HEAT CONDUCTION AND MASS DIFFUSION

McGraw-Hill Series in Mechanical Engineering

Consulting Editors

Jack P. Holman, *Southern Methodist University*
John R. Lloyd, *Michigan State University*

Anderson: *Modern Compressible Flow: With Historical Perspective*
Arora: *Introduction to Optimum Design*
Bray and Stanley: *Nondestructive Evaluation: A Tool for Design, Manufacturing, and Service*
Culp: *Principles of Energy Conversion*
Dally: *Packaging of Electronic Systems: A Mechanical Engineering Approach*
Dieter: *Engineering Design: A Materials and Processing Approach*
Eckert and Drake: *Analysis of Heat and Mass Transfer*
Edwards and McKee: *Fundamentals of Mechanical Component Design*
Gebhart: *Heat Conduction and Mass Diffusion*
Heywood: *Internal Combustion Engine Fundamentals*
Hinze: *Turbulence*
Howell and Buckius: *Fundamentals of Enginering Thermodynamics*
Hutton: *Applied Mechanical Vibrations*
Juvinall: *Engineering Considerations of Stress, Strain, and Strength*
Kane and Levison: *Dynamics: Theory and Applications*
Kays and Crawford: *Convective Heat and Mass Transfer*
Kelly: *Fundamentals of Mechanical Vibrations*
Kimbrell: *Kinematics Analysis and Synthesis*
Martin: *Kinematics and Dynamics of Machines*
Modest: *Radiative Heat Transfer*
Norton: *Design of Machinery*
Phelan: *Fundamentals of Mechanical Design*
Raven: *Automatic Control Engineering*
Reddy: *An Introduction to the Finite Element Method*
Rosenber and Karnopp: *Introduction to Physics*
Schlichting: *Boundary-Layer Theory*
Shames: *Mechanics of Fluids*
Sherman: *Viscous Flow*
Shigley: *Kinematic Analysis of Mechanisms*
Shigley and Mischke: *Mechanical Engineering Design*
Shigley and Uicker: *Theory of Machines and Mechanisms*
Stiffler: *Design with Microprocessors for Mechanical Engineers*
Stoecker and Jones: *Refrigeration and Air Conditioning*
Ullman: *The Mechanical Design Process*
Vanderplaats: *Numerical Optimization: Techniques for Engineering Design, with Applications*
White: *Viscous Fluid Flow*
Zeid: *CAD/CAM Theory and Practice*

HEAT CONDUCTION AND MASS DIFFUSION

Benjamin Gebhart

University of Pennsylvania

McGraw-Hill, Inc.

New York   St. Louis   San Francisco   Auckland   Bogotá
Caracas   Lisbon   London   Madrid   Mexico   Milan   Montreal
New Delhi   Paris   San Juan   Singapore   Sydney   Tokyo   Toronto

This book was set in Times Roman by Science Typographers, Inc.
The editors were John J. Corrigan and John M. Morriss;
the production supervisor was Kathryn Porzio.
The cover was designed by John Hite.
R. R. Donnelley & Sons Company was printer and binder.

HEAT CONDUCTION AND MASS DIFFUSION

2 3 4 5 6 7 8 9 0 DOC DOC 9 0 9 8 7 6 5 4 3

ISBN 0-07-023151-6

Library of Congress Cataloging-in-Publication Data

Gebhart, Benjamin.
    Heat conduction and mass diffusion/Benjamin Gebhart.
       p.   cm.
    Includes bibliographical references and index.
    ISBN 0-07-023151-6
     1. Heat—Conduction. 2. Mass transfer. I. Title.
QC321.G42   1993   92-35803
621.402′2—dc20

ABOUT THE AUTHOR

Benjamin Gebhart was born in Cincinnati and grew up in Miamisburg, Ohio. After a short stint as an apprentice tool maker, he joined the Marine Corps and spent two and a half years in the Pacific, in World War II. The GI Bill supported the BSE (ME) and MSE (ME) degrees at the University of Michigan. The Ph.D. degree was obtained at Cornell University and he remained there for twenty-four years. In 1975 he went to the State University of New York at Buffalo as a Leading Professor and became Chairman of Mechanical Engineering. In 1980 he was appointed the Samuel Landis Gabel Professor of Mechanical Engineering at the University of Pennsylvania. On leaves, he has been a professor at the University of California at Berkeley, the University of Marseille, Oregon State University, at the Ecole des Mines in France, and the Naval Post Graduate School, as the NAVSEA Research Chair Professor. He has also been associated widely in Europe, Scandinavia, Germany, and in the former USSR in research interactions.

The research over this period has been in many areas of buoyancy induced flows and transport. Other investigations have concerned mixed convection, mass diffusion, melting and freezing, flow interactions, density extrema effects, transport in pure and saline water, and microconfigured surface radiation and phase change processes. Earlier books were *Heat Transfer* (1961) and (1971) and *Buoyancy-Induced Flow and Transport*, with Professor Y. Jaluria, Professor R. L. Mahajan, and Dr. B. Sammakia. He is on the editorial boards of many archive journals and is listed in eight biographical references. He is a fellow of the ASME.

An interest for over three decades has been the assembly of land parcels into a large nature sanctuary near Ithaca, N.Y. Reforestation, wildlife habitat improvement, and soil restoration remain a continuing and strenuous activity there.

To Lorna and Raïssa

CONTENTS

Appendixes

PREFACE

This book is a result of the author's activity in teaching conduction, with an added element of mass transfer. It provides a suitable text in conduction heat transfer and also includes much material concerning many kinds of mass diffusion processes. It demonstrates heat conduction and mass diffusion processes in terms of many and diverse engineering applications. Many excellent books have concerned the fundamental aspects of both conduction mechanisms and analysis. There are also a number of books which principally concern mass diffusion. These also include good coverage of additional and more diverse fundamental mechanisms which commonly arise in mass diffusion. A principal purpose here is to link analysis, application, and heat and mass diffusion.

Heat conduction and mass diffusion are very closely interrelated in both applied science and in technology. Many of the physical processes and methodologies are very similar. Their consideration together is well known to be an effective teaching strategy. The appropriate rate laws are often similar. Many direct analogies are commonly used in calculating transport. Considering heat and mass transfer together is an economy. It is also an advantage for students whose work will span these fields.

This book brings heat conduction and mass diffusion into a common treatment. The level is appropriate for a first advanced course. The total content of this book is beyond coverage in a single course. Choices are to be made in terms of the objectives of the particular instructor.

Both the content and presentations in this book are often different than common in past texts. Several objectives are followed throughout. A most important matter is a careful description of the physics of the many fundamental processes. For example, diffusion mechanisms are discussed in detail, in terms of the constituent microscopic processes.

Most solutions demonstrated here were chosen as characteristic of different applications. The separation of variables, the Laplace transform, and the Duhamel methods are given, along with some particularly useful solutions. Most graduate students are receiving instruction in applied mathematics. Therefore, the amount of such analysis, and the number of solutions, are considerably reduced here. This has made possible a broader representation of the diverse mechanisms and processes which are of increasing importance in our field. Abundant references also are given to the more specialized treatments available in many other excellent books and in the literature.

This book includes useful coverage of many typical and currently important processes and applications. Examples are composite materials, composite insulation, contact resistance, catalysis, moving sources, phase fronts, noncontinuum diffusion, effects of thermal stress, and shape factors. Combined and simultaneous processes of heat and mass diffusion, such as in transpiration cooling and porous region drying, are also treated.

Chapters 3, 4, and 5 concern steady-state and single and multidimensional unsteady-state processes. The results are commonly in terms of heat conduction bounding conditions, since many of the mechanisms do not have simple or direct mass diffusion analogs. Examples are contact resistance, moving phase fronts and moving sources.

The simplest treatments of the numerical methods used to generate approximate solutions are given in Chaps. 3, 4, and 5. These apply, in turn, to multidimensional steady-state, one-dimensional transients, and multidimensional transient processes. This divided treatment is used to emphasize the several distinct aspects of numerical modeling. Section 3.9 concerns the subdivision of a continuous region into the numerical simulation of regular boundary conditions, in steady state. Section 4.9 is the first consideration of the numerical simulation of an evolving transient response, including questions of calculational stability. Section 5.3 then combines these concepts, for multidimensional transients. Other numerical formulations and methodologies are thereafter considered in more detail in Chap. 9. Depending on the objectives of the instructor, and the previous preparation of the students, the material in Secs. 3.9, 4.9, and 5.3 may serve only as an initial review of some of the basic aspects of numerical representation.

Chapter 6 concerns many important mass diffusion mechanisms and processes which do not have very common or direct analogs in heat conduction. Section 6.1 concerns the equation transformations characteristic of mass diffusion, the concept of porous region permeability and the fundamental diffusion mechanisms of chemical species in gases, liquids, and solids. Section 6.2 examines the effects of changing mass diffusivity, D, over a region, due to its sensitivity to both local concentration and to region inhomogeneity. Section 6.3 treats transients, variable diffusivity, surface processes, and the measurement of diffusivity. Section 6.4 concerns interfaces and moving boundaries, in terms of surface region processes, such as surface oxidation layers and moving locations, or fronts, of abruptly changing diffusivity. Section 6.5 concerns distributed internal chemical reactions. Both chemically irreversible and reversible reactions are analyzed, to determine sorption and desorption rates. Section 6.6 concerns several kinds of combined heat transfer and mass diffusion mechanisms. These include the frequent thermal effects of internal species adsorption, chemical species diffusion in a flame front and the transpiration cooling of a hot surface. Non-Fickian transport, due to both noncontinuum and to surface-diffusion effects, is also analyzed.

The formulations and analyses in this book do not include either the Soret or Dufour effects. The Soret effect is the concentration gradient which sometimes arises from an imposed temperature gradient, as in saline water. The Soret effect would then cause saline diffusion. The Dufour effect is the inverse kind of process, thermal diffusion arising from an imposed concentration gradient. Similar simpler kinds of coupled processes arise in the mechanisms discussed in Sec. 6.6.

Chapter 7 concerns composite regions. Thermal transport contact resistance is quantified in detail, in recognition of its importance in many and diverse applications. It is also a good physical example of the interaction of parallel and series processes. The conductivity mechanisms of composite materials are discussed. The last section concerns composite insulation, and superinsulation in particular.

Chapter 8 is related to a group of important conduction-mediated applications. Extended surface heat transfer is treated in Sec. 8.2. Section 8.3 concerns welding, in terms of the internal conduction responses to concentrated moving energy sources. Thermal stresses, which arise in many processes, are also considered for several simple examples, in Sec. 8.4. Section 8.5 concerns the average heat conduction rate in randomly disturbed conduction environments. Section 8.6 concerns fluid flows in which the convection transport field is actually analyzed as a purely conductive process. The examples are liquid films and internal flows.

The treatment of numerical analysis in Chap. 9 presumes the background given in Secs. 3.9, 4.9, and 5.3. Section 9.1 contrasts the finite difference and finite element methods. Then finite difference formulations are given, in Sec. 9.2, along with the resulting errors due to truncation, discretization, and round-off. Examples of general higher-order estimates are also given. Section 9.3 concerns truncation errors and considerations of stability in transients. Common numerical methods are summarized. Section 9.4 treats important additional aspects of calculation techniques, including numerical iteration and the treatment of irregular boundaries. Section 9.5 concerns the effects of variable properties. Section 9.6 formulates the finite-element method of numerical analysis and gives a simple example.

This book covers a relatively wide diversity of heat and mass transfer processes, over a broad range of applications. The objective is to bring these considerations together in a consistent way, in a book of reasonable size. On the other hand, this material includes the treatment of many important mechanisms not commonly treated in text material concerning the heat and mass transfer mechanisms. Examples include moving phase interfaces, contact resistance, cryogenic insulation, composites, chemical processes, thermal stress, and random conduction effects.

The result here has been that the coverage of some commonly important matters is less detailed. One example is the material concerning finite-difference and finite-element implementation, in Chap. 9. Another is the relatively smaller scope given to classical analytical solution techniques. However, these treatments, along with many others, are widely available in the other literature referenced here.

A large number of problems are given. These cover most aspects of the material in the book. Appendix A gives conversion factors. Appendixes B, C, and D concern thermal properties, and E tabulates mass diffusivities for gases, liquids, and solids. Appendixes F, G, and H concern the error function, Laplace transform pairs, and piping and tubing dimensions. Both English and SI units are used in this book, since practice continues to indicate that the wide use of both systems is a continuing reality in our field.

McGraw-Hill and the author would like to thank the following reviewers for their many helpful comments and suggestions: Douglas Baines, University of Toronto; Christopher Beckermann, University of Iowa; Ralph Greif, University of California–Berkeley; Yogesh Jaluria, Rutgers–The State University of New Jersey; David Lilley, Oklahoma State University; John Lloyd, Michigan State University; Richard Pletcher, Iowa State University; and J. R. Thomas, Virginia Polytechnic Institute.

Benjamin Gebhart

HEAT CONDUCTION AND MASS DIFFUSION

CHAPTER
1

INTRODUCTION

1.1 DIFFUSION PROCESSES IN SOLIDS

The study of heat and mass transport continues to be an increasingly intense concern in technology and in the earth sciences. In every field concerned with energy production and exchanges, the need to understand, to predict, and to optimize has led to increasingly detailed study of how thermal energy and chemical species are carried, distributed, and diffused in and by material.

This book concerns the rate of heat and chemical species diffusion through materials, by the mechanism called conduction or diffusion. This is idealized as a stationary region of material through which heat or chemical species diffuse. A common example of heat conduction is heating an object in an oven or furnace. The material remains stationary throughout, neglecting thermal expansion, as the heat diffuses inward to increase its temperature. A comparable chemical species diffusion process arises when a dry fibrous material is placed in a humid environment, to increase the water content of the fibers by inward vapor diffusion.

The common feature of both kinds of processes is that both the thermal and the mass diffusion often take place without an important effect on the configuration or volume of the material through which the diffusion occurs. These are simple conduction modes. However, there are many other transport processes in which both thermal and mass diffusion arise, wherein the material is flowing or in relative motion. These are convection processes. These are covered by the formulations developed here only for flow processes in which the motion does not influence the thermal or mass diffusion. This sometimes occurs when the flow is essentially parallel and also normal to the imposed gradient of temperature or concentration. Then the motion may not affect the diffusion process.

1.1.1   Heat Conduction and Thermal Conductivity

The rate of heat conduction through a material may be proportional to the temperature difference across the material and to the area perpendicular to heat flow and inversely proportional to the length of the path of heat flow between the two temperature levels. This dependence was established by Fourier and is analogous to the relation for the conduction of electricity, called Ohm's law. The constant of proportionality in Fourier's law, denoted by k, is called the thermal conductivity. It is a property of the conducting material and of its state. With the notation indicated in Fig. 1.1.1, Fourier's law is

$$q = \frac{kA}{L}(t_1 - t_2) \qquad (1.1.1)$$

where kA/L is called the conductance of the geometry.

The thermal conductivity k, which is analogous to electrical conductivity, is a property of the material. It is equivalent to the rate of heat transfer between opposite faces of a unit cube of the material which are maintained at temperatures differing by 1°. In engineering units in the English system, k is expressed in

$$\frac{\text{Btu}}{\text{hr ft}^2\,°\text{F}/\text{ft}} = \text{Btu}/\text{hr ft }°\text{F}$$

in SI units, k is expressed as W/m^2 K/m, or W/m K.

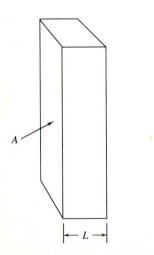 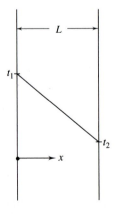

FIGURE 1.1.1
One-dimensional steady-state heat conduction.

The conduction equation (1.1.1) may also be written as the heat transfer rate per unit area normal to the direction of heat flow, q'', as

$$\frac{q}{A} = q'' = \frac{k}{L}(t_1 - t_2) = k\left[-\frac{(t_2 - t_1)}{L}\right] = -k\frac{dt}{dx} \qquad (1.1.2)$$

The quantity q'' is very useful and is hereafter called the heat flux. Note that the quantity in brackets is minus the temperature gradient through the material, that is, $-dt/dx$.

The transport property, thermal conductivity, varies over a wide range for the various substances commonly encountered. For example, at 25°, the values for air and water are 0.26×10^{-3} and 6.11×10^{-3} W/cm K, respectively. The values for iron, silicon, aluminum, copper, and silver are 0.80, 1.49, 2.37, 4.01, and 4.29. For diamonds of types I, IIa, and IIb, the values are 9.9, 23.2, and 13.6 W/cm K. The ratio for diamond IIa to air is 89,230 to 1. Among materials, gases generally have the lowest thermal conductivities, followed by good insulators, nonmetallic liquids, nonmetallic solids, liquid metals, metal alloys, and, finally, the best group of conductors, pure metals.

The need for the long-term storage of cryogenic liquids has led to the development of much better insulators than either gases or a vacuum alone. These superinsulators amount to a very highly evacuated layer lightly filled with material, to trap thermal radiation. Values of effective conductivity k_e, down to 4×10^{-7} W/cm K are commonly achieved. See Sec. 7.5.1. The resulting ratio of diamond IIa to k_e is then 58×10^6. A selection of thermal-conductivity data is given in Tables B, C, and D in the Appendix.

The thermal conductivity of a given material depends upon its state and may vary with temperature, pressure, and so on. For moderate pressure levels the effect of pressure is small. However, for many substances the effect of temperature upon k is not negligible. Therefore, the thermal conductivity in Eq. (1.1.1) is meant to be some average value if k is a function of temperature or location. This ambiguity concerning the definition of thermal conductivity may be avoided if the relationship between the heat flux and temperature difference is written for a very thin conduction region. Another and equivalent procedure is to write the local heat flux in terms of the local value of k and the local temperature gradient in the direction of heat flow:

$$q''_n = -k_n\frac{\partial t}{\partial n} \qquad (1.1.3)$$

This is the conduction heat transfer rate per unit area in terms of the local temperature gradient in the direction of heat flow. The coordinate in the direction of heat flow is denoted by n. The thermal conductivity k_n applies for heat conduction in the direction of n. Many materials have different conduction characteristics in different directions. For examples, wood and other ordered fibrous materials have higher thermal conductivities parallel to the grain than perpendicular to it. In multidimensional temperature fields there will be temperature gradients in each coordinate direction. Equation (1.1.3), with the proper value of k, may then apply in any direction. In general, a more complicated tensor conductivity, defined in Chap. 2, applies for all processes.

The minus sign appears in Eq. (1.1.3) because it is customary to denote heat flux, q_n'', in the direction of increasing n, that is, in the positive direction, as positive. However, positive heat flow results if t decreases with n, that is, if $\partial t/\partial n$ is negative. Therefore, a minus sign is required.

Equation (1.1.3) is called Fourier's law of conduction, and indeed it is a law if k is taken as dependent only upon the local state of the material and upon the direction in which n is measured. To a good approximation, the thermal conductivity is a function only of the local material and state for many circumstances common in engineering practice. That is, k is essentially independent of the local temperature gradient and of direction. Then the analysis of heat conduction is based upon Eq. (1.1.3), where k is considered to be a transport property.

TRANSIENT CONDUCTION. A great majority of the conduction processes of interest are transients. That is, due to changing boundary temperatures and other effects, the local temperature in the conduction region depends on time as well as on location. An example is given in Fig. 1.1.2. A plane layer of material, originally uniformly at t_i throughout, has its temperature on the left side, at $x = 0$, increased to $t_a > t_i$ in Fig. 1.1.2(a) or decreased to $t_b < t_i$ in Fig. 1.1.2(b), at time $\tau = 0$. The temperature in the region, at x, changes with time τ. That is, $t = t(x, \tau)$ in the region $0 < x < L$.

If the temperature at $x = L$ is maintained at t_i thereafter, $t(x, \tau)$ increases with time in Fig. 1.1.2(a) and decreases in Fig. 1.1.2(b). The instantaneous distributions of $t(x, \tau)$ are shown at later times, τ_1, τ_2, and τ_3 and in the eventual steady state, as $\tau \to \infty$. Also shown in Fig. 1.1.2 are the local slopes of the instantaneous temperature distributions at time τ_2 at $x = L/4$ and $L/2$.

In Fig. 1.1.2(a) the heat flow is toward the right, a positive heat flux $q_a''(x, \tau_2)$. The slope, or temperature gradient, is greater in magnitude at $L/4$ than at $L/2$. That is, $q''(L/2, \tau_2)$ is therefore greater than $q''(L/4, \tau_2)$. This difference is the rate of thermal energy being absorbed at time τ_2 in the layer of material between $x = L/4$ and $x = L/2$. This rate of energy absorption heats this material up from its temperature level at time τ_2 toward the distribution shown at the later time τ_3. Cooling the left surface, as seen in Fig. 1.1.2(b), produces the equal but opposite effect of cooling the material.

This example shows that temperature increases with time occur in regions of local instantaneous temperature distributions, $t(x, \tau)$, which are concave upward. That is, for $\partial^2 t/\partial x^2 > 0$, as in Fig. 1.1.2($a$), $t(x, \tau)$ is increasing. Decreases of $t(x, \tau)$ with time occur for $\partial^2 t/\partial x^2 < 0$, as in Fig. 1.1.2($b$). This is a reliable generalization when k is constant and uniform across the conduction layer and when there is no additional energy source in the layer. Such an energy source might arise, for example, from internal chemical reaction or from

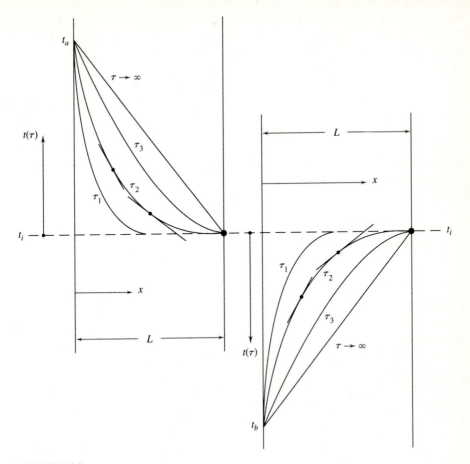

FIGURE 1.1.2
Transient conduction in a solid plane layer of material.

internal heating as by microwave radiation. The resulting volumetric rate of energy generation is written as q'''.

1.1.2 Thermal Conduction Mechanisms

Theoretical predictions and measurements have been made of the value of thermal conductivity, k, for many types of substances. In gases, heat is conducted (i.e., thermal energy is diffused) by the random motion of molecules. Higher-velocity molecules from higher-temperature regions move about randomly, and some reach regions of lower temperature. By a similar random process, lower-velocity molecules from lower-temperature regions reach higher-temperature regions. Thereby, net energy is exchanged between the two regions.

The thermal conductivity depends upon the space density of molecules, upon their mean free path, and upon the magnitude of the molecular velocities. The net result of these effects, for gases having very simple molecules, is a dependence of k upon $\sqrt{T}$, where T is the absolute temperature. This results from the kinetic theory of gases.

In considering conduction processes in general, including purely conduction heat transfer in gases as well as in liquids and solids, additional and different microscopic transport mechanisms must be taken into account. A classification of the common and relatively well-understood mechanisms is listed as follows together with the materials in which each mechanism is important in determining the overall conduction transport:

1. Classical gas, molecular diffusion (gases, porous solids, electrons in semiconductors)
2. Degenerate electron gas diffusion (metals)
3. Elastic phonon diffusion (all solids)
4. Thermal radiation, electromagnetic photons (internal emission and reabsorption, perhaps continuous spectra)
5. Classical theory for common liquids

The macroscopically observed conduction characteristic of any material may be the sum of several or all of the previous effects, as well as of others which may arise.

The preceding microscopic mechanisms are well understood in several types of materials. Their transport properties are most easily discussed in terms of a single general model, as follows. Consider the x direction in a material in which an energy gradient (dt/dx) exists. Points are laid off, as in Fig. 1.1.3, a distance λ apart, where λ is equal to the distance over which an exchange by any mechanism is essentially complete. This would be the mean free path in a gas, the absorption length for radiation, and so on, and λ is assumed small compared to the thickness of the region in this continuum model.

Energies per unit volume at two adjacent points are denoted as e_1 and e_2. The characteristic velocity of energy transport is taken as v. The net rate of energy flux in the positive x direction, that is, the difference between transport

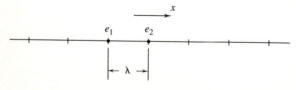

FIGURE 1.1.3
Model for microscopic conduction mechanisms.

to the right and transport to the left, is

$$q''_x = e_1 v - e_2 v = -v(e_2 - e_1)$$

$$= -v\frac{de}{dx}\lambda = -\lambda v\frac{de}{dt}\frac{dt}{dx}$$

$$= -c\lambda\frac{dt}{dx} \tag{1.1.4a}$$

where $c = de/dt$ is the specific heat of the material per unit volume.

Comparing this result with the definition of thermal conductivity k in Eq. (1.1.3),

$$k = c\lambda v \tag{1.1.4b}$$

This model has exchanged the unknown k for the product of the quantities c, λ, and v. These quantities are given for mechanisms 1 to 5 in Table 1.1.1.

TABLE 1.1.1
Transport quantities for the basic conduction mechanisms

Mechanism	c	λ	v	Temperature dependence of k
1. Gas	$3/2nk'$	$\dfrac{1}{ns}$	$\left(\dfrac{k'T}{m}\right)^{1/2}$	$\sqrt{T}$
2. Electron	cT	$f(T)$	v_f, velocity at the Fermi energy level	$Tf(T)$
3. Phonon	$\left(A\dfrac{k'T^3}{v_s h}k'\right)^*$	$e^{\theta/T}$	v_s, velocity of sound	$T^3 e^{\theta/T}$
4. Radiation	$\left(A\dfrac{k'T^3}{v_l h}k'\right)^{**}$	$\dfrac{l}{\alpha}$	v_l, velocity of light	T^3
5. Liquids	$3nk'$	$\dfrac{1}{n\bar{\lambda}^2}$	v_s, velocity of sound	

The various quantities are defined as follows:

k'	Boltzmann constant
n, m	particle number density and mass
s	scattering cross section
T	absolute temperature
θ	Debye temperature
h	Planck constant
l	length
α	absorption characteristic
$\bar{\lambda}$	mean free path
*	at very low temperatures
**	for continuous spectra

The product of the three columns, in each mechanism, indicates how transport by the mechanism depends upon physical constants and upon temperature. The last column indicates the net temperature dependence of thermal conductivity. The relative importance of the several mechanisms which may simultaneously occur in a given material and temperature level may be determined by comparison. Other discussions of conduction mechanisms are given by Bird et al. (1960) and Eckert and Drake (1972).

In spite of the relevant microscopic mechanisms, the macroscopic formulation is used. That is, Fourier's law of conduction, Eq. (1.1.3), is applied to the analysis of conduction processes, without regard to the way in which the thermal conductivity arises for various substances. Equation (1.1.3) is employed in writing an energy balance for a small element of material in a conduction region. This energy balance equates the rate of gain of energy by conduction to the rate of change of stored energy in the element. The resulting relation is a partial differential equation which must be satisfied by the temperature distribution in the conduction region. The most representative differential equations are derived in Sec. 1.3.

1.1.3 Mass Diffusion and Diffusivity

Mass diffusion through a region occurs by the motion of quantities of mass or chemical species through the material. Many physical mechanisms arise. A simple model is that in which atoms or molecules, of local concentration C, move through the structure of the material. They move in the direction of decreasing concentration. For example, the drying of wood by heating releases vapor. The vapor moves out through the structure by diffusion. Ripening of some fruits is accomplished by placing them in an atmosphere of a gas which diffuses inward to cause a chemical effect which amounts to ripening. Hardening the surface layer of low-carbon-steel parts may be accomplished by heating them in a kiln, while packed in a high-carbon solid material. The carbon diffuses into the steel surface regions. Subsequent quenching of the part produces a hardened surface layer which will take a high finish and also resist abrasion.

The diffusion mechanism in a plane layer of material is formulated in Fig. 1.1.4. A layer L thick and of area A has concentration levels C_1 and C_2 maintained at the two boundaries, at $x = 0$ and $x = L$. In this example it is assumed that there is no local adsorption (or retention) or release of stored or bound diffusing material C in the region $x = 0$ to L. The mass diffusion coefficient D is also taken as uniform, that is, constant across the region.

The rate of mass diffusion through the layer in steady state, m, will be proportional to the concentration difference across the material, $(C_1 - C_2)$, and to the area perpendicular to the mass diffusion, A. It will be inversely proportional to the length of the path L of mass diffusion between the two imposed concentration levels C_1 and C_2. Then the mass flow rate through the region, m,

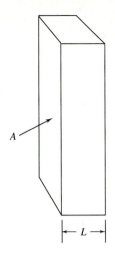

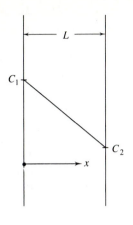

FIGURE 1.1.4
Imposed surface concentrations C_1 and C_2 at $x = 0$ and L, for a diffusing material.

of species C is given by

$$m = \frac{DA}{L}(C_1 - C_2) \tag{1.1.5}$$

where DA/L is the "mass" conductance across the region. This result is completely analogous to Eq. (1.1.1) for heat flow, which follows from the Fourier law of conduction. The mass diffusion formulation given in Eq. (1.1.5) is a form of Fick's first law.

Equation (1.1.5) is also written in terms of a mass flux, that is, the diffusion rate in mass per unit cross-sectional area, per unit time, as m'':

$$\frac{m}{A} = m'' = \frac{D}{L}(C_1 - C_2) = D\left[-\frac{(C_2 - C_1)}{L}\right] = -D\frac{dC}{dx} \tag{1.1.6}$$

The quantity in brackets is minus the concentration gradient through the material, that is, $-dC/dx$.

Several consistent systems of dimensions and units are used for the physical quantities in Eq. (1.1.6). A common practice expresses concentration in mass per unit volume, M/L^3, and m'' in mass flux per unit area and time, that is, M/L^2T. Then D has the dimensions of L^2/T. The units may be cm^2/s or ft^2/hr, as in Tables E.1 to E.7 in the Appendix, or as m^2/s in SI units. Other commonly used measures and terms of composition and concentration include number density or fraction, partial density, mass or mole fraction, and molar concentration.

The magnitude of the diffusion coefficient depends strongly on both the material through which diffusion occurs and the diffusing species. For a static region of air at 25°C, $D = 0.712$ and 0.066 cm^2/s for the diffusion of the gases H_2 and naphthalene, respectively. This process is the diffusion of one gas through another, due to local molecular mixing at the length scale of the molecular mean free path. Fick's law is also a good approximation for gas diffusion at low pressures.

For water at 20°C, $D = 5.13 \times 10^{-5}$ and 0.45×10^{-5} cm^2/s for the diffusion of H_2 and sucrose, respectively. This rate is far slower, since the diffusion process now occurs at the far shorter length scale of the spacing of water molecules. For gas or vapor diffusion through passive porous solids, D is merely the permeability constant P. This varies over a wide range for any diffusion material. It depends on the nature of the porosity.

Basic physical mechanisms of mass diffusion are discussed in detail in Sec. 6.1.4. The models are for gases, liquids, and crystalline solids. For gases, the kinetic theory models species diffusion, in terms of the mean free path, as for thermal conductivity in Sec. 1.1.2. Liquid molecules are much more closely spaced. Then the diffusion is between adjacent molecules in thermal motion. For crystalline solids, the similar process is internal to the lattice structure and through vacancies in the structure. Diffusion coefficients in gases, liquids, and solids are commonly on the order of 10^{-1}, 10^{-5}, and 10^{-7} cm^2/s. See Sec. 6.1.4 and Kirkaldy and Young (1987) and Cussler (1985) for other related aspects.

As for heat conduction, as discussed in Sec. 1.1.1, Eq. (1.1.6) indicates that both the gradient of concentration dC/dx and the mass flux $m''(x)$ are constant across the region. This follows from the condition of steady state when D is uniform across the region and no local adsorption or desorbtion of species C occurs in the region.

TRANSIENTS. The similarity between heat conduction, q'', and mass diffusion, m'', seen in Eqs. (1.1.2) and (1.1.6), is the basis of the frequent close similarity between these two processes. Thus D is the counterpart of k, as dC/dx is of dt/dx. Mass diffusion processes may also be transient, as $C(x, \tau)$ in one dimension. For D uniform, the preceding discussion, related to Fig. 1.1.2 in terms of $t(x, \tau)$ and its concavity, applies directly to mass diffusion. The later discussion there, concerning distributed internal energy sources, q''', also applies here in terms of a distributed internal species source, as C'''. Such effects may arise from changes in the local concentration of adsorped material.

MECHANISMS OF MASS DIFFUSION. Several different aspects of mass diffusion arise as distinctive characteristics which must be taken into account. The simple example given previously makes two assumptions. One is that mass diffusivity D is the same throughout the region $x = 0$ to L. The other is that the local mass flux m'' is also the same throughout. That is, there is no adsorption or desorption across the region.

Since diffusivity D arises as the ability of species C to diffuse, it depends upon the specific mechanisms whereby the diffusion occurs. For example, the diffusion of one gas through another arises as a result of random atomic or molecular thermal motions. The resulting local mixing at the molecular level results in the net migration of any species in the direction of a decreasing local concentration gradient of that species. In diffusion through a solid, with some porosity effect, the thermal motion of species C results in the same kind of mechanism.

However, the net local diffusion of species C through the material is sometimes the sum of several simultaneous and parallel physical mechanisms. This often arises with simultaneous water vapor and water diffusion in fibrous materials and in soils. Then both phases may move through the porosity of the structure. Water may also diffuse on internal surface layers, due to capillary and other effects. Vapor may move through the porosity in either a continuum or noncontinuum gas diffusion mode. It may also be adsorbed locally on the surface. It may also move by particle surface diffusion processes.

The processes are commonly absorption, adsorption, desorption, and chemical and physical interaction with the material. Simple absorption amounts to a diffusion material simply penetrating the region, as a solvent into a polymer or vapor diffusion into a porous material. However, in many processes the diffusing material becomes chemically or physically bound locally into the structure. The carburization of steel amounts to a chemical bond. This kind of effect is sometimes called chemisorption. Physisorption follows from physical bonding, as by thin layers of the diffusing material bonding with the internal structure, as with liquid water layers diffusing through soil or SO_2 molecules bonding to catalytic surfaces. Adsorption is a common general term which refers to the local holding of the diffusing material, by any of the preceding mechanisms. Desorption is the release. The terms used commonly here are adsorption, and desorption.

The previous diffusion and reaction mechanisms are very common effects. For example, consider a continuing adsorption of species C in the material across the region $x = 0$ to L, as shown in Fig. 1.1.4. then m'', the flux of C, would depend on x, as $m''(x)$. With adsorption, $C''' < 0$, the flux $m''(x)$ would decrease from $m''(0)$ to $m''(L)$. The local concentration variation would then be a curve of decreasing slope; see Eq. (1.1.6). That is, it would be concave upward. On the other hand, desorption, for $C''' > 0$, would result in the reverse effect.

The basic diffusion equations, governing the simplest diffusion mechanisms, as in Eq. (1.1.6), are developed in Sec. 1.3. These equations result from a balance of the processes involving the species C. Resulting mass diffusion analysis and results appear in Chap. 2 as analogs to heat conduction. This analogy applies most simply where both k and D are constant and uniform across the region of diffusion. Chapter 6 considers the more complicated diffusion mechanisms most characteristic of many other important mass diffusion processes.

1.2 HEAT CONDUCTION EQUATIONS

Heat conduction is increasingly important in modern technology, in the earth sciences, and in many other evolving areas of thermal analysis. The specification of temperatures, of heat sources, and of heat flux, in regions of material in which conduction occurs, gives rise to analyses of temperature distributions, heat flows, and conditions of thermal stressing. The importance of such conditions has led to an increasingly highly developed field of analysis in which sophisticated mathematical and increasingly powerful numerical techniques are used. All such analysis proceeds from the equations which result from the basic physical formulations of the phenomena relevant to conduction.

This section considers the basic physical mechanisms which commonly arise in heat conduction. General and specific equations are derived, in terms of various physical effects. In Sec. 1.4, characteristic kinds of temperature boundary conditions are given, along with typical initial conditions in time-varying processes. The following chapters then consider many different kinds of heat conduction processes, as well as applications, including phase change, conduction, composite materials, and contact resistance.

Since the analytical or numerical formulation which is to be applied to a particular conduction circumstance depends primarily upon the detailed nature of the conduction process, a classification system is used. Steady state means that the conditions (i.e., temperature, density, etc.) at all points of the conduction region are independent of time. Unsteady state implies a change with time, usually only of the temperature. There are two distinct types of unsteady state. In pure periodics, the temperature variation with time at all points in the region is periodic. In transients, the temperature changes with time are not periodic. An example of periodic conduction may be the conduction of energy in a thick concrete roof slab due to daily solar effects. The immersion of a hot piece of metal in a cold quenching bath is an example of transient conduction. Transient periodics also commonly arise.

Another important aspect of classification is the minimum number of space coordinates (or dimensions) necessary to describe the temperature field. Three coordinates suffice in all circumstances, but many configurations are simpler because of the geometry of the conduction region or because of the symmetries of the temperature distribution. One- and two-dimensional configurations are common.

Many applications also involve the additional feature of heat generation internal to the conduction region. This may be due, for example, to a chemical reaction or to an energy dissipation process. The source of energy may be concentrated, as in a small-diameter current-carrying electrical conductor inside thick insulation. Or the source may be distributed, as in the elements or shield of a nuclear reactor. Then the generation of energy is due to the deceleration and absorption of both high-energy particles and radiation throughout the solid material. Another very common example in use is electrical induction heating throughout a material as, for example, in a microwave oven.

The classification of a problem according to the distinctions enumerated previously indicates the particular approach to be used for its solution. It also suggests the amount of difficulty that will be encountered in obtaining the solution, that is, the internal temperature distribution. Commonly, an analytic solution is not possible. Approximate information is then obtained by numerical or other approximate analyses.

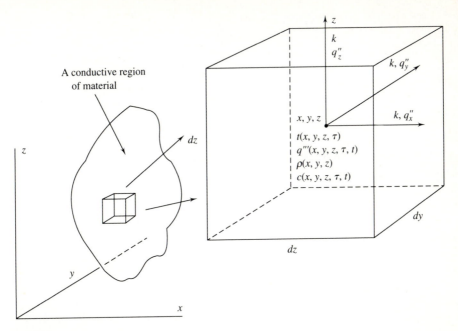

FIGURE 1.2.1
Differential element for heat conduction analysis.

The "solution" of a conduction problem amounts to the functional dependence of temperature upon location in the conduction region, and upon time in unsteady-state circumstances. Obtaining a solution means determining a temperature distribution which is consistent with the conditions on the boundaries and also consistent with any specified constraints internal to the region.

The temperature distribution in the region must also be consistent with the relation among heat flux, thermal conductivity, and temperature gradient embodied in Fourier's law of conduction. That is, the temperature distribution must satisfy any specified conditions and Fourier's law simultaneously. However, the law of conduction is not applied in the form of, for example, Eq. (1.1.3). Instead, a differential equation is derived from this relation. This equation constrains the form of the temperature distribution, by specifying a relation between the space and time derivatives of this distribution. The general differential equations for conduction in a material of invariant density are developed in the following discussion. A Cartesian differential volume, $dx\, dy\, dz$, is used, as shown in Fig. 1.2.1. This is done here for clarity in distinguishing all of the independent energy effects. The other common method uses a global summation of individual effects, over a finite control volume. See, for example Özisik (1980).

Many different kinds of effects arise in applications, such as variable conductivity and distributed energy sources. The most general equations are derived first. They are then specialized to the simpler forms which apply to particular applications. These equations are then the basis of the analyses given in subsequent chapters. The differential equations, which express the physics of the conduction process, are also the basis of the numerical analyses of conduction behavior. They are converted to the required finite-difference forms, for particular applications, in Secs. 3.9, 4.9, and 5.3. The accuracy, uses, and other properties of numerical formulations are discussed in much greater detail in Chap. 9.

1.2.1   The Conduction Equation

Consider a differential element of volume $dx\,dy\,dz$ of a material subject to a conduction process, as shown in Fig. 1.2.1. The differential equation of conduction is derived by writing an instantaneous energy balance for this element. The net rate of energy gain across the six faces, by conduction, is added to the rate of energy generation in the element. This sum is set equal to the time rate of change of energy stored in the mass inside the volume element, as indicated by its changing temperature level $t(x, y, z, \tau)$.

The conduction rates are related to the relevant thermal conductivities and temperature gradients by, for example, Fourier's law of conduction, Eq. (1.1.3). The time rate of change of stored energy is equal to the thermal capacity of the mass in $dx\,dy\,dz$, $\rho c\,dx\,dy\,dz$, times the time rate of change of the local temperature, where ρ and c are the local density and specific heat. The conduction and storage terms are the only ones required in these terms of the energy balance only if there is no movement of mass across the six faces of the element $dx\,dy\,dz$. This condition is invoked by specifying that internal material velocities are zero, as in the absence of flow. This also implies and requires that there is no change in mechanical strain with time in the material of the conduction region. That is, the density $\rho(x, y, z)$ is taken to remain constant at each location in the region.

The midpoint of the differential element is denoted as x, y, z. At time τ the temperature and specific heat at this point are t and c. The density $\rho(x, y, z)$ is taken independent of time. The energy generation is assumed due to a source distributed throughout the material. Its strength q''' is the rate of energy generation at x, y, z at time τ per unit volume of the material. This rate may also depend upon local temperature, as with a chemical reaction. Therefore, in general, $q''' = q'''(x, y, z, t, \tau)$, and has the units of energy per unit time and volume. The thermal conductivity k may also depend on location, temperature, and time, due to changing local properties, as $k(x, y, z, t, \tau)$.

In the general analysis, the thermal conductivity at point x, y, z will be considered to depend upon the orientation of the surface for which Fourier's law of conduction is written. The simplest thermal-conductivity characteristic (isotropic) is that in which k is the same in all directions, as in gases, most liquids, and in amorphous solids. Then

$$q''_{x_i} = -k \frac{\partial t}{\partial x_i} \qquad (1.2.1)$$

where x_i is either x, y, or z. The heat flux is then a vector $\bar{q}''$. This is written in terms of the gradient $\bar{\nabla}$ of the temperature field, $\bar{\nabla}t$, as

$$\bar{q}'' = -k\,\bar{\nabla}t \tag{1.2.2}$$

In a more general formulation of conduction (anisotropic), it is postulated that the component of the heat flux in any direction, for example, q_x'' in the x direction, depends upon the temperature gradients in each of the three coordinate directions. That is,

$$-q_x'' = k_{11}\frac{\partial t}{\partial x} + k_{12}\frac{\partial t}{\partial y} + k_{13}\frac{\partial t}{\partial z}$$

Three conductivity coefficients may also arise in the y and z coordinate directions. Then the thermal conductivity becomes the following second order tensor quantity:

$$k_{ij} = \begin{bmatrix} k_{11} & k_{12} & k_{13} \\ k_{21} & k_{22} & k_{23} \\ k_{31} & k_{32} & k_{33} \end{bmatrix} \tag{1.2.3a}$$

Carslaw and Jaeger (1959) summarize the specific form of this tensor for several kinds of crystalline systems. The heat flux component in the x_i direction, q_{x_i}'', is then

$$q_{x_i}'' = -k_{ij}\frac{\partial t}{\partial x_j} \tag{1.2.3b}$$

Recall that one sums on any repeated subscript, for example j in the previous equation. Then, choosing $i = 1, 2, 3$, all three heat flux components are generated. The general anisotropic formulation includes the isotropic case if $k_{ij} = 0$ for $i \neq j$ and if the remaining $k_{ii} = k$; that is, if all nondiagonal terms in the tensor formulation given previously are zero and all the remaining diagonal ones are equal.

Another simple special circumstance often arises in ordered materials such as wood, fibrous materials, and numerous crystalline substances. This occurs when the surfaces upon which boundary conditions are to be applied are normal to the principal axes of conductivity for the anisotropic mechanism. Then the coordinate axes coincide with the principal axes, $k_{ij} = 0$ for $i \neq j$, and the remaining diagonal terms k_{ii} are written as k_i. Such conduction is called orthotropic. The heat flux components then are

$$q_x'' = -k_x\frac{\partial t}{\partial x} \qquad q_y'' = -k_y\frac{\partial t}{\partial y} \qquad q_z'' = -k_z\frac{\partial t}{\partial z} \tag{1.2.4}$$

That is, the directions x, y, and z specify simple directional characteristics of the material.

This method, however, may not be used to simplify all problems. In the event that boundary conditions must be applied on surfaces other than those normal to the coordinate axes, difficulties arise at these surfaces. These matters are discussed in more detail in Carslaw and Jaeger (1959).

The following derivation of the general conduction equation will assume the orthotropic mode of conduction. The isotropic result is then obtained by taking $k_x = k_y = k_z = k$. Subsequently, the general anisotropic form will be inferred from these results. Recall that the general k_{ij} may be functions of location in a nonhomogeneous material and functions of temperature level in any material. They may also be functions of time, either through a dependence on local temperature or through a change in local material state or condition with time. That is,

$$k_{ij} = k_{ij}(x, y, z, t, \tau)$$

Consider an orthotropic material. The rate of heat conduction, per unit area at the point x, y, z in Fig. 1.2.1, and across a surface perpendicular to the x axis at that point, is

$$q_x'' = -k_x \frac{\partial t}{\partial x} \tag{1.2.5}$$

If differentials of higher order than the first are neglected, the instantaneous difference in the rate of heat conduction per unit area across the two surfaces of the differential element $dx\, dy\, dz$, and perpendicular to the x axis, is

$$-\frac{\partial}{\partial x}(q_x'')\, dx = -\frac{\partial}{\partial x}\left(-k_x \frac{\partial t}{\partial x}\right) dx = \frac{\partial}{\partial x}\left(k_x \frac{\partial t}{\partial x}\right) dx$$

This is the change in q_x'' from $x - dx/2$ to $x + dx/2$. The minus sign is introduced to yield a minus quantity if the conduction rate out of the right face is greater than the conduction rate into the left face. That will result in a heat gain in the material region under consideration being a positive quantity.

The net rate of gain in the element $(dx\, dy\, dz)$ due to conduction in the x direction is the net rate, per unit area, times the conduction area normal to the x direction, $dy\, dz$:

$$\frac{\partial}{\partial x}\left(k_x \frac{\partial t}{\partial x}\right) dx\,(dy\, dz)$$

Similar relations are written for the y and z directions as

$$\frac{\partial}{\partial y}\left(k_y \frac{\partial t}{\partial y}\right) dy\,(dx\, dz) \quad \text{and} \quad \frac{\partial}{\partial z}\left(k_z \frac{\partial t}{\partial z}\right) dz\,(dx\, dy)$$

The net rate of energy gain by conduction in volume $dx\, dy\, dz = dV$ is, therefore,

$$\left[\frac{\partial}{\partial x}\left(k_x \frac{\partial t}{\partial x}\right) + \frac{\partial}{\partial y}\left(k_y \frac{\partial t}{\partial y}\right) + \frac{\partial}{\partial z}\left(k_z \frac{\partial t}{\partial z}\right)\right] dx\, dy\, dz$$

The rate of energy generation in this volume, $dV = dx\, dy\, dz$, is equal to the strength of the distributed source per unit volume, $q'''(x, y, z, \tau, t)$, times the volume of the element:

$$q'''(dx, dy\, dz)$$

The time rate of change of stored energy in the element dV, due to a changing temperature level, is equal to the product of its mass, specific heat, and the time rate of change of temperature:

$$(\rho\, dx\, dy\, dz)c\frac{\partial t}{\partial \tau}$$

where, in general, c may vary as $c(x, y, z, \tau, t)$.

The general conduction equation is obtained by equating the sum of the net rates of energy gain by conduction and generation to the time rate of increase of stored energy. The result is

$$\frac{\partial}{\partial x}\left(k_x \frac{\partial t}{\partial x}\right) + \frac{\partial}{\partial y}\left(k_y \frac{\partial t}{\partial y}\right) + \frac{\partial}{\partial z}\left(k_z \frac{\partial t}{\partial z}\right) + q''' = \rho c \frac{\partial t}{\partial \tau} \qquad (1.2.6)$$

This quite general conduction equation is applied to specific circumstances by reducing it to the simplest form consistent with the conditions which must be imposed. Several of the more important of these simpler forms have been given specific names and are presented in the following discussion.

However, at this point the form of the general conduction equation applicable to an anisotropic material will be written. Recall that the individual conduction terms in Eq. (1.2.6) are the gradients of heat flux components in each of the three coordinate directions. For a general anisotropic material the heat flux components are

$$q''_{x_i} = -k_{ij}\frac{\partial t}{\partial x_j} \qquad (1.2.7)$$

Therefore, the typical conduction term (in direction x_i) is

$$\frac{\partial}{\partial x_i}\left(k_{ij}\frac{\partial t}{\partial x_j}\right)$$

The general conduction equation then becomes

$$\frac{\partial}{\partial x_i}\left(k_{ij}\frac{\partial t}{\partial x_j}\right) + q''' = \rho c \frac{\partial t}{\partial \tau} \qquad (1.2.8)$$

This relation reduces to the simpler orthotropic and isotropic conduction mechanisms for the simpler conditions on k_{ij} discussed previously. Other discussions and derivations of various of the equations given in this section are found in Jakob (1949), Schneider (1955), and Carslaw and Jaeger (1959).

1.2.2   Simpler Forms

Consider now the form of Eq. (1.2.8) applicable to an isotropic material, $k_x = k_y = k_z = k$. The subscripts x, y, z are suppressed in Eq. (1.2.6). However, the thermal conductivity k may not be factored out of the three conduction terms unless it is independent of x, y, and z at each time, that is, unless k is uniform over the conduction region at each time. This results in a conduction region consisting of a homogeneous material whose thermal conductivity is independent of temperature. For such a thermal conductivity condition, the differential equation becomes

$$\frac{\partial^2 t}{\partial x^2} + \frac{\partial^2 t}{\partial y^2} + \frac{\partial^2 t}{\partial z^2} + \frac{q'''}{k} = \frac{\rho c}{k}\frac{\partial t}{\partial \tau} = \frac{1}{\alpha}\frac{\partial t}{\partial \tau} \qquad (1.2.9)$$

where the quantity $k/\rho c$, the instantaneous uniform thermal conductivity divided by the thermal capacity ρc, is denoted by α, called the thermal diffusivity.

If, in addition to the conditions necessary for Eq. (1.2.9), there is no distributed source, that is, if $q''' = 0$, and α is constant, the conventional Fourier equation results:

$$\nabla^2 t = \frac{1}{\alpha}\frac{\partial t}{\partial \tau} \qquad (1.2.10)$$

This equation then applies to three-dimensional unsteady-state conduction in an isotropic region of uniform thermal conductivity. It also applies in an isotropic region when α is only a function of time.

If the conditions for Eq. (1.2.10) are met and if the conduction circumstance occurs in steady state, the time derivative is zero. The resulting relation is called the Laplace equation:

$$\nabla^2 t = 0 \qquad (1.2.11)$$

Another common circumstance is steady-state conduction in an isotropic region of uniform thermal conductivity which is subject to a distributed source of energy. The strength of the generation process, q''', may vary over the region as $q'''(x, y, z, t)$. The resulting relation is called Poisson's equation:

$$\nabla^2 t + \frac{q'''}{k} = 0 \qquad (1.2.12)$$

The similarity between Eqs. (1.2.10)–(1.2.12) and those which characterize various other physical phenomena is an interesting and important matter. It means, for example, that the solution of a heat transfer problem may also be the solution of interesting problems involving completely different phenomena. This similarity of equations is also the basis of the electrical and soap-film analogs which have been used to determine solutions of conduction problems. See Gebhart (1971).

Equations (1.2.9)–(1.2.12) apply for an isotropic material for which k is a simple constant throughout the region, or at any time in an unsteady process.

Each equation is linear in temperature t. Equations (1.2.10) and (1.2.11) are also homogeneous, which leads to important simplifications in analysis. In addition, if q''' in Eqs. (1.2.9) and (1.2.12) is directly proportional to temperature, they may also be made homogeneous by a transformation.

However, there are many practical circumstances in which the temperature variation and temperature changes in the conduction region cause an appreciable change in thermal conductivity, if it is temperature dependent, that is, if $k = k(t)$. The conduction region may also be inhomogeneous, with k varying with location. In this event, for an isotropic material, Eq. (1.2.6) becomes

$$\bar{\nabla} \cdot k \bar{\nabla} t + q''' = \rho c \frac{\partial t}{\partial \tau} \tag{1.2.13}$$

where q''' may depend on location, temperature, and time. The first term in this equation is nonlinear.

1.2.3 Transformations

Under some circumstances the nonlinear characteristic of Eq.(1.2.13) may be removed, by the Kirchhoff transformation, for $k = k(t)$. The dependent variable t is transformed to a new quantity V, defined as follows:

$$dV = \frac{k(t)}{k(t_0)} dt = \frac{k}{k_0} dt \tag{1.2.14a}$$

where k_0 is the value of k at some temperature level t_0. The typical derivatives in Eq. (1.2.13) are written in terms of V as follows:

$$\frac{\partial V}{\partial \tau} = \frac{dV}{dt} \frac{\partial t}{\partial \tau} = \frac{k}{k_0} \frac{\partial t}{\partial \tau} \quad \text{and} \quad \frac{\partial^2 V}{\partial x^2} = \frac{1}{k_0} \frac{\partial}{\partial x} \left(k \frac{\partial t}{\partial x} \right)$$

Then Eq. (1.2.13) becomes

$$\nabla^2 V + \frac{q'''}{k_0} = \frac{\rho c}{k} \frac{\partial V}{\partial \tau} = \frac{1}{\alpha} \frac{\partial V}{\partial \tau} \tag{1.2.14b}$$

This equation is linear in V only if the temperature dependence of c and k are proportional to each other. This is a quite special condition. However, in steady-state problems the term on the right-hand side is absent and the transformation to V achieves a general linearized form. See Sec. 9.5 for a numerical analysis of conduction with variable physical properties.

Another interesting transformation of Eq. (1.2.6) removes the complexity of directional conductivities k_x, k_y, and k_z for conduction in orthotropic materials. It also applies in anisotropic materials when principal axes may be used. All thermal-conductivity coefficients are taken as constant and uniform in the region. Introducing the scaling of the x, y, and z coordinates in terms of new space coordinates X, Y, and Z,

$$X = x \left(\frac{k_a}{k_x} \right)^{1/2} \qquad Y = y \left(\frac{k_a}{k_y} \right)^{1/2} \qquad Z = z \left(\frac{k_a}{k_z} \right)^{1/2} \qquad (1.2.15a)$$

results in the following equation, where k_a may be chosen as $\sqrt[3]{(k_x k_y k_z)}$ in interpreting any solution; see Chao (1963):

$$\frac{\partial^2 t}{\partial X^2} + \frac{\partial^2 t}{\partial Y^2} + \frac{\partial^2 t}{\partial Z^2} + \frac{q'''}{k_a} = \frac{\rho c}{k_a} \frac{\partial t}{\partial \tau} \qquad (1.2.15b)$$

Equation (1.2.15) reduces to the special forms in Eqs. (1.2.10)–(1.2.12) when $k_x = k_y = k_z = k$ is both uniform and constant over the conduction region.

In some processes, k may vary only with time, as $k(\tau)$. Then Eq. (1.2.13) becomes Eq. (1.2.9). A new time τ' is defined as

$$d\tau' = \frac{k(\tau)}{\rho c} \, d\tau \quad \text{and} \quad \tau' = \int_0^\tau \frac{k(\tau)}{\rho c} \, d\tau \qquad (1.2.16)$$

For $q''' = 0$, Eq. (1.2.9) becomes

$$\nabla^2 t = \frac{\partial t}{\partial \tau'} \quad \text{where} \quad t = t(x, y, z, \tau') \qquad (1.2.17)$$

This is the same kind of relation as the Fourier equation, for k constant and uniform.

Conductivity is often variable across a conduction region, because of inhomogeneity of the substance or because of varying porosity. The resulting Poisson equation in terms of, for example, $k = k(x, y, z)$, is

$$\nabla \cdot k \, \nabla t + q''' = 0 \qquad (1.2.18a)$$

or

$$\nabla \cdot k \nabla \left(\frac{\sqrt{k} \, t}{\sqrt{k}} \right) + q''' = 0 \qquad (1.2.18b)$$

This form suggests a transformation in terms of $\sqrt{k} \, t = y$, as

$$\nabla \cdot \left(k \nabla (y / \sqrt{k}) \right) + q''' = 0 \qquad (1.2.18c)$$

This result is shown by Munoz and Burmeister (1988) to result in

$$\nabla^2 y - \frac{\nabla^2 \sqrt{k}}{\sqrt{k}} + \frac{q'''}{\sqrt{k}} = 0 \qquad (1.2.19)$$

Solutions of this relation are found for circular geometries in which k varies in appropriate ways. Also see Bellman (1953) and Luikov (1971). These studies reduce the complexity of solutions which account for spatially variable conductivity.

1.2.4 Other Coordinates

The equations developed previously are largely in terms of a Cartesian coordinate system. For many applications the cylindrical or spherical systems are more natural and convenient. For example, the equations for k constant and uniform may be converted to the cylindrical or spherical forms by the relations between coordinates in these different systems, as defined in Fig. 1.2.2. The Fourier equation (1.2.10), in cylindrical coordinates (r, θ, z), is

$$\frac{\partial^2 t}{\partial r^2} + \frac{1}{r}\frac{\partial t}{\partial r} + \frac{1}{r^2}\frac{\partial^2 t}{\partial \theta^2} + \frac{\partial^2 t}{\partial z^2} + \frac{q'''}{k} = \frac{1}{\alpha}\frac{\partial t}{\partial \tau} \tag{1.2.20}$$

where $x = r\cos\theta$ and $y = r\sin\theta$. In polar spherical coordinates (r, θ, ϕ), it becomes

$$\frac{\partial^2 t}{\partial r^2} + \frac{2}{r}\frac{\partial t}{\partial r} + \frac{1}{r^2\sin\theta}\frac{\partial}{\partial\theta}\left(\sin\theta\frac{\partial t}{\partial\theta}\right)$$

$$+ \frac{1}{r^2\sin^2\theta}\frac{\partial^2 t}{\partial\phi^2} + \frac{q'''}{k} = \frac{1}{\alpha}\frac{\partial t}{\partial \tau} \tag{1.2.21}$$

where $x = r\sin\theta\cos\phi$, $y = r\sin\theta\sin\phi$, and $z = r\cos\theta$.

For variable conductivity, $k(x, y, z, t, \tau)$, in the conduction region, Eq. (1.2.13) applies. The resulting forms of $\overline{\nabla} \cdot k\,\overline{\nabla}t$ in cylindrical and spherical

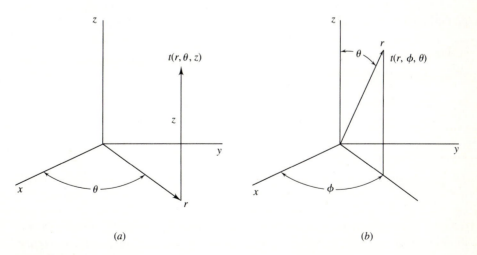

(a)   (b)

FIGURE 1.2.2
The cylindrical and spherical coordinate systems.

coordinates result in

$$\frac{1}{r}\frac{\partial}{\partial r}\left(rk\frac{\partial t}{\partial r}\right) + \frac{1}{r^2}\frac{\partial}{\partial \theta}\left(k\frac{\partial t}{\partial \theta}\right) + \frac{\partial}{\partial z}\left(k\frac{\partial t}{\partial z}\right) + q''' = \rho c\frac{\partial t}{\partial \tau} \qquad (1.2.22)$$

$$\frac{1}{r^2}\frac{\partial}{\partial r}\left(r^2 k\frac{\partial t}{\partial r}\right) + \frac{1}{r^2 \sin \theta}\frac{\partial}{\partial \theta}\left(k \sin \theta\frac{\partial t}{\partial \theta}\right)$$

$$+ \frac{1}{r^2 \sin^2 \theta}\frac{\partial}{\partial \phi}\left(k\frac{\partial t}{\partial \phi}\right) = \rho c\frac{\partial t}{\partial \tau} \qquad (1.2.23)$$

in terms of the coordinates r, θ, and ϕ defined in Fig. 1.2.2.

SOLUTIONS. The preceding collection of conduction differential equations is adequate to most needs. The meaning of the particular equation applicable to any given physical process is that it specifies the form the temperature distribution must have in order that the requirement of conservation of energy is met at every point in the conduction region at all times.

The particular information required in conduction analysis depends upon the use to be made of the results. Occasionally, the temperature distribution itself is wanted as a guide for material selection, as in a nuclear-reactor core or rocket nozzle. Sometimes the analysis is intended instead to yield heat flow rate information, for example, in walls and other structural members. Another frequent need is information concerning the temperature distribution and its change with time in order to evaluate thermal stressing effects. These may arise because of differential thermal expansion over the conduction region. For all of these needs, the same initial approach is followed. The solution is determined. That is, the temperature distribution $t(x, y, z, \tau)$ is found throughout the conduction region.

Admissible forms of such solutions are dictated by the relevant differential equation. The particular and definite result is determined by any imposed boundary conditions. For example, these may simply be the temperatures on surfaces that the solution must also satisfy. In addition, in a transient problem, the solution must also satisfy, at some specified time τ, an initially known temperature distribution. This is called the initial condition. Thus there are boundary-condition and initial-condition circumstances. These and other bounding conditions are discussed in detail in Sec. 1.4.

1.3   MASS DIFFUSION EQUATIONS

The importance of mass diffusion processes continues to increase with developing technology in materials treatment, processing, and fabrication. Mass diffusion is also increasingly important in determining the effective life of many devices. Many of these depend, for their continuing successful operation, on initially built-in concentration differences and gradients. These tend to diffuse away with time. There are also increasing applications in evaluating diffusion effects in biological materials and in chemical separation, as in reverse osmosis.

1.3.1 The Diffusion Coefficient

Over the span of such processes many different kinds of mechanisms arise. One basic consideration is the kind of mass diffusion coefficient, D, which will properly describe the process. As written most simply, in Eq. (1.1.6), D is a constant. It may also be a function of location, temperature, concentration, and time, as $D(x, y, z, t, C, \tau)$ in general. This is the isotropic formulation. The mass flux component in the n direction m''_n, where n may be x, y, or z, may be written as

$$m''_n = -D\frac{\partial C}{\partial n} \quad \text{or} \quad \bar{m}'' = -D\bar{\nabla}C \tag{1.3.1}$$

This is analogous to Eq. (1.2.2) for conduction.

However, in some materials D may be different in different directions, as in water vapor migration through wood having an important grain structure. Then D might be expressed as D_x, D_y, D_z. This is an orthotropic formulation. This is like that for k_{x_i}, as k_x, k_y, k_z in Eq. (1.2.1). This leads to Eq. (1.2.6), which governs the temperature distribution $t(x, y, z, \tau)$.

The most general diffusion formulation is that for an anisotropic region, as in Eq. (1.2.3a) for conduction. That is,

$$D_{ij} = \begin{bmatrix} D_{11} & D_{12} & D_{13} \\ D_{21} & D_{22} & D_{23} \\ D_{31} & D_{32} & D_{33} \end{bmatrix} \tag{1.3.2a}$$

The mass diffusion rates in the three coordinate directions m''_i, that is, m''_x, m''_y, m''_z, become, as in Eq. (1.2.3b) for thermal diffusion,

$$m''_{x_i} = -D_{ij}\frac{\partial C}{\partial x_j} \tag{1.3.2b}$$

where the sum over $j = 1, 2, 3$ results in m''_{x_i} and $i = 1, 2, 3$ gives the three directional components m''_i.

The orthotropic circumstance is recovered in the general formulation Eq. (1.3.2b) if the D_{ij} for $i \neq j$ are zero. Then D has three components D_x, D_y, D_z. If, in addition, $D_x = D_y = D_z$, the diffusion process is isotropic, as in Eq. (1.3.1). In general, the diffusion coefficient D_{ij} may depend on location, direction, local temperature t, concentration C, and time.

Also, as indicated in Sec. 1.1.3, diffusion coefficients at any location may be the sum of several local parallel processes which diffuse the species C through the material. For example, a vapor may diffuse through the voids in a porous material, whereas the liquid phase flows by capillary forces along the solid internal surfaces of the structure. These more complicated matters are considered in Chap. 6 and may also be referred to in Crank (1979), Schewmon (1963), and Bird et al. (1960). In this section it is assumed that only one kind of diffusion process arises, as characterized by Eq. (1.3.2b), in general.

OTHER MASS DIFFUSION EFFECTS. In addition to the frequent effects of local concentration C and temperature on the diffusion coefficient, other mechanisms arise which affect the local internal concentration level. One mechanism is the adsorption or reaction of the diffusing species locally in the material in the region. An example of adsorption arises in putting a dry piece of material, such as wood, in a humid environment. The water vapor diffuses inward, due to the concentration, or partial pressure difference. The diffusing vapor is locally partially adsorbed as it diffuses inward to the drier interior. If this wood is then put in a dry environment, the direction of diffusion reverses. Vapor arises locally and diffuses outward as the wood dries.

These two processes amount to a local sink and source of species C, H_2O vapor in this example. The volumetric rate of adsorption or release of vapor is formulated as $C'''(x, y, z, C, t, \tau)$ or as $m'''(x, y, z, C, t, \tau)$. The units of C''' are commonly mass per unit volume of region per unit of time, M/L^3T. This quantity is analogous to q''', the strength of a distributed energy source, $q'''(x, y, z, t, \tau)$, in Sec. 1.2, in W/L^3T.

Such a local internal disappearance or appearance of species C often arises from a chemical effect or reaction. This effect may be either exothermic or endothermic. Then a positive or negative contribution to q''' also arises. The rate may depend on C. This must be included in the energy equation analysis for the region, Eq. (1.2.8) in general. Chemical reactions are usually rate processes, in terms of a rate constant. The forward rate is assumed, for example, to consume species C locally. This may be countered by an opposing reaction rate, which produces species C.

These processes may or may not be in local equilibrium, at each local concentration and temperature level arising across the diffusion region. Also, these chemical processes are in part determined by the temperatures, t_s, and concentrations, C_s, imposed or resulting on the surfaces of the conduction–diffusion region. Therefore, the formulation to determine both the temperature and concentration distributions is coupled. It must then include both the chemical rate constants and the net distributed source terms C''' and q'''.

1.3.2   The Mass Diffusion Equation

This relation is written in terms of a balance of a diffusing species, C, in the differential control volume $dx\,dy\,dz = dV$ in Fig. 1.3.1. The concentration is $C(x, y, z, \tau)$ at x, y, z, where any species production or adsorption rate is $C'''(x, y, z, t, C, \tau)$. The diffusion region will be assumed to be orthotropic with mass diffusivities D_x, D_y, and D_z in the coordinate directions. The local material density may vary across the region, as in an inhomogeneous porous region, as $\rho(x, y, z)$. However, it is assumed that mass diffusion effects do not distort the material with time. This is the equivalent of the condition of no change in mechanical strain, invoked in deriving the conduction equations in Sec. 1.2.

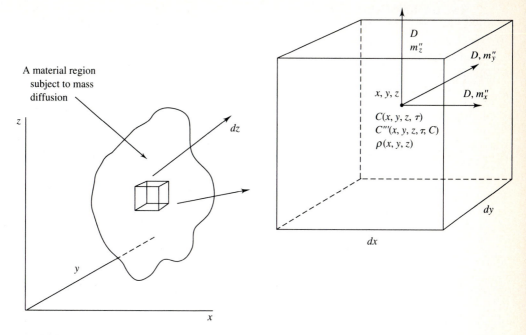

FIGURE 1.3.1
Differential element for mass diffusion analysis.

The balance begins by summing the rates of contributions of species C to the volume $dV = dx\,dy\,dz$, by the fluxes m''_x, m'_y, and m''_z and the local source C'''. This rate is then set equal to the rate of increased storage of species C in $dy\,dy\,dz$, that is, $(\partial C/\partial\tau)\,dV$.

Consider first the diffusion flux component m''_x at x, in the x direction. The local diffusion rate is $m''_x\,dy\,dz$. The change from $x - dx/2$ to $x + dx/2$ is the derivative of the local diffusion rate:

$$\frac{\partial}{\partial x}(m''_x\,dy\,dz)\,dx = \left[\frac{\partial m''_x}{\partial x}\right]dV = \left[\frac{\partial}{\partial x}\left(-D_x\frac{\partial C}{\partial x}\right)\right]dV$$

If the preceding quantities in brackets, the local gradient of the mass flux, are positive, the loss of C at $x + dx/2$ is greater than the gain at $x - dx/2$. This amounts to a rate of loss. Therefore, a gain is positive if the preceding term is taken as negative, as follows:

$$-\frac{\partial}{\partial x}(m''_x\,dy\,dz)\,dx = \frac{\partial}{\partial x}\left(D_x\frac{\partial C}{\partial x}\right)$$

Three such terms for diffusion are summed, with $C'''\,dV$ taken positive for a source of species C. This sum is set equal to the rate of increasing storage, to give the following equation for $C(x, y, z, \tau)$:

$$\frac{\partial}{\partial x}\left(D_x \frac{\partial C}{\partial x}\right) + \frac{\partial}{\partial y}\left(D_y \frac{\partial C}{\partial y}\right) + \frac{\partial}{\partial z}\left(D_z \frac{\partial C}{\partial z}\right) + C''' = \frac{\partial C}{\partial \tau} \qquad (1.3.3)$$

For D_x, D_y, and D_z constant and uniform in the region,

$$D_x \frac{\partial^2 C}{\partial x^2} + D_y \frac{\partial^2 C}{\partial y^2} + D_z \frac{\partial^2 C}{\partial z^2} + C''' = \frac{\partial C}{\partial \tau} \qquad (1.3.4)$$

For isotropic diffusion, $D = D_x = D_y = D_z$, where D may still be variable, as $D(x, y, z, C, \tau)$, the equation becomes

$$\frac{\partial}{\partial x}\left(D \frac{\partial C}{\partial x}\right) + \frac{\partial}{\partial y}\left(D \frac{\partial C}{\partial y}\right) + \frac{\partial}{\partial z}\left(D \frac{\partial C}{\partial z}\right) + C''' = \bar{\nabla} \cdot D \bar{\nabla} C + C''' = \frac{\partial C}{\partial \tau}$$

$$(1.3.5)$$

If, in addition, D is uniform over the region at all times,

$$D\left(\frac{\partial^2 C}{\partial x^2} + \frac{\partial^2 C}{\partial y^2} + \frac{\partial^2 C}{\partial z^2}\right) + C''' = D \nabla^2 C + C''' = \frac{\partial C}{\partial \tau} \qquad (1.3.6)$$

These relations, which may be collectively called Fick's second law, are completely analogous to those for thermal conduction. Compare Eqs. (1.3.3) and (1.2.6) for orthotropic regions, and Eqs. (1.3.5) and (1.2.13) for isotropic regions. The Fourier, Laplace, and Poisson equations, like Eqs. (1.2.10)–(1.2.12), again arise here, from Eq. (1.3.6), for $C''' = 0$, C''' and $\partial C/\partial \tau = 0$, and $\partial C/\partial \tau = 0$, respectively. See also the result in Crank (1979) for anisotropic processes.

The previous mass diffusion balance and resulting equations are formulated for species diffusion through a fixed and stationary matrix. This approximation is reasonable for a very large proportion of the processes considered here. However, for appreciable simultaneous counterdiffusion, as for the interdiffusion of two gases, the preceding formulation is not always a sufficiently accurate procedure. Then other flux formulations become necessary. These are reviewed by Edwards et al. (1979). A frequently adequately accurate improvement is to formulate the local diffusion mass flux rate relative to the local mass average velocity instead.

1.3.3   Distributed Sources and Sinks

The local species source or sink of diffusing species C is expressed as $C'''(x, y, z, t, C, \tau)$, per unit volume and time. This is the rate at which the diffusing material is mobilized, $C''' > 0$, or immobilized, $C''' < 0$, by the physical or chemical structure of the material. These effects are often called physisorption and chemisorption. Common examples of mobilization occur in layers of plastics which solidify by a chemical reaction which releases chemical species or solvents. Immobilization occurs, for example, in the carburization of steel at high temperature. Then the diffusing carbon atoms are locally adsorbed in the steel. Another example of adsorption is the humidification of dry wood placed in a humid environment, as water vapor diffuses through it. Desorption, $C''' > 0$, occurs as wood is dried in air at low humidity.

Such effects arise as transient processes when the immobilized concentration C has not yet reached an equilibrium distribution, as when carburizing steel, that is, $C''' < 0$, or in drying wood, $C''' > 0$. These effects may also occur in steady state in a process in which the species C is continuously consumed in a local chemical process. An example of this is the oxygen diffusion into living material, to support metabolic and other chemical processes. In steady state, the absence of local processes which create or consume C results in $C''' = 0$.

The evaluation of C''' in other circumstances often depends upon the physics and chemistry of the local processes. Several general and characteristic mechanisms arise. If the chemical species is merely created or adsorbed uniformly in the region, as in the aging of a material, C''' is uniform and positive and negative, respectively.

However, in many important processes the local concentration level of diffusing species C interacts directly with the release or adsorption of that particular species in the material. That is, more is adsorbed at high local concentrations of diffusing species C. If the reaction which adsorps or desorps species C locally is very fast compared to the diffusion rate $\bar{m}''$, then the instantaneous local level of material immobilization, or adsorption, $c(x, y, z, \tau)$, may be proportional to the local diffusing concentration, C, as follows, where R_r is a constant of proportionality. Here, R_r and D are assumed not affected by any temperature effects in the region:

$$c = R_r C \quad \text{and} \quad \frac{\partial c}{\partial \tau} = R_r \frac{\partial C}{\partial \tau} \tag{1.3.7}$$

For example, if the local diffusing concentration C increases, the adsorbed material, as c, increases at a comparable rate. This amounts to a sink, or $C''' = -\partial c/\partial \tau$. Therefore, Eq. (1.3.6) becomes

$$D \nabla^2 C - \frac{\partial c}{\partial \tau} = \frac{\partial C}{\partial \tau} \tag{1.3.8a}$$

or

$$\frac{D}{(1 + R_r)} \nabla^2 C = D' \nabla^2 C = \frac{\partial C}{\partial \tau} \tag{1.3.8b}$$

Thus this effect is very simply accommodated by using the effective diffusivity $D' = D/(1 + R_r) < D$ in the analysis. The physical result of this adsorption mechanism is a decreased mass diffusion flux, as R_r increases. That is, the very mobile local storage mechanism of species C damps imposed transient changes of concentration conditions applied at the boundaries of the region.

Other kinds of processes arise which require nonlinear relations such as $c = R_r C^n$, $n \neq 1$. Also, R may be temperature dependent. In other processes, adsorption and desorption may be the net rate of simultaneous forward and reverse chemical reactions. These additional considerations are discussed in detail in Sec. 6.5.

1.3.4 Equation Transformations

As seen for heat conduction, there are also several transformations which will reduce the complexity of the preceding differential equations, under certain restricted physical conditions. First, for Eq. (1.3.5), where D may be a function of local concentration C, as $D(C)$, the Kirchhoff transformation is again used. The method is to transform $C(x, y, z, \tau)$ into a new variable $W(x, y, z, \tau)$, defined as follows:

$$dW = \frac{D(C)}{D(C_0)} dC = \frac{D}{D_0} dC \qquad (1.3.9a)$$

where D_0 is a reference value of D at C_0. Typical terms in Eq. (1.3.5) are then written in terms of W as

$$\frac{\partial W}{\partial \tau} = \frac{dW}{dC} \frac{\partial C}{\partial \tau} = \frac{D}{D_0} \frac{\partial C}{\partial \tau} \quad \text{and} \quad \frac{\partial^2 W}{\partial x^2} = \frac{1}{D_0} \frac{\partial}{\partial x} \left(D \frac{\partial D}{\partial x} \right)$$

Then Eq. (1.3.5) becomes

$$\nabla^2 W + \frac{C'''}{D_0} = \frac{1}{D} \frac{\partial W}{\partial \tau} \qquad (1.3.9b)$$

This equation is not linear because of $D(C)$ on the right side. However, in steady-state processes this term disappears, yielding a linear equation in W of the Poisson form.

Another transformation applies to transients in orthotropic regions, with D_x, D_y, and D_z unequal but constant and uniform, Eq. (1.3.4). The x, y, and z coordinates are then scaled as follows:

$$X = x \left(\frac{D_a}{D_x} \right)^{1/2} \qquad Y = y \left(\frac{D_a}{D_y} \right)^{1/2} \qquad Z = z \left(\frac{D_a}{D_z} \right)^{1/2}$$

to yield

$$\frac{\partial^2 C}{\partial X^2} + \frac{\partial^2 C}{\partial Y^2} + \frac{\partial^2 C}{\partial Z^2} + \frac{C'''}{D_a} = \frac{1}{D_a} \frac{\partial C}{\partial \tau} = \frac{\partial C}{\partial (D_a \tau)} \qquad (1.3.10)$$

where D_a may be taken as $\sqrt[3]{D_1 D_2 D_3}$.

A sometimes convenient transformation arises when a uniform diffusivity D in the region changes with time, as $D(\tau)$. This could arise in a changing physical state through time, as by material aging. A new time τ' is defined as

$$d\tau' = D(\tau) \, d\tau$$

Then Eqs. (1.3.5) or (1.3.6), with $C''' = 0$, become

$$\nabla^2 C = \frac{\partial C}{\partial \tau'} \quad \text{where } C = C(x, y, z, \tau')$$

(1.3.11)

and

$$\tau' = \int_0^\tau D(\tau)\, d\tau$$

(1.3.12)

Thereby, the relation between τ' and τ may be determined for the variation $D(\tau)$ which applies.

1.4 BOUNDING CONDITIONS AND ANALYSIS

Sections 1.2 and 1.3 developed the differential equations of both heat and mass transfer for a wide range of the physical conditions encountered in practice. The role of these equations is to impose limits on the forms of the solutions, $t(x, y, z, \tau)$ and $C(x, y, z, \tau)$, over the conduction and diffusion regions. These distributions must be consistent with the laws they embody. The equations indicate the admissible ways in which the solutions may depend on x, y, z, and τ.

However, to completely specify a solution for any particular physical circumstance, the bounding conditions these solutions must also satisfy must be completely specified. These include conditions on or at the boundaries of the region. Common kinds of such boundary conditions are: specified surface temperature or concentration levels in t_s or C_s; imposed flux levels of q_s'' or m_s''; and convective processes between the surface and a surrounding environment at some temperature or concentration level, t_e or C_e. Thermal radiant exchange may also occur at a boundary. Commonly there are also other surface loading effects such as penetrating solar or microwave radiation and fission product irradiation. These latter mechanisms may cause internal generation, as q''' or C'''.

In steady-state processes such boundary conditions are commonly the only additional considerations which arise to determine the solution for $t(x, y, z)$ or $C(x, y, z)$, interior to the region. However, if the process in the region is transient, the solution also requires the internal temperature or concentration distribution at some specified time, usually taken as $\tau = 0$. That is, the initial condition is $t_i(x, y, z, 0)$ or $C_i(x, y, z, 0)$. Transients also require boundary conditions. These sometimes also depend upon time.

An important special kind of a time-dependent process is the steady periodic. This is defined as a transient in which the temperature or concentration at each point in the region varies periodically with time. That is, the average value of t or C at each point in the region does not change through time. The initial condition is then replaced by the periodic boundary or internal condition which drives the process.

A periodic model approximates the common example of the variation of the temperature in the surface layer of exposed soil, driven by the daily cycle of atmospheric conditions. It has a period of one day. This example also has an additional feature, the conduction of the geothermal heat q_g'' up from far below. The surface process is still a periodic. However, it has the feature that the average value of t increases downward. This process then amounts to a steady periodic, superimposed on a one-dimensional conduction process. A comparable example is the periodic temperature variation in the cylinder wall of a reciprocating engine, water cooled on the outside. A steady periodic may also arise as a combination of a collection of constituent individual steady harmonics. An example is the annual variation of soil surface temperature.

1.4.1   Temperature Boundary Conditions

These are formulated as they apply to a three-dimensional region, with the surface area locations of the region designated as s. The six following conditions (T.1 to T.6) include: specified surface temperature, t_s; imposed surface heat flux, q_s''; surface convection, h; a combination of surface flux, convection, and surface irradiation at a rate of q_s''; a radiation interaction with an adjacent environment consisting of a group of surfaces at different temperature levels, t_i; and surface irradiation at a rate Q_s'' by a penetrating energy effect, such as microwaves. Several of these conditions are expressed in terms of conduction region heat flux imposed at the surface. This is accommodated by inward conduction into the region as $-k(\partial t/\partial n)_s$. Then the local coordinate n is inward and normal to the surface, over its extent, where $n = 0$ at the surface.

T.1. AN IMPOSED SURFACE TEMPERATURE. $t_s(x, y, z, \tau)$ is imposed over the surface which bounds the region. This is sometimes called a Dirichlet condition. Some forms of this condition are:

1. The local t_s independent of time, $t_s(x, y, z)$.
2. Uniform t_s over the surface, but changing with time, $t_s(\tau)$.
3. Uniform and constant t_s; a homogeneous boundary condition results, for $t(x, y, z, \tau)$, written as $\theta(x, y, z, \tau) = t(x, y, z, \tau) - t_s$. Then θ over the boundary is 0.

T.2. AN IMPOSED SURFACE HEAT FLUX CONDITION. Over the surface, this is $q_s''(x, y, z, \tau)$. This is sometimes called a von Neumann condition. The region surface condition is

$$q_s''(x, y, z, \tau) = -k(\partial t/\partial n)_s \tag{1.4.1}$$

The conduction inward, into the region, $-k(\partial t/\partial n)_s$, balances the imposed flux q_s''. A special form of this condition arises for $q_s'' = 0$, a perfectly insulating environment. Then $(\partial t/\partial n)_s = 0$.

T.3. AN IMPOSED CONVECTION PROCESS AT THE SURFACE. An adjacent ambient medium, locally at $t_e(x, y, z, \tau)$, with a local convection coefficient $h(x, y, z, \tau, t_e, t_s) = h$ at the surface:

$$h[t_e(x, y, z, \tau) - t_s(x, y, z, \tau)] = -k\left(\frac{\partial t}{\partial n}\right)_s \qquad (1.4.2)$$

For h and t_e constant and uniform, again

$$h[t_e - t_s(x, y, z, \tau)] = -k\left(\frac{\partial t}{\partial n}\right)_s \qquad (1.4.3)$$

T.4. A COMBINATION OF CONVECTION AND FLUX EFFECTS. Here a given surface flux loading, q_s'', as by radiation, is applied where t_e, h, t_s, and q_s'' may or may not depend on (x, y, z, τ) over the surface:

$$h(t_e - t_s) + q_s'' = -k\left(\frac{\partial t}{\partial n}\right)_s \qquad (1.4.4)$$

Each of the last three conditions amount to requiring that the solution in the region, $t(x, y, x, \tau)$, has a certain temperature gradient approaching the surface. This must balance, by internal conduction, the sum of the energy effects imposed at the surface.

T.5. RADIATION SURFACE LOADING $A_s = A_j$. The equivalent surrounding enclosure, shown in Fig. 1.4.1, consists of n gray and opaque radiating surfaces of area A_i, at T_i. A conduction region surface temperature $T_s = T_j$ is shown. The total enclosure, is an assembly of n surfaces, as idealized in Fig. 1.4.1. The net rate of radiant loss from this typical surface area of such an enclosure, $A_j = A_s$, is

$$q_j = W_j A_j - \sum_{i=1}^{n} B_{ij} W_i A_i \qquad (1.4.5)$$

where $W_j = \epsilon_j \sigma T_j^4$ is the hemispherical emission rate per unit area of $A_s = A_j$, $\epsilon_j = \alpha_j = 1 - \rho_j$ is the emittance, α_j is the absorptance, and ρ_j is the reflectance. The surface A_j is taken to be the boundary of any particular conduction region of interest. Then $t_s = t_j$ becomes the absolute temperature $T_j = T_s$ and the surface flux q_s'', formulated in T.2, becomes $-q_j/A_j$. Recall that q_s'' is considered positive in T.2 if it is an energy addition to the region, whereas q_j is defined in Eq. (1.4.5) as positive if it is a loss.

The first term in Eq. (1.4.5) is the gross emission rate of surface A_j. Each term in the sum is the net absorption rate, at A_j, of the gross emission of each of the n surfaces, A_{ij} including A_j. The absorbtion factors B_{ij} are, therefore, the fraction of the total emission rate of surface A_i, eventually absorbed at A_j. This summation is to include all effects of reflectance ρ_i, at all n surfaces. The

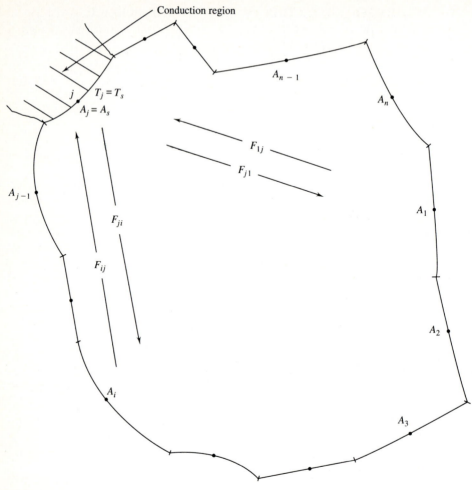

FIGURE 1.4.1
A radiant enclosure, including the surface area of extent A_s of a conduction region.

equations for the B_{ij} are given as follows; see also Gebhart (1957):

$$\sum_{i=1}^{n} (\alpha_{pi} - \delta_{pi}) B_{ij} + F_{pj}\epsilon_j = 0 \qquad (1.4.6)$$

where $\alpha_{pi} = F_{pi}\rho_i$, δ_{pi} is Kronecker's delta, and F_{pj} is the angle factor from A_p to A_j. This formulation, in terms of B_{ij}, applies for diffuse emission and reflection at each surface. It also assumes uniform irradiation over each surface, from each other surface. Then reflections may also be distributed in terms of the F_{ij}.

The angle factors in Eq. (1.4.6) are the fraction of the total emission of surface $A_{p,i}$ which irradiates surface A_j directly, that is, not including any intervening reflections from any surface. The angle factors, for example the F_{ij}, are calculated in terms of the geometric relation between surfaces A_i and A_j. See Siegel and Howell (1981), Howell (1982), Rohsenow et al. (1985), and Gebhart (1971). For an enclosure, as in Fig. 1.4.1,

$$\sum_{j=1}^{n} F_{ij} = 1 \tag{1.4.7}$$

Also, in general, the F_{ij} and F_{ji} are related by a reciprocity relation, such as $F_{ij}A_i = F_{ji}A_j$, for diffuse emission and reflection. There are n^2 values of F in an n-surfaced enclosure.

There are also n^2 values of B_{ij}. These are given by Eq. (1.4.6) upon summing on $p = 1, 2, \ldots, n$ and then on $j = 1, 2, \ldots, n$. The sum on p results in n independent equations for $B_{1j}, B_{2j}, \ldots, B_{nj}$. This is the jth column in the B_{ij} matrix. This information is sufficient to calculate q_j, for any j, from Eq. (1.4.6). Summing on p and j generates the whole B_{ij} matrix. Equation (1.4.6) may also be written as

$$\sum_{i=1}^{n} (\alpha_{pi} - \delta_{pi})K_{ij} + \frac{F_{pj}}{A_j} = 0 \tag{1.4.8}$$

where $K_{ij} = B_{ij}/\epsilon_j A_j = K_{ji} = B_{ji}/\epsilon_i A_i$. This is the K_{ij} matrix $\overline{K}$. It is symmetric, as is the F_{ij}/A_j matrix, $\overline{F}$. See Gebhart (1971). The matrix form of Eq. (1.4.8), in terms of the identity matrix $\overline{I}$, becomes

$$\overline{K} = -(\overline{\alpha} - \overline{I})^{-1}\overline{F}\overline{A}^{-1} \tag{1.4.9}$$

or
$$\overline{K}^{-1} = -\overline{A}\overline{\rho} + \overline{A}\overline{F}^{-1} \tag{1.4.10}$$

This calculation procedure for the B_{ij} is very simple, after the F_{ij} are known. However, the exact geometric determinations of F_{ij} are usually very demanding, except for very simple geometric relationships. See Siegel and Howell (1981) for a large collection of information. Approximation methods are often used for large relative spacing of the surfaces.

The preceding formulation is also very simply applied as a uniform flux surface radiant dissipation rate, as q_s''. The index j in Eq. (1.4.5) may be denoted as s. Then $A_j = A_s$ and $T_j = T_s$. The value of q_s'' is

$$q_s'' = \frac{q_s}{A_s} = \frac{W_s A_s - \sum_{i=1}^{n} B_{is}W_i A_i}{A_s}$$

$$= W_s - \sum_{i=1}^{n} B_{is}W_i a_i \tag{1.4.11}$$

where $a_i = A_i/A_s$ is the dimensionless area of surface A_i. Note that the preceding analysis assigns only one temperature level to each surface, this is to be some appropriate average value. Increased accuracy may be achieved by subdividing areas which have a large temperature variation over their extent. The previous general formulation may also be adapted to insulated surfaces, and to openings and windows in the enclosure; see Gebhart (1971). See also Siegel and Howell (1981), Sect. 8-3.2, for a matrix inversion technique similar to that in Eqs. (1.4.9) and (1.4.10).

The preceding general result of surface radiation flux loading, q_s'' in Eq. (1.4.11), may be approximated in simpler form in many applications. If the surface area A_s is actually in radiation energy exchange with an extensive environment, $A_e \gg A_s$, uniformly at T_e, then $q_s'' = A_s \alpha_s (T_e^4 - T_s^4)$, if the extent of surface A_s is at a uniform temperature of T_s.

T.6. INTERNAL REGION IRRADIATION. This is taken to arise from energy effects coming from the surroundings, which penetrate the region. These effects may be absorbed and dissipated internally. Examples are microwave heating, fission product irradiation and deceleration, and solar energy penetration. The incident energy intensity at the surface is taken as Q_s''. At some distance n inside the region, it is diminished by absorption to $Q''(n)$. We assume that the "absorption length" for this energy source is short, compared to any radius of curvature of the surface. The absorption occurs inside a region, as modeled in Fig. 1.4.2.

In Fig. 1.4.2, consider the thin internal layer of material, dn, in which amount dQ'' is absorbed, where dQ'' is taken as proportional to the local intensity $Q''(n)$, as follows:

$$dQ'' = -\mu Q''(n)\, dn \quad \text{or} \quad \frac{dQ''}{Q''(n)} = -\mu\, dn$$

where μ is the absorption coefficient. The solution for $Q''(n)$ is then an

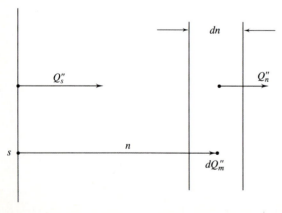

FIGURE 1.4.2
Energy absorption in the conduction region.

exponential decay in local intensity, as

$$\frac{Q''(n)}{Q''_s} = e^{-\mu n} \qquad (1.4.12)$$

The energy absorbed per unit volume locally in the region, q''', is

$$q''' = -\frac{dQ''_n}{dn} = \mu Q(n) = \mu Q''_s e^{-\mu n} \qquad (1.4.13)$$

This quantity is the local strength of the distributed energy source to be used in the relevant differential equation.

SUMMARY. The preceding conditions cover most circumstances which arise. However, some others are often too complicated or nonlinear to permit a mathematical determination of the conduction region temperature distribution. Then recourse is to numerical or other approximate analysis. In such techniques the previous complexities may usually be accommodated quite simply. Other interesting discussions of bounding conditions are to be found in Carslaw and Jaeger (1959), Arpaci (1966), and Özisik (1980).

1.4.2 Concentration Boundary Conditions

Several commonly occurring such conditions are similar to temperature conditions T.1 and T.3 given previously.

C.1. AN IMPOSED SURFACE CONCENTRATION LEVEL. This is for $C_s(x, y, x, \tau)$ applied over the surface which bounds the diffusion region. Three similar special forms commonly arise:

1. $C_s(x, y, z)$ independent of time.
2. C_s uniform, over the surface, but changing with time, as $C_s(\tau)$.
3. C_s uniform and constant, again a homogeneous condition for the new variable $C(x, y, z, \tau) - C_s$, which is zero over the boundary.

C.2. CONVECTION OF SPECIES C TO THE SURFACE. This arises from an adjacent ambient medium. The ambient has concentration levels locally of $C_e(x, y, x, \tau)$, with a surface mass transfer convection coefficient $h_m(x, y, z, \tau, C_e, C_s) = h_m$:

$$h_m[C_e(x, y, z, \tau) - C_s(x, y, z, \tau)] = -D\left(\frac{\partial C}{\partial n}\right)_s \qquad (1.4.14)$$

For h_m and C_e constant and uniform, again

$$h_m[C_e - C_s(x, y, z, \tau)] = -D\left(\frac{\partial C}{\partial n}\right)_s \qquad (1.4.15)$$

There are common circumstances of mass diffusion which also amount to a specified surface flux of a species at the surface, as $m''_s(x, y, z, \tau)$. This is the analog of q_s in conditions T.2 and T.4. However, these conditions often follow from more complicated physical mechanisms. Therefore, this formulation is deferred to Chap. 6, after several other necessary concepts have been clarified.

Also the kind of penetrating surface irradiation process which resulted in condition T.6 and Eq. (1.4.13) has implications related to mass transfer. Penetrating energy or particle effects may also cause time-dependent changes in internal diffusion mechanisms. For example, the resulting local thermal generation process modeled in T.6, as Eq. (1.4.13), would affect mass diffusion by an adsorption mechanism similar to that modeled in Eq. (1.3.8). A fission product irradiation may also produce other species, in local collisions and interactions. This would thereby generate additional diffusing species and mechanisms. Some similar mechanisms are considered in Chap. 6.

1.4.3 Initial Conditions

In conduction and mass diffusion, initial conditions commonly arise as the initial temperature and concentration distributions in the region. These distributions change after the time at which the bounding values are changed to different levels. The initial conduction condition over a region is $t(x, y, x, 0)$. This develops, through later time, as $t(x, y, x, \tau)$, subject to the new boundary conditions. The changed conditions might also include the beginning of a distributed internal energy source, q''', as by beginning irradiation, microwave, or induction heating. Mass diffusion initial conditions are often equally simple. Later chapters indicate in detail the ways in which initial conditions are accommodated, by several methods, to determine transient response.

1.4.4 Subsequent Analysis

This chapter has set forth the physical processes of heat conduction and mass diffusion transport. These were the basis of the generation of quite general equations governing both the instantaneous temperature and concentration distributions in regions. These equations were then reduced to the simpler forms adequate for a very large majority of applications. The boundary and initial conditions given also anticipate quite complicated circumstances. However, they reduce simply to forms suitable for the more common circumstances.

The chapters which follow develop the results which stem from the preceding formulations. Chapter 2 demonstrates many of the mechanisms, in the simplest application, steady state in one dimension. Chapter 3 concerns the particular methods used and the resulting behavior in multidimensional steady-state processes. Chapters 4 and 5 then consider transients, first for one and then for multidimensional regions. The Laplace and separation of variables methods of solution are also demonstrated and applied. Chapters 3 and 4 also consider analysis and approximation methods which are sometimes more effective in analyzing such processes. The results in Chaps. 2 through 5 are similar to comparable treatments of mass diffusion processes. These simplest numerical methods are first developed in Chaps. 3 through 5 for both steady and transient processes. Chapter 9 concerns more advanced aspects of numerical and other approximate methods.

Chapter 6 considers mass diffusion mechanisms and results which are more complicated than the analogous heat conduction processes in Chaps. 2 through 5. These include parallel local diffusion modes, general formulations of local chemical reaction, the coupling of local reactions with temperature, and simultaneous heat conduction effects.

Chapter 7 concerns contact resistance between adjacent conduction regions, as in a composite heat transfer barrier such as a multi-ply wall. Conduction in composite materials, such as a solid with embedded geometries of a different material, is considered. The heat transfer mechanisms in insulation are treated.

Chapter 8 considers a group of applications in which the heat transfer rate, determined by a conduction mechanism, dominates the processes. These include finned surfaces, welding, thermal stresses, random conductive effects, liquid films, and conductive internal flows. The following subsection develops the particular parameters which very commonly arise to characterize conduction regimes in terms of dimensionless parameters.

1.4.5 Similarity Analysis

The differential equations developed in Secs. 1.2 and 1.3 are to be solved, in principle, with the appropriate bounding conditions. The result, for example, is $t(x, y, z, \tau)$. However, particular solutions are not usually given in the form of temperature or concentration. Instead, they are written nondimensionally. The solutions may then be expressed in terms of nondimensional parameters whose values vary from one circumstance or time level to another. Each such parameter which arises is characteristic of a basic feature of the process. For example, the time level in a transient, τ, becomes $\alpha\tau/L^2$ where L is the characteristic dimension of the conduction region. The resulting functional dimensionless dependencies are developed in the following discussion for both temperature and convection surface conditions.

A SURFACE TEMPERATURE CONDITION. Figure 1.4.3 shows an arbitrary conduction region geometry. This example of that particular geometry has a characteristic size, or dimension, L. The region is initially uniformly at some temperature t_i inside. The temperature of the surface then changed to a uniform level t_s, which is held constant thereafter. The temperature t_s is to apply over the surface, whose location in x, y, z is described by some function $f_s(x, y, z, L) = 0$. For example, if the geometry is a sphere of radius R, then $f_s = x^2 + y^2 + z^2 - R^2 = 0$, for a centered origin.

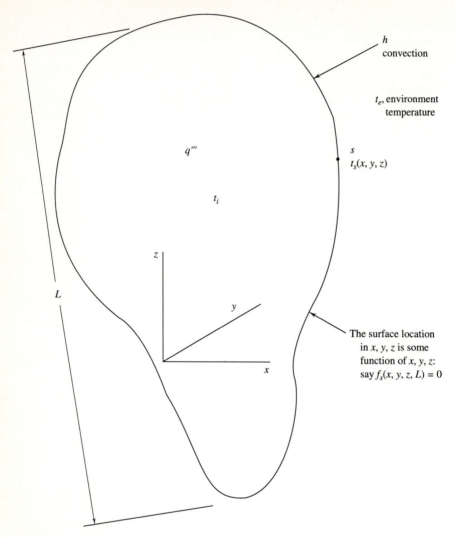

FIGURE 1.4.3
A transient conduction region.

The transient response $t(x, y, z, \tau)$ is to be determined, assuming that k is initially uniform inside the region and will remain constant. The differential equation and boundary conditions which determine $t(x, y, z, \tau)$ are

$$\frac{\partial^2 t}{\partial x^2} + \frac{\partial^2 t}{\partial y^2} + \frac{\partial^2 t}{\partial z^2} + \frac{q'''}{k} = \frac{1}{\alpha}\frac{\partial t}{\partial \tau} \qquad (1.4.16)$$

at $\tau = 0$, $t = t_i$
over $f_s(x, y, z, L) = 0$, $t = t_s$

The coordinates x, y, z, the function $f_s(x, y, z, L) = 0$, and the temperature are nondimensionalized as follows:

$$X = \frac{x}{L} \qquad Y = \frac{y}{L} \qquad Z = \frac{z}{L} \qquad \frac{t - t_i}{t_s - t_i} = \phi(x, y, z)$$

where $f_s(x, y, z, L) = 0$ becomes nondimensional as $F_s(X, Y, Z) = 0$ and ϕ varies from 0 to 1.0. Substituting the preceding transformation into Eq. (1.4.16) results in the following differential equation:

$$\frac{(t_s - t_i)}{L^2} \left[\frac{\partial^2 \phi}{\partial X^2} + \frac{\partial^2 \phi}{\partial Y^2} + \frac{\partial^2 \phi}{\partial Z^2} \right] + \frac{q'''}{k} = \frac{(t_s - t_i)}{\alpha} \frac{\partial \phi}{\partial \tau} \qquad (1.4.17)$$

Multiplying through $L^2/(t_s - t_i)$ yields

$$\frac{\partial^2 \phi}{\partial X^2} + \frac{\partial^2 \phi}{\partial Y^2} + \frac{\partial^2 \phi}{\partial Z^2} + \frac{q''' L^2}{k(t_s - t_i)} = \frac{L^2}{\alpha} \frac{\partial \phi}{\partial \tau} = \frac{\partial \phi}{\partial(\alpha\tau/L^2)} \qquad (1.4.18a)$$

The bounding conditions are

$$\text{at } \tau = 0, \ \phi = 0 \qquad (1.4.18b)$$
$$\text{over } F_s(X, Y, Z) = 0, \ \phi = 1 \qquad (1.4.18c)$$

Equations (1.4.18) indicate that the solution ϕ will depend only on $X, Y, Z, q''' L^2/k(t_s - t_i)$, time as $\alpha\tau/L^2$, and on the form of the nondimensional function $F_s(X, Y, Z) = 0$. That is,

$$\phi = \frac{(t - t_i)}{(t_s - t_i)} = \phi\left[X, Y, Z, \left(\frac{q''' L^2}{k(t_s - t_i)} \right), \alpha\tau/L^2, \text{Geometry} \right] \quad (1.4.19)$$

Therefore, the only aspects of any particular conduction situation which affects the solution, when expressed as ϕ, for any given geometry are

$$\frac{q''' L^2}{k(t_s - t_i)} \quad \text{and} \quad \frac{\alpha\tau}{L^2} \qquad (1.4.20)$$

where $\alpha\tau/L^2 \equiv \text{Fo}$, the Fourier number. This relates energy conduction, and storage. The dimensionless parameter $q''' L^2/k(t_s - t_i)$ is the ratio of the internal generation rate to a conduction rate and is a generation number, say Q_g.

A CONVECTION SURFACE CONDITION. The preceding similarity analysis used a surface temperature boundary condition, for the simplest first use of this method. Had the formulation been instead for a convection process at the surface, h, the boundary condition would have been T.3, Eq. (1.4.2), as follows, where t_e is taken as constant and uniform:

$$\text{over } f_s(x, y, z, L) = 0 \quad \text{and} \quad h(t_e - t_s) = -k\left(\frac{\partial t}{\partial n} \right)_s$$

The characteristic temperature difference is now $(t_e - t_i)$ and ϕ becomes

$(t - t_i)/(t_e - t_i)$. The boundary condition, generalized as $N = n/L$, is

$$\text{over } F_s(X, Y, Z) = 0 \quad \text{and} \quad h\phi(t_e - t_s) = -k\frac{(t_e - t_s)}{L}\left(\frac{\partial\phi}{\partial N}\right)_s \quad (1.4.21a)$$

$$\text{or} \qquad \frac{hL}{k}\phi = -\left(\frac{\partial\phi}{\partial N}\right)_s \qquad (1.4.21b)$$

The quantity $hL/k \equiv \text{Bi}$ is the Biot number. It is the ratio of the convection coefficient h to a characteristic conductance of the region k/L. Thus the dependence of ϕ is now

$$\phi = \frac{t - t_i}{t_e - t_i} = \phi(X, Y, Z, Q_g, \text{Fo}, \text{Bi}, \text{Geometry}) \qquad (1.4.21c)$$

The additional parameter is Bi.

Succeeding chapters are concerned with finding both analytical and numerical solutions, usually in terms of ϕ, under many different conditions. Solutions are very commonly given in terms of ϕ. This is often called a temperature excess ratio. This is because it is a ratio of the amounts by which t and t_s or t_e are in excess of t_i.

PROBLEMS

1.1.1. Plot the logarithm of thermal conductivity vs. temperature (as possible) in the range 0 to 500°C for the following substances:

air	liquid sodium	brass
low-pressure steam	concrete	steel
carbon dioxide gas	building brick	tin
water	glass	copper
mercury	ice	silver

Put all curves on a single logarithmic plot.

1.1.2. For the gases CO_2 and O_2 at atmospheric pressure and steam at zero pressure, calculate the Prandtl number at 200°C.

1.1.3. For water at 60°F the molecular spacing and sonic velocity are 1.2×10^{-8} in. and 5000 fps. From these values, compute the thermal conductivity and compare the result with the value tabulated in the Appendix.

1.2.1. Prove, for steady-state conduction in a slab whose surfaces are maintained at t_1 and t_2, that the temperature distribution is a straight line if k is uniform throughout the slab. Indicate the shape of the temperature distribution if k increases with t, if $t_2 > t_1$.

1.2.2. Using the relations between Cartesian and cylindrical coordinates, derive Eq. (1.2.18) from Eq. (1.2.9).

1.2.3. Using cylindrical coordinates, derive directly, from Fourier's law, the steady-state differential equation for a distributed source in a material of uniform conductivity, where temperature is a function only of radius.

1.2.4. Repeat Prob. 1.2.3 for k variable over the region, as $k(r, t)$.

1.2.5. Repeat Prob. 1.2.3 for a region described in spherical coordinates.

1.2.6. Repeat Prob. 1.2.4 for spherical coordinates.

1.2.7. A round rod of radius b has the following initial instantaneous temperature distribution, at time τ:

$$t = A\left[8b^4 + (x^4 + y^4) - 3b^2(x^2 + y^2)\right] \quad \text{for } r < b$$

(a) At that time, determine the portion of the rod cross section in which the temperature is decreasing with time.

(b) At what location in the rod is the temperature changing most rapidly at this time?

1.4.1. Consider boundary condition T.2. The solar radiant flux on a stone concrete wall surface is 400 W/m^2. For a surface absorptance of 0.80, determine the temperature gradient and its sign, in the concrete, immediately adjacent to the surface at $x = 0$, neglecting other energy effects.

1.4.2. For the surface in Prob. 1.4.1, consider instead a convective surface effect to an ambient at 60°F, at a time when the surface temperature is 80°F. For a convection coefficient of $h = 5$ Btu/hr ft^2 °F, determine the surface temperature gradient and its sign.

1.4.3. For the concrete wall surface in Prob. 1.4.1, combine the radiant and convection surface condition in Probs. 1.4.1 and 1.4.2.

(a) Determine the gradient and its sign at $x = 0$.

(b) If the concrete wall is replaced by a "perfect" insulator, determine its surface temperature.

1.4.4. A 0.5-cm-thick pane of tinted glass is subject to a solar flux of 400 W/m^2. The front and back surface reflectivities are 0.10 and 0, respectively. The value of μ is 0.5 per centimeter.

(a) Determine the absorption rate in the glass.

(b) Calculate the intensity of the transmitted flux.

1.4.5. Consider a flat, steel surface having an absorptance of 0.8 and a surface temperature of 100°C, in a large radiant environment 300°C.

(a) Calculate the resulting surface net heat flux loading and the internal gradient at the surface.

(b) If convection with the environment also occurs, with $h = 80$ W/m^2 K, determine the net flux and the gradient in the steel, at the surface.

1.4.6. A steel sphere of 1-in. diameter at an initial temperature of 100°F is suddenly placed in an environment at 50°F. A convection coefficient of $h = 20$ Btu/ft^2 hr °F applies, over the surface. Determine the Biot number and the dependence of the Fourier number on the time, in seconds.

REFERENCES

Arpaci, V. S. (1966) *Conduction Heat Transfer*, Addison-Wesley, Reading, MA.

Bellman, R. (1953) *Stability Theory of Differential Equations*, McGraw-Hill, New York.

Bird, R. B., W. E. Stewart, and E. N. Lightfoot (1960) *Transport Phenomena*, John Wiley and Sons, New York.

Carslaw, H. S., and J. C. Jaeger (1959) *Conduction of Heat in Solids*, 2nd ed., Clarendon Press, Oxford.

Chao, B. T. (1963) *Appl. Sci. Res.* **A12**, 134.

Crank, J. (1979) *The Mathematics of Diffusion*, 2nd ed., Clarendon Press, Oxford.

Cussler, E. L. (1985) *Diffusion Mass Transfer in Fluid Systems*, Cambridge Univ. Press.

Eckert, E. R. G., and R. M. Drake, Jr. (1972) *Analysis of Heat and Mass Transfer*, McGraw-Hill, New York.

Edwards, D. K., V. E. Denny, and A. F. Mills (1979) *Transfer Processes*, 2nd ed., McGraw-Hill, New York.

Gebhart, B. (1957) Unified treatment for thermal radiation processes—gray, diffuse radiators and absorbers, Paper 57-A-34, ASME, New York. See also: *Trans. ASHRAE*, **65**, 321, 1959; *Heat Transfer*, 2nd ed., McGraw-Hill, New York, 1971.

Gebhart, B. (1971) *Heat Transfer*, McGraw-Hill, New York.

Howell, J. R. (1982) *Radiation Configuration Factors*, McGraw-Hill, New York.

Jakob, M. (1949) *Heat Transfer*, Vol. 1, John Wiley and Sons, New York.

Kirkaldy, J. S., and D. J. Young (1987) *Diffusion in the Condensed State*, The Institute of Metals, London.

Luikov, A. V. (1971) *Heat Transfer-Soviet Research*, **3**, 1.

Munoz, A. V., and L. C. Burmeister (1988) *Trans. ASME, J. Heat Transfer*, **110**, 778.

Özisik, M. N. (1980) *Heat Conduction*, John Wiley and Sons, New York.

Rohsenow, W. M., J. P. Hartnett, and E. N. Ganic (1985) *Handbook of Heat Transfer Fundamentals*, McGraw-Hill, New York.

Schneider, P. J. (1955) *Conduction Heat Transfer*, Addison-Wesley, Reading, MA.

Schewmon, P. G. (1963) *Diffusion in Solids*, McGraw-Hill, New York.

Siegel, R., and J. R. Howell (1981) *Thermal Radiation Heat Transfer*, 2nd ed., McGraw-Hill, New York.

CHAPTER
2

STEADY-STATE PROCESSES IN ONE DIMENSION

2.1 PLANE, CYLINDRICAL, AND SPHERICAL REGIONS

This chapter concerns the simplest of conduction mechanisms. The temperature variation in the region is described by one variable alone: x in plane regions and r in cylindrical and spherical regions. The plane region one-dimensional approximation applies for flat surfaces whose face dimensions in each direction along the surface are very large compared to the region thickness L, and when a uniform boundary condition is applied to each surface. The approximation for cylindrical geometries applies when the axial length is very large compared to the maximum conduction region radius. An exception to these requirements arises when the edge conditions for the plane region or the end conditions for the cylindrical geometry result in uniform temperature gradients in the y and z and in the r directions, respectively. The only requirement that arises for the spherical geometry is that a uniform condition applies to each concentric surface which bounds the region.

2.1.1 The Differential Equations

The general equations which apply for the three geometries are given in the following discussion. The formulation given includes the possibility that k is a function of x or r, due to material inhomogeneity. Also k may be a function of

43

local temperature t. Finally, k may also be a function of both location and temperature, as $k(x, t)$. The symbol q''' in the equations remains the generation rate per unit volume. The three differential equations, from the general forms given in Eqs. (1.2.13), (1.2.20), and (1.2.21), are

$$\frac{d}{dx}\left(k\frac{dt}{dx}\right) + q''' = 0 \quad \text{(plane region)} \tag{2.1.1}$$

$$\frac{1}{r}\frac{d}{dr}\left(rk\frac{dt}{dr}\right) + q''' = 0 \quad \text{(cylindrical region)} \tag{2.1.2}$$

$$\frac{1}{r^2}\frac{d}{dr}\left(r^2k\frac{dt}{dr}\right) + q''' = 0 \quad \text{(spherical region)} \tag{2.1.3}$$

Steady-state mass transfer is also considered in this chapter. The spatial conditions under which one-dimensional mass transport results are the same as set forth previously for conduction heat transfer. The relevant equations given in the following discussion are completely similar and the diffusion coefficient, D, sometimes also depends on location as $D(x)$, on concentration as $D(C)$, or on both as $D(x, C)$. The source term, C''', in mass diffusion, may arise from spontaneous chemical species generation locally. It may also arise from equilibrium adsorption and desorption when D is replaced by $D' = D/(1 + R_r)$, as discussed in Sec. 1.3. The three differential equations, from the general form given in Eq. (1.3.5), are

$$\frac{d}{dx}\left(D\frac{dC}{dx}\right) + C''' = 0 \quad \text{(plane region)} \tag{2.1.4}$$

$$\frac{1}{r}\frac{d}{dr}\left(rD\frac{dC}{dr}\right) + C''' = 0 \quad \text{(cylindrical region)} \tag{2.1.5}$$

$$\frac{1}{r^2}\frac{d}{dr}\left(r^2D\frac{dC}{dr}\right) + C''' = 0 \quad \text{(spherical region)} \tag{2.1.6}$$

The remainder of this section analyzes heat and mass diffusion for regions having uniform k and D. This includes a layer of a single material, regions which are a composite of layers of several materials, and infinite regions. Contact resistance is also discussed, as it relates to conduction through composite layers. Section 2.2 considers conditions wherein k may vary over the region as $k(t)$, $k(x)$, and $k(t, x)$, or in terms of r for the other two geometries. Section 2.3 applies these kinds of variations of k in composite barriers. The effect of internal energy generation q''' are analyzed in Sec. 2.4. Section 2.5 concerns mass transfer processes which are similar to the heat transfer analyses given in preceding sections.

2.1.2   Uniform Conductivity and Diffusivity

Plane, cylindrical, and spherical layers are shown in Fig. 2.1.1, each with boundary conditions t_1 and t_2 at the two bounding surfaces. In the absence of a distributed source, Eqs. (2.1.1)–(2.1.3) become the following. Analogous forms for mass diffusion will follow from Eqs. (2.1.4)–(2.1.6). Boundary conditions are included in the following discussion.

$$\frac{d^2t}{dx^2} = 0$$

$$\text{at } x = 0, t = t_1; \text{ at } x = L, t = t_2 \quad (2.1.7)$$

$$\frac{1}{r}\frac{d}{dr}\left(r\frac{dt}{dr}\right) = \frac{d^2t}{dr^2} + \frac{1}{r}\frac{dt}{dr} = 0$$

$$\text{at } r = r_1, t = t_1; \text{ at } r = r_2, t = t_2 \quad (2.1.8)$$

$$\frac{1}{r^2}\frac{d}{dr}\left(r^2\frac{dt}{dr}\right) = \frac{d^2t}{dr^2} + \frac{2}{r}\frac{dt}{dr} = 0$$

$$\text{at } r = r_1, t = t_1; \text{ at } r = r_2, t = t_2 \quad (2.1.9)$$

PLANE LAYERS. The solution follows and the heat flux q'' and q are calculated as

$$k\frac{d^2t}{dx^2} = 0 \rightarrow k\frac{d}{dx}\left(\frac{dt}{dx}\right) = 0 \rightarrow k\frac{dt}{dx} = B_1 \rightarrow t = B_1x + B_2 \quad (2.1.10)$$

where $-B_1$ is the heat flux q''. The boundary conditions determine $t(x)$, from which q'' is then calculated

$$t(x) = t = (t_2 - t_1)\frac{x}{L} + t_1 \quad \text{or} \quad \phi = \frac{t - t_2}{t_1 - t_2} = 1 - \frac{x}{L} \quad (2.1.11)$$

where ϕ is called the temperature excess ratio and

$$q''(x) = q'' = -k\frac{dt}{dx} = -k\frac{(t_2 - t_1)}{L} = \frac{(t_1 - t_2)}{R_t} \quad (2.1.12)$$

$$q = Aq'' = kA\frac{(t_1 - t_2)}{L} \quad (2.1.13)$$

where k/L is the conductance C_t and L/k is the resistance R_t, both per unit area.

For mass diffusion the differential equation and boundary conditions are the same as in Eq. (2.1.7) with t replaced by C. The resulting concentration

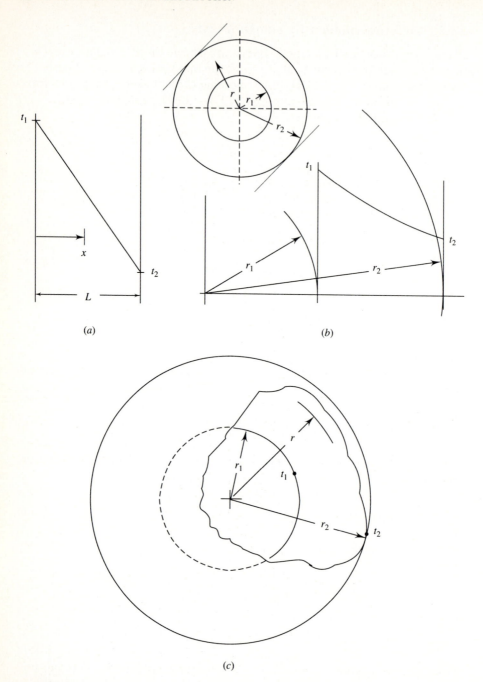

(a)

(b)

(c)

FIGURE 2.1.1
One-dimensional regions: (a) plane; (b) cylindrical; (c) spherical.

distribution and mass flux are

$$C(x) = C = (C_2 - C_1)\frac{x}{L} + C_1 \quad \text{or} \quad \tilde{C} = \frac{C - C_2}{C_1 - C_2} = 1 - \frac{x}{L} \quad (2.1.14)$$

$$m''(x) = m'' = D\frac{(C_1 - C_2)}{L} \quad \text{and} \quad m = DA\frac{(C_1 - C_2)}{L} \quad (2.1.15)$$

This result is the same as for conduction, as will also be shown in Sec. 2.5 for other typical results in Secs. 2.1 through 2.4.

CYLINDRICAL TUBES. Figure 2.1.1(b) (see page 46) indicates the geometry and Eq. (2.1.8) the equation and boundary conditions. The solution is simply found by replacing dt/dr in the equation by the function p. Then

$$\frac{dp}{dr} + \frac{p}{r} = 0 = r\,dp + p\,dr = d(pr) \quad (2.1.16)$$

and

$$pr = r\frac{dt}{dr} = B_1 \quad \text{and} \quad t(r) = B_1 \ln r + B_2 \quad (2.1.17)$$

The constants are evaluated and the solution is

$$t = \frac{1}{\ln(r_2/r_1)}\left[(t_2 - t_1)\ln r + t_1 \ln r_2 - t_2 \ln r_1\right] \quad (2.1.18a)$$

or

$$\phi = \frac{t - t_2}{t_1 - t_2} = \frac{\ln(r_2/r)}{\ln(r_2/r_1)} \quad (2.1.18b)$$

This distribution is seen to be logarithmic and is plotted in Fig. 2.1.1(b) for $t_1 > t_2$. The slope decreases in magnitude from r_1 to r_2, indicating that the heat flux per unit area, q'', decreases as r increases. This is consistent with the condition that the total heat flow per unit cylinder length must be independent of r and that the area for conduction increases with r. The heat transfer rate per unit area at r and the total conduction rate per unit cylinder length are calculated as

$$q''(r) = -k\frac{dt}{dr} = -\frac{k(t_2 - t_1)}{r\ln(r_2/r_1)}$$

$$= \frac{k(t_1 - t_2)}{r\ln(r_2/r_1)} \quad (2.1.19)$$

$$\frac{q}{L} = q' = q''2\pi r = \frac{2\pi k(t_1 - t_2)}{\ln(r_2/r_1)} = C_t(t_1 - t_2) = \frac{(t_1 - t_2)}{R_t} \quad (2.1.20)$$

The conductance C_t of the cylindrical wall is the coefficient of $(t_1 - t_2)$ and the resistance is R_t, each per unit length in the axial direction.

The expression for the heat flow rate for a cylinder is different from that for a flat plate. However, for small values of $(r_2 - r_1)/r_1$, or for r_2/r_1 near 1, this conductance becomes essentially equal to $A_m k/\Delta r$, where A_m is the conduction area at the average radius, per unit length. Jakob (1949) introduced a shape factor F as follows:

$$q' = \frac{2\pi k(t_1 - t_2)}{\ln(r_2/r_1)} \approx \frac{A_m k(t_1 - t_2)}{F \Delta r} \tag{2.1.21}$$

Calculations indicated that $F \leq 1.01$ for $r_2/r_1 \leq 1.4$. That is an error of 1% which arises when taking $F = 1$ in Eq. (2.1.21).

SPHERICAL SHELLS. Figure 2.1.1(*c*) (see page 46) is the geometry and the equations and boundary conditions are Eq. (2.1.9). This solution is found by again replacing dt/dr by the function p. Then

$$\frac{d^2 t}{dr^2} + \frac{2}{r}\frac{dt}{dr} = 0 = r^2\, dp + p(2r\, dr) = d(r^2 p) \tag{2.1.22}$$

and

$$r^2 p = B_1 \quad \text{and} \quad t(r) = -\frac{B_1}{r} + B_2 \tag{2.1.23}$$

The constants are evaluated and the solution is

$$t = \frac{1}{1/r_1 - 1/r_2}\left(\frac{t_1 - t_2}{r} + \frac{t_2}{r_1} - \frac{t_1}{r_2}\right) \tag{2.1.24a}$$

or

$$\phi = \frac{t - t_2}{t_1 - t_2} = \frac{1}{1 - (r_1/r_2)}\left[\frac{r_1}{r} - \frac{r_1}{r_2}\right] \tag{2.1.24b}$$

The heat transfer rate per unit area, q'', and the total heat transfer rate, q, are found

$$q'' = -k\frac{dt}{dr} = \frac{k(-B_1)}{r^2} = \frac{k(t_1 - t_2)}{r^2(1/r_1 - 1/r_2)} \tag{2.1.25}$$

$$q = 4\pi r^2 q'' = \frac{4\pi k(t_1 - t_2)}{1/r_1 - 1/r_2} = C_t(t_1 - t_2) = \frac{(t_1 - t_2)}{R_t} \tag{2.1.26}$$

The conductance of the spherical shell is the coefficient of $(t_1 - t_2)$ in Eq. (2.1.26). For thin shells, that is, with $(r_2 - r_1)/r_1$ very small, the conductance becomes $A_m k/\Delta r$, where A_m is the conduction area at the average radius. The error in this procedure is less than 1% if $r_2/r_1 \leq 1.2$.

2.1.3   Composite Regions

In the majority of processes of practical interest, the heat conduction or mass diffusion region is composed of successive layers. This assembly may also include surfaces or internal spaces, subject to convection, radiation, and other processes. An example, composed of plane layers, is shown in Fig. 2.1.2. Comparable cylindrical and spherical composite barriers are similarly described

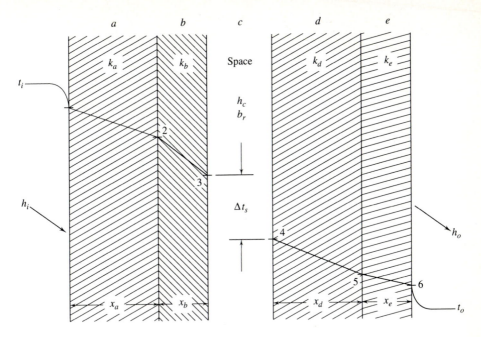

FIGURE 2.1.2
A composite plane barrier.

in terms of radii of the interfaces of the shells, as r_1, r_2, and so on. This geometry is shown in Fig. 2.1.3.

The simplest way to determine the heat or mass transfer through a barrier is to sum the resistances in series from the first resistance on one side, through the last resistance on the other side, as $R_T = \sum_i R_i$.

A PLANE COMPOSITE. The structure is seen in Fig. 2.1.2. The solid layer resistances are $R_t = (x/k)$, and the left and right surface resistances are $1/h_i$ and $1/h_o$. The air layer resistance is composed of two resistances in parallel, R_c and R_r. The resulting layer resistance R_L is $1/(h_c + h_r)$, where h_c is convection and h_r is the equivalent radiation conductance. This kind of formulation for radiant transport applies only for $t_3 - t_4 \ll T_3$ and T_4 in Fig. 2.1.2, where T is the absolute temperature. The total resistance is

$$R_T = \frac{1}{h_i} + \left(\frac{x}{k}\right)_a + \left(\frac{x}{k}\right)_b + \frac{1}{h_c + h_r} + \left(\frac{x}{k_c}\right) + \left(\frac{x}{k_d}\right) + \frac{1}{h_o} \quad (2.1.27)$$

The flux through the barrier is then

$$q'' = \frac{(t_i - t_o)}{R_T} \quad (2.1.28)$$

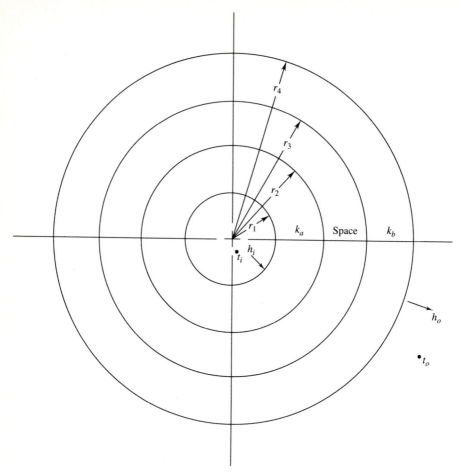

FIGURE 2.1.3
A composite cylindrical or spherical barrier.

A CYLINDRICAL COMPOSITE. A structure having the same kind of elements of resistance as in Fig. 2.1.2 is shown in Fig. 2.1.3. The individual components of resistance for a cylindrical geometry are to be calculated per unit of axial length. Then, when summed, they will determine the heat flow through the barrier per unit of length q/L. For the two conduction layers, the conduction resistances per unit length are seen from Eq. (2.1.20) to be

$$R_a = \frac{\ln(r_2/r_1)}{2\pi k_a} \quad \text{and} \quad R_b = \frac{\ln(r_4/r_3)}{2\pi k_b} \tag{2.1.29}$$

The surface and air layer resistances, R_L, per unit length, are

$$\frac{1}{2\pi h_i r_1}, \quad \frac{1}{2\pi h_o r_4}, \quad \text{and} \quad R_L^{-1} = 2\pi r_2 h_c + 2\pi r_2 h_r$$

where the parallel convection and radiation coefficients in the air layer are based on r_2, that is, $A_2/L = 2\pi r_2$. The total resistance and heat flow are then

$$R_T = \frac{1}{2\pi h_i r_1} + \frac{\ln(r_2/r_1)}{2\pi k_a} + \frac{1}{2\pi r_2(h_c + h_r)} + \frac{\ln(r_4/r_3)}{2\pi k_b} + \frac{1}{2\pi h_o r_4}$$

$$(2.1.30)$$

$$\frac{q}{L} = \frac{(t_i - t_o)}{R_T} \tag{2.1.31}$$

A SPHERICAL COMPOSITE. Figure 2.1.3 may also be interpreted as a spherical composite barrier. The heat flow through each region is the same, q. The total resistances of the two conduction regions are seen from Eq. (2.1.26) to be

$$R_a = \frac{1/r_1 - 1/r_2}{4\pi k_a} \quad \text{and} \quad R_b = \frac{1/r_3 - 1/r_4}{4\pi k_b} \tag{2.1.32}$$

The two surface and the air layer resistances are

$$R = \frac{1}{4\pi h_i r_1^2}, \quad \frac{1}{4\pi h_o r_4^2}, \quad \text{and} \quad \frac{1}{4\pi r_2^2(h_c + h_r)} \tag{2.1.33}$$

The total resistance R_T is the sum of the previous five components and the heat flow is

$$q = \frac{t_i - t_o}{R_T} \tag{2.1.34}$$

2.1.4 Contact Resistance

The analysis of conduction through composite layers in the preceding section assumed that all adjacent layers of solid material were in perfect thermal contact for conduction heat transfer. For example, for the plane composite barrier in Fig. 2.1.2, layers a and b and c and d were taken to be in perfect contact at locations 2 and 5. This condition was incorporated in assuming that each surface had the same temperature at its common plane of contact. Layer c was in air layer which had a thermal resistance of $R_L = 1/(h_c + h_r)$. This resistance R_L may be thought of as a contact resistance, as R_c. It is seen to be the inverse of the sum of two conductance components, h_c and h_r.

In actually assembling such a barrier in a conventional way, there would be imperfect contact between both layers a and b and d and e at locations 2 and 5, respectively. This usually arises because such layers are not perfectly flat

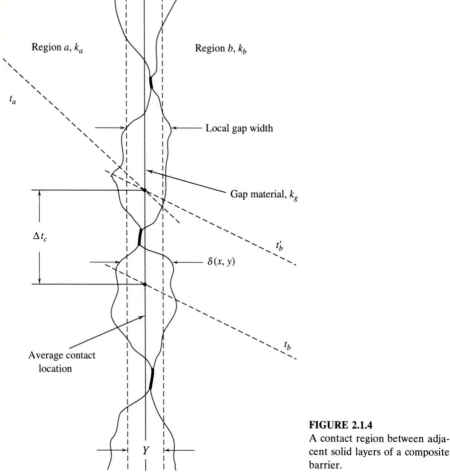

FIGURE 2.1.4
A contact region between adjacent solid layers of a composite barrier.

or smooth and are not commonly bonded together as by a highly conductive adhesive. Then there is a contact region, idealized in Fig. 2.1.4, at high magnification.

This contact region differs from the air layer c in Fig. 2.1.2 in that there is solid contact at various locations along the contact region. This region is also usually much thinner than gaps intentionally left in a composite barrier. However, the formulation of contact region conductance is often in terms of the convective, radiative, and conductive process across the contact region. The conductive mechanism between regions a and b arises across the various local areas at which they are in actual physical contact. The number and size of these contact areas will increase as the pressure p, applied to push regions a and b together, increases. This results from mechanical deformation of both materials.

The effects of the contact resistance on conduction and heat flux are shown in Fig. 2.1.4. The two dashed vertical lines are at about the average locations of the interfaces of each region a and b. The average gap thickness is Y. The temperature distributions shown inside of each region a and b, approaching the contact region, are linear. The different slopes shown assume that $k_a < k_b$. Recall that the flux q'' was and still is uniform across the whole structure, in steady state.

With perfect thermal bounding of the two regions, that is, no contact resistance, the temperature distributions in regions a and b are t_a and t'_b. These distributions would then meet, as shown, at the average contact location. However, with added resistance in the contact region, there is a larger temperature gradient in this immediate region. The resulting difference in the extrapolations of t_a and the changed t_b is shown in Fig. 2.1.4, as Δt_c, the contact region temperature drop. This is positive for $t_a > t_b$. The extrapolation is to the average contact location. This difference Δt_c is the effect which would be detected upon examining the t_a and t_b distributions in regions a and b, relatively far away from the contact region. Thus Δt_c resulted from the contact resistance R_c and is calculated from the conduction heat flux normal to the barrier, q''_n, as

$$\Delta t_c = R_c q''_n \qquad (2.1.35)$$

The immediate local effect of R_c is to interpose a temperature drop at each contact resistance across any composite layer. This increases the total resistance across the layer, for example, R_T, in Eqs. (2.1.27) and (2.1.28) for a plane composite.

Similar contact effects also arise at interfaces in mass diffusion processes. The mechanism is often the same as that in heat conduction. The two regions a and b, as in Fig. 2.1.4, may have mass diffusivities D_a and D_b which are different from that of the gap material D_g. This would impose an added contact region resistance if D_g is less than D_a and D_b. However, for the low diffusivities common in solids, as discussed in Sec. 6.1.4, the gap material of average thickness Y may reduce the overall mass diffusion resistance of the structure, for D_g large. The gap resistance $R_{c,D} = Y/D_g$ would be less for a higher diffusivity layer.

The same formulation applies for other than plane contact region geometries, when allowance is made for any appreciable geometric changes which arise due to the added thickness caused by a contact region. The general circumstance is shown in Fig. 2.1.5, where n is the local normal to the contact region. The general relation between the conductivities and gradients on the two sides of the contact region, and Δt_c and R_c, is

$$-k_a \left(\frac{dt}{dn} \right)_a = -k_b \left(\frac{dt}{dn} \right)_b = \frac{\Delta t_c}{R_c} \qquad (2.1.36)$$

where Δt_c is again the extrapolated difference of t_a and t_b to some average contact location, defined as shown in Fig. 2.1.4.

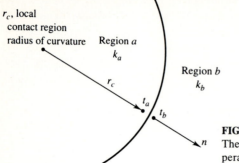

r_c, local
contact region
radius of curvature   Region a
                           k_a

                               Region b
                                   k_b

r_c

t_a

t_b

n

FIGURE 2.1.5
The relation of contact resistance R_c to the temperature field in the conduction region.

TABLE 2.1.1
Contact region conditions, for the variations of contact conductance shown in Fig. 2.1.6

Curve	Material pair	rms surface finish (in. $\times$ 10^{-6})	Gap material	Mean contact temperature (°F)
1	Aluminum (2024-T3)	48–65	Vacuum‡	110
2	Aluminum (2024-T3)	8–18	Vacuum‡	110
3	Aluminum (2024-T3)	6–8*	Vacuum‡	110
4	Aluminum (75S-T6)	120	Air	200
5	Aluminum (75S-T6)	65	Air	200
6	Aluminum (75S-T6)	10	Air	200
7	Aluminum (2024-T3)	6–8*	Lead foil (0.008 in.)	110
8	Aluminum (75S-T6)	120	Brass foil (0.001 in.)	200
9	Stainless (304)	42–60	Vacuum‡	85
10	Stainless (304)	10–15	Vacuum‡	85
11	Stainless (416)	100	Air	200
12	Stainless (416)	100	Brass foil (0.001 in.)	200
	Magnesium (AZ-31B)	50–60†	Vacuum‡	85
14	Magnesium (AZ-31B)	8–16†	Vacuum‡	85
15	Copper (OFHC)	7–9	Vacuum‡	115
16	Stainless/aluminum	30–65	Air	200
17	Iron/aluminum	—	Air	80
18	Tungsten/graphite	—	Air	270

*Not flat.

†Oxidized.

‡10^{-4} mm Hg pressure = 10^{-4} Torr.

CONTACT RESISTANCES. These are both calculated and measured. Taking the example shown in Fig. 2.1.4, with air, a very thin gap will suppress convection. Then the modes of heat transfer in the contact region are conduction through the gas, radiation across the gap, and conduction through the direct contact areas. These are the three parallel components of the total flux $q'' = q''_g + q''_r + q''_{dc}$. The first two components are approximated as

$$q''_g = \left(\overline{\frac{1}{\delta}}\right) k_g (t_a - t_b) \quad \text{and} \quad q''_r = \frac{\sigma\left(T_a^4 - T_b^4\right)}{[1/\epsilon_a + 1/\epsilon_b - 1]} \qquad (2.1.37)$$

where $(\overline{1/\delta})$ may be taken as a suitable average of $1/\delta$, over a typical part of the contact region. The radiation effect is calculated as between two parallel surfaces, of regions a and b, having emissivities ϵ_a and ϵ_b. However, the direct contact flux, q''_{dc} must take into account mechanisms which are strongly dependent on surface flatness and finish, on the applied assembly pressure level p, and on material elasticity and ductility.

In many critical applications, particularly in computer circuit cooling, many techniques have been developed to reduce R_c at the necessary material interfaces between the heat-dissipating circuit elements and the heat sink. Soft materials may be placed between the regions. Soft material and heat-conductive greases are often inserted during assembly. Layers of soft and deformable material are also often deposited on the contacting surfaces before assembly.

CONTACT RESISTANCE AND CONDUCTANCE VALUES. Detailed mechanisms are discussed in Secs. 7.2 and 7.3. Here a selection of contact conductance information is given. These results describe several of the more important specific aspects in applications with metallic regions. The listing of contact region conditions, in Table 2.1.1, is for contact regions between metals under vacuum or containing air and foils. The measured contact conductances, C_c, as a function of mating pressure, p, are shown in Fig. 2.1.6. The dashed line at the lower left is the trend of the component of C_c due to the direct contact areas over the unit area of the contact region. The exponent, 2/3, applies for elastic deformation. See Sec. 7.3.2.

Curves 1 and 2 show the improved conductance which results from flatter surfaces, as do curves 9 and 10. Curves 4, 5, and 6 show the improvement with air in the gaps and the more effective radiation transfer at a higher temperature level. Curve 7 is an improvement over curve 3, due to a softer foil between the two surfaces. Curve 8, compared to curve 4, shows the effect of a harder brass foil. Curve 11 again shows the air and radiation effects. The brass foil between the harder stainless, curve 12, improves conductance. Curves 16, 17, and 18 show the effect of dissimilar materials. High values of C_c result from a combination of effects, such as greater deformation and differential thermal expansion. Several other related aspects are discussed at the end of Sec. 7.3.

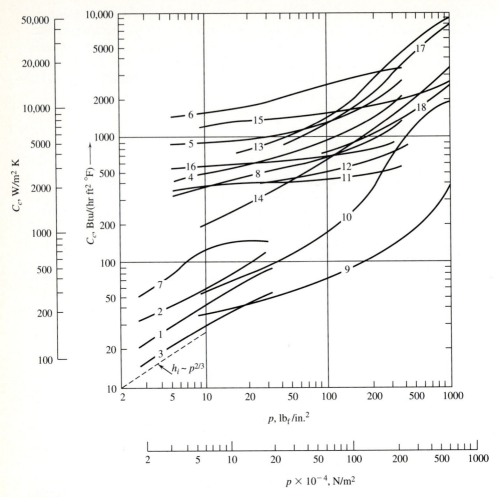

FIGURE 2.1.6
Contact region conductance $C_c = R_c^{-1}$ as a function of mating pressure p, for the contact region conditions shown in Table 2.1.1. [From Rohsenow et al. (1985).]

2.1.5 Infinite-Region One-Dimensional Conduction

In many applications there is a localized heating condition over a surface embedded in a very extensive surrounding conduction region. This circumstance is sketched in Fig. 2.1.7 for an extensive flat surface, a long cylinder, and a sphere. Each surface is at t_1 and in a region whose distant temperature is t_∞. The one-dimensional conduction in the regions results in the three temperature distributions, given in Eqs. (2.1.10), (2.1.17), and (2.1.23), as follows:

$$t(x) = B_1 x + B_2, \ t(r) = B_1 \ln r + B_2 \text{ and } t(r) = (B_1/r) + B_2 \quad (2.1.38)$$

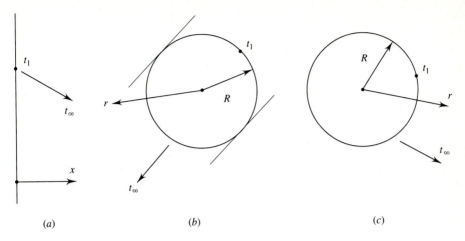

FIGURE 2.1.7
Plane, cylindrical, and spherical geometries embedded in very extensive regions.

The boundary conditions are

plane:	$x = 0$, $t = t_1$ and $t \to t_\infty$ as $x \to \infty$
cylindrical:	$r = R$, $t = t_1$ and $t \to t_\infty$ as $r \to \infty$
spherical:	$r = R$, $t = t_1$ and $t \to t_\infty$ as $r \to \infty$

Applying the first boundary condition for each geometry determines the three values of B_2 in terms of t_1 and B_1. However, applying the condition in x or in $r \to \infty$ may not be used to determine the constant B_1, for either the plane or the cylindrical geometries. There is no solution which will satisfy this condition. Another interpretation is that no steady-state heat flow may arise from these two configurations embedded in an infinite medium.

However, for the sphere, $B_1 = t_\infty$ and $B_2 = R(t_1 - t_\infty)$. The resulting solution and heat rejection rate are

$$\frac{t - t_\infty}{t_1 - t_\infty} = \frac{R}{r} \tag{2.1.39a}$$

$$q = 4\pi R^2 k \left(-\frac{dt}{dr} \right)_R = 4\pi R k (t_1 - t_\infty) \tag{2.1.39b}$$

Therefore, the conduction coefficient per unit area, h, from a spherical surface into an infinite region at t_∞, is

$$h = \frac{q}{A(t_1 - t_\infty)} = \frac{q}{4\pi R^2 (t_1 - t_\infty)} = \frac{k}{R} \tag{2.1.40}$$

or

$$\frac{hR}{k} = 1 \tag{2.1.41}$$

This last quantity is analogous to the Nusslet number, common to convection processes. The preceding nonzero values of q and h are also useful in convection as the limiting condition of convective transport, as the velocity in the moving surrounding fluid goes to zero. There is no such asymptotic value available in the other two geometries. This result may be interpreted as due to their conduction region being too constrained, as one and two dimensional, respectively, for the plane and cylindrical sources of infinite area.

2.2 VARIABLE CONDUCTIVITY

In many processes the material of the region is not homogeneous and the conductivity (and mass diffusivity) vary across the region. This would be as $k = k(x)$ or $k(r)$, for one-dimensional conduction in plane, cylindrical, and spherical regions. Also, the imposed temperature conditions are often sufficiently different that the conductivity varies enough with temperature, $k = k(t)$, that this effect must also be taken into account. There are also common circumstances in which both effects arise together, as in an inhomogeneous material with a highly temperature-dependent thermal conductivity, as $k(x, t)$ or $k(r, t)$.

The relevant equations in one dimension in steady state are Eqs. (2.1.1)–(2.1.6), where k and D may be variable. The analysis is quite simple when q''' and C''' are taken as zero, as first considered here. Each equation is then homogeneous in the quantities in parentheses in Eqs. (2.1.1)–(2.1.6). Results are given in Sec. 2.2.1 for $k = k(t)$ in the three geometries. The general procedure for $k(x)$, or $k(r)$, then appears in Sec. 2.2.2. Section 2.2.3 discusses dependence on both location and temperature, for example, as $k(t, x)$. A solution is also given for a special such circumstance.

2.2.1 Temperature-Dependent Conductivity

Many kinds of variation occur and a linear variation is often assumed. However, this is not necessary to obtain a general result, as will be seen.

PLANE REGION. The region is bounded by surfaces at t_1 and t_2, as in Fig. 2.1.1(a) (see page 46). The equation and boundary conditions become

$$\frac{d}{dx}\left(k \frac{dt}{dx} \right) = 0$$

$$\text{at } x = 0, t = t_1; \text{ at } x = L, t = t_2 \quad (2.2.1)$$

and
$$k \frac{dt}{dx} = B_1 \quad \text{or} \quad -k \frac{dt}{dx} = q'' = -B_1$$

This again confirms that the flux q'' is constant across the layer.

$$k \frac{dt}{dx} = B_1 = -q''$$

The minus sign in the preceding equation arises because a positive heat flux, q'', arises from a negative temperature gradient. Integration from $x = 0$ to some location x gives

$$\int_{t_1}^{t} k \, dt = B_1 \int_0^x dx = B_1 x = -q'' x \tag{2.2.2}$$

B_1 is evaluated at $x = L$, where $t = t_2$, as

$$\int_{t_2}^{t_1} k \, dt = C_1 L \quad \text{or} \quad q'' = -B_1 = -\frac{1}{L} \int_{t_1}^{t_2} k \, dt \tag{2.2.3}$$

An average, or effective conductivity, k_m, is defined as

$$k_m = \frac{1}{(t_2 - t_1)} \int_{t_1}^{t_2} k \, dt \tag{2.2.4}$$

This is the integrated average value of $k(t)$ over the temperature range t_1 to t_2. This is in general not equal to $[k(t_1) + k(t_2)]/2$, for a nonlinear variation of $k(t)$.

Eliminating the integral between Eqs. (2.2.3) and (2.2.4) gives the heat flux as

$$q'' = \frac{k_m}{L}(t_1 - t_2) = \frac{(t_1 - t_2)}{R_t} \tag{2.2.5}$$

where R_t is the resistance of the layer. The temperature distribution $t(x)$ is determined from Eq. (2.2.2) and (2.2.5) as

$$\int_{t_1}^{t} k \, dt = k_m(t_2 - t_1) \frac{x}{L} \tag{2.2.6}$$

Postulating the functional dependence of $k(t)$, k_m is determined from Eqs. (2.2.4). Then the integral in Eq. (2.2.6) may be evaluated for the dependence of t on x.

For $k(t)$ linear in t, or $k(t) = k(t_2) + \beta(t - t_2)$, $k_m = [k(t_1) + k(t_2)]/2$ and the heat flux is again given by Eq. (2.1.12) as

$$q'' = \frac{k_m}{L}(t_1 - t_2) \tag{2.2.7}$$

CYLINDRICAL REGION. This geometry is shown in Fig. 2.1.1(b) (see page 46), and Eq. (2.1.2) for $q''' = 0$, which follows, is then integrated once

$$\frac{1}{r} \frac{d}{dr} \left(rk \frac{dt}{dr} \right) = 0$$

$$\text{at } r_1, t = t_1; \text{ at } r_2, t = t_2 \tag{2.2.8}$$

$$rk \frac{dt}{dr} = B_1 = -\frac{q'}{2\pi} \tag{2.2.9}$$

This result confirms that the conduction rate q' is conserved at all radial locations. The next integration yields

$$\int_{t_1}^{t} k\, dt = B_1 \int_{r_1}^{r} \frac{dr}{r} = B_1 \ln \frac{r}{r_1} = -\frac{q'}{2\pi} \ln \frac{r}{r_1} \qquad (2.2.10)$$

Then B_1 is evaluated at $r = r_2$, where $t = t_2$, as

$$\int_{t_1}^{t_2} k\, dt = B_1 \ln \frac{r_2}{r_1} \quad \text{or} \quad -\frac{q'}{2\pi} = \frac{1}{\ln(r_2/r_1)} \int_{t_1}^{t_2} k\, dt \qquad (2.2.11)$$

The average conductivity is defined as before as

$$k_m = \frac{1}{(t_2 - t_1)} \int_{t_1}^{t_2} k\, dt \qquad (2.2.12)$$

Eliminating the integral between Eqs. (2.2.11) and (2.2.12) gives the heat flow as

$$q' = \frac{2\pi k_m(t_1 - t_2)}{\ln(r_2/r_1)} = \frac{(t_1 - t_2)}{R_t} \qquad (2.2.13)$$

The temperature distribution $t(r)$ is then evaluated from Eqs. (2.2.10) and (2.2.13):

$$\int_{t_1}^{t} k\, dt = k_m(t_2 - t_1) \frac{\ln(r/r_1)}{(r_2/r_1)} \qquad (2.2.14)$$

SPHERICAL REGION. Equation (2.1.3) applies and for $q''' = 0$ it becomes

$$\frac{1}{r^2} \frac{d}{dr}\left(r^2 k \frac{dt}{dr}\right) = 0$$

$$\text{at } r_1, t = t_1; \text{ at } r_2, t = t_2 \quad (2.2.15)$$

The integrations give the following, where q is the constant conduction rate at all r:

$$r^2 k \frac{dt}{dr} = B_1 = -\frac{q}{4\pi} \qquad (2.2.16)$$

Then

$$\int_{t_1}^{t} dt = B_1 \int_{r_1}^{r} \frac{dr}{r^2} = -B_1\left(\frac{1}{r} - \frac{1}{r_1}\right) = \frac{q}{4\pi}\left(\frac{1}{r} - \frac{1}{r_1}\right) \qquad (2.2.17)$$

Then B_1 is evaluated at r_2 where $t = t_2$:

$$\int_{t_1}^{t_2} k\, dt = -B_1\left(\frac{1}{r_2} - \frac{1}{r_1}\right) = \frac{q}{4\pi}\left(\frac{1}{r_2} - \frac{1}{r_1}\right) \qquad (2.2.18)$$

Defining k_m as before, the heat flow becomes

$$q = \frac{4\pi k_m(t_1 - t_2)}{1/r_1 - 1/r_2} = \frac{(t_1 - t_2)}{R_t} \qquad (2.2.19)$$

and the temperature distribution is evaluated from Eqs. (2.2.17) and (2.2.19) as

$$\int_{t_1}^{t} k \, dt = k_m(t_2 - t_1)\left(\frac{r_1}{r} - 1\right)\bigg/\left(1 - \frac{r_1}{r_2}\right) \qquad (2.2.20)$$

SUMMARY. For each of the preceding geometries, the heat transfer is calculated from k_m. The temperature distribution is determined for a particular $k(t)$ for the geometry of interest, as in Eq. (2.2.20) for a spherical shell. The integration determines $t(r)$. Such results are given in Jakob (1949) for k varying linearly with temperature, for all three geometries. This effect is also discussed in detail by Özisik (1980) and also by Carslaw and Jaeger (1959), Arpaci (1966), Schneider (1955), and Eckert and Drake (1972). This effect is considered in more detail here, in Sec. 9.5, in relation to numerical analysis. Myers (1971) also gives a formulation and calculation procedure, also for variable area in the direction of conduction.

The physical effect of the variation of thermal conductivity as $k(t)$ is often very similar to the mass diffusion analog, in terms of mass diffusivity D as a function of local concentration C, as $D = D(C)$. Accounts of those effects are given by, for example, Crank (1975) and Kirkaldy and Young (1987). These matters are considered in detail in Sec. 6.2.1.

2.2.2 Location-Dependent Conductivity

This condition commonly arises. Material inhomogeneity may result during processing or fabrication due to mechanical, chemical, density, or many other effects. Examples of such materials are heat-treated metals, castings, extruded forms, and chemically aged or irradiated materials. Other common circumstances include a material which is a mixed composite of several materials, as with small particles or voids of varying spatial density, embedded in another uniform material. Recall the general transformation in Sec. 1.2.3. The equations are still (2.1.1)–(2.1.3), with $q''' = 0$. The difference from Sec. 2.2.1 is that k is now dependent on x, and r, for the three geometries.

PLANE REGIONS. Equation (2.1.1), with $q''' = 0$, $k = k(x)$, and the boundary conditions, is

$$\frac{d}{dx} k(x) \frac{dt}{dx} = 0$$

$$\text{at } x = 0, t = t_1; \text{ at } x = L, t = t_2 \qquad (2.2.21)$$

Thus the local flux $-k(x)\,dt/dx$ is still constant at q''. However, the gradient is not, since $k(x)$ varies, as in Sec. 2.2.1 due to $k = k(t)$. The first and second integrations of Eq. (2.2.21) give

$$k(x)\frac{dt}{dx} = B_1 = -q'' \quad \text{or} \quad \frac{dt}{dx} = \frac{B_1}{k(x)}$$

and
$$\int_{t_1}^{t} dt = t - t_1 = B_1 \int_0^x \frac{dx}{k(x)} \tag{2.2.22}$$

Then B_1 is evaluated using $t = t_2$ at $x = L$ as

$$B_1 = -q'' = (t_2 - t_1) \Big/ \int_0^L \frac{1}{k(x)} dx \tag{2.2.23}$$

The flux is

$$q'' = \left(\frac{(t_1 - t_2)}{L}\right) \Big/ \frac{1}{L} \int_0^L \frac{1}{k(x)} dx = \frac{(t_1 - t_2)}{R_t} \tag{2.2.24}$$

The denominator above is merely the average of $1/k(x)$ over the region from $x = 0$ to $x = L$, as

$$\left(\frac{1}{k(x)}\right)_a = \frac{1}{L} \int_0^L \frac{1}{k(x)} dx \tag{2.2.25}$$

and
$$q'' = (t_1 - t_2)/L(1/k)_a \tag{2.2.26}$$

On the other hand, $k(x)$ in Eq. (2.1.1) may be written in terms of $1/k(x) = R(x)$, where $R(x)$ is the equivalent resistivity to heat conduction. Then Eq. (2.2.24) is instead

$$q'' = \frac{(t_1 - t_2)}{L} \Big/ \frac{1}{L} \int_0^L R(x) \, dx \tag{2.2.27}$$

The denominator above is now simply the spatial average of $R(x)$, as of $1/k(x)$, called here R_a. Then

$$q'' = \frac{(t_1 - t_2)}{LR_a} \tag{2.2.28}$$

The temperature distribution $t(x)$ is simply found by combining Eqs. (2.2.22) and (2.2.23), in terms of $k(x)$, as follows:

$$\frac{t(x) - t_1}{t_2 - t_1} = \frac{1}{LR_a} \int_0^x R(x) \, dx \tag{2.2.29}$$

The preceding formulation accommodates any variation of $R(x)$ or of $1/k(x)$, which may be summed over the region $x = 0$ to L. It is noted here that many variations of $k(x)$ are not of the form which will result in the heat flux calculation being in terms of an average value of $k(x)$ over L, as k_m, in the form $q'' = k_m(t_1 - t_2)/L$.

CYLINDRICAL REGIONS. The equation and boundary conditions yield results similar to those given previously:

$$\frac{1}{r}\frac{d}{dr}[rk(r)]\frac{dt}{dr} = 0$$

$$\text{at } r = r_1, t = t_1; \text{ at } r = r_2, t = t_2 \quad (2.2.30)$$

$$rk(r)\frac{dt}{dr} = B_1 = -\frac{q'}{2\pi} \quad (2.2.31)$$

$$\int_{t_1}^{t} dt = t - t_1 = B_1 \int_{r_1}^{r} \frac{dr}{rk(r)}$$

Then B_1 is evaluated at $r = r_2$, $t = t_2$ as

$$B_1 = (t_2 - t_1) \bigg/ \int_{r_1}^{r_2} \frac{dr}{rk(r)} = -\frac{q'}{2\pi} \quad (2.2.32)$$

The heat flow per unit length is then

$$q' = 2\pi(t_1 - t_2) \bigg/ \int_{r_1}^{r_2} \frac{dr}{rk(r)}$$

$$= \frac{2\pi(t_1 - t_2)}{r_2 - r_1} \bigg/ \frac{1}{(r_2 - r_1)} \int_{r_1}^{r_2} \frac{dr}{rk(r)} = \frac{(t_1 - t_2)}{R_t} \quad (2.2.33)$$

where the denominator in Eq. (2.2.33) is the average of $1/rk(r) = (1/rk)_a$ over the region from r_1 to r_2. The heat flow is then

$$q' = \frac{2\pi(t_1 - t_2)}{(r_2 - r_1)} \bigg/ \left(\frac{1}{rk}\right)_a = \frac{(t_1 - t_2)}{R_t} \quad (2.2.34)$$

SPHERICAL REGIONS. The same procedure again applies as follows:

$$\frac{1}{r^2}\frac{d}{dr}\left(r^2 k(r)\frac{dt}{dr}\right) = 0$$

$$\text{at } r = r_1, t = t_1; \text{ at } r = r_2, t = t_2 \quad (2.2.35)$$

$$r^2 k(r)\frac{dt}{dr} = B_1 = -\frac{q}{4\pi} \quad (2.2.36)$$

$$\int_{t_1}^{t} dt = t - t_1 = B_1 \int_{r_1}^{r} \frac{dr}{r^2 k(r)}$$

$$B_1 = (t_2 - t_1) \bigg/ \int_{r_1}^{r_2} \frac{1}{r^2 k(r)} = -\frac{q}{4\pi} \quad (2.2.37)$$

and
$$q = \frac{4\pi(t_1 - t_2)}{(r_1 - r_2)} \bigg/ \frac{1}{(r_1 - r_2)} \int_{r_1}^{r_2} \frac{dr}{r^2 k(r)} = \frac{(t_1 - t_2)}{R_t} \qquad (2.2.38)$$

where the denominator is the average of $1/r^2 k(r) = (1/r^2 k)_a$ over the region from r_1 to r_2. Then the heat flow is

$$q = \frac{4\pi(t_1 - t_2)}{(r_2 - r_1)} \bigg/ \left(\frac{1}{r^2 k}\right)_a = \frac{(t_1 - t_2)}{R_t} \qquad (2.2.39)$$

SUMMARY. The previous three heat flow equations, for $k = k(x)$ or $k(r)$, are completely similar in that the characteristic thickness is L or $(r_2 - r_1)$. The variable conductivity effect is incorporated as the averages of $1/k(x)$, $1/rk(r)$, and $1/r^2 k(r)$, respectively. This method may be applied for any $k(x)$ or $k(r)$ for which these averages may be calculated. This method may be used for discontinuities in $k(x)$ or $k(r)$ and across composite barriers, as well, when there is no contact resistance between the layers. The following section gives some results for the thermal conductivity varying both with temperature and with location across the region.

2.2.3 Temperature- and Location-Dependent Conductivity

Sections 2.2.1 and 2.2.2 consider the thermal conductivity as temperature and location dependent, respectively, across a conduction region, bounded by conditions of t_1 and t_2. A dependence on location typically arises from a difference in material structural arrangement or chemical condition across the region. The temperature dependence usually results from differences in the fundamental conduction transport mechanisms which arise at locations of different temperature levels across the region.

PLANE REGIONS. If the location and temperature dependence of $k(t, x)$ are independent of each other, the conductivity may be written as

$$k(t, x) = K(t)f(x) \qquad (2.2.40a)$$

where $f(x)$ is the location dependence of k, written in nondimensional form. An example of Eq. (2.2.40a) is a linear dependence of k on both t and x. This may be written as

$$K(t)f(x) = K_0[1 + \beta(t - t_0)][1 + \gamma(x - x_0)] \qquad (2.2.40b)$$

where t_0 and x_0 are reference values, β and γ have units of $°C^{-1}$ and m^{-1}, respectively, and $K_0 = k(t_0, x_0)$.

In general, for the form of k in Eq. (2.2.40), and for $q''' = 0$, Eq. (2.1.1) is given and integrated as follows:

$$\frac{d}{dx} k(t, x) \frac{dt}{dx} = \frac{d}{dx} [K(t)f(x)] \frac{dt}{dx} = 0$$

$$\text{at } x = 0, t = t_1; \text{ at } x = L, t = t_2 \quad (2.2.41)$$

$$K(t)f(x)\frac{dt}{dx} = B_1 = -q'' \quad (2.2.42)$$

or

$$K(t)\, dt = B_1 \frac{dx}{f(x)}$$

Then

$$\int_{t_1}^{t} K(t)\, dt = B_1 \int_0^x \frac{dx}{f(x)} \quad (2.2.43)$$

and

$$\int_{t_1}^{t_2} K(t)\, dt = B_1 \int_0^L \frac{dx}{f(x)} = -q'' \int_0^L \frac{dx}{f(x)} \quad (2.2.44)$$

Defining a mean value $k_m(t)$ of the temperature-dependent component of $k(t, x)$, that is, of $K(t)$, as follows:

$$K_m = \frac{1}{t_2 - t_1} \int_{t_1}^{t_2} K(t)\, dt \quad (2.2.45)$$

Eqs. (2.2.44) and (2.2.45) give

$$q'' = \frac{K_m(t_1 - t_2)}{L} \Big/ \frac{1}{L}\int_0^L \frac{dx}{f(x)} = \frac{(t_1 - t_2)}{R_t} \quad (2.2.46)$$

where R_t is now defined in this equation in terms of $[1/f(x)]_a$.

This is a very simple result. It amounts to the result in Eq. (2.2.5) for k dependent only on t, modified by the term in the denominator allowing for $f(x)$. If $f(x) = 1$, Eq. (2.2.46) is, of course, the same as Eq. (2.2.5). Therefore, this formulation, for given functions $K(t)$ and $f(x)$, amounts to determining K_m and the average value of $[1/f(x)]$ across the conduction region, $[1/f(x)]_a$. Then, if the temperature distribution is desired, $B_1 = -q''$ is used in Eq. (2.2.43), with the given function $K(t)$ to calculate $t(x)$. A numerical calculation is indicated.

CYLINDRICAL AND SPHERICAL REGIONS. In circumstances wherein the conductivity $k(t, r)$ may again be separated as $k(t, r) = K(t)f(r)$, the same kind of analysis applies, as for the plane layer geometry given previously. There, the temperature and location variations are separated. The temperature effect is accommodated in terms of the temperature average of $K(t)$, as K_m, as in Sec. 2.2.1. The separable effect of location dependence, $f(x)$, appears in the allowance for $k(x)$ in the formulations in Sec. 2.2.2. These appear in Eqs. (2.2.34) and (2.2.39), for the cylindrical and spherical geometries, respectively.

SUMMARY. The preceding formulation and results for $k = k(t, x,$ or $r)$ assumes that the temperature, t, and location effects, in x or r, are separable as $k(t, x$ or $r) = K(t)f(x)$ or $K(t)f(r)$. This idealization applies when the location variation of k, as $f(x)$, is inherent to the mechanical, crystalline, chemical, or other aspects of the conduction region, and independent of the temperature distribution. Then K_0 is the relevant value of $K(t)$ at some temperature level and location, t_0, x_0.

This idealization also requires that the $K(t)$ component of thermal conductivity is only temperature dependent throughout the region. That is, $K(t)$ does not depend, in addition, on location. This may be a reasonable approximation in some materials. For example, in a material of small but variable porosity, the resulting conduction over the region may depend primarily on the solid phase, consisting of a particular material. The conductivity mechanisms of this solid phase may continue to depend only on the solid material conductivity, independent of porosity. Then $K(t)$ is unchanged and $f(x)$ depends only on the local porosity. This measure may also be a good representation when the region consists of a microscopic composite of several crystalline materials which vary in concentration over the region.

In processes when a separation of $k(t, x)$ is not an accurate procedure, the general relation, for example, Eq. (2.2.41) for a plane layer, applies. The difficulty for analysis is seen in the following result from the first integration:

$$k(t, x)\frac{dt}{dx} = B_1$$

In general, for such forms of $k(t, x)$, a numerical calculation is indicated. See Sec. 9.5 for a numerical analysis of conduction with variable physical properties.

2.3   VARIABLE CONDUCTIVITY IN COMPOSITE BARRIERS

The analysis in Secs. 2.2.1 through 2.2.3 determines the steady-state heat flow through plane, cylindrical, and spherical layers of a single material of possibly variable thermal conductivity, k, across the conduction region. This section applies these results, which are for a single conduction layer, to plane, cylindrical, and spherical composite layers. Composite barrier analysis, for layers having uniform conductivity, is given in Sec. 2.1.3. The three kinds of variation of k considered are

1. $k = k(t)$, a function a local temperature.
2. $k = k(x)$ or $k(r)$, a function of location in a plane layer, or in cylindrical and spherical layers.
3. $k = k(t, x)$ or $k(t, r)$, a function both of temperature and of location, due to varying local material physical condition.

For each of the three kinds of dependence shown previously, there are three geometries. In Secs. 2.2.1 through 2.2.3 the heat flow rates are given for each of these geometries, for each kind of conductivity dependence. The heat flow quantity is q'', q', and q for the three geometries, respectively. The conductive resistance in each geometry is defined, in terms of the temperatures t_1 and t_2, imposed at the two surfaces as

$$q'', q', \text{ and } q = \frac{(t_1 - t_2)}{R_t}$$

The values of R_t are given in Table 2.3.1 for the analyses in Secs. 2.2.1 through 2.2.3. In all conditions of a temperature dependence of k, the effective layer conductivity is k_m and K_m, defined in Table 2.3.1 and written in Eqs. (2.2.4) and (2.2.45).

Using these results, the composite barrier heat transfer rate is easily written for each of the geometries in Figs. 2.1.2 and 2.1.3. The surface convection and air layers included in these figures have resistances as formulated in Eqs. (2.1.27), (2.1.30), and (2.1.33). These are formulated in the same way with variable conductivity. The resulting total resistance R_T is written as follows, as an example, for the plane composite, when $k = k(t, x) = K(t)f(x)$ in each of the four conduction layers:

$$q'' = \frac{(t_1 - t_2)}{R_T} \tag{2.3.1}$$

$$R_T = \frac{1}{h_i} + \left[\frac{\int_0^L \frac{dx}{f(x)}}{K_m} \right]_a + \left[\frac{\int_0^L \frac{dx}{f(x)}}{K_m} \right]_b + \frac{1}{h + h_r}$$

$$+ \left[\frac{\int_0^L \frac{dx}{f(x)}}{K_m} \right]_c + \left[\frac{\int_0^L \frac{dx}{f(x)}}{K_m} \right]_d + \frac{1}{h_o} \tag{2.3.2}$$

where the K_m, $f(x)$, and L quantities in each term are for the particular region noted.

This result, and those in the material in Sec. 2.2, indicate that formulating the effects of variable conductivity is a simple matter. Even complicated variations of $K(t)$ and $f(x)$ are accommodated in the formulation, to determine the same kind of layer resistance which arises for uniform k.

However, in a composite barrier, subject only to bounding temperatures like t_i and t_o, as in Figs. 2.1.2 and 2.1.3, the surface temperatures, say t_1 and t_2, of any particular layer are not initially known. For purely x or r dependence of k, known directly from the material composition and its variation, the resistances in the second column of Table 2.3.1 apply.

On the other hand, if $k = k(t)$ or $k(t, x \text{ or } r)$, the value of the averages, k_m or K_m, are required, as seen in the first and third columns. However, the temperature limits of integration, for averaging, are required to calculate k_m or

TABLE 2.3.1
The conductive resistance, R_t, of plane, cylindrical, and spherical layers having variable conductivities as $k(t)$, $k(x)$ or $k(r)$, and $k(t, x) = K(t)f(x)$

Geometry	Conductivity variation		
	$k(t)$	$k(x)$ or $k(r)$	$k(t, x) = K(t)f(x)$
Plane	$\left[\dfrac{L}{k_m}\right]^*$ Eq. (2.2.5)	$\left[\displaystyle\int_0^L \dfrac{dx}{k(x)}\right]$ Eq. (2.2.24)	$\left[\dfrac{\displaystyle\int_0^L \dfrac{dx}{f(x)}}{K_m}\right]^*$ Eq. (2.2.46)
Cylindrical	$\left[\dfrac{\ln(r_2/r_1)}{2\pi k_m}\right]^*$ Eq. (2.2.13)	$\left[\dfrac{\displaystyle\int_{r_1}^{r_2} \dfrac{dr}{rk(r)}}{2\pi}\right]$ Eq. (2.2.33)	
Spherical	$\left[\dfrac{(1/r_1 - 1/r_2)}{4\pi k_m}\right]^*$ Eq. (2.2.19)	$\left[\dfrac{\displaystyle\int_{r_1}^{r_2} \dfrac{dr}{r^2 k(r)}}{4\pi}\right]$ Eq. (2.2.38)	

$^*k_m$ and K_m are the average of k and K, respectively, over the imposed temperature range, from t_1 to t_2.

K_m. Thus an iteration is required. This is the same procedure which may be followed when the k variation is not large, yet representative values are wanted for each conduction layer. The iteration procedure for highly variable $k(t)$ and $K(t)$ will generally require more steps, for the same level of accuracy.

2.4 INTERNAL ENERGY GENERATION

Many processes occur in a region of material subjected to a distributed internal dissipation process which releases thermal energy throughout the region. This is the distributed source of strength q''' in Eqs. (2.1.1)–(2.1.3). Examples are microwave, or induction heating, in a microwave oven. Another occurrence is irradiation of a material by high-energy particles, which give up energy in internal collisions. Also, an electrical current dissipates energy over a resistive region. In these examples, $q''' > 0$. Distributed chemical reactions in, for example, hydration or curing, also release or absorb energy throughout the region. Then q''' will be positive or negative. These last processes may also release or absorb a chemical species, C, as well. Then $C''' \neq 0$ and mass diffusion may also arise in the region. The equations for this effect are then (2.1.4)–(2.1.6), in the three geometries.

A distributed source, $q''' > 0$, or sink, $q''' < 0$, alters the internal temperature distribution which would result with only conduction, subject to the same imposed boundary conditions. The strength of the source is sometimes uniform over the region. However, many processes arise in which the source strength varies across the region, as with temperature-dependent effects.

There are two typical mechanisms which result in spatially varying internal generation. These are distinguished by the input which results in the distributed source q'''. Consider an effect which amounts to a surface loading, which penetrates the region and results in internal energy generation. Examples of this are surface irradiation by high-energy particles and by microwaves. The energy absorption and resulting generation rate is highest near the surface. It decreases inward as the remaining local intensity decreases across the region. Recall Eq. (1.4.12). Therefore, $q''' = q'''(x, y, z)$ in general, even in steady state.

Another common mechanism leading to variable source strength arises when the rate depends on the total temperature level, t, as $q'''(t)$. This arises in internal mechanisms in which the rate of the process causing the energy source is, itself, also temperature dependent. An example of this mechanism is a distributed chemical reaction, which is either exothermic or endothermic. Another circumstance is electric current in a region whose local resistance is temperature dependent.

The following four sections consider these mechanisms, q''' uniform and both location and temperature dependent. Results are given in Sec. 2.4.1, for q''' uniform, in plane, cylindrical, and spherical regions. In Sec. 2.4.2, a spatial dependence of the distributed source is analyzed, as $q'''(x)$, for a plane region. The effect of source strength temperature dependence, $q'''(t)$, is given in Sec. 2.4.3, for the plane, cylindrical, and spherical geometries. Section 2.4.4 considers the effects of variable conductivity $k(t, x)$, in conjunction with a uniformly distributed source q'''. Distributed source effects in transients and in multidimensional regions are considered in Chaps. 3 through 5.

2.4.1 A Uniform Internal Distributed Energy Source, q'''

The plane, cylindrical, and spherical regions in Fig. 2.1.1 are considered, with the boundary conditions shown. Equations (2.1.1)–(2.1.3) apply, as follows, for uniform conductivity over the region:

$$\frac{d^2t}{dx^2} + \frac{q'''}{k} = 0$$

$$\text{at } x = 0, t = t_1; \text{ at } x = L, t = t_2 \quad (2.4.1)$$

$$\frac{d^2t}{dr^2} + \frac{1}{r}\frac{dt}{dr} + \frac{q'''}{k} = 0$$

$$\text{at } r = r_1, t = t_1; \text{ at } r = r_2, t = t_2 \quad (2.4.2)$$

$$\frac{d^2t}{dr^2} + \frac{2}{r}\frac{dt}{dr} + \frac{q'''}{k} = 0$$

$$\text{at } r = r_1, t = t_1; \text{ at } r = r_2, t = t_2 \quad (2.4.3)$$

They are seen to be inhomogeneous in the term q'''/k. This results in the added temperature effect produced by the source. The three regions are analyzed, in turn, in the following discussion.

PLANE REGIONS. Equation (2.4.1) is integrated and the constants are determined as follows to give the solution $t(x, q''')$, for surface temperatures t_1 and t_2:

$$\frac{dt}{dx} + \frac{q'''}{k}x = B_1$$

$$t + \frac{q'''}{k}\frac{x^2}{2} = B_1 x + B_2 \qquad (2.4.4)$$

$$B_2 = t_1 \qquad B_1 = \frac{t_2 - t_1}{L} + \frac{q'''L}{2k}$$

$$t = \frac{q'''L^2}{2k}\left[\frac{x}{L} - \left(\frac{x}{L}\right)^2\right] + (t_2 - t_1)\frac{x}{L} + t_1 \qquad (2.4.5)$$

The temperature distribution reduces to Eq. (2.1.11) for $q''' = 0$. Equation (2.4.5) may be interpreted as that linear distribution, with the following parabolic term added:

$$\frac{q'''}{k}\frac{L^2}{2}\left[\frac{x}{L} - \left(\frac{x}{L}\right)^2\right]$$

The distribution is rewritten in generalized form as

$$\frac{t - t_2}{t_1 - t_2} = \frac{q'''L^2}{2k(t_1 - t_2)}\left(1 - \frac{x}{L}\right)\frac{x}{L} + \left(1 - \frac{x}{L}\right)$$

$$= B\frac{x}{L}\left(1 - \frac{x}{L}\right) + \left(1 - \frac{x}{L}\right) \qquad (2.4.6)$$

This relation is plotted in Fig. 2.4.1, for the various values of B given previously.

The local heat flux is equal to minus the product of k and the local temperature gradient:

$$q''(x) = -k\frac{dt}{dx} = -k\left(B_1 - \frac{q'''}{k}x\right) = -k\left(\frac{t_2 - t_1}{L} + \frac{q'''L}{2k} - \frac{q'''x}{k}\right)$$

or $\quad q''(x) = k\left[\frac{t_1 - t_2}{L} - \frac{q'''L}{k}\left(\frac{1}{2} - \frac{x}{L}\right)\right] \qquad (2.4.7)$

For $q''' = 0$ this expression reduces to Eq. (2.1.12), for the plane region without generation. The second term in Eq. (2.4.7) is the influence of generation on heat flux. For example, for $(t_1 - t_2)$ and q''' positive, sufficiently large values of q''' can reverse the direction of q''. This can be seen in Fig. 2.4.1, from the changes of the slope of the curves as B increases.

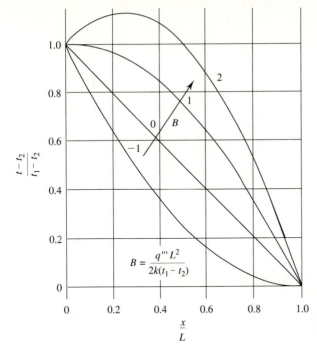

FIGURE 2.4.1
Temperature distribution in a plane region with a uniform distributed energy source q'''.

The preceding results for $t(x)$ and q'' have a number of interesting characteristics. The first term inside the brackets in Eq. (2.4.7) is the simple conduction for the boundary conditions t_1 and t_2. This is the curve $B = 0$ in Fig. 2.4.1. The second term, due to q''', is symmetric about $x = L/2$. That is, the effect beyond pure conduction goes equally each way.

Another feature of this result is that it is also the solution for a region of thickness $2L$ having both surface temperatures the same. This is merely a symmetric parabolic distribution and is obtained from the preceding solution by taking $B = 1$. Then the distribution in the right half is as seen in Fig. 2.4.1. This solution, from Eq. (2.4.6), when L is measured each way from $x = 0$, is

$$\frac{t - t_2}{t_1 - t_2} = 1 - \frac{x^2}{L^2} \tag{2.4.8a}$$

where the following characteristic temperature difference arises:

$$t_1 - t_2 = q'''L^2/2k = \Delta t_{ch} \tag{2.4.8b}$$

CYLINDRICAL REGIONS. The temperature distribution in a tube wall or in a rod is determined from Eq. (2.4.2). Replacing dt/dr by p and multiplying through by $r\,dr$ gives

$$r\,dp + p\,dr + \frac{q'''}{k}r\,dr = d(rp) + \frac{q'''}{2k}d(r^2) = 0$$

and

$$rp + \frac{q'''}{2k}r^2 = B_1$$

This may be written as

$$dt = -\frac{q'''}{2k}r\,dr + B_1\frac{dr}{r}$$

and, therefore,

$$t = -\frac{q'''}{4k}r^2 + B_1 \ln r + B_2 \qquad (2.4.9)$$

Consider first a tube of inside and outside radii a and b with inside and outside surface temperatures maintained at t_1 and t_2. The constants are evaluated and substituted into Eq. (2.4.9) to yield:

$$\frac{t - t_2}{t_1 - t_2} = \frac{q''b^2}{4k(t_1 - t_2)}\left[\left(1 - \frac{a^2}{b^2}\right)\frac{\ln(r/b)}{\ln(b/a)} - \left(1 + \frac{r^2}{b^2}\right)\right] - \frac{\ln(r/b)}{\ln(b/a)}$$

$$(2.4.10)$$

The local heat flow rate is determined from the temperature gradient.

The general solution, Eq. (2.4.9), is next applied to a solid rod of radius b maintained at t_b on its surface. One boundary condition is

$$\text{At } r = b: t = t_b$$

Two conditions are necessary for the second-order differential equation. There are two constants, B_1 and B_2, to be determined. The other condition is evident from the general solution, Eq. (2.4.9). Since, for uniform generation, the temperature at $r = 0$ is finite, the coefficient of $\ln r$ must be zero. The constant B_2 is then evaluated, and the temperature distribution is

$$t - t_b = \theta = \frac{q'''b^2}{4k}\left(1 - \frac{r^2}{b^2}\right) \qquad (2.4.11)$$

This is written as follows as the ratio of $t - t_0$ to $t(0) - t_0$, where $t(0)$ is the temperature at $r = 0$:

$$\frac{t - t_b}{t(0) - t_b} = 1 - \frac{r^2}{b^2} \qquad (2.4.12)$$

where the characteristic temperature difference Δt_{ch} is

$$t(0) - t_b = q'''b^2/4k = \Delta t_{ch} \qquad (2.4.13)$$

SPHERICAL REGIONS. Equation (2.4.3) applies and is again transformed by taking $p = dt/dr$. The result is

$$d(r^2 p) + \frac{q''' r^2\, dr}{k} = 0 \qquad (2.4.14)$$

$$t(r) = -\frac{q''' r^2}{6k} - \frac{B_1}{r^3} + B_2 \qquad (2.4.15)$$

The following two sets of boundary conditions apply for a spherical shell of inner and outer radius a and b and for a solid sphere of radius b:

$$\text{at } r = a,\ t = t_1;\ \text{at } r = b,\ t = t_2 \qquad (2.4.16)$$

$$\text{a bounded solution at } r = 0;\ \text{at } r = b,\ t = t_b \qquad (2.4.17)$$

The temperature distribution through the shell material is

$$\frac{t - t_2}{t_1 - t_2} = \frac{1}{1 - a/b}\left(\frac{a}{r} - \frac{a}{b}\right)$$

$$+ \frac{q''' b^2}{6k(t_1 - t_2)}\left[\left[1 - \frac{r^2}{b^2}\right] + \left(1 + \frac{a}{b}\right)\left[\frac{a}{b} - \frac{a}{r}\right]\right] \qquad (2.4.18)$$

where the first term is the pure conduction effect. The distribution in the shell, in terms of $t - t_b$, and the value of $t(r)$ at $r = 0$, $t(0)$, are

$$\frac{t - t_b}{t(0) - t_b} = \left(1 - \frac{r^2}{b^2}\right) \qquad (2.4.19)$$

where

$$t(0) - t_b = \frac{q''' b^2}{6k} = \Delta t_{\text{ch}} \qquad (2.4.20)$$

SUMMARY. Comparing this last relation with the results for the rod and plane layer, Eqs. (2.4.8) and (2.4.11), the temperature distribution is seen to be the same. The maximum temperature differences are different only as 2, 4, and 6 in the denominator. A similarity also results in the average temperature levels, t_m, internal to these three geometries. The results are, respectively,

$$\frac{t_m - t_b}{t(0) - t_b} = \frac{2}{3}, \frac{1}{2}, \text{ and } \frac{2}{5} \qquad (2.4.21)$$

2.4.2 Spatially Varying Distributed Internal Energy Sources, q'''

Such a source may arise as a surface load which penetrates the region. The flux level gradually diminishes in intensity by absorption, which releases energy. Examples are surface irradiation by fission products, as high-energy particles, or by microwave heating. These effects may cause different mechanisms of energy release and scatter in the conduction region. Particle collisions typically result in a large amount of scatter, whereas electromagnetic radiation may penetrate more uniformly, depending on the material of the region. Visible radiation in clear glass is an extreme sample of small scatter.

Nevertheless, whenever scatter is an appreciable part of the propagation process, the internal effect will in general be multidimensional in all but very extensive plane regions. That geometry is the only mechanism considered here. Equation (2.4.1) applies, where $q''' = q'''(x)$, as

$$\frac{d^2t}{dx^2} + \frac{q'''(x)}{k} = 0 \quad \text{or} \quad t = -\int\int \frac{q'''(x)}{k}(dx)^2 + B_1 x + B_2 \quad (2.4.22)$$

A particular solution will be given for a circumstance which approximates particle absorption in a radiation shield. An extensive plate of thickness L, having inside and outside surface temperatures t_1 and t_2, is assumed subjected to a gamma-ray intensity at its inside surface. The rate is q_0'' energy units per unit time. A uniform attenuation coefficient μ is assumed, and secondary radiation effects are ignored.

Recall the boundary condition T.6 in Sec. 1.4.1, for a penetrating energy flux. The local generation rate is first determined from the attenuation dq'' as follows:

$$\frac{dq''}{q''} = -\mu \, dx$$

$$\int_{q_0''}^{q''} d \ln q'' = -\mu \int_0^x dx$$

$$\ln \frac{q''}{q_0''} = -\mu x \quad \text{or} \quad \frac{q''}{q_0''} = e^{-\mu x}$$

$$q'''(x) = -\frac{dq''}{dx} = \mu q_0'' e^{-\mu x}$$

The temperature distribution is determined from

$$\frac{d^2t}{dx^2} + \frac{\mu q_0''}{k} e^{-\mu x} = 0$$

$$\text{at } x = 0, t = t_1; \text{ at } x = L, t = t_2$$

$$\frac{dt}{dx} = \frac{q_0''}{k} e^{-\mu x} + B_1$$

$$t = -\frac{q_0''}{\mu k} e^{-\mu x} + B_1 x + B_2$$

and
$$t - t_1 = \frac{q_0''}{\mu k}\left[(1 - e^{-\mu x}) - \frac{x}{L}(1 - e^{-\mu L})\right] - (t_1 - t_2)\frac{x}{L} \quad (2.4.23)$$

The local conduction heat flux may be determined from the first derivative of the temperature distribution as

$$q''(x) = \frac{k}{L}(t_1 - t_2) + \frac{q_0''}{\mu L}\left[\mu L e^{-\mu x} + (1 - e^{-\mu L})\right] \qquad (2.4.24)$$

where the first term is again conduction, in the absence of a distributed source. Recall from Sec. 2.1.4 that there is no solution for an infinite solid for any imposed plane energy input.

2.4.3 Temperature-Dependent Energy Generation, $q'''(t)$

This very commonly arises, when the local temperature level interacts with the mechanism causing the generation rate q'''. Examples are internal chemical reactions and electrical energy dissipation in regions where the electrical resistance changes with temperature.

The equations in the three regions are (2.4.1)–(2.4.3), where $q''' = q'''(t)$ is a function of temperature. Results are given here for a plane layer, a cylindrical rod, and a solid sphere. The temperature dependence of q''' is taken in the following simple form, as $q'''(t)$ a linear function of local temperature t:

$$q'''(t) = q_s'''[1 + \beta(t - t_s)] = q_s'''(1 + \beta\theta) \qquad (2.4.25)$$

where $\theta = t - t_s$ and q_s''' is the local distributed source strength at the surface of the region. This form is also found in Jakob (1949).

PLANE REGIONS. For simplicity, $x = 0$ is taken as the midplane of the plane region, as seen in Fig. 2.4.2(a), where s is the half-thickness of the layer. Therefore, the temperature field is symmetric around $x = 0$. The surface temperature on each side, at $x = s$ and $-s$, is taken as t_s. Then q''' in Eq. (2.4.25) is the generation at $x = s$, and $\beta q_s'''$ is the rate of change of q''', with local temperature t. The level of q''' results from the temperature t_s imposed at $x = s$ and $-s$, at the two surfaces.

Since both surfaces are at t_s, the internal temperature is symmetric about the midplane at $x = 0$, in the plane region of thickness $2s$. Therefore, $dt/dx = 0$ at $x = 0$, and the equation and boundary conditions are

$$\frac{d^2\theta}{dx^2} + \frac{q_s'''}{k}(1 + \beta\theta) = 0$$

$$\text{at } x = 0, \frac{dt}{dx} = 0; \text{ at } x = s, t = t_s \qquad (2.4.26a)$$

or

$$\frac{d^2\theta}{dx^2} + n + p\theta = 0$$

$$\text{at } x = 0, \frac{d\theta}{dx} = 0; \text{ at } x = s, \theta = 0 \qquad (2.4.26b)$$

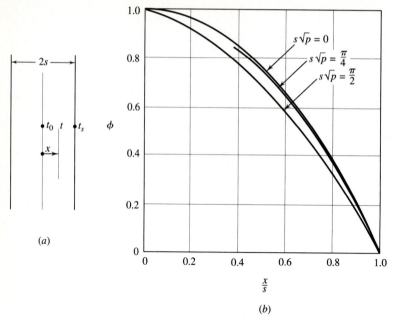

FIGURE 2.4.2
Temperature-dependent energy generation, $q''' = q'''(t)$, in a plane region.

where $\theta = t - t_s$ and

$$n = \frac{q_s'''}{k} \quad \text{and} \quad p = \beta \frac{q_s'''}{k}$$

Equation (2.4.26) is linear in θ, since q''' was taken as linear in θ in Eq. (2.4.25a). The further transformation $\theta' = n + p\theta$ results in

$$\frac{d^2\theta'}{dx^2} + \frac{1}{p}\theta' = 0$$

The solution is

$$\theta' = n + p\theta = B_1' \cos\left(x\sqrt{p}\right) + B_2' \sin\left(x\sqrt{p}\right)$$

$$\theta = B_1 \cos\left(x\sqrt{p}\right) + B_2 \sin\left(x\sqrt{p}\right) - \frac{n}{p}$$

$$\frac{d\theta}{dx} = -B_1\sqrt{p} \sin\left(x\sqrt{p}\right) + B_2\sqrt{p} \cos\left(x\sqrt{p}\right)$$

$$\left(\frac{d\theta}{dx}\right)_0 = 0 = B_2\sqrt{p} \quad \text{and} \quad B_2 = 0$$

$$\theta_s = 0 = B_1 \cos\left(L\sqrt{p}\right) - \frac{n}{p} \quad \text{and} \quad B_1 = \frac{n}{p \cos\left(s\sqrt{p}\right)}$$

From these results

$$\theta(x) = \frac{1}{\beta}\left[\frac{\cos(x\sqrt{p})}{\cos(s\sqrt{p})} - 1\right] \tag{2.4.27}$$

This temperature distribution expression has several interesting features. The temperature excess is a maximum at $x = 0$ and decreases to zero at $x = s$. However, for $s\sqrt{p} = \pi/2$, the temperature excess is infinite for all x. Since s is the measure of the temperature effect upon the local generation rate, the condition $s\sqrt{p} < \pi/2$ sets a limit on temperature dependence. Beyond this condition, arbitrarily high temperatures and destruction of the plane region would result. This circumstance can be interpreted as the consequence of a temperature dependence of sufficient magnitude that the temperature rise necessary to cause the generated energy to be conducted to the surface causes an increase in generation rate which is too large to be conducted away. This limitation in terms of q_s''' and β is

$$s\sqrt{p} = s\sqrt{\frac{\beta q_s'''}{k}} < \frac{\pi}{2}$$

or

$$\beta q_s''' < \left(\frac{\pi}{2}\right)^2 \frac{k}{s^2}$$

Also for $\theta \geq 0$, for $q_s''' \geq 0$, the range of x is limited by $\cos(x\sqrt{p}) > \cos(s\sqrt{p}.)$ This is the range $0 \leq x \leq s$.

For a temperature dependence less than the limiting value, the midplane temperature excess θ_0, the mean temperature excess ratio ϕ_m, and the temperature excess ratio ϕ distribution are

$$\theta_0 = \frac{1}{\beta}\left[\frac{1}{\cos(s\sqrt{p})} - 1\right] \tag{2.4.28}$$

$$\phi_m = \frac{\theta_m}{\theta_0} = \frac{\sin(s\sqrt{p}) - s\sqrt{p}\,\cos(s\sqrt{p})}{s\sqrt{p}\left[1 - \cos(s\sqrt{p})\right]} \tag{2.4.29}$$

$$\phi = \frac{\theta}{\theta_0} = \frac{\cos(x\sqrt{p}) - \cos(s\sqrt{p})}{1 - \cos(s\sqrt{p})} \tag{2.4.30}$$

The temperature distribution, expressed as an excess ratio, ϕ, varies between 0 and 1. The form of the distribution is relatively insensitive to $s\sqrt{p}$. Curves for various values of $s\sqrt{p}$ are plotted in Fig. 2.4.2(b) (see page 76).

CYLINDRICAL AND SPHERICAL REGIONS. These are diagrammed in Fig. 2.4.3, where s is the region radius, t_s is the surface temperature, $\theta = t - t_s$, and the variable generation rate q''' is again linear in temperature as in Eq. (2.4.25). The

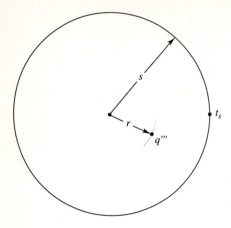

FIGURE 2.4.3
The solid cylinder and sphere, with $q''' = q'''(t)$.

two equations and apparent boundary conditions are

$$\frac{d^2\theta}{dr^2} + \frac{1}{r}\frac{d\theta}{dr} + \frac{q'''_s}{k}(1 + \beta\theta) = 0 \tag{2.4.31}$$

$$\frac{d^2\theta}{dr^2} + \frac{2}{r}\frac{d\theta}{dr} + \frac{q'''_s}{k}(1 + \beta\theta) = 0 \tag{2.4.32}$$

$$\text{at } r = s, \ \theta = 0; \text{ at } r = 0, \ \frac{d\theta}{dr} = 0 \tag{2.4.33}$$

Solutions were given by Jakob (1949), for each geometry, in terms of the midpoint temperature $\theta_0 = \theta(s, \beta, p)$, the local temperature excess $\phi(r, s, p)$, and the mean value $\phi_m(s, p)$. For the solid rod these are

$$\theta_0 = \frac{1}{\beta}\left[\frac{1}{J_0(s\sqrt{p})} - 1\right] \tag{2.4.34}$$

$$\phi_m = \frac{\theta_m}{\theta_0} = \frac{J_0(r\sqrt{p}) - J_0(s\sqrt{p})}{1 - J_0(s\sqrt{p})} \tag{2.4.35}$$

$$\phi = \frac{\theta}{\theta_0} = \frac{1}{s\sqrt{p}}\frac{2J_1(s\sqrt{p}) - (s\sqrt{p})J_0(s\sqrt{p})}{1 - J_0(s\sqrt{p})} \tag{2.4.36}$$

where J_0 and J_1 are the Bessel functions of first order and first kind. The upper limit of $s\sqrt{p}$ for a finite temperature across a plane region was seen to be $s\sqrt{p} < \pi/2$. For the solid rod, the limit was found to be $s\sqrt{p} < 2.4048$. The proper solution also requires that $J_0(r\sqrt{p}) > J_0(s\sqrt{p})$.

For the spherical region the results are given as follows in the same form:

$$\theta_0 = \frac{1}{\beta}\left(\frac{s\sqrt{p}}{\sin(s\sqrt{p})} - 1\right) \tag{2.4.37}$$

$$\phi_m = \frac{\theta_m}{\theta_0} = \frac{3\sin(s\sqrt{p}) - 3(s\sqrt{p})\cos(s\sqrt{p}) - (s^2p)\sin(s\sqrt{p})}{(s^2p)[(s\sqrt{p}) - \sin(s\sqrt{p})]} \tag{2.4.38}$$

$$\phi = \frac{\theta}{\theta_0} = \frac{(s/r)\sin(r\sqrt{p}) - \sin(s\sqrt{p})}{s\sqrt{p} - \sin(s\sqrt{p})} \tag{2.4.39}$$

The limits for the solid sphere are $s\sqrt{p} < \pi$ and $\sin(r\sqrt{p})/(r\sqrt{p}) > \sin(s\sqrt{p})/(s\sqrt{n})$.

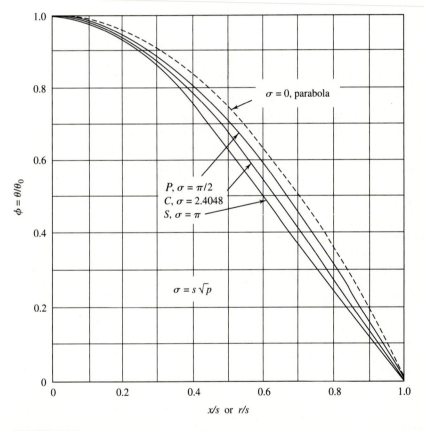

FIGURE 2.4.4

A comparison of the temperature distributions ϕ in a plane layer vs. (x/s), in a cylindrical rod and solid sphere vs. (r/s). The dashed curve is for no temperature dependence of q'''. The solid curves are the limiting solutions, that is, those which apply at the maximum permissible value in Eq. (2.4.25).

SUMMARY. The results for the cylindrical and spherical regions are similar to those shown for the plane region in Fig. 2.4.2(b). The three solutions are compared in Fig. 2.4.4. Curves are shown there fore the two limiting conditions in the geometry. The first is $s\sqrt{p} = 0$. That is, $p = q_s'''\beta/k = 0$, from $\beta = 0$. These are then the uniform generation conditions given in Sec. 4.2.1. The other curves shown are the three limiting solutions, for $\sigma = s\sqrt{p} = \pi/2$, 2.4048, and π.

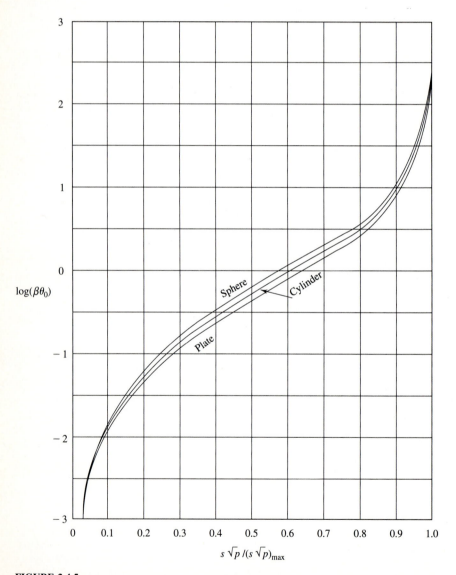

FIGURE 2.4.5
Central temperatures θ_0 as a function of $s\sqrt{p}$ normalized by $(s\sqrt{p})_{max} = \pi/2$, 2.4048, and π for the plane, cylindrical, and spherical regions, respectively.

The curves indicate two separate conclusions. First, for each geometry, a variation of $s\sqrt{p}$ over the whole range $0 = s\sqrt{p} = (s\sqrt{p})_{max}$ causes a maximum difference in these distributions, when normalized by θ_0, of 20% or less. The second conclusion, comparing the three geometries at $s\sqrt{p} = 0$ and at $(s\sqrt{p})_m$, is that the differences are only about 5% and 15%, respectively. This shows that the parameters and variables of the formulation are very appropriate for these mechanisms. Figure 2.4.5 shows the central temperature for each of the three geometries, in terms of $\beta\theta_0$ versus $s\sqrt{p} = \sqrt{\beta s^2 q_s'''/k}$, normalized by the relevant maximum value.

2.4.4 Conductivity and Source Strength Both Variable with Temperature and Location

Variable conductivity, as $k(t, x)$, was discussed in Sec. 2.2.3, with no generation, that is, with $q''' = 0$. In Sec. 2.4.2, q''' was considered as variable with location, and then variable with temperature in Sec. 2.4.3. There are many circumstances in which conductivity and q''' both vary significantly over the material region of interest. Examples in which k is to be taken as $k(t, x)$ were discussed in Sec. 2.2.3. The mechanisms which cause q''' to vary over the region were discussed previously.

Considering a plane region, the equation is given as follows, formulating $k(t, x)$ as $K(t)f(x)$ as in Eq. (2.2.40). In an analogous way q'''/k is written as $Q(t)g(x)$ as a function of t and x. That is, $q'''(t, x)$ is assumed separable in t and x. Recall the discussion in Sec. 2.2.3. Here, $Q(t)$ has the dimensions of q''' and $g(x)$ is dimensionless. For a plane region of thickness L with imposed surface temperatures of t_1 and t_2:

$$\frac{d}{dx}k\frac{dt}{dx} + q''' = \frac{d}{dx}[K(t)f(x)]\frac{dt}{dx} + Q(t)g(x) = 0$$

$$\text{at } x = 0, t = t_1; \text{ at } x = L, t = t_2 \quad (2.4.40)$$

or
$$d[K(t)f(x)]\frac{dt}{dx} + Q(t)g(x)\,dx = 0 \quad (2.4.41)$$

Since t is a function of x, the second term may not be evaluated in the first integration. A numerical method is indicated. However, if $q''' = G(x)$, that is, no temperature dependence of q''', the following results:

$$K(t)f(x)\frac{dt}{dx} + \int_0^x G(x)\,dx = B_1$$

Then

$$\int_{t_1}^t K(t)\frac{dt}{dx} + \int_0^x \frac{1}{f(x)}\left[\int_0^x G(x)\,dx\right]dx = B_1 x \quad (2.4.42)$$

and
$$\int_{t_1}^{t_2} K(t)\,dt + \int_0^L \frac{1}{f(x)}\left[\int_0^x G(x)\,dx\right]dx = B_1 L \quad (2.4.43)$$

Equation (2.4.43) determines B_1 when $K(t)$, $f(x)$, and $G(x)$ are known, in terms of t_1, t_2, and L. Then Eq. (2.4.42), in terms of $K(t)$, is used to determine $t(x)$. Then $q''(x)$ may also be calculated.

2.5   MASS TRANSFER IN ONE-DIMENSION

Mass transfer mechanisms are considered in Sec. 1.3. The equations were developed in general form and specialized to simpler circumstances. For one-dimensional steady-state processes, the general relations are Eqs. (2.1.4)–(2.1.6). These are completely similar to those for conduction heat transfer, Eqs. (2.1.1)–(2.1.3). This similarity stems from the similarity of the Fourier law of conduction, Eq. (1.1.3), and the Fick law of diffusion, Eq. (1.3.1). This similarity also extends to the meaning of the distributed sources q''' and C'''.

The preceding considerations are the basis of the frequent analogy between heat conduction and mass diffusion. As a result, virtually all of the results given in this chapter for conduction effects, in terms of temperature distributions and heat flux, may be transformed into comparable mass diffusion predictions, in terms of local concentration and mass flux.

This was done for a plane layer of uniform conductivity in Sec. 2.1.1. The same procedure applies to composite barriers and other geometries. The analogs of the temperature, location, and time dependence of thermal conductivity, k, are the same kinds of variations of the mass diffusivity D. Internal generation as q''' becomes C'''.

However, this analogy has very strict limits. These arise because several additional common kinds of mechanisms in mass diffusion have no direct analogy in thermal conduction. For example, k for conduction in a uniform material may be taken, with good accuracy, to depend only on the temperature. However, in mass diffusion, D commonly depends on multiple and largely independent processes. Often the net diffusion amounts to the combined effects of several phases of the species C, for example, vapor and liquid. These different component processes may depend on several independent properties of both the two phases and the physical nature of the region.

Another aspect arises from the local chemical adsorption, desorption, and chemical reactions which commonly arise. A simple example of this is discussed in Sec. 1.3.3. If the local chemical reaction is very fast, then the local level of adsorption c depends only on the local concentration C, as $c = R_r C$, where R_r is a constant of proportionality. It is shown in Sec. 1.3.3 that this effect is accommodated in the analysis by replacing D by $D/(1 + R_r) = D'$, as in Eq. (1.3.8b). However, in reaction-rate-controlled local reactions, a more elaborate formulation is required. These and other aspects and special characteristics are discussed in Sec. 6.5, as a basis for analysis there.

Four one-dimensional steady-state mass transport processes are analyzed here. These demonstrate the preceding methods in this chapter, as applied to conditions more typical of mass diffusion. These all apply to steady-state processes in a plane layer of thickness L. The four examples are as follows:

1. A concentrated plane source of chemical species input at the center of the layer, at a rate of M, per unit area, at $x = L/2$.
2. Liquid diffusion in the plane layer, with evaporation at one surface.
3. An internal distributed species source C''', as with the aging of a plastic or due to the loss of a solvent.
4. An internal species source, with local adsorption or desorption.

The equation and boundary conditions which include all of the foregoing effects are

$$\frac{d^2C}{dx^2} + \frac{C'''}{D} = 0$$

$$\text{at } x = 0,\, C = C_1;\, \text{at } x = L,\, C = t_2 \quad (2.5.1)$$

For case 1 there is an added condition at $x = L/2$ for the concentrated layer of input at a rate of M. For case 2 there is, in effect, a plane concentration sink at the surface of evaporation.

1. A CONCENTRATED PLANE INTERNAL INPUT. In the absence of a distributed source, Eq. (2.5.1) becomes

$$\frac{d^2C}{dx^2} = 0 \quad \text{or} \quad C(x) = Ax + B \quad (2.5.2)$$

That is, $C(x)$ is linear in both subregions. Several characteristic distributions of $C(x)$ are shown in Fig. 2.5.1(a), for $M > 0$, $M = 0$, and $M < 0$. It is seen that the level of the midplane concentration $C(L/2)$, with respect to the imposed conditions, C_1 and C_2, is strongly dependent on the sign and magnitude of M. The boundary condition at $x = L/2$ is written as follows:

$$M = +D\left(\frac{dC}{dx}\right)_L - D\left(\frac{dC}{dx}\right)_R \quad (2.5.3)$$

where L and R are the left and right regions. The concentration distributions in the left and right regions are

$$C_L(x) = [C(L/2) - C_1]\frac{2x}{L} + C_1 \quad (2.5.4a)$$

and

$$C_R(x) = \frac{2x}{L}[C_2 - C(L/2)] + 2C(L/2) - C_2 \quad (2.5.4b)$$

Calculating gradients from Eq. (2.5.4), Eq. (2.5.3) yields the following value of

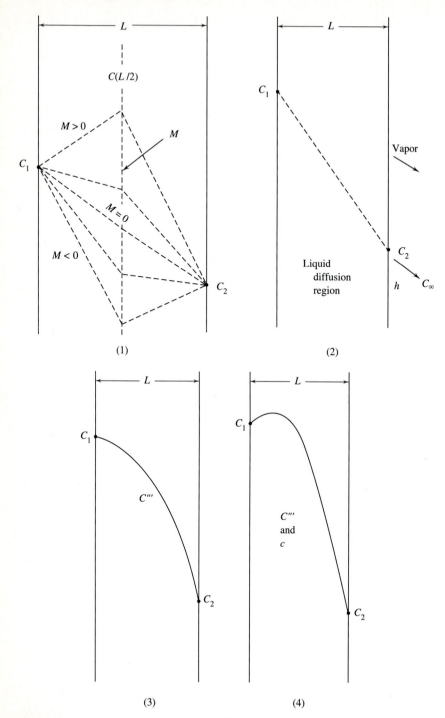

FIGURE 2.5.1
Mass diffusion in plane layers of thickness L.

the midplane concentration level $C(L/2)$:

$$C(L/2) = \frac{ML}{4D} + \frac{(C_1 + C_2)}{2} \qquad (2.5.5)$$

This simple result shows that $C(L/2)$ is the average of the bounding conditions, C_1 and C_2, plus the effect of the plane source, as $ML/4D$. The preceding analysis also applies if the two regions have different mass diffusivities, D_L and D_R, in Eq. (2.5.3). The result then is

$$C(L/2) = \frac{LM}{2(D_L + D_R)} + \frac{D_L C_1 + D_R C_2}{(D_L + D_R)} \qquad (2.5.6)$$

2. LIQUID DIFFUSION AND SURFACE EVAPORATION. Figure 2.5.1(b) (see page 84) shows the liquid diffusion layer. The evaporation process at the right face and the convection process h to an environment at C_∞ are shown. The liquid diffusion and convective rates are as follows:

$$m'' = -D\frac{dC}{dx} = -D\frac{(C_2 - C_1)}{L} \quad \text{and} \quad m''_v = h(C_2 - C_\infty) \qquad (2.5.7)$$

where D applies for the liquid diffusion through the porous solid layer. The bounding concentrations in this configuration are C_1 and C_∞. If C_1 is to be adjusted so that the diffusing liquid flux at $x = L$, m'', is to be removed at the same rate by the convective evaporation process, then $m'' = m''_v$, and

$$m'' = -D\frac{(C_2 - C_1)}{L} = h(C_2 - C_\infty) = m_v \qquad (2.5.8)$$

When this condition is met, the concentration C_2 will be

$$C_2 = \frac{C_1 - \text{Bi}\, C_\infty}{1 + \text{Bi}} \qquad (2.5.9)$$

where $\text{Bi} = hL/D$ is the mass diffusion Biot number. This number is the ratio of the convection conductance h to the diffusion conductance through the region, D/L.

3. A UNIFORM DISTRIBUTED SOURCE, C'''. Species C is assumed to be formed at a constant rate and uniformly across the region. The equations and boundary conditions are Eq. (2.5.1). The solutions to the completely analogous conduction circumstances are Eqs. (2.4.6) and (2.4.7). For concentration $\tilde{C}$, the distribution is

$$\tilde{C} = \frac{C - C_2}{C_1 - C_2} = B_c \frac{x}{L}\left(1 - \frac{x}{L}\right) + \left(1 - \frac{x}{L}\right) \qquad (2.5.10)$$

where $\tilde{C}$, an excess ratio, is analogous to ϕ and

$$B_c = \frac{C'''L^2}{2D(C_1 - C_2)} \tag{2.5.11}$$

The local mass flux is

$$m'''(x) = D\left[\frac{C_1 - C_2}{L} + \frac{C'''L}{D}\left(\frac{1}{2} - \frac{x}{L}\right)\right] \tag{2.5.12}$$

Again the second term in Eq. (2.5.10) is simple mass diffusion. The first term is the effect of the source. The distribution in Fig. 2.5.1(b) applies for some particular value of B_c.

For a discussion of other applications of this result and of the solution forms for cylindrical and spherical regions, see Sec. 2.4.1. The mass diffusion solutions for variable C''', as a function of location, local concentration C, and of both of these, are of the same form as those given in Secs, 2.4.2 through 2.4.4.

4. A CONTINUING INTERNAL SOURCE, C''', WITH LOCAL ADSORPTION. This result is similar to the immediately preceding one. Now D there is replaced by $D' = D/(1 + R_r)$. Recall from Eq. (1.3.7) that the local level of adsorption for very rapid rates is $c = R_rC$. That is, the total local concentration of species C is the sum of the diffusing and adsorbed concentrations, as $C + c = C + R_rC = C(1 + R_r)$. The general governing relation is Eq. (1.3.8b) in one dimension, with an added source term C''':

$$\frac{D}{(1 + R_r)}\frac{d^2C}{dx^2} + C''' = D'\frac{d^2C}{dx^2} + C''' = \frac{\partial C}{\partial \tau} \tag{2.5.13}$$

where the chemical mechanism for the release, or distributed source rate, C''', is assumed independent of the diffusion, and the adsorption mechanism, as $c = R_rC$.

In the eventual steady state in this process, $C = C(x)$, and c is, as assumed in developing Eq. (2.5.13), always in equilibrium with C, as $c = R_rC$. Therefore, C''' merely diffuses through the material, independent of the local adsorption level. The results are then determined from Eq. (2.5.1) and are those given in Eqs. (2.5.10)–(2.5.12). However, in an unsteady process, $C = C(x, \tau)$, the local adsorption level c would also be x and τ dependent. Then C''' would interact with both of these effects, as formulated in Eq. (2.5.13).

PROBLEMS

2.1.1. A 2-by-3-m concrete slab 15 cm thick, having a thermal conductivity of 2.5 W/m °C, has its two surfaces maintained at 70 and 20°C, respectively. For steady-state conduction, find the heat transfer rate through the slab and the slab resistance.

2.1.2. The spherical pressure vessel of a small homogeneous reactor is to have a 1-m ID and 1.5-m OD and to be made of a steel having a thermal conductivity of 41 W/m °C. The outside surface of the vessel is to be sufficiently well insulated so that the heat loss through the insulation will be 375 W/m^2 of area, based upon the outside surface of the pressure vessel. If the inside surface of the pressure vessel is to be at 400°C, find the outside surface temperature. Calculate the total conduction rate through the spherical shell and the shell resistance.

2.1.3. A wall is constructed of a 6-in. layer of concrete with a a 2-in. layer of lime plaster on the inside and 2-in. limestone facing on the outside.

(a) Find the conductance of each layer per unit area, and for steady-state conduction, the ratio of the temperature differences across the plaster and limestone facing to that across the concrete.

(b) For a total temperature difference of 30°F across the wall, find each of the temperature differences and the heat flux.

2.1.4. A 3/4-in. schedule 80 steel pipe has inside and outside temperatures of 20 and 160°F. Find the heat flow per unit pipe length and the flux per unit inside area and per unit outside area.

2.1.5. Consider a long tube of inside and outside diameters a and b, with a uniform thermal conductivity. Under what conditions is the following expression for heat flow per unit tube length accurate within 4%?

$$\frac{q}{L} = \frac{2\pi a k}{b - a}(t_1 - t_2)$$

2.1.6. A sandwich construction consists of two extensive flat plates, 1 and 2, of the same material, each of thickness L. They are separated by a very thin electrically conducting sheet. Both plates are in good thermal contact with the sheet and are not themselves electrically conducting. The thin sheet dissipates thermal energy internally uniformly at the rate of q'' per unit area and time. The other surfaces of the left and right plates are maintained at t_1 and t_2.

(a) Sketch the steady-state temperature distributions across the assembly.

(b) Determine the temperature at the midplane of the geometry.

(c) Calculate the ratio of the heat flow through the left plate to the total input q'', in terms of t_1, t_2, and other quantities.

2.1.7. For the sandwich construction referred to in problem 2.1.6, the other surface of the left plate is maintained at t_1. The right plate is at $t_2 < t_1$.

(a) Sketch the temperature distribution through the assembly.

(b) Calculate the midplane temperature t_c and the ratio of the heat flow through the left plate to the total input q'', in steady state, in terms of t_1, t_2 and other quantities.

(c) Repeat part (a) if $k_1 = 2k_2$ and $t_1 = t_2$ and comment on the resulting heat flux in each direction.

2.1.8. A lake surface is covered by a 4-in. layer of ice on a day when the air temperature is 14°F. A thermocouple embedded very near the upper surface of the ice layer indicates a temperature of 25°F. Assuming steady-state conduction in the ice and no water supercooling, find the heat loss from the lake per acre of surface. What is the conductance of the ice layer? One acre = 43,560 ft^2.

2.1.9. Consider a composite barrier to consist of only two layers, a and b, of equal thickness L and thermal conductivity k. Take $h_i = h_o$. Consider the contact between the two solid layers to be perfect, case a, and to have a contact resistance R_c, case b.

(a) Calculate the overall resistances for each circumstance, for $k = L = h = R_c = 1$.

(b) Find the ratio of the overall resistances in cases a and b.

2.2.1. Consider a cylindrical element of inside and outside diameters a and b with the inside and outside temperature levels t_1 and t_2. For a linear variation of k with t, as $k_0 = k(1 + \beta t)$, show that Eq. (2.1.19) is correct if k is replaced by its average value.

2.2.2. A wall of thickness L is made of a material whose thermal conductivity varies with temperature as follows: $k = k_0 t^2$.

(a) Find an expression for the steady heat conduction through the wall per unit area, that is, q'', if the two surfaces are maintained at t_1 and t_2.

(b) If we wish to write q'' as the product of the temperature difference and a mean thermal conductivity divided by L, at what temperature must this conductivity be calculated so that such an equation will give the right result?

(c) Plot the temperature distribution for $t_1 > t_2$ in suitable nondimensional terms.

2.2.3. An extensive slab of solid material of thickness L has one face maintained at t_1 and the other at t_2. The thermal conductivity of the material is temperature dependent, as follows:

$$k = k_0\left(t^2 - t_0^2\right)$$

Find (a) an expression in terms of t_1 and t_2, for the heat flux through the slab, and (b) the temperature t_r at which an appropriate "average" thermal conductivity k_m should be calculated.

2.2.4. (a) For a plate of thickness L having surface temperatures of t_1 and t_2 and a linear variation of k with t, that is,

$$k = k_0(1 + \beta t)$$

find the expression for the temperature distribution.

(b) For a concrete slab 20 cm thick having $t_1 = 50°C$, $t_2 = 15°C$, $k_0 = 1.8$ W/m °C, and $\beta = 0.005°F^{-1}$, compute the heat flux and plot the actual temperature distribution and the distribution that would apply for a uniform k, equal to the average value.

2.2.5. The two surfaces of a 2-by-3-m plate 5 cm thick are at 70 and 30°C. Find the total heat flow rate if the thermal conductivity has the following dependence upon t:

$$k = 5.4(1 + 0.003t + 0.6 \times 10^{-4}t^2) \quad \text{in W/m}°C$$

2.3.1. A plane wall of thickness L consists of a material whose conductivity varies as $k(t, x)$. The conduction is largely by the radiative phonon mode 4, and the temperature dependence of k arises from that effect. There is also a compositional variation of k across the region. It amounts to an x dependence of k. Both effects are approximated by independent linear components of $k(t, x)$, around the location $x = 0$. Take $t(0) = t_1$, and $t(L) = t_2$.

(a) Determine the appropriate approximation of the temperature effect.

(b) Calculate the temperature distribution.

(c) Determine the heat flux.

2.4.1. Two large steel plates at temperatures of 100 and 60°C are separated by a steel rod 1 m long and 3 cm in diameter. The rod is welded to each plate. The space between the plates is filled with insulation, which also insulates the lateral faces of the rod. Because of a voltage difference between the two plates, current flows through the rod, dissipating electrical energy at a rate of 80 W. The maximum temperature in the rod and the heat flux at each end are to be determined. For the rod, $k = 44.2$ W/m °C.

2.4.2. One method of determining the inside surface convection coefficients for dielectric fluids flowing through tubes involves supplying the heat by passing an electric current through the tube wall material between two electrodes welded to the tube at different points. The strength of the Joulean heating is assumed uniform at q'''. The outside surface is insulated. Therefore, the inside heat flux to the fluid is computed from q'''. The outside surface temperature is measured with thermocouples and, from this, the inside surface temperature is computed, knowing the generation rate q''' and tube dimensions.

(a) Write the governing differential equation and boundary conditions.

(b) Derive the relation for the inside surface temperature t_i in terms of the measured outside surface temperature t_o and other variables, for a tube of inside and outside diameter of a and b, respectively.

(c) Compute the average inside surface temperature for an experimental run using a 2.5-cm ID, 0.30-cm wall thickness and 2-m-long stainless-steel tube for which the following data were obtained: outside temperature, 550°C; voltage difference between electrodes, 100; current, 192 A; and $k = 17$ W/m °C for the tube material.

2.4.3. A copper rod 0.3 cm in diameter and 1 m long runs between two large bus bars. The rod is insulated on its lateral surface against the flow of heat and electric current. The bus bars will be at 20°C. What is the maximum current the rod may carry if its temperature is not to exceed 120°C at any point? Assume that the electrical resistivity of copper is constant at 1.72×10^{-6} Ω cm.

2.4.4. A small dam, which may be idealized as an extensive plate of 1.5 m thick, is to be completely poured in a short period of time. The hydration of the fresh concrete, initially at 20°C, results in the equivalent of a distributed source of constant strength, 40 W/m³. If both dam surfaces are kept at 20°C, find the maximum temperature to which the concrete will be subjected, in steady state, assuming a simple conduction process, $k = 1.2$ W/m °C.

2.4.5. For a plate of thickness L having a distributed source of uniform strength q''' and equal surface temperatures t_o, derive an expression from Poisson's equation for the maximum temperature in the plate. Find an expression for $(t - t_0)/(t_m - t_0)$, where t_m is the maximum temperature in the plate.

2.4.6. An electrical conductor consists of a copper wire 0.5 cm in diameter covered with a 12-cm layer of an insulating material having a thermal conductivity of 0.16

W/m °C. The conductor passes through a pool of liquid which maintains the outside surface of the insulation at essentially 20°C. If the insulation temperature is not to exceed 80°C at any point, what is the maximum current the conductor may carry? Assume the resistivity of copper uniform at 1.72×10^{-6} Ω cm.

2.4.7. For the maximum permissible current in Prob. 2.4.6, what is the highest temperature level in the copper conductor?

2.4.8. Consider a solid sphere of radius a in an ambient medium at t_e. A convection coefficient h applies for transport between the sphere surface and the environment. The sphere material is subjected to a uniform and constant distributed energy source q'''. We wish to calculate the steady-state temperature distribution in the sphere material.
(a) Write the equations and boundary conditions in complete form, using $\theta(r) = t - t_e$.
(b) Solve for the general solution $\theta(r)$. It is not required to evaluate the constants of integration.

2.4.9. A metal rod of diameter D and length $2L$ is insulated against radial heat flow. Its ends are maintained at equal temperature t_0. A distributed source, q''', is present in the left half of the rod.
(a) Find the temperature distribution along the rod, in steady state.
(b) Find the position of the maximum temperature and sketch the temperature distribution.
(c) Find the fraction of the total heat input Q which goes out the left end of the rod.

2.4.10. The surface of a long solid cylindrical rod of radius R is maintained uniformly at t_0. The interior of the rod is subject to a distributed energy source whose strength is zero at $r = 0$ and increases linearly with distance from the cylinder's axis.
(a) Write the complete statement for determining the temperature distribution.
(b) Calculate this distribution.
(c) What is the resulting characteristic temperature difference?

2.4.11. A circular rod of length L, thermal conductivity k, and cross-sectional area A has the left and right ends maintained at temperatures of t_1 and t_2, respectively. A distributed source results in a generation rate which increases linearly from zero at the left end, that is, as $q''' = bx$. The lateral surface of the rod is insulated.
(a) Derive the equation for the temperature distribution in the rod.
(b) Develop the conditions under which a temperature maximum will appear between the ends of the rod. Sketch such a distribution and the resulting heat flux distribution for $t_1 = t_2$.

2.4.12. Repeat Prob. 2.4.3 for a temperature-dependent resistivity for copper of

$$R = R_o[1 + \beta(t - t_0)]$$

$$= 1.724 \times 10^{-6}[1 + 0.0040(t - t_0)] \ \Omega \ \text{cm}$$

where $t_0 = 16°C$. What is the upper limit of current for the conditions specified?

2.4.13. An extensive slab of material of thickness L is insulated on the left surface and is in contact with a fluid at t_e on the right one. The resulting convection coefficient is h. The slab material is subject to a spatial nonuniformly distributed energy source q''' whose strength varies from zero at the left surface as $q''' = Cx^2/L^2$; x

is the distance into the material from the left surface. The thermal conductivity of the material varies as $k = k_0(t - t_e)$.

(a) Write the equation and all boundary conditions for the steady-state temperature distribution through the slab material.

(b) Sketch the expected temperature distribution, for C both positive and negative.

(c) Calculate the temperature distribution in the slab.

2.4.14. If the geometry in Prob. 2.4.11 had been a long solid cylinder of radius R, with $q''' = Qr^2/R^2$, and a surface temperature condition t_0 instead:

(a) Write the governing equation.

(b) From the boundary condition at $r = R$, determine the first constant of integration.

(c) Examine the question of the other necessary boundary condition.

2.4.15. The iron thermal shield of a nuclear reactor may be idealized as an infinite flat plate 12 cm thick. A neutron flux at the inner surface of 7×10^{12} cm^{-2} s^{-1} produces a net heat release per unit volume which decreases exponentially from 6 W/cm^3 at the inside surface to half that value 2.5 cm from the inside surface. If both surfaces of the shield are to be maintained at 50°C, find the location of the maximum temperature in the shield and its value.

2.4.16. Consider the pressure shell of a nuclear reactor as being a large flat metal plate of thickness s. The gamma-ray heating rate within the plate per unit volume and time will vary exponentially with x, where the origin for x is at the inside face:

$$q''' = q'' \mu e^{-\mu x}$$

The constants μ and q'' are respectively, the absorption coefficient and the gamma-ray heating rate at the inside surface.

(a) If the inside surface of the shell is adiabatic and the outside surface is maintained at t_s, find the temperature excess distribution through the shell.

(b) Calculate the heat flux at the outside shell surface due to gamma-ray absorption within the shell material.

2.4.17. Determine the steady-state temperature distributions in a layer of thickness $2s$, subject to a uniform distributed source of q'''. The surfaces are at t_0 and

(a) $k = k(t)$.

(b) $k = k_0[1 + \gamma(t - t_0)]$.

2.4.18. A circular rod of length L and cross-sectional area A has the left and right ends maintained at t_1 and t_2, respectively. A distributed source is present which increases linearly from a value of zero at the left end, as $q''' = bx$. The lateral surface is insulated.

(a) Determine the temperature distribution in the rod.

(b) Find the conditions under which a temperature maximum will occur in the rod material.

(c) For $t_1 = t_2$ sketch a temperature distribution and the associated local heat flux.

REFERENCES

Arpaci, V. S. (1966) *Conduction Heat Transfer*, Addison-Wesley, Reading, MA.

Carslaw, H. S., and J. C. Jaeger (1959) *Conduction of Heat in Solids*, 2nd ed, Oxford Univ. Press.

Crank, J. (1975) *The Mathematics of Diffusion*, 2nd ed., Oxford Science Publications, Clarendon Press, Oxford.

Eckert, E. R. G., and R. M. Drake, Jr. (1972) *Analysis of Heat and Mass Transfer*, McGraw-Hill, New York.

Jakob, M. (1949) *Heat Transfer*, John Wiley and Sons, New York.

Kirkaldy, J. S., and D. J. Young (1987) *Diffusion in the Condensed State*, The Institute of Metals, London.

Myers, G. E. (1971) *Analytical Methods in Conduction Heat Transfer*, McGraw-Hill, New York.

Özisik, N. M. (1980) *Heat Conduction*, Wiley Interscience, New York.

Rohsenow, W. M., J. P. Hartnett, and E. Ganic, Eds. (1985) *Handbook of Heat Transfer Fundamentals*, McGraw-Hill, New York.

Schneider, P. J. (1955) *Conduction Heat Transfer*, Addison-Wesley, Reading, MA.

CHAPTER
3

STEADY-STATE PROCESSES IN SEVERAL DIMENSIONS

3.1 METHODS OF ANALYSIS

Steady-state one-dimensional plane, cylindrical, and spherical processes were analyzed in Chap. 2. Included were the characteristics of limited and infinite regions of one material, composite barriers, contact resistance, temperature and spatially dependent conductivity, and internal energy generation. Mass diffusion was also analyzed for mechanisms comparable to those considered for heat conduction.

Many important applications are at least approximately one dimensional. Those results indicated many very important mechanisms. However, many applications are inherently multidimensional. The detailed mechanisms must be perceived in that way as related to many different geometrical configurations.

This chapter considers steady-state processes in two and three dimensions, in rectangular, cylindrical, and spherical regions. These model conduction processes in geometries in which the boundary conditions of a region may be conveniently and simply expressed in terms in these coordinates.

Many of the mechanisms considered in the following sections are important in diverse applications. Rectangular, cylindrical, and spherical geometries are treated. Calculations are given for different kinds of boundary conditions. The effects of distributed energy sources and of temperature-dependent properties are analyzed. Both the shape factor and the simplest forms of approximate numerical methods are given. This collection of examples and techniques provides a general background.

The first consideration in this section is an example of the use of the method of separation of variables. It is applied in rectangular coordinates in two dimensions. The solution, $t(x, y)$, is expressed as a product of functions of x and y as $t(x, y) = X(x)Y(y)$. This formulation is then applied to obtain the solutions of two classical problems in conduction. Then this method is applied to processes described in rectangular, cylindrical, and spherical coordinates. The results are the algebraic forms of the temperature dependence, which satisfy the Laplace relations in Eq. (1.2.11), in terms of the relevant space coordinates.

Sections 3.2 through 3.4 then give solutions of some important problems in the three geometries. Section 3.5 relates to the use of geometric shape factors to estimate conduction in other multidimensional geometries. Section 3.6 then includes distributed internal energy sources, q''', as analyzed in Sec. 2.4 for one-dimensional processes. The equation then used is the Poisson equation, Eq. (1.2.12).

Section 3.7 then considers several examples of variable conductivity. Section 3.8 extends the consideration of contact resistance, in Sec. 2.1, to multidimensional conduction fields. In Sec. 3.9 the formulation is given for the numerical calculation of steady-state multidimensional conduction. Several kinds of boundary conditions are formulated and the effects of internal distributed sources are included.

Chapter 4 then considers one-dimensional unsteady-state processes in plane, cylindrical, and spherical regions. Then, for example, $t(x, \tau)$ becomes $X(x)T(\tau)$, again as determined by the method of separation of variables. Both transients and periodic processes are analyzed, over a wide range of important applications. Numerical formulations for transients are given and the conditions required for stability in such calculations are determined. Chapter 5 then combines multidimensional effects with transients, in both analysis and in the simplest numerical formulation.

The analysis and solutions given in Chaps. 3 through 5 are in terms of temperature and heat conduction, as described by the Laplace, Poisson, and Fourier equations, and their simple extensions. Therefore, most of the results given also apply to mass diffusion in the similar Fickian formulations in Eqs. (1.3.3)–(1.3.6), (1.3.8), (1.3.10), and (1.3.11). For brevity in the presentation, the equivalent results for mass diffusion are not given. Instead, consideration of mass diffusion is largely deferred to Chap. 6. There, other and distinctive mass diffusion mechanisms are formulated and then applied to characteristic circumstances. Chapters 7 and 8 consider applications. Chapter 9 is a general discussion of finite-difference numerical analysis and of the formulation of the finite-element method.

3.1.1 Steady-State Processes and Example Solutions

The differential equations governing the temperatures field in a conduction region, and the mass concentration field in a diffusion region, are developed in Chap. 1. Transient and internal generation effects are included in the most general forms, Eqs. (1.2.13) and (1.3.5). In this chapter only steady-state processes are considered and Secs. 3.1 through 3.6 assume uniform properties. Equations (1.2.13) and (1.3.5) then become

$$\nabla^2 t + \frac{q'''}{k} = 0 \qquad (3.1.1)$$

$$\nabla^2 C + \frac{C'''}{D} = 0 \qquad (3.1.2)$$

These are inhomogeneous, in the second term. Such distributed sources are considered in Sec. 3.6.

The conduction equations, for $q''' = 0$ and $C''' = 0$, are then in the simplest form, the Laplace equation. Equation (3.1.1) will first be applied to the infinite strip of material in the z direction, shown in Fig. 3.1.1, with the boundary conditions shown:

$$\frac{\partial^2 t}{\partial x^2} + \frac{\partial^2 t}{\partial y^2} = 0 \qquad (3.1.3)$$

$$\text{at } x = 0 \text{ and } L, t = t_0; \text{at } y = 0, t = t_1$$

Four boundary conditions are needed. The additional one will arise in considering the behavior of $t(x, y)$ as $y \to \infty$.

The temperature field, $t(x, y)$, is assumed to consist of separable x and y effects, as

$$t(x, y) = X(x)Y(y) \qquad (3.1.4)$$

where X and Y depend only on x and y, respectively. The solutions of linear and homogeneous equations such as the Laplace and Fourier equations have this property for many kinds of boundary and initial conditions. Equation (3.1.4) is substituted into Eq. (3.1.3):

$$\frac{\partial^2 t}{\partial x^2} + \frac{\partial^2 t}{\partial y^2} = \frac{\partial^2 (XY)}{\partial x^2} + \frac{\partial^2 (XY)}{\partial y^2} = Y \frac{d^2 X}{dx^2} + X \frac{d^2 Y}{dy^2} = 0$$

or
$$\frac{1}{X} \frac{d^2 X}{dx^2} = -\frac{1}{Y} \frac{d^2 Y}{dy^2}$$

This last equation has an interesting feature. Apparently, the left-hand side may be a function only of x and the right-hand side only of y. However, x and y may vary independently of each other over the region. Therefore, the two sides of the equation would vary independently of each other if they were, respectively, functions of x and y. Thus they are not functions of x and y but

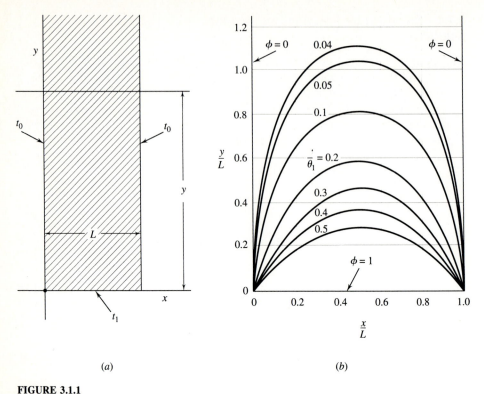

(a)   (b)

FIGURE 3.1.1
Conduction in a narrow strip of finite length: (a) the strip and the boundary conditions; (b) the location of isotherms in the conduction region.

are equal to the same constant, say $-\lambda^2$. The result is

$$\frac{d^2X}{dx^2} + \lambda^2 X = 0$$

$$\frac{d^2Y}{dy^2} - \lambda^2 Y = 0$$

(3.1.5)

These ordinary differential equations may be integrated to give

$$X = C_1 e^{i\lambda x} + C_2 e^{-i\lambda x}$$

$$Y = C e^{\lambda y} + D e^{-\lambda y}$$

Using the relation $e^{\pm i\lambda x} = \cos \lambda x \pm i \sin \lambda x$, the solution becomes

$$X = A \cos \lambda x + B \sin \lambda x$$

The general solution is

$$t(x, y) = XY = (A \cos \lambda x + B \sin \lambda x)(C e^{\lambda y} + D e^{-\lambda y}) \qquad (3.1.6)$$

This result is next applied in the two particular circumstances seen in Figs. 3.1.1 and 3.1.2.

Note that a constant $+\lambda^2$ is equally acceptable in the separation, Eq. (3.1.5). Then the roles of x and y are then merely interchanged. The two forms are capable of satisfying different boundary conditions in x and y.

CONDUCTION IN A STRIP. Consider the conduction region shown in Fig. 3.1.1, which occupies the region $0 \le x \le L$ and $y \ge 0$. It is assumed that there is no conduction in the z direction. This condition would result, for example, if the region were a thin strip insulated on the top and bottom surfaces. This might also be thought of as a section in the edge region of a plate which is very large in the y direction, and in the z direction, normal to the plane shown. The solution will be given for the two edges, or plate surfaces, maintained at t_0 and the bottom maintained at a temperature t_1. The differential equation and boundary conditions are written in terms of temperature excess $\theta = t - t_0$ as

$$\frac{\partial^2 \theta}{\partial x^2} + \frac{\partial^2 \theta}{\partial y^2} = 0 \tag{3.1.7}$$

1. $x = 0$ and $x = L$; $\theta = t - t_0 = 0$.
2. $y = 0$, $0 < x < L$; $\theta_1 = t_1 - t_0$.
3. $\lim\limits_{y \to \infty} \theta = 0$.

Equation (3.1.6) is a solution of Eq. (3.1.7). The boundary conditions are used to determine A, B, C, and D in Eq. (3.1.6). Boundary condition 1 gives $A = 0$ because $\cos 0 = 1$ and θ must be zero at $x = 0$. Similarly, $\sin \lambda x$ must be zero at $x = L$; that is, λL must be an integral multiple of π, where λ is taken as positive.

$$\lambda L = n\pi \quad \text{or} \quad \lambda = \frac{n\pi}{L}$$

Condition 3 is satisfied by $C = 0$. Equation (3.1.6) is then

$$\theta = BDe^{-(n\pi/L)y} \sin\left(\frac{n\pi}{L}x\right) = Ee^{-n(\pi/L)y} \sin\left(\frac{n\pi}{L}x\right)$$

This expression satisfies the differential equation for any integral value of n equal to or greater than zero, and the result for zero is trivial. Therefore, since the sum of any two solutions of a linear differential equation is also a solution, the general solution is obtained by summing all possible solutions:

$$\theta = \sum_{n=1}^{\infty} E_n e^{-(n\pi/L)y} \sin\left(\frac{n\pi}{L}x\right)$$

Boundary condition 2 is used to evaluate the E_n. At $y = 0$ the expression is

$$\theta_1 = \sum_{n=1}^{\infty} E_n \sin\left(\frac{n\pi}{L}x\right) \qquad 0 < x < L$$

This is recognized as the Fourier sine series expansion of a constant θ_1 in the interval 0 to L. The constants E_n are the Fourier coefficients for such an expansion and are given by

$$E_n = \frac{2}{L} \int_0^L \theta_1 \sin\left(\frac{n\pi}{L}x\right) dx$$

$$= \frac{2\theta_1}{L} \int_0^L \sin\left(\frac{n\pi}{L}x\right) dx = -\frac{2\theta_1}{L}\frac{L}{n\pi}\cos\left(\frac{n\pi}{L}x\right)\Big|_0^L$$

$$= \frac{4\theta_1}{n\pi} \qquad \text{for } n = 1, 3, 5, \ldots$$

$$= 0 \qquad \text{for } n = 2, 4, 6, \ldots$$

The solution is then

$$\theta = \frac{4\theta_1}{\pi} \sum_{n=1,3,5}^{\infty} \frac{e^{-(n\pi/L)y}}{n} \sin(n\pi x/L) \tag{3.1.8}$$

This may be written in closed form as

$$\theta = \frac{2\theta_1}{\pi} \arctan \frac{\sin[(\pi/L)x]}{\sinh[(\pi/L)y]} \tag{3.1.9}$$

where

$$2\sinh\left(\pi\frac{y}{L}\right) = \left(e^{(\pi/L)y} - e^{-(\pi/L)y}\right)$$

The nature of the solution may be visualized by noting that isothermals are given by

$$\frac{\sin[(\pi/L)x]}{\sinh[(\pi/L)y]} = \tan\left(\frac{\pi}{2}\frac{\theta}{\theta_1}\right) = \text{constant} = C$$

$$\sin\left(\frac{\pi}{L}x\right) = \frac{C}{2}\left(e^{\pi y/L} - e^{-\pi y/L}\right)$$

The isothermals are plotted in Fig. 3.1.1(b) for various values of θ/θ_1. This multidimensional example is one of the simplest kinds.

The preceding solution requires a Fourier series expansion to satisfy the condition $\theta = \theta_1 = t_1 - t_0$ at $y = 0$. In effect, the constant θ_1 is represented as a sum of periodic modes, $\sin(n\pi x/L)$, each times an amplitude E_n. In initially solving this conduction circumstance, Fourier encountered this need and thereby generated the Fourier series formulation.

This solution has another interesting characteristic, seen as follows. The conduction heat transfer across the bottom of the strip, at $y = 0$ and for $x = 0$

to L, is defined as $Q(0)$. This same heat flow rate leaves the strip across the two vertical edges, at $x = 0$ and L, equally. Recall Fig. 3.1.1(a). The local conductive heat flow rate upward in the region, as a function of y, is defined as $Q(y)$. It is calculated from Eq. (3.1.8) as follows:

$$Q(y) = \int_0^L q_y''(x, y) \, dx = -k \int_0^L \left(\frac{\partial t}{\partial y} \right) dx = -k \int_0^L \left(\frac{\partial \theta}{\partial y} \right) dx$$

From Eq. (3.1.8):

$$\frac{\partial \theta}{\partial y} = \frac{4\theta_1}{L} \sum_{n=1,3,5}^{\infty} e^{-(n\pi/L)y} \sin(n\pi x/L)$$

Therefore,

$$Q(y) = -\frac{4k\theta_1}{L} \int_0^L \sum_{n=1,3,5}^{\infty} e^{-(n\pi/L)y} \sin(n\pi x/L) \, dx$$

$$= \frac{4k\theta_1}{L} \sum_{n=1,3,5}^{\infty} e^{-(n\pi/L)y} \frac{L}{n\pi} \cos(n\pi x/L) \Big|_0^L$$

$$= \frac{8k\theta_1}{\pi} \sum_{n=1,3,5}^{\infty} \frac{e^{-(n\pi/L)y}}{n} \tag{3.1.10}$$

In the limit, as $y \to 0$, the preceding result, an infinite sum in $1/n$, $n = 1, 3, 5, \ldots$, is divergent. The consequence is $Q(0) \to \infty$. This results from the strong singularities in heat flux at $x = 0$ and L at $y = 0$. These arise in the boundary conditions following Eq. (3.1.7). They assign a temperature discontinuity at each of those locations. This feature commonly arises in such conduction solutions. However, the solutions away from these locations are often very reliable. Of course, in an actual physical conduction region, large local fluxes would arise to moderate the actual boundary conditions at such locations, to give a bounded total heat flow rate.

This kind of singularity persists for corner angles other than 90° in Fig. 3.1.1; see Nansteel et al. (1986). See also Bassani et al. (1987) for adjacent insulated and flux loading surface conditions. Although a local singularity occurs, the integrated heat flow is bounded.

Both of these kinds of results suggest uncertainties in numerical analysis results, in some circumstances. For the discontinuous temperature condition, as in Fig. 3.1.1, a finite heat flow results. However, its magnitude is dependent on the choice of grid spacing in the numerical analysis. For adjacent adiabatic and heat flux conditions, along a boundary, the flux is bounded. However, the magnitude of the integrated heat flow would again depend upon the grid spacing. An accurate solution of the contribution of the local flux singularity is necessary to properly evaluate its effect on the overall conduction heat transfer calculation.

CONDUCTION IN A LONG RECTANGULAR ROD. For a long rod in the z direction and for assigned surface temperatures which are not z dependent, the temperature distribution in the material depends only on x and y, that is, $t = t(x, y)$. Consider the rod in Fig. 3.1.2 which is H high (y) and L wide (x) with a temperature t_1 on the upper face and t_0 on each of the other three faces. The equation and boundary conditions, in terms of $\phi = (t - t_0)/(t_1 - t_0)$, are

$$\frac{\partial^2 \phi}{\partial x^2} + \frac{\partial^2 \phi}{\partial y^2} = 0$$

1. $\phi = 0$, $x = 0$
2. $\phi = 0$, $x = L$
3. $\phi = 0$, $y = 0$
4. $\phi = 1$, $y = H$

The solution is again Eq. (3.1.6). The constants are evaluated from the boundary conditions as follows:

$$\text{1 and 2:} \quad A = 0 \quad \text{and} \quad \lambda = \frac{n\pi}{L}$$
$$\text{3:} \quad C + D = 0 \quad C = -D$$

We have then, summing over all n,

$$\phi = \sum_n E_n \left(e^{(n\pi y/L)} - e^{-(n\pi y/L)} \right) \sin\left(n\pi \frac{x}{L}\right)$$

$$= \sum_n 2E_n \sinh\left(n\pi \frac{y}{L}\right) \sin\left(n\pi \frac{x}{L}\right)$$

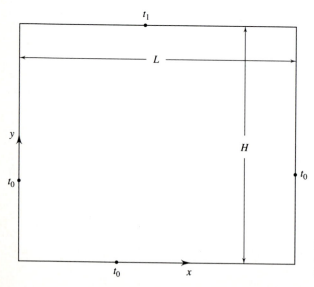

FIGURE 3.1.2
A long rectangular rod and the surface boundary conditions.

The last boundary condition is now used to evaluate $2E_n \sinh(n\pi y/L)$ at $y = H$, as the coefficients of a sine series expansion in x, of the boundary condition $\phi = 1$. The resulting solution is

$$\phi = \frac{4}{\pi} \sum_{n=1,3,5} \frac{1}{n} \frac{\sinh[n\pi(y/L)]}{\sinh[n\pi(H/L)]} \sin\left(n\pi\frac{x}{L}\right) \qquad (3.1.11)$$

Conduction for this geometry may also be analyzed for all four surface temperatures being different and even for each being variable over the surface. This solution would be obtained simply by adding four solutions, each found for one of the surfaces at some temperature, while the other three surfaces are held at, say, $0°$. The preceding result would be one of the four solutions; the other three solutions can be obtained from Eq. (3.1.11) by permuting y, H and x, L and by translating the x and y axes. This procedure is considered in Sec. 3.2.

SUMMARY. Here it is noted that the preceding two solutions, Eqs. (3.1.9) and (3.1.11), for the temperature distributions, θ or $\phi = \theta/\theta_1$, do not depend on the thermal conductivity, k, of the region. These solutions satisfy uniform temperature boundary conditions t_1 and t_0. The orthogonal isothermals and heat flow lines are purely geometric. The characteristic temperature is simply $\theta_1 = t_1 - t_0$.

On the other hand, an imposed flux surface condition, q'', is often imposed, as in the third condition in Fig. 3.2.1(a). Then the characteristic temperature, Δt_{ch}, depends on q'', as in the example in Eq. (3.2.9b). Similarly, with a distributed source q''' in a region, Δt_{ch} may depend on q''', as in Eq. (2.4.8b).

3.1.2 Separation of Variables for Rectangular, Cylindrical, and Spherical Regions

The separation of $t(x, y)$ in Eq. (3.1.5) resulted in a single separation constant $-\lambda^2$, where λ was taken as a real number in such a way as to satisfy the imposed boundary conditions. The separation process may be more complicated in three-dimensional conduction. Also, different forms of the product functions arise in the cylindrical and spherical coordinate systems. The three-dimensional results, for the Laplace equations in the three coordinate systems, are given in the following discussion.

RECTANGULAR COORDINATES. In three dimensions, Eq. (3.1.4) becomes instead

$$t(x, y, z) = X(x)Y(y)Z(z) \qquad (3.1.12)$$

Then $\nabla^2 t = 0$ is evaluated in terms of X, Y, and Z:

$$YZ \nabla^2 X + XZ \nabla^2 Y + XY \nabla^2 Z = 0$$

or

$$\frac{1}{X}\frac{d^2X}{dx^2} + \frac{1}{Y}\frac{d^2Y}{dy^2} + \frac{1}{Z}\frac{d^2Z}{dz^2} = 0 \qquad (3.1.13)$$

Each of the three terms may involve only one independent variable, x, y, and z, respectively. Therefore, each must be a constant. Again taking the first term as $-\lambda^2$ and the succeeding two as $-\gamma^2$ and $-\delta^2$, the following relations result:

$$\lambda^2 + \gamma^2 + \delta^2 = 0 \tag{3.1.14a}$$

$$\frac{1}{X}\frac{d^2X}{dx^2} = -\lambda^2 \qquad \frac{1}{Y}\frac{d^2Y}{dy^2} = -\gamma^2 \qquad \frac{1}{Z}\frac{d^2Z}{dz^2} = -\delta^2 \tag{3.1.14b}$$

The equations for X, Y, and Z become

$$\frac{d^2X}{dx^2} + \lambda^2 X = 0 \tag{3.1.15a}$$

$$\frac{d^2Y}{dy^2} + \gamma^2 Y = 0 \tag{3.1.15b}$$

$$\frac{d^2Z}{dz^2} + \delta^2 Z = 0 \tag{3.1.15c}$$

The solutions of these equations, $X(\lambda, x)$, $Y(\gamma, y)$, and $Z(\delta, z)$, may be expressed in several equivalent functional forms.

CYLINDRICAL COORDINATES, (r, θ, z). The temperature distribution $t(r, \theta, z)$ is governed by Eq. (1.2.18), with $q''' = 0$:

$$\frac{\partial^2 t}{\partial r^2} + \frac{1}{r}\frac{\partial t}{\partial r} + \frac{1}{r^2}\frac{\partial t}{\partial \theta^2} + \frac{\partial^2 t}{\partial z^2} = 0 \tag{3.1.16}$$

Then

$$t(r, \theta, z) = \bar{R}(r)\Theta(\theta)Z(z) \tag{3.1.17}$$

and

$$\frac{1}{\bar{R}}\left(\frac{d^2\bar{R}}{dr^2} + \frac{1}{r}\frac{d\bar{R}}{dr}\right) + \frac{1}{r^2\Theta}\frac{d^2\Theta}{d\theta^2} + \frac{1}{Z}\frac{d^2Z}{dz^2} = 0 \tag{3.1.18}$$

Taking the second and third terms as $-\gamma^2/r^2$ and $-\delta^2$, respectively, Eq. (3.1.18) becomes

$$\frac{1}{\bar{R}}\left(\frac{d^2\bar{R}}{dr^2} + \frac{1}{r}\frac{d\bar{R}}{dr}\right) - \frac{\gamma^2}{r^2} - \delta^2 = 0$$

Therefore, the relations for Θ, Z, and $\bar{R}$ become

$$\frac{d^2\Theta}{d\theta^2} + \gamma^2\Theta = 0 \tag{3.1.19}$$

$$\frac{d^2Z}{dz^2} + \delta^2 Z = 0 \tag{3.1.20}$$

$$\frac{1}{\bar{R}}\left(\frac{d^2\bar{R}}{dr^2} + \frac{1}{r}\frac{d\bar{R}}{dr}\right) - \left(\delta^2 + \frac{\gamma^2}{r^2}\right) = 0 \tag{3.1.21}$$

The functions $\Theta(\gamma, \theta)$ and $Z(\delta, z)$ are nominally in terms of sines and cosines. The solutions of Eq. (3.1.21) for $\bar{R}_\gamma(\delta, r)$ are modified Bessel functions of order γ, of the first and second kind, respectively. This added complexity in $\bar{R}_\gamma$ arises because of the cylindrical effects in the dR/dr and Θ terms in Eq. (3.1.18).

SPHERICAL POLAR COORDINATES, (r, ϕ, θ). The temperature distribution $t(r, \phi, \theta)$, where θ is the polar angle, is determined from Eq. (1.2.19), for $q''' = 0$:

$$\frac{\partial^2 t}{\partial r^2} + \frac{2}{r}\frac{\partial t}{\partial r} + \frac{1}{r^2 \sin\theta}\frac{\partial}{\partial \theta}\left(\sin\theta \frac{\partial t}{\partial \theta}\right) + \frac{1}{r^2 \sin^2\theta}\frac{\partial^2 t}{\partial\phi^2} = 0 \quad (3.1.22)$$

Further consideration of the properties of this relation is more convenient when θ is replaced by the variable

$$\mu = \cos\theta \quad (3.1.23)$$

Equation (3.1.22) then becomes

$$\frac{\partial^2 t}{\partial r^2} + \frac{2}{r}\frac{\partial t}{\partial r} + \frac{1}{r^2}\frac{\partial}{\partial \mu}\left[(1 - \mu^2)\frac{\partial t}{\partial \mu}\right] + \frac{1}{r^2(1 - \mu^2)}\frac{\partial^2 T}{\partial\phi^2} = 0 \quad (3.1.24)$$

The solution is expressed as follows:

$$t(r, \mu, \theta) = \bar{R}(r)M(\mu)\Phi(\phi) \quad (3.1.25)$$

The three component equations then become

$$\frac{d^2\Phi}{d\phi^2} + m^2\Phi = 0 \quad (3.1.26)$$

$$\frac{d^2\bar{R}}{dr^2} + \frac{2}{r}\frac{d\bar{R}}{dr} - \frac{n(n+1)}{r^2}\bar{R} = 0 \quad (3.1.27)$$

$$\frac{d}{d\mu}\left[(1 - \mu^2)\frac{dM}{d\mu}\right] + \left[n(n+1) - \frac{m^2}{1 - \mu^2}\right]M = 0 \quad (3.1.28)$$

The function $\Phi(m, \phi)$ is nominally in terms of sines and cosines. $\bar{R}(n, r)$ is in terms of exponentials r^n and $r^{-(n+1)}$. $M(\mu, n, m, \mu)$ satisfies Legendre's associated differential equation. These have solutions $P_n^m(\mu)$ and $Q_n^m(\mu)$, which are called associated Legendre functions of degree n and order m of the first and second kind.

SUMMARY. The previous forms are the basis for the analysis of multidimensional transport results given in Secs. 3.2 through 3.4, in both two and three dimensions. These results also apply to comparable mass diffusion. Various examples among the boundary conditions in Secs. 1.4.1 and 1.4.2 are used. The particular examples given are intended to represent some of the most typical and important mechanisms. In Sec. 3.6 the presence of distributed sources q''' and C''' are included and their effects are evaluated.

3.2   RECTANGULAR REGIONS

In two- and three-dimensional geometries in steady state, many solutions may be generated for different kinds of surface boundary conditions. Most of these follow for region boundaries which each correspond to fixed or limiting values of either x, y, or z. The two examples given in Figs. 3.1.1 and 3.1.2 have specified conditions at $x = 0$ and L and at $y = 0$. This first example has a limiting value of $\phi = 0$ as $y \to \infty$. The rod has a fixed value, $\phi = 1$ at $y = H$.

Further examples of such solutions will be given for the following boundary conditions discussed in Sec. 1.4.1: specified surface temperature conditions, T.1; specified surface heat flux, T.2; and a surface convection condition, T.3. The two-dimensional geometries in Sec. 3.2.1 are the half-infinite and infinite strips. Rectangular rods are considered in Sec. 3.2.2. The three-dimensional results are for rectangular parallelepipeds and are given in Sec. 3.2.3. These geometries are shown in Fig. 3.2.1. These results are chosen as charateristic of the very large amount of such material available in the literature.

3.2.1   Thin Strips

The region in Fig. 3.2.1(a) is like that in Fig. 3.1.1, which has boundary conditions t_1 and t_0. Here three other boundary conditions are used. For each one $\theta(L, y) = 0$. For the first, the conditions are as in Fig. 3.1.1, except that at $y = 0$, $\theta(x, 0) = t(x, 0) - t_0 = f(x)$. This is the first condition shown in Fig. 3.2.1(a). The second circumstance shown is $\theta(x, 0) = 0$ and $\theta = t(0, y) - t_0 = f(y)$ at $x = 0$, as seen in Fig. 3.2.1(a). The third condition is $\theta = 0$ at $x = 0$ and L and a uniform flux q'' is imposed at $y = 0$. The fourth result given here is in Fig. 3.2.1(b), an infinite strip with $\theta(0, y) = t(0, y) - t_0 = f(y)$ and $\theta = 0$ at $x = L$.

HALF-INFINITE STRIPS. For the three conditions in Fig. 3.2.1(a), the general solution is again Eq. (3.1.6). Applying the boundary conditions for the first example, $\theta = 0$ at $x = 0$ and L and $\theta \to 0$ as $y \to \infty$, yields again

$$\theta(x, y) = \sum_{n=1}^{\infty} E_n e^{-(n\pi/L)y} \sin \frac{n\pi}{L} x$$

The changed boundary condition at $y = 0$, $\theta = f(x)$, is satisfied by the expansion

$$f(x) = \sum_{n=1}^{\infty} E_n \sin \frac{n\pi}{L} x$$

This is a sine series expansion of $f(x)$ and the solution becomes

$$t(x, y) - t_0 = \frac{2}{L} \sum_{n=1,3,5}^{\infty} e^{-(n\pi/L)y} \sin \frac{n\pi x}{L} \int_0^L f(x) \sin \frac{n\pi x}{L} \, dx \quad (3.2.1)$$

For $f(x) = t_1 - t_0 = \theta_1$, Eq. (3.1.8) is again found.

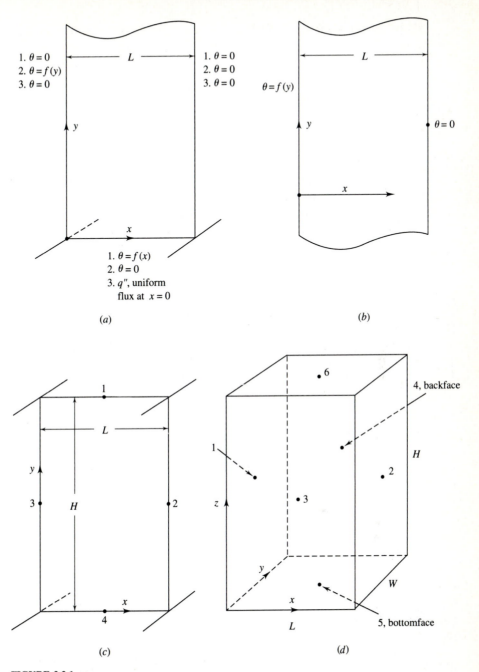

FIGURE 3.2.1
Rectangular geometries: (*a*) Half-infinite strip; (*b*) infinite strip; (*c*) long rectangular rod;
(*d*) rectangular parallelepiped.

There are forms of $f(x)$ which would not result in the infinite heat flow rate $Q(y)$ at $y = 0$, as found in Eq. (3.1.10) for the previous solution, Eq. (3.1.8). For example, $f(x) = \sin \pi x / L$ would remove the singularities at both $x = 0$ and L.

The second circumstance also applies to Fig. 3.2.1(a), with the boundary conditions: at $x = L$, and at $y = 0$, $t = t_0$. At $x = 0$, $\theta(0, y) = t(0, y) - t_0 = f(y)$. The general solution is again Eq. (3.1.6). Since $\theta = 0$ at $x = L$, $A \cos \lambda x + B \sin \lambda x = 0$ and $A \cos \lambda L = -B \sin \lambda L$. The two other conditions yield the following result; see Carslaw and Jaeger (1959), page 165:

$$t(x, y) - t_0$$

$$= \frac{1}{2L} \sin \frac{\pi x}{L} \int_{y'=0}^{\infty} f(y') \left[\frac{1}{\cos[\pi(L-x)/L] + \cosh[\pi(y-y')/L]} \right.$$

$$\left. - \frac{1}{\cos[\pi(L-x)/L] + \cosh[\pi(y+y')/L]} \right] dy' \quad (3.2.2)$$

The third circumstance in Fig. 3.2.1(a) is like the first in that $t = t_0$, or $\theta = 0$, at $x = 0$ and L. However, at $y = 0$ a uniform surface flux q'' is imposed. The solution, before applying the flux condition, is again

$$\theta(x, y) = \sum_{i=1}^{\infty} E_n e^{-(n\pi/L)y} \sin \frac{n\pi x}{L}$$

at $y = 0$:

$$q'' = -k \frac{\partial t}{\partial y} = -k \left(\frac{\partial \theta}{\partial y} \right)_0 = k \sum_{i}^{n} E_n \frac{n\pi}{L} \sin \frac{n\pi}{L} x$$

Therefore,

$$kE_n \frac{n\pi}{L} = \frac{2}{L} \int_0^L q'' \sin \frac{n\pi}{L} x \, dx = -\frac{4q''}{n\pi} \qquad n = 1, 3, \ldots$$

and

$$E_n = \frac{4Lq''}{k\pi^2} \frac{1}{n^2}$$

and

$$\theta(x, y) = \frac{4Lq''}{k\pi^2} \sum_{n=1,3}^{\infty} \frac{1}{n^2} e^{-(n\pi/L)y} \sin \frac{n\pi}{L} x$$

where $4Lq''/k\pi^2$ is the characteristic temperature difference Δt_{ch} discussed in Sec. 3.1.1.

A singularity would exist in the first solution at $x = 0$ and L unless $f(x)$ at $x = 0$ and L is zero. The second solution would have a singularity at $x = 0$, and $y = 0$, unless $f(y)$ is zero there. Also, in the second solution, $f(y)$ must be bounded for a physically reasonable solution. The general requirement would be that conduction across the surface at $x = 0$, for $y = \Delta$, some distance, be bounded. The third solution has no such limitations.

INFINITE STRIP. The region in Fig. 3.2.1(b) has $\theta = t(L, y) - t_0 = 0$ and $\theta = f(y)$ at $x = 0$. The solution is

$$t(x, y) - t_0 = \frac{1}{2L} \sin\frac{\pi x}{H} \int_{-\infty}^{\infty} \frac{f(y')\, dy'}{\cos[\pi(L - x)/L] + \cosh[\pi(y - y')/L]}$$

$$(3.2.3)$$

Consideration of the physical reasonableness of the form of $f(y)$ again applies, as for the preceding first two half-infinite-strip configurations.

SUMMARY. Although the preceding solutions have interesting characteristics, they do sometimes have unreasonable consequences, as do many such idealized conduction formulations. These features stem from the imposition of temperature boundary conditions which are often discontinuous at locations where boundaries meet. Imposed flux formulations, as at $y = 0$ in Fig. 3.2.1(a), often do not have this property.

3.2.2 Long Rectangular Rods

The geometry is shown in Fig. 3.2.1(c) (see page 105). This is the same geometry as in Fig. 3.1.2. It was analyzed in Sec. 3.1.1. for $t = t_0$ at $x = 0$, at $x = L$, and at $y = 0$, with $t = t_1$ at $y = H$. This section considers the same geometry. Five different kinds and combinations of boundary conditions are applied to boundaries numbered 1, 2, 3, and 4 in Fig. 3.2.1(c). The different circumstances are shown in Fig. 3.2.2, where (a) is the solution in Sec. 3.1.1., leading to Eq. (3.1.11).

The boundary conditions are uniform surface temperature, uniform imposed heat flux q'', an insulation covering approximated as $q'' = 0$, assigned variable surface temperature as $f(x)$ and/or $f(y)$, and convection h to an ambient at t_e. The solutions for conditions (b), (c), (d), (e), and (f) are given in the following discussion.

FOUR ASSIGNED SURFACE TEMPERATURES. This result is obtained by the general method called superposition. Given the linear nature of the differential equation, solutions in a given geometry with different homogeneous boundary conditions may be added together. This sum is then also the solution of the differential equation, subject to the resulting sum of the boundary conditions at each surface. The procedure is diagrammed in Fig. 3.2.3 for the conditions in Fig. 3.2.2(b) and (c). The four formulations on the right sum to be the desired solution on the left.

Therefore, the solution of the circumstance in Fig. 3.2.2(b) is constructed of the four components seen in Fig. 3.2.3(a). The first component, t_1 in

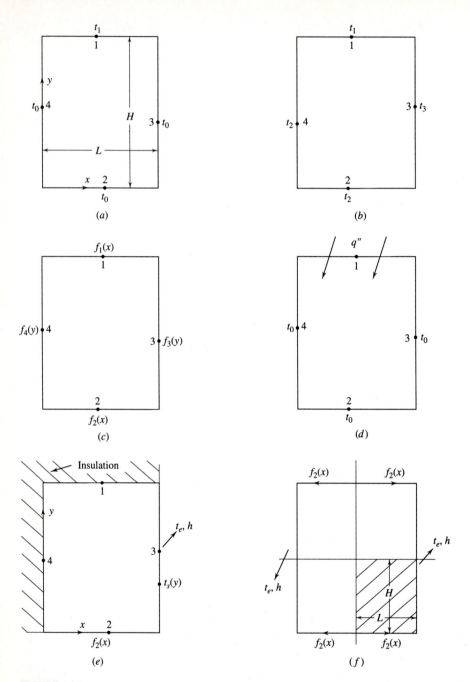

FIGURE 3.2.2
Long rectangular rods, L by H, with five different kinds of boundary conditions at the four surfaces 1, 2, 3, and 4. (a) Surface 1 at $\theta_1 = t_1 - t_0$, the others at t_0. (b) Surfaces at t_1, t_2, t_3, and t_4. (c) Surfaces at $f_1(x)$, $f_2(x)$, $f_3(y)$, and $f_4(y)$. (d) Uniform imposed flux over surface 1, the other surfaces at t_0. (e) Surfaces 1 and 4 insulated, surface 2 at $t - t_e = f(x)$, and convection on surface 3. (f) The circumstance in part (e), as a component of a larger geometry.

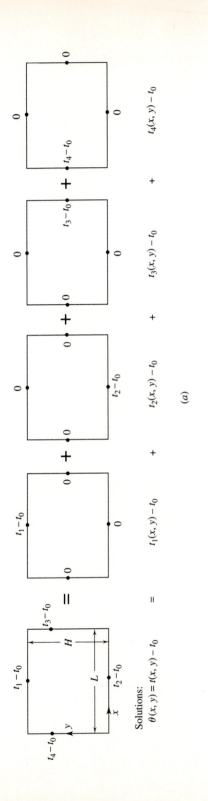

Solutions:

$$\theta(x, y) = t(x, y) - t_0 = t_1(x, y) - t_0 + t_2(x, y) - t_0 + t_3(x, y) - t_0 + t_4(x, y) - t_0$$

(a)

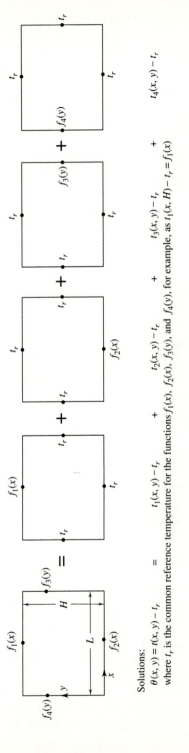

Solutions:

$$\theta(x, y) = t(x, y) - t_r = t_1(x, y) - t_r + t_2(x, y) - t_r + t_3(x, y) - t_r + t_4(x, y) - t_r$$

where t_r is the common reference temperature for the functions $f_1(x)$, $f_2(x)$, $f_3(y)$, and $f_4(y)$, for example, as $t_1(x, H) - t_r = f_1(x)$

(b)

FIGURE 3.2.3
Superposition solutions: (a) and (b) for conditions shown in Fig. 3.2.1(b) and (c). Temperature boundary conditions are, for example, $t_1 - t_0$ in (a) and $t_1 - t_r = f_1(x)$ in (b).

Fig. 3.2.2, is Eq. (3.1.11) written as follows in terms of $t_1(x, y) - t_0$:

$$\phi_1 = \frac{t(x, y) - t_0}{t_1 - t_0} = \frac{4}{\pi} \sum_{n=1,3,5}^{\infty} \frac{1}{n} \frac{\sinh[n\pi y/L)]}{\sinh[n\pi H/L)]} \sin\left(n\pi \frac{x}{L}\right) \quad (3.1.11)$$

This is to be summed with the three other equations, for $t_2(x, y) - t_0$, $t_3(x, y) - t_0$, and $t_4(x, y) - t_0$. For example, $t_3(x, y) - t_0$ is obtained by interchanging y, L, H, and t_1 in Eq. (3.1.11) with x, H, L, and t_3, to obtain

$$\phi_3 = \frac{t_3(x, y) - t_0}{t_3 - t_0} = \frac{4}{\pi} \sum_{n=1,3,5}^{\infty} \frac{1}{n} \frac{\sinh[n\pi(x/H)]}{\sinh[n\pi(L/H)]} \sin\left(n\pi \frac{y}{H}\right) \quad (3.2.4)$$

The remaining two equations, for $t_2(x, y) - t_0$ and $t_4(x, y) - t_0$, are obtained from Eqs. (3.1.11) and (3.2.4) by the translation and rotation of the coordinate axes shown in Fig. 3.2.3(a).

ASSIGNED SURFACE TEMPERATURE DISTRIBUTIONS. The analysis for the more general boundary conditions shown in Fig. 3.2.3(b) is more complicated. First, a solution is necessary for $t_1(x, y) - t_r$, in terms of $f_1(x)$, the analog of Eq. (3.1.11) for t_1 uniform. The separation of variables form is Eq. (3.1.6). Applying the other surface boundary conditions apparent in Fig. 3.2.3(b), the form is

$$t_1(x, y) - t_r = \sum_{n=1}^{\infty} E_{n,1} [e^{n\pi(y/L)} - e^{-n\pi(y/L)}] \sin n\pi(x/L)$$

$$= \sum_{n=1}^{\infty} 2 E_{n,1} \sinh n\pi(y/L) \sin n\pi(x/L) \quad (3.2.5)$$

The condition on surface 1 is applied to determine the $E_{n,1}$:

$$t_1(x, H) - t_r = f_1(x) = \sum_{n=1}^{\infty} 2 E_{n,1} \sinh n\pi(H/L) \sin n\pi(x/L)$$

$$2 E_{n,1} \sinh n\pi(H/L) = \frac{2}{L} \int_0^L f_1(x) \sin n\pi(x/L) \, dx$$

and $t_1(x, y) - t_r$

$$= \frac{2}{L} \sum_{n=1}^{\infty} \frac{\sinh n\pi(y/L) \sin n\pi(x/L)}{\sinh n\pi(H/L)} \int_0^L f_1(x) \sin n\pi(x/L) \, dx$$

$$(3.2.6)$$

The procedure for assembling $t(x, y) - t_r$ is seen in Fig. 3.2.3(b). For example, $t_3(x, y) - t_r$ may be found by interchanging y, L, H, and f_1 in

Eq. (3.2.6), to obtain

$$t_3(x, y) - t_0$$

$$= \frac{2}{L} \sum_{n=1}^{\infty} \frac{\sinh n\pi(x/H) \sin n\pi(y/L)}{\sinh n\pi(L/H)} \int_0^H f_3(y) \sin n\pi(y/H) \, dy$$

$$(3.2.7)$$

The functions $t_2(x, y)$ and $t_4(x, y)$ may then be obtained from the two preceding equations by translation and rotation of the axes.

AN IMPOSED SURFACE FLUX. This circumstance is seen in Fig. 3.2.2(d) (see page 108). The temperature level variable is $\theta = t(x, y) - t_0$. The flux boundary condition at $y = H$ results in the temperature distribution having a uniform value of the slope in the y direction, $\partial t/\partial y$, immediately inside the boundary at $y = H$. This conducts the imposed uniform flux into the interior of the region.

Given the three other boundary conditions, at t_0, the general form of the solution is the same as for the conditions on $t_1(x, y) - t_r$ in Fig. 3.2.3(b). Here, for $t(x, y)$, this is Eq. (3.2.5), where $t_1(x, y) - t_r$ now becomes $t(x, y) - t_0 = \theta(x, y)$:

$$\theta(x, y) = \sum_{n=1}^{\infty} 2E_n \sinh n\pi(y/L) \sin n\pi(x, L) \qquad (3.2.8)$$

The flux q'' is determined from Eq. (3.2.8) as

$$q'' = -k \left(\frac{\partial \theta}{\partial y} \right)_H = - \sum_{n=1}^{\infty} 2kE_n \frac{n\pi}{L} \cosh n\pi(H/L) \sin n\pi(x/L)$$

Therefore,

$$-\frac{2kE_n n\pi}{H} \cosh n\pi(H/L) = \frac{2}{L} \int_0^L q'' \sin n\pi(x/L) \, dx$$

$$= \begin{cases} -\dfrac{4q''}{n\pi} & \text{for } n = 1, 3, \ldots \\ 0 & \text{for } n = 2, 4, \ldots \end{cases}$$

The result is

$$\theta(x, y) = \frac{4Lq''}{k\pi^2} \sum_{n=1,3}^{\infty} \frac{1}{n^2} \frac{\sinh n\pi(y/L)}{\cosh n\pi(H/L)} \sin n\pi(x/L) \qquad (3.2.9a)$$

In this kind of a conduction circumstance, there is no imposed temperature difference, as $t_1 - t_0$ in the circumstance in Fig. 3.2.2(a), which led to Eq. (3.1.11). However, Eq. (3.2.9a) indicates that there is a characteristic temperature difference, $\Delta t_{ch} = 4Hq''/k\pi^2$. Thus ϕ might be written as

$$\phi = \frac{\theta(x, y)}{\Delta t_{ch}} = \sum_{n=1,3}^{\infty} \frac{1}{n^2} \frac{\sinh n\pi(y/L)}{\cosh n\pi(H/L)} \sin n\pi(x/L) \qquad (3.2.9b)$$

This characterization in terms of Δt_{ch} usually arises in conditions when the conduction transport is driven by an imposed heat flux condition.

INSULATING AND CONVECTIVE SURFACE CONDITIONS. The example considered here is Fig. 3.2.2(e) (see page 108). Surfaces 1 and 4 are perfectly insulated. Surface 2 is at $f_2(x) = t_2(x,0) - t_e$. Surface 3 has a convective condition T.3 in Eq. (1.4.2). For an environment uniformly at t_e, condition T.3 at $x = L$ for $0 \le y \le H$ is

$$h(t_e - t_s) = -k\left(\frac{\partial t}{\partial x}\right)_s \tag{3.2.10}$$

where t_3 in Fig. 3.2.2(e) is an unknown resulting function $t_s(y)$ at $x = L$. The other boundary conditions are

$$\frac{\partial t}{\partial x} = 0 \quad \text{at } x = 0$$

$$\frac{\partial t}{\partial y} = 0 \quad \text{at } y = H \tag{3.2.11}$$

$$t(x,0) - t_e = f_2(x)$$

Starting again from Eq. (3.1.6), the first two conditions in Eq. (3.2.11) are applied

$$t(x,y) - t_e = \theta(x,y) = (A\cos\lambda x + B\sin\lambda x)(Ce^{\lambda y} + De^{-\lambda y}) \tag{3.1.6}$$

First

$$\frac{\partial t}{\partial x} = \frac{\partial\theta}{\partial x} = (-A\lambda\sin\lambda + B\lambda\cos\lambda x)(Ce^{\lambda y} + De^{-\lambda y})$$

This is zero at $x = 0$ for $B = 0$. The condition at $y = H$ is used.

$$\frac{\partial t}{\partial y} = \frac{\partial\theta}{\partial y} = A\cos\lambda x(C\lambda e^{\lambda y} - D\lambda e^{-\lambda y})$$

This is zero at $y = H$ for $C = De^{-2\lambda H}$. Therefore,

$$\theta(x,y) = AD\cos\lambda x[e^{\lambda y - 2\lambda H} + e^{-\lambda y}]$$

At $x = L$, the convection condition, Eq. (3.2.10), is applied

$$\left(\frac{\partial t}{\partial x}\right)_L = \left(\frac{\partial\theta}{\partial x}\right)_L = -AD\lambda\sin\lambda L[e^{\lambda y - 2\lambda H} + e^{-\lambda y}]$$

$$= -\frac{h}{k}\theta(H,y) = -\frac{h}{k}AD\cos\lambda L[e^{\lambda y - 2\lambda H} + e^{-\lambda y}]$$

The result is

$$\lambda\sin\lambda L = \frac{h}{k}\cos\lambda L \quad \text{or} \quad (\lambda L)\tan(\lambda L) = \frac{hL}{k} = \text{Bi} \tag{3.2.12}$$

where Bi is the Biot number, a ratio of the two conductances h and k/L.

This transcendental equation has the roots $(\lambda L)_1, (\lambda L)_2, \ldots, (\lambda L)_n, \ldots,$ which depend on Bi. Taking $(AD)_n = E_n$, the result is now

$$\theta(x, y) = \sum_{n=1}^{\infty} E_n \cos[(\lambda L)_n(x/L)][e^{(\lambda L)_n(y/L) - 2(\lambda L)_n(H/L)} e^{-(\lambda L)_n(y/L)}]$$

The last condition, at $y = 0$, gives

$$t(x, 0) - t_e = \theta(x, 0) = f_2(x) = \sum_{n=1}^{\infty} E_n[\cos(\lambda L)_n(x/L)][e^{-2(\lambda L)_n(H/L)} + 1]$$

$$= \sum_{n=1}^{\infty} B_n \cos[(\lambda L)_n(x/L)]$$

This is an expansion of $f_2(x)$ and B_n becomes

$$B_n = \frac{2\left[\mathrm{Bi}^2 + (\lambda L)_n^2\right]}{\left[\mathrm{Bi}^2 + \mathrm{Bi} + (\lambda L)_n^2\right]\cosh(\lambda L)_n(H/L)}$$

$$\times \int_0^L f_2(x) \cos[(\lambda L)_n(x/L)]\, d(x/L) \qquad (3.2.13)$$

The solution is then

$$t(x, y) - t_e = \theta(x, y) = \sum_{n=1}^{\infty} B_n \cos[(\lambda L)_n(x/L)] \cosh\{(\lambda L)_n[(H - y)/L]\}$$

$$(3.2.14)$$

This solution is complicated by the $(\lambda L)_n$, to be determined from Eq. (3.2.12), as dependent on Bi. The lower limiting value of Bi is zero, for $h/k = 0$. The roots are then $(\lambda L)_n = 0, \pi, 2\pi, \ldots$. This is an insulating condition at $x = L$. Then there is no heat supply or loss from the region, other than that caused by a nonuniform assigned condition $f_2(x)$ along $y = 0$. The other limit of Bi is for $h/k \to \infty$. Then the $(\lambda L)_n$ are $\pi/2, 3\pi/2, \ldots$. The result is $\theta(L, y) = 0$, or $t_3 = t_e$. That is, the surface at $x = L$ is isothermal at t_e.

Another special form of this solution is that for $\theta(x, 0) = t_2 - t_e$, an isothermal surface. The result is

$$\theta(x, y) = t(x, y) - t_e$$

$$= \frac{2hL^2(t_2 - t_e)}{kH} \sum_{n=1}^{\infty} \frac{\cos[(\lambda L)_n(x/L)] \cosh\{(\lambda L)_n[(H - y)/L]\}}{\left[\mathrm{Bi}^2 + \mathrm{Bi} + (\lambda L)_n^2\right]\cos(\lambda L)\cosh(\lambda H)}$$

$$(3.2.15)$$

This result has an important application. Perfectly insulated surfaces may also often be interpreted as planes of symmetry in a larger geometry. Thus the element in Fig. 3.2.2(e) is only the lower right-hand region of the geometry in Fig. 3.2.2(f) (see page 108). Thereby, Eq. (3.2.15) gives the solution for the larger geometry, for the boundary conditions shown.

3.2.3   Rectangular Parallelepipeds

Fully three-dimensional processes may also be constructed, using the separation of variables result in Eq. (3.1.15). The notation is shown in Fig. 3.2.1(d) (see page 105), including the surface numbering 1 through 6. Results are given for four representative combinations of surface conditions:

1. Surface 1, at $x = 0$, at t_1; surface 2, at $x = L$, at t_2; the other surfaces at t_0 and $\theta(x, y, z) = t - t_0$, or $\phi(x, y, z) = (t - t_0)/(t_1 - t_0)$.
2. Same conditions on surfaces 1 and 2 as given previously, but with convection on the other four surfaces, to an ambient medium at t_e. Then $\theta(x, y, z) = t - t_e$ and $\phi(x, y, z) = (t - t_e)/(t_1 - t_e)$.
3. Surface 1 at t_1 and convection at the other five surfaces to an ambient at t_e, with θ and ϕ defined as in 2.
4. Region *extends infinitely* in the x direction, from $x = 0$. Surface 1, at $x = 0$, at t_1; convection at the other four surfaces, with θ and ϕ defined as in 2.

The separation of variables form is given for the first solution. The four specific solutions are then given. For the additional intermediate steps, see Carslaw and Jaeger (1959).

TEMPERATURE SURFACE CONDITIONS, 1. A solution of $\nabla^2\theta = 0$ is the following, in terms of n and m:

$$\theta(x, y, z) = t - t_0$$

$$= \frac{(t_1 - t_0)\sinh u(L - x) + (t_2 - t_0)\sinh ux}{\sinh uL}\sin\frac{m\pi y}{W}\sin\frac{n\pi z}{H}$$

$$(3.2.16a)$$

where

$$u^2 = \pi^2\left(\frac{m^2}{W^2} + \frac{n^2}{H^2}\right) \qquad (3.2.16b)$$

The following solution is in terms of summations in q and p, defined in Eq. (3.2.17b).

$$\phi(x, y, z) = \frac{t - t_0}{t_1 - t_0}$$

$$= \frac{16}{\pi^2}\sum_{p=0,1}^{\infty}\sum_{q=0,1}^{\infty}\frac{[\sinh u(L - x) + \phi_2\sinh ux]}{(2p + 1)(2q + 1)\sinh uL}$$

$$+ \frac{\sin[(2p + 1)\pi y/W]\sin[(2q + 1)\pi z/H]}{(2p + 1)(2q + 1)\sinh uL} \qquad (3.2.17a)$$

where $\quad u^2 = \dfrac{(2p+1)^2\pi^2}{W^2} + \dfrac{(2q+1)^2\pi^2}{H^2}\quad$ and $\quad \phi_2 = \dfrac{t_2 - t_0}{t_1 - t_0}\quad$ (3.2.17b)

The effect of the second temperature condition, t_2, appears in terms of ϕ_2 in Eq. (3.2.17a).

TWO SURFACE TEMPERATURE CONDITIONS AND CONVECTION, 2. This is the same as solution 1 but with convection on the remaining four surfaces instead. See the boundary condition T.3, as used in Eq. (3.2.10). The appropriate form is similar to Eq. (3.2.16), which now may satisfy condition T.3. The solution is

$$\phi(x, y, z) = \frac{t - t_e}{t_1 - t_e}$$

$$= \sum_{r=1}^{\infty} \sum_{s=1}^{\infty} \frac{4\,\mathrm{Bi}_L^2 \sinh v(L - x) + \phi_2 \sinh vx \big] \cos \alpha_r y \cos \beta_s z}{C_r C_s \cos \alpha_r W \cos \beta_s H \sinh vL}$$

(3.2.18a)

where

$$C_r = \left(WL\alpha_r^2 + \mathrm{Bi}_{LW}^2 + \mathrm{Bi}_L\right) \qquad (3.2.18b)$$

$$C_s = \left(HL\beta_s + \mathrm{Bi}_{HL}^2 + \mathrm{Bi}_L\right) \qquad (3.2.18c)$$

$$\mathrm{Bi}_L = hL/k \qquad \mathrm{Bi}_{LW}^2 = h^2LW/k^2 \qquad \mathrm{Bi}_{LH}^2 = h^2LH/k^2 \quad (3.2.18d)$$

$$v^2 = \alpha_r^2 + \beta_s^2 \quad \text{and} \quad \phi_2 = (t_2 - t_e)/(t_1 - t_e) \qquad (3.2.18e)$$

α_r and β_s are roots of

$$\alpha \tan \alpha/W = h/k \quad \text{and} \quad \beta \tan \beta H = h/k \qquad (3.2.18f)$$

The conditions in Eq. (3.2.18f) are analogous to the convection result in Eq. (3.2.12) for the rectangular region. However, three different forms of the Biot number arise in Eqs. (3.2.18d).

ONE SURFACE TEMPERATURE CONDITION AND CONVECTION ELSEWHERE, 3. This circumstance, having $t = t_1$ at $x = 0$ and convection on the other five surfaces, has the following solution:

$$\phi(x, y, z) = \frac{t - t_e}{t_1 - t_e}$$

$$= \sum_{r=1}^{\infty} \sum_{s=1}^{\infty} \frac{4\,\mathrm{Bi}_L \cos \alpha_r y \cos \beta_s z}{C_r C_s(\mathrm{Bi}_L \sinh vL + v \cosh vL) \cos(\alpha_r W) \cos(\beta_s H)}$$

$$+ \frac{\left[\mathrm{Bi}_L^2 \sinh v(L - x) + vL \cosh v(L - x)\right]}{C_r C_s(\mathrm{Bi}_L \sinh vL + v \cosh vL) \cos(\alpha_r W) \cos(\beta_s H)}$$

(3.2.19)

where C_r and C_s remain as defined in Eqs. (3.2.18b) and (3.2.18c) and where α_r, β_s, and the Biot numbers are as defined in Eqs. (3.2.18d)–(3.2.18f). This result is a model of a square fin of base area HW, extending out a distance L from a surface at t_1, into an environment at t_e.

END REGION OF AN INFINITELY LONG BAR H BY W, 4. This configuration is the same as in solution 3, except that the square region extends from $x = 0$ to infinity. This actually amounts to the end region of a long rod of cross-sectional are HW, with the end face, $x = 0$, maintained at t_1. Convective losses from the lateral faces reduce the temperature difference $t - t_e$ to zero at sufficiently large x. The solution is

$$\phi(x, y, z) = \frac{t - t_e}{t_1 - t_e}$$

$$= 4\,\mathrm{Bi}_L^2 \sum_{r=1}^{\infty} \sum_{s=1}^{\infty} \frac{e^{-vx}\cos\alpha_r y \cos\beta_s z}{C_r C_s \cos\alpha_r W \cos\beta_s H} \qquad (3.2.20)$$

where again α_r, β_s, v, and the Biot numbers are as in Eq. (3.2.18a). The preceding result indicates that the absence of a sixth surface simplifies the result.

SUMMARY. The foregoing examples quantify the effects of characteristically different kinds and arrangements of boundary conditions in three dimensions. Such solutions are easily assembled for the boundaries and imposed conditions which lead to multidimensional solutions.

3.3 CYLINDRICAL REGIONS

Section 3.2 gave several solutions in two and three dimensions in Cartesian coordinates. They included finite and infinite regions with surface temperature, flux loading, and convection conditions. The techniques of analysis often used Fourier series to satisfy a particular bounding condition.

In this section similar kinds of solutions are given for conduction in geometries described in cylindrical coordinates. These examples are all two-dimensional temperature fields. The first is a short rod subject to uniform temperature and then to convection surface conditions. The second is a circumferential variation of surface temperature. The last is a disk-shaped temperature condition on the surface of a semiinfinite region. The results have applications concerning fins and contact resistance, as discussed in detail in Secs. 7.2, 7.3, and 8.2.

3.3.1 Short Cylinders

The geometry and notation are seen in Fig. 3.3.1(a). The solution for t_s uniform or convection over $r = R$, and for any $t = f(r)$ applied at $z = 0$, is $t(r, z) =$

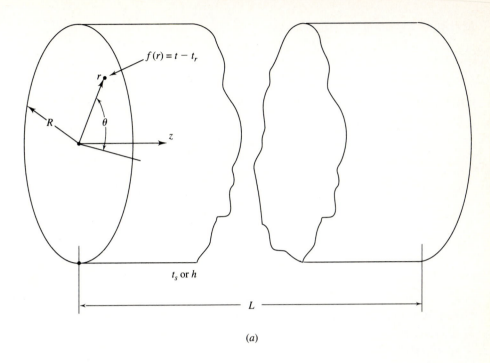

(a)

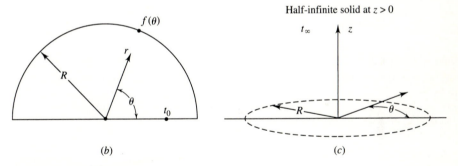

(b) (c)

FIGURE 3.3.1
Cylindrical conduction regions: (a) short cylinder; (b) circumferential temperature variation; (c) a disk-shaped input boundary condition.

$R(r)Z(z)$. Equation (3.1.16) becomes

$$\frac{\partial^2 t}{\partial r^2} + \frac{1}{r}\frac{\partial t}{\partial r} + \frac{\partial^2 t}{\partial z^2} = 0 \tag{3.1.16}$$

and the product solutions for $Z(z)$ and $\bar{R}(r)$ are

$$\frac{d^2 Z}{dz^2} + \delta^2 Z = 0 \tag{3.1.20}$$

$$\frac{1}{\bar{R}}\left(\frac{d^2\bar{R}}{dr^2} + \frac{1}{r}\frac{d\bar{R}}{dr}\right) - \frac{\gamma^2}{r^2} = 0 \tag{3.1.21}$$

Equation (3.1.16) has solutions in terms of the products of Bessel and exponential functions or of modified Bessel and trigonometric functions.

Figure 3.3.1(a) approximates a pin-shaped fin, attached at $z = 0$ to a heat transfer surface. The two examples given here are

1. t_s is uniform over $r = R$ and $z = L$ and $t - t_s = f(r)$ is applied at $z = 0$.
2. Convection, h, over the surfaces at $r = R$ and $z = L$ to an ambient medium at t_e. At $z = 0$, $t - t_e = f(r)$.

The first solution is given as follows, in terms of the $f(r)$ applied at $z = 0$:

$$t - t_s = \frac{2}{R^2}\sum_{n=1}^{\infty}\frac{J_0(r\alpha_n)\sinh(L-z)\alpha_n}{J_1^2(R\alpha_n)\sinh L\alpha_n}\int_0^R rf(r)J_0(r\alpha_n)\,dr \tag{3.3.1}$$

where J_0 and J_1 are Bessel functions of the first kind of order zero and one and the α_n are the positive roots of the relation $J_0(R\alpha_n) = 0$. The second solution is

$$t - t_e = \sum_{n=1}^{\infty}A_n J_0(r\alpha_n)\frac{\alpha_n\cosh\alpha_n(L-z) + (k/h)\sinh\alpha_n(L-z)}{\alpha_n\cosh\alpha_n L + (k/h)\sinh\alpha_n L} \tag{3.3.2a}$$

where

$$A_n = \frac{2\alpha_n^2}{\left[\mathrm{Bi}_R^2 + (\alpha R)_n^2\right]J_0^2(R\alpha_n)}\int_0^R rf(r)J_0(r\alpha_n)\,dr \tag{3.3.2b}$$

and the α_n are the roots of $\alpha J_0'(R\alpha_n) + (h/k)J_0(R\alpha_n) = 0$, where J_0' is given by J_1.

Carslaw and Jaeger (1959), Sec. 8.3, also give solutions for many other boundary conditions. Schneider (1955), in Sec. 6.6, discusses the result, Eq. (3.3.1), for t uniform at $z = 0$. The temperature fields for several values of L/R indicated the increasing value of L/R for which the conduction heat transfer becomes independent of the end condition, at $z = L$.

3.3.2 Circumferential Temperature Variation

The long half-cylinder in Fig. 3.3.1(b) (see page 117) is considered, for $t(R, \theta)$ $- t_0 = f(\theta)$. Here $t(r, \theta) = \bar{R}(r)\Theta(\theta)$. The Euler equation results for $\bar{R}$. The other relation is that for Θ. The solutions are also given

$$r^2 \frac{d^2 \bar{R}}{dr^2} + r \frac{d\bar{R}}{dr} - n^2 \bar{R} = 0 \tag{3.3.3a}$$

$$\frac{d^2 \Theta}{d\theta^2} + n^2 \Theta = 0 \tag{3.3.4}$$

$$\bar{R}(r) = C_1 r^n + C_2 r^{-n} \tag{3.3.3b}$$

$$\Theta = C_3 \cos n\theta + C_4 \sin n\theta$$

For $t - t_0$ to be bounded as $r \to 0$, $C_2 = 0$. Therefore,

$$t(r, \theta) - t_0 = r^n (A \cos n\theta + B \sin n\theta)$$

where θ is defined in Fig. 1.2.2.

The boundary conditions are $t(r, 0) - t_0 = 0$ and $t(r, \pi) - t_0 = 0$. Therefore, $A = 0$, $n = 1, 2, 3 \ldots$, and

$$t(r, \theta) - t_0 = \sum_{n=1}^{\infty} B_n r^n \sin n\theta$$

The condition at $r = R$, $t(R, \theta) - t_0 = f(\theta)$, is next applied to determine the B_n:

$$t(R, \theta) - t_0 = f(\theta) = \sum_n B_n R^n \sin n\theta$$

$$B_n R^n = \frac{2}{\pi} \int_0^{\pi} f(\theta) \sin n\theta \, d\theta$$

A specific result follows from a specification of $f(\theta)$. For $t(R, \theta) = t_1$, a constant, $f = t_1 - t_0$ and the solution is

$$\frac{t(r, \theta) - t_0}{t_1 - t_0} = \frac{t - t_0}{t_1 - t_0} = \phi = \frac{4}{\pi} \sum_{n=1,3}^{\infty} \frac{1}{n} \left(\frac{r}{R} \right)^n \sin n\theta \tag{3.3.5}$$

Recall the effect of the discontinuities at $x = 0$ and L in the strip geometry shown in Fig. 3.1.1. Similar behavior would not arise here for more reasonable values of $f(\theta)$ at $\theta = 0$ and π.

3.3.3 A Disk-Shaped Input Condition at a Region Boundary

This is shown in Fig. 3.3.1(c) (see page 117). The condition may be applied as either flux or a temperature condition over the surface of radius R. It may be on the bottom surface of a half-infinite solid above, or between two half-infinite regions, one above and one below. Here $t = t(r, z) = \bar{R}(r)Z(r)$. There are several important circumstances in which these configurations arise. One is with a single half-infinite region, remotely at t_∞, with either a temperature or flux condition maintained over the disk area. The rest of the surface, at $r > R$, may be insulated. Then the condition, upward in z, spreads outward in r. Also, $\partial t/\partial z = 0$, at $z = 0$, for $r > R$. Another process is with the two semiinfinite regions in direct contact only over $r \le R$, with either a temperature or flux condition maintained over $r \le R$. Several interesting solutions are given in the following discussion. See also Carslaw and Jaeger (1959).

A HALF-INFINITE REGION, A TEMPERATURE CONDITION. The temperatures over the disk area and at $z \to \infty$ are taken as t_1 and t_∞, respectively. The distribution $t(r, z)$ and the resulting total heat flow rate Q, for $\partial t/\partial z = 0$ for $r > R$, become

$$\phi = \frac{t(r, z) - t_\infty}{t_1 - t_\infty} = \frac{2}{\pi} \int_0^\infty e^{-\lambda z} J_0(\lambda r) \sin \lambda R \frac{d\lambda}{\lambda}$$

$$= \frac{2}{\pi} \sin^{-1} \frac{2R}{\left[(r - R)^2 + z^2\right]^{1/2} + \left[(r + R)^2 + z^2\right]^{1/2}} \quad (3.3.6)$$

and
$$Q = 4kR(t_1 - t_\infty) \quad (3.3.7)$$

The preceding process amounts to a heat flow, Q, from a small circular area, at t_1, into a large region remotely at t_∞. The driving temperature difference is $t_1 - t_\infty$. The conductive resistance, R_c, may be calculated from

$$Q = (t_1 - t_\infty)/R_c \quad \text{or} \quad R_c = (t_t - t_\infty)/Q$$

From Eq. (3.3.7) it is

$$R_x = 1/4kR \quad (3.3.8)$$

This process is a model in evaluating the contact resistance of a single contact between two rough surfaces. See Sec. 2.1.4 and Fig. 2.1.4.

A HALF-INFINITE REGION, FLUX CONDITION. If the condition over the disk is an imposed uniform heat flux q'' instead, with $\partial t/\partial z = 0$ for $r > R$, the solution is

$$t - t_\infty = \frac{Rq''}{k} \int_0^\infty e^{-\lambda z} J_0(\lambda r) J_1(\lambda R) \frac{d\lambda}{\lambda} \quad (3.3.9)$$

The resulting average temperature $\overline{t - t_\infty}$ over the disk surface is

$$\overline{(t - t_\infty)} = \frac{8Rq''}{3\pi k} \quad (3.3.10)$$

The characteristic temperature is seen in Eq. (3.3.9) to be Rq''/k and the conductive resistance, in terms of $(t - t_\infty)$, is $8/3\pi^2 kR$.

TWO HALF-INFINITE REGIONS. The regions are assumed to contact each other only over the disk, of radius R. At $r > R$, $\partial t / \partial z = 0$. Regions 1 and 2 have conductivities k_1 and k_2 and remote temperatures of $t_{1,\infty}$ and $t_{2,\infty}$, respectively. The local region temperatures are $t_1(r, z)$ and $t_2(r, z)$, respectively. The disk is assumed to be at a uniform temperature t_d. The solutions in the two regions, as ϕ_1 and ϕ_2, are written in terms of

$$\phi_1 = \frac{t_1(r, z) - t_{1,\infty}}{t_d - t_{1,\infty}} \quad \text{and} \quad \phi_2 = \frac{t_2(r, z) - t_{2,\infty}}{t_d - t_{2,\infty}} \tag{3.3.11}$$

The solution for each of these distributions, for t_d uniform, is then the form of Eq. (3.3.6).

This applies for the disk being of uniform temperature over its extent. The two solutions are joined by the condition that the total heat flow on each side of the disk surface has the same magnitude. This procedure is not a strictly correct matching of the two temperature fields. The differing convergence of the heat flow lines, as sketched in Fig. 7.2.1, for $k_b > k_a$ indicate a nonuniform gradient $\partial t / \partial z$ across the region of the disk-shaped contact area.

The total thermal resistance, $R_t = R_{1c} + R_{2c}$, between the remote regions at $t_{1,\infty}$ and $t_{2,\infty}$, is found from $R_t = (t_{1,\infty} - t_{2,\infty})/Q$ as

$$R_t = \frac{(k_1 + k_2)}{2Rk_1k_2} = \frac{(k_1 + k_2)}{R(2k_1k_2)} = \frac{1}{Rk_s} \quad k_s = \frac{2k_1k_2}{k_1 + k_2} \tag{3.3.12}$$

where R is the contact region radius, used in Sec. 7.3.3 to calculate constriction resistances, and k_s is sometimes called the harmonic mean of k_1 and k_2. It is twice the product, divided by the sum.

SUMMARY. The preceding resistances, in Eq. (3.3.12) and that for a single region in Eq. (3.3.8), are called constriction resistances. They arise as the heat flow paths are forced to constrict to pass across the small area of the direct contact disk area. This is the model for the evaluation of the two constriction components, R_{ac} and R_{bc}, of the total contact resistance mechanism. Constriction is pictured in Fig. 7.2.1. The effect appears in the circuit of the total contact resistance R_c, shown in Fig. 7.3.1(b), as $R_{ac,i}$ and $R_{bc,i}$.

3.4 SPHERICAL REGIONS

Two examples are given of conduction in spherical coordinates as defined in Fig. 1.2.2(b). The first is a hollow sphere, Fig. 3.4.1(a), maintained at a uniform temperature t_0 on the outside surface at R_o. The inside surface, at R_i, has a temperature which varies with the angle downward from the upper pole, as $t(R_i, \theta) - t_0 = f(\theta)$. That is, $t(R_i, \theta) - t_0$ is taken as independent of ϕ. Therefore, $t = t(r, \theta)$. The second geometry is a solid hemisphere of radius R_H, Fig. 3.4.1(b), having a uniform base temperature t_0. The surface is at $t(R_H, \theta) - t_0 = f(\theta)$. Again $t = t(r, \theta)$.

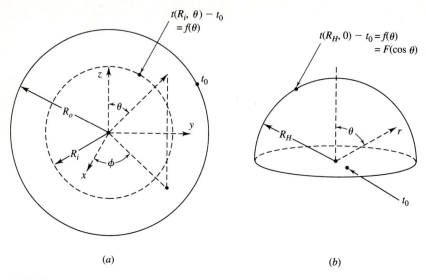

(a)   (b)

FIGURE 3.4.1
Spherical geometries: (*a*) a hollow sphere; (*b*) a solid hemisphere.

HOLLOW SPHERE. Separation of variables resulted in Eqs. (3.1.27) and (3.1.28) for $t(r, \theta) = \bar{R}(r)M(\mu)$, where in the following equations $\mu = \cos \theta$:

$$\frac{d^2 \bar{R}}{dr^2} + \frac{2}{r} \frac{d\bar{R}}{dr} - \frac{n(n+1)}{r^2} \bar{R} = 0 \qquad (3.1.27)$$

$$\frac{d}{d\mu}\left[(1 - \mu^2)\frac{dM}{d\mu}\right] + n(n+1)M = 0 \qquad (3.1.28)$$

where $M = M(\mu) = M(\cos \theta)$. The solution of the Euler equation is

$$R(r) = C_1 r^m + C_2 r^{-(m+1)} \qquad (3.4.1)$$

The solutions of Eq. (3.1.28) are Legendre polynomials $P_m(\mu) = P_m(\cos \theta)$ and the form of the solution is

$$t(r, \theta) - t_0 = \sum_m \left(A_m r^m + B_m r^{-(m+1)}\right) P_m(\cos \theta)$$

Boundary conditions are applied and the $f(\theta)$ is expressed as an expansion in terms of Legendre polynomials, to give the following result. See Schneider (1955), Sec. 6.7, for further details.

$$t - t_0 = \sum_{m=0,1}^{\infty} \frac{p}{2}\left[\frac{R_o^p - r^p}{R_o^p - R_i^p}\right]\left(\frac{R_i}{r}\right)^{p+1/2} P_p(\cos \theta) \int_0^\pi P_m(\cos \theta) F(\cos \theta) \sin \theta \, d\theta$$

$$(3.4.2)$$

Here $p = 2m + 1$, for convenience. The preceding reference gives a specific solution and a plot of isotherms in the shell, for $R_o/R_i = 2$ for $t(R_i, \theta) - t_0 = f(\theta) = 100 \cos^2 \theta$.

HEMISPHERICAL SOLID. The conditions in Fig. 3.4.1(b) are similar to those in Fig. 3.4.1(a), and also in Fig. 3.3.1(b), for the long half-cylinder. In all three, the boundary condition variation is in θ and $t(r, \theta) = \overline{R}(r)\Theta(\theta)$.

The present procedure is very similar to that for the hollow sphere given previously, except that the boundary condition at $r = R_i$ is replaced, in the region $\theta = 90°$ and $-90°$, by $t(r, \theta) - t_0 = 0$, for r in the range 0 to R_H. Using these boundary conditions and $C_2 = 0$ in Eq. (3.3.3b), the form of the solution is

$$t(r, \mu) - t_0 = \sum_{n=1}^{\infty} B_n r^n P_n(\mu) \qquad \mu = \cos \theta \qquad (3.4.3)$$

The solution, in terms of $F(\cos \theta) = F(\mu)$ is

$$t(r, \mu) - t_0 = \sum_{n=0,2}^{\infty} (2n + 1) \left(\frac{r}{R_H}\right)^n P_n(\mu) \int_0^1 F(\mu) P_n(\mu)\, d\mu \quad (3.4.4)$$

This result is similar in form to that given in Sec. 3.3.2 for the long half-cylinder.

3.5 CONDUCTION SHAPE FACTORS

The foregoing sections of this chapter indicate the kinds of steady-state solutions which follow in rectangular, cylindrical, and spherical regions. The procedures and results given depend on boundary conditions each being applied at specific or limiting values of single coordinates among (x, y, z), (r, θ, z), and (r, ϕ, θ), depending on the geometry.

It is far more common in practical circumstances that a particular boundary condition must be simultaneously expressed in terms of several of the space coordinates. Several characteristic examples are shown in Fig. 3.5.1. In Fig. 3.5.1(a) the Laplace equation in the region is in terms of r and θ. However, the surface boundary condition $t = t_2$ applies at $\cos \theta = D/R'$. In Fig. 3.5.1(b) each condition, t_1 and t_2, applies at four different locations, in each x and y. In Fig. 3.5.1(c), if the origin is taken at the center of the internal circular passage, the condition t_2 applies at some $F(r, \theta)$ over the outer surface. In Fig. 3.5.1(d) the situation is even more complicated.

Some such complications may be overcome in analysis. However, the great diversity in practice has led to other alternatives. These include direct numerical analysis, graphical methods, and quite approximate parallel–series one-dimensional conduction models, as seen in Sec. 7.4.3, applied to composite materials.

Experimental methods have also been used, often in an analogous phenomenon in which experimentation is less difficult than it commonly is in making heat flow measurements. An example is electrical conduction through an electrolitic bath of the same shape as the conduction region of interest. This

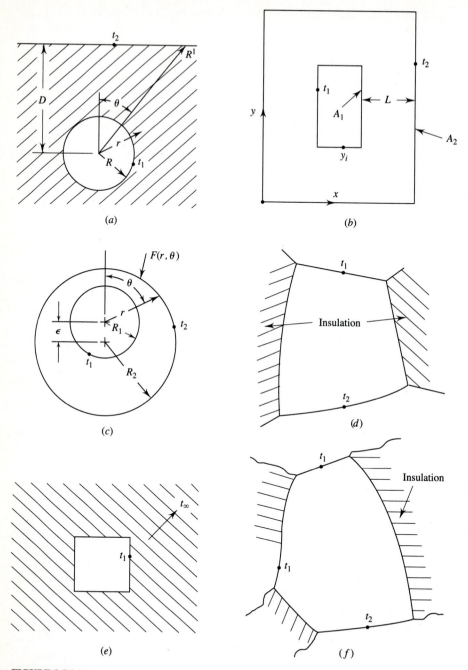

FIGURE 3.5.1
More complicated boundary conditions: (a) a long buried cylinder; (b) a thick-walled duct; (c) an off-centered internal passage in a tube; (d) an irregular rod; (e) a square rod in an infinite medium distantly at t_2; (f) a more complicated region.

models the Laplace equation, which is common to both phenomena. Another simple experimental analog is the use of soap films spread on region boundaries. Boundary heights above a reference plane simulate different temperature boundary conditions. The addition of an inflation pressure under the soap film simulates the distributed source in the Poisson equation, Eq. (1.2.12).

This section discusses a very widely used practical alternative for estimating the conduction heat transfer through irregular geometries. It is called the shape factor method. The method most simply applies to conduction regions of uniform and isotropic conductivity, k, subject to only two temperature boundary conditions, for example, t_1 and t_2, on surfaces. Any other surfaces present must be adiabatic, called insulated. All of the geometries in Fig. 3.5.1 satisfy these requirements, even those in Fig. 3.5.1(e) and (f).

All of the uniform temperature boundary condition solutions in Secs. 3.2 through 3.4 also satisfy the preceding conditions. However, many of those solutions are unrealistic, because of discontinuous temperature boundary conditions. These often result in infinite heat flow rates. An example was given for the region in Fig. 3.1.1. This gives the result for $Q(y)$ in Eq. (3.1.10). This is unbounded as $y \rightarrow 0$. Such solutions are not useful in evaluating shape factors.

3.5.1 Shape Factor Formulation

For the kinds of geometries and conditions specified previously, the heat conduction rate through any region, q, is proportional to the imposed temperature difference $(t_1 - t_2)$ across it as

$$q = (t_1 - t_2)/R_t = C_t(t_1 - t_2) \tag{3.5.1}$$

where R_t and C_t are the resistance and conductance of the region, between the imposed temperature conditions t_1 and t_2. The region may be of any geometry and q denotes the total heat conduction rate between the surfaces at t_1 and t_2.

In these linear processes $R_t \propto k^{-1}$ and $C_t \propto k$. Therefore, Eq. (3.5.1) may be written in terms of $C_t = Sk$, as

$$q = C_t(t_1 - t_2) = Sk(t_1 - t_2) \tag{3.5.2}$$

or
$$S = q/k(t_1 - t_2) \tag{3.5.3}$$

where S is the general conductance shape factor. Also, S^{-1} is the resistance shape factor. For a plane region of surface area A, as in Fig. 1.1.1, the value of S, from Eq. (1.1.1), is $S = A/L$, with the dimension of length. For a sphere at t_1 embedded in an extensive medium distantly at t_∞, Eq. (2.1.39b) compared with Eq. (3.5.2) shows that the shape factor S is $4\pi R$. For most kinds of cylindrical geometries, Eq. (3.5.2) is written in terms of q' in Eq. (2.1.20). Then S becomes the shape factor per unit length and is dimensionless. For a spherical region, in terms of q as in Eq. (2.1.26), S has the units of length. Similar considerations allow the formulation of S for any conduction process where the motive temperature difference is simply $(t_1 - t_2)$.

In the preceding postulate, S is a function only of geometry if k is uniform throughout the region. If, in a given application, k is a function of local region temperature or of location, some average value would apply, as in the regular one-dimensional geometries in Sec. 2.2. However, those formulations of proper average values are based on integrals over regular geometries. Therefore, determining estimates in regions such as those in Fig. 3.5.1 introduces additional approximations.

AN EXAMPLE OF SHAPE FACTORS. Consider the geometry in Fig. 3.5.1(b), as a thick-walled rectangular box, instead of a duct. It has inside and outside areas A_1 and A_2. An estimate of the conduction through the six walls of thickness L, between the inside and outside surfaces at t_1 and t_2, would be

$$q \approx \frac{(A_1 + A_2)}{2L} k(t_1 - t_2) = \frac{A_m}{L} k(t_1 - t_2) = Sk(t_1 - t_2) \quad (3.5.4)$$

where A_m is a simple average conduction area. This estimate implies that the region is equivalent to a flat region of area A_m, L thick. Then $S \approx A_m/L$.

This procedure does not represent a systematic estimate of the conduction effects through the twelve edges and eight corner regions of this geometry. Consider boxes in which L becomes large, compared, for example, to the three characteristic edge lengths inside, $L_i = L_1, L_2, L_3$. There are twelve edges and possibly three different values among the L_i. For L much greater than all the L_i, the edge and corner conduction effects would dominate. The results which follow show that this effect becomes increasingly large as L/L_i increases.

Langmuir et al. (1913) determined the heat transfer through thick-walled rectangular boxes, from analog measurements in electrolytes. The results are written in terms of an effective area A_e divided by the uniform thickness of the walls, L, as

$$q = \frac{A_e k}{L}(t_1 - t_2) \quad (3.5.5)$$

The shape factor is $S = A_e/L$ given previously. It has units of length and was given by Langmuir et al. (1913) in terms of the relation of the uniform wall thickness L to the three lengths L_i of the edges of the inside volume.

The values of A_e/L were correlated in terms of four regimes, based on the relationship between L and the L_i. The effect which determines the regimes is the relative importance of conduction through the twelve edges and eight corner regions, compared to that through the six walls. For example, the data for the thinnest walls studied were correlated by

$$S = \frac{A_e}{L} = \frac{A_1}{L} + 0.54 \sum y + 1.2L = S_1 + S_2 + S_3 \quad (3.5.6)$$

where A_1 is the inside area and $\sum y$ is the sum of the twelve inside edge

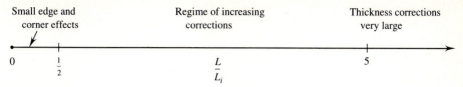

FIGURE 3.5.2
The conduction regimes, in L/L_i, for thick-walled boxes.

lengths, as

$$\sum y = 4 \sum_{i=1,2,3} L_i$$

Equation (3.5.6) applies for the L/L_i between $1/2$ and 5. For $L/L_i < 1/2$ the A_m approximation, in Eq. (3.5.4), applies approximately. In Eq. (3.5.6) the first approximation is $S_1 = A_1/L$, plus the additional two for the edges and corners.

The range of L/L_i is seen in Fig. 3.5.2, from relatively thin walls, for $L/L_i < 1/2$, to very thick ones, for $L/L_i > 5$. The three other equations given by Langmuir et al. (1913) are for one, two, and three of the distinct values of the L_i being less than $L/5$. The equations are

$$\frac{A_e}{L} = \frac{A_1}{L} + 0.465 \sum y + 0.35L \tag{3.5.7}$$

$$\frac{A_e}{L} = \frac{2.78 L_{i,m}}{\log(A_2/A_1)} \tag{3.5.8}$$

$$\frac{A_e}{L} = 0.79\sqrt{A_1 A_2/L^2} \tag{3.5.9}$$

where $L_{i,m}$ is the length of the edge which is less than $L/5$.

The second and third terms in the first relation, Eq. (3.5.6), amount to the edge and the corner conduction effects, for thinner walls, $L/L_i < 2$. The edge shape factor correction, S_2, is seen to average 0.54 per unit of inside edge length. The corner correction $S_3 = 1.2L$ averages $0.15L$ for each corner. This component evaluation procedure may be used for other three-dimensional geometries and also for intersecting regions of thick walls in two dimensions. See, for example, Schneider (1955) and Jakob and Hawkins (1957).

3.5.2 Some Shape Factor Solutions

Many of the shape factor determinations have been for buried geometries such as spheres and cylinders. These approximate applications such as geothermal deposits, underground storage volumes, and buried cables and pipelines.

Some results relate to the interaction of several deeply buried elements at different temperatures, as shown in Fig. 3.5.3(a). The temperatures t_1 and t_2 are then the boundary conditions. If the distant medium is at some different temperature, t_3, this process would not fulfill the requirement for a shape factor in Eq. (3.5.2).

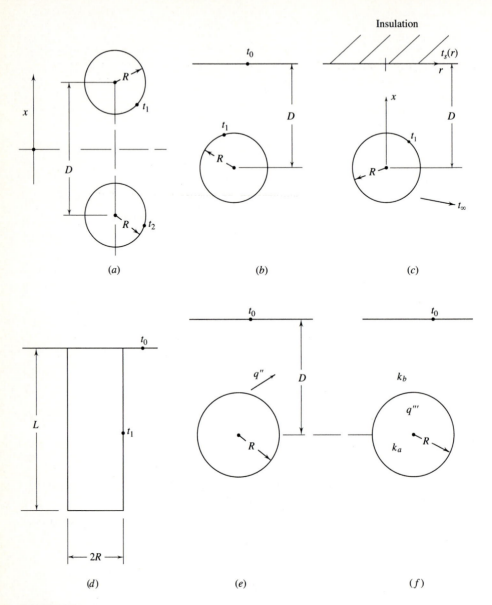

FIGURE 3.5.3

Examples of typical heat transfer configurations with buried surfaces: (a) two elements in an infinite region; (b) a sphere or cylinder at t_1, D below a surface at t_0; (c) a sphere or cylinder at t_1, below an insulated surface; (d) a rod at t_1 where it is submerged at and below a surface at t_0; (e) a sphere dissipating uniform surface flux below a surface at t_0; (f) a sphere with uniform internal energy generation, adjacent to a surface at t_0.

However, most applications concern a single geometry at some heating or cooling condition, buried near or at the surface of the surrounding medium, as in Fig. 3.5.3(b) and (c). Two kinds of boundary conditions simulate a wide range of applications. One is the distant surface at a uniform temperature t_0 and the buried geometry at a surface temperature of t_1. Then heat flow is between the geometry and the surface. The other is an insulating surface, as seen in Fig. 3.5.3(c). Then steady-state heat flow occurs only if the distant ambient is at $t_\infty \neq t_1$.

There are also interesting applications for the bounding surface at t_0 with the buried geometry dissipating a uniform heat flux q'' at its surface, as in Fig. 3.5.3(e), or internally at a rate q''', as in Fig. 3.5.3(f).

This section includes consideration of results for each of the geometries shown in Fig. 3.5.3. Shape factors are given for both temperature and insulating conditions. Other sources of shape factor information are then given in Sec. 3.5.3.

TWO SPHERES. With surface temperatures given on both spheres, the only conductivity of importance in determining the heat transfer rate is that of the infinite medium, in Fig. 3.5.3(a). A source–sink analysis by Hahne and Grigull (1973) applies at a good approximation at larger spacings, for $D/R \geq 5$. The resulting heat flow and shape factor are

$$q = k(t_1 - t_2)\frac{2\pi R}{(1 - R/D)} \tag{3.5.10}$$

$$S = \frac{2\pi R}{(1 - R/D)} \tag{3.5.11}$$

There are several interesting consequences of the conduction field in Fig. 3.5.3(a). First, as $x \to \infty$ and as $x \to -\infty$, the remote region temperatures approach t_1 and t_2, respectively. Also, the symmetric dashed location is isothermal, at $t = (t_1 + t_2)/2$. Note that for $R/D \to 0$ the result from Eq. (3.5.11) is

$$S = 2\pi R \tag{3.5.12}$$

This applies for an infinite distance between two spheres.

Recall that, following Eq. (3.5.3), it was shown that the conductance shape factor for a sphere in an infinite medium is $4\pi R$. The preceding example, $R/D \to 0$, is merely two such resistances in series, between spheres 1 and 2. Adding the two resistances, each of $1/4\pi R$, gives the result in Eq. (3.5.12).

Hahne and Grigull (1974) also report the results of Attwood (1949) for spacings in the range $2 < D/R < 5$ as

$$S = 2\pi R\left[1 + \frac{R}{D} + \left(\frac{R}{D}\right)^2 + \left(\frac{R}{D}\right)^3 + 2\left(\frac{R}{D}\right)^4 + 3\left(\frac{R}{D}\right)^5 + \cdots\right] \tag{3.5.13}$$

This is within about 1%, in the range $D/R \geq 2$, of the improved estimation of S given previously, over the simplest model, which resulted in Eq. (3.5.12).

A BURIED SPHERE NEAR THE SURFACE. Two surface conditions are shown in Fig. 3.5.3(b) and (c), isothermal at t_0 and insulated. The isothermal surface condition is the same as for the region $x \leq 0$ in Fig. 3.5.3(a). The dashed line at $x = 0$ is an isothermal, due to symmetry. However, the distance over which $t_1 - t_0$ occurs is half the length in which $t_1 - t_2$ occurs, between two spheres. Therefore, the heat transfer and conductance shape factor is twice as great and Eq. (3.5.11) becomes, for $D/R \geq 5$:

$$S = 4\pi D/2(1 - R/D) \qquad (3.5.14)$$

where D is still as defined in Fig. 3.5.3(a).

This configuration was analyzed in more detail by Weihs and Small (1978), over the range of $(D - R)/R$ down to 10^{-2}, using a bispherical formulation. The results were compared with those from the earlier approximate results in Eqs. (3.5.13) and (3.5.14). These results are all in close agreement for $(D - R)/R$ greater than approximately 0.5. At $(D - R)/R = 10^{-2}$, the approximate results were about 35% below the bispherical calculation. Recall that this solution is comparable to that between two buried spheres of equal sizes.

For the insulating condition at $x = D$, in Fig. 3.5.3(c), the surface temperature at $x = D$, $t_s(r)$, is a maximum directly above the sphere and decreases radially away from that location. All of the heat flow is eventually downward toward the remote region temperature t_∞. Hahne and Grigull (1974) gave an approximate shape factor, as that given in Eq. (3.5.14). The insulated surface temperature variation $t_s(r)$ was also determined by Weihs and Small (1978). For example, for $D/R = 1.005$, $t_s(r) = t_1$ over most of the surface immediately above. For $D/R \approx 3.3$, the maximum value of $t_s(0)$ decreased by 50% of $(t_1 - t_\infty)$.

A LONG CYLINDER AT UNIFORM DEPTH. This geometry models many applications with buried conduits. There is often also convective heat transfer and energy generation internal to the conduit. Mass diffusion processes may also arise outside in the ambient region. The result given here is for conduction from an isothermal cylinder at t_1 at depth D below a surface at t_0, as in Fig. 3.5.3(b) for a sphere. The shape factor for a long cylindrical geometry is in terms of q/L, as defined initially in Eq. (2.1.20):

$$\frac{q}{L} = q' = Sk(t_1 - t_0) \qquad (3.5.15)$$

where S is dimensionless.

An analysis, based on a mirror-image method, is given by Eckert and Drake (1972). The result is

$$S = \frac{2\pi}{\ln\left[D/R + \sqrt{(D/R)^2 - 1} \right]} \qquad (3.5.16)$$

For $D/R \gg 1$ the result is $2\pi/\ln(2D/R)$.

OFF-CENTERED PASSAGE IN A TUBE. The geometry is shown in Fig. 3.5.1(c). The shape factor, given by Eckert and Drake (1974), is again dimensionless:

$$S = \frac{2\pi}{\cosh^{-1}\left[(8R_1^2 + 8R_2^2 - \epsilon^2)/4R_1R_2\right]} \qquad (3.5.17)$$

where ϵ is the eccentricity and R_1 and R_2 are the two radii.

A VERTICAL ROD BURIED FLUSH WITH THE SURFACE. The rod of length L and radius R, Fig. 3.5.3(d) (see page 128), has a surface temperature of t_1. The surface of the ambient region is at t_0. The result is

$$S = \frac{2\pi L}{\ln(2L/R)} \qquad (3.5.18)$$

The evaluation method results in a definite value of S, even though there is a temperature discontinuity around the upper rod edge, at t_1 over its surface, and the surface temperature elsewhere, of t_0. Eckert and Drake (1974) also provide a table of additional results.

IMPOSED HEAT FLUX AND INTERNAL ENERGY GENERATION. The uniform surface flux condition in Fig. 3.5.3(e) was applied to the surface of a long buried horizontal cylinder by Thiyagarajan and Yovanovich (1974). Bicylindrical coordinates were used to analyze the temperature distribution in the ambient region. The resistance from the cylinder to the flat surface was determined, in terms of the boundary temperature t_0 and the average cylinder surface temperature t_a. The result, expressed as a conductive shape factor per unit length, where $\cosh A = (D/R)$, is

$$S = \left[\frac{A}{2\pi} + \frac{1}{\pi}\sum_{n=1}^{\infty}\frac{e^{-2nA}}{n}\tanh(nA)\right]^{-1} \qquad (3.5.19)$$

The results given, in terms of D/R, span the range $1.001 \leq D/R \leq 10$, the bounding values being $A = 0.04$ and 2.99.

The other example here is the sphere, in Fig. 3.5.3(f), containing a uniformly distributed source of strength q''', for example, in W/m^3. The temperature of the spherical surface varies and the thermal conductivity of both regions, k_a and k_b, must be taken into account in the analysis. Simultaneous solutions inside the sphere and in the ambient region must be matched at their interface.

The analysis by Bau (1982) compares the results of a bispherical and an approximate source–sink analysis, for $k_b/k_a = 0.1$, 1, and 10. Both results were in very close agreement, for $\cosh^{-1}(D/R) > 0.8$, or $D/R > 1.34$. The shape factor is calculated in terms of the average temperature, t_a, over the spherical region as

$$S = Q/k_b(t_a - t_0) \qquad Q = \frac{4}{3}\pi R^3 q''' \tag{3.5.20}$$

where $(t_a - t_0)$ is determined in the analysis. The resulting shape factor is

$$S = \frac{4\pi R}{\dfrac{k_b}{5k_a} - \dfrac{R}{2D} + 1} \tag{3.5.21}$$

SUMMARY. The results in this section are for buried spheres, cylinders, and for several other simple shapes. Most of the analysis has been based on approximations which become inaccurate for close spacings of the surfaces at which the boundary conditions are applied. More accurate analysis has shown the limitations of some of these results and has extended the range of accurate information.

The preceding results are very valuable in many applications and in characterizing the mechanisms related to shape factor determination. However, many applications do not involve extensive ambient media or the simplest boundary conditions and geometries. The following section summaries sources of additional information relevant to other geometries and bounding conditions.

3.5.3   Other Shape Factor Information

The foregoing section indicates many of the basic considerations in determining conduction and shape factors. The geometries are simple and the results have many applications. This section gives a brief summary of some of the additional shape factor information which is available.

Hahne and Grigull (1974) obtained shape factor information, from earlier solutions, related to the kinds of geometries which originate from point sources, embedded in an infinite medium. These include ellipsoids, disks, ribbon strips, short wires, and cylinders. The variation of the shape factor for each geometry was determined. The results are given as a function of the aspect ratio of the geometry, over a very wide range.

Hahne and Grigull (1975) presented shape factors for some 50 different kinds of geometries. Most are for regions of limited extent. The examples include multiple adjacent surfaces, surfaces embedded in a small geometry, disks, rings, and conduits. Most of these geometries simulate a very important conduction effect characteristic of a practical application.

Yovanovich (1973) discusses the methodology for calculating shape factors in geometries conveniently described in cylindrical, spherical, elliptic cylindrical, bicylindrical, and oblate and prolate spheroidal coordinates. Resistances are calculated for geometric examples in each of these coordinate systems. Solutions are also given for strips on the surface of extensive regions. Yovanovich (1977) also gives further results for some similar shapes.

A compendium of conduction shape factors also appears in Rohsenow et al. (1985), (4-162). Most of the geometries are single and multiple isothermal shapes associated with infinite regions. Several examples of passages are also given. The information is in terms of q, k, and Δt, rather than the shape factor S.

Another source of shape factor information is in steady-state solutions such as those given in Secs. 3.2.3 and 3.2.4. Consider any given solution, in a region, which depends only on two different boundary conditions. If these conditions are merely t_1 and t_2, and q as a function of conductivity k is calculated from the solution, then S may be found from Eq. (3.5.3). However, many solutions, such as that for the region in Fig. 3.1.1, have an infinite q. Then S will be infinite. Nevertheless, there are many other analytical solutions for which reasonable values of q result.

The shape factor procedure may also be used for some other pairs of boundary conditions. For example, a solution for an imposed heat flux surface condition, q'', permits the calculation of the resulting surface temperature along that boundary, as well as its average value, t_a. Also q is known from the q'' boundary condition. Then S may be calculated. An example of such a solution is Eq. (3.2.9).

Different forms of assigned surface temperature variations, for example, $f(x)$ and $f(r)$, may also be averaged to calculate the resulting shape factors. Even with a linear surface process such as convection, the shape factor procedure may be useful. The average temperature of the convective surface may be calculated from the solution. The shape factor then will also depend on the Biot number.

3.6 DISTRIBUTED ENERGY SOURCES

There are many circumstances in which a region is subject to a dissipation of energy in its interior. Commonly important causes are penetration by solar, microwave, and particle irradiation and internal electrical energy dissipation or chemical reaction. The general formulation is Eq. (1.2.8) where the distributed source strength, q''', may be a function of location in the region, as it usually is with irradiation. See Sec. 2.4.2. It also may vary with local material properties, with temperature, and with time, in transients and in periodics.

Distributed sources and sinks are also very common in mass diffusion processes. They may arise from local adsorption or desorption of the diffusing species of local concentration C. The local rates of such effects are expressed as C'''. The comparable relevant formulation for this effect is Eqs. (1.3.6) and (1.3.8). The more complicated mass diffusion effects, such as rate reactions, are considered in Chap. 6.

One-dimensional uniform and varying distributed sources are considered in Sec. 2.4. Several two-dimensional and one three-dimensional example are given in this section. These demonstrate the method used and indicate the properties of such conduction and diffusion processes. The first example is a long square rod with a source of uniform strength. Then spatially variable q''' is discussed. A cubical region with a uniform internal source is then analyzed.

The last configuration is a finite flat plate region with three plate edges at a uniform temperature t_0. The other edge is at $\theta_1 = t_1 - t_0 = f(x)$. This geometry is in an environment at t_0. The distributed source effect is simulated by convection from the two plate surfaces to an environment also at t_0. This last example approximates a finlike geometry, wherein the surface loss on the two sides is simulated using a distributed sink of local strength $q'''(x, y)$.

3.6.1   Multidimensional Distributed Energy Sources

The half-lengths of the sides of the long square rod in Fig. 3.6.1(a) are taken as L, for convenience in the analysis. The temperature excess, defined as $\theta(x, y) = t(x, y) - t_0$, satisfies the following Poisson equation. The boundary conditions are also given.

$$\frac{\partial^2 \theta}{\partial x^2} + \frac{\partial^2 \theta}{\partial y^2} + \frac{q'''}{k} = 0 \tag{3.6.1}$$

$$\theta(\pm L, \pm L) = 0 \tag{3.6.2}$$

The solution $\theta(x, y)$ is postulated as

$$\theta(x, y) = \theta_1(x, y) + \theta_2(x, y) \tag{3.6.3}$$

where $\theta_1(x, y)$ and $\theta_2(x, y)$ are to be solutions of the Poisson and the Laplace equations, respectively. Thereby, the solution for a distributed source will be transformed into a solution for $\theta_2(x, y)$, with different boundary conditions.

The preceding formulation is first shown to satisfy Eq. (3.6.1). Then the boundary conditions for each function θ_1 and θ_2 are given. These collectively satisfy the conditions in Eq. (3.6.2):

$$\nabla^2 \theta + \frac{q'''}{k} = \nabla^2 \theta_1 + \nabla^2 \theta_2 + \frac{q'''}{k} = 0$$

This relation is satisfied if θ_2 satisfies the Laplace equation $\nabla^2 \theta_2 = 0$ and if θ_1 is a solution of

$$\nabla^2 \theta_1 + \frac{q'''}{k} = 0 \tag{3.6.4}$$

UNIFORMLY DISTRIBUTED SOURCE q''' IN A LONG ROD. Both $\theta_1(x, y) = -q'''x^2/2k$ and $-q'''y^2/2k$ satisfy Eq. (3.6.4). Taking either form is sufficient. Choosing the first, the boundary condition on $\theta(x, y)$ at $x = \pm L$ becomes

$$\theta(\pm L, y) = 0 = \theta_1(\pm L, y) + \theta_2(\pm L, y)$$

$$= -\frac{q'''L^2}{2k} + \theta_2(\pm L, y) \tag{3.6.5}$$

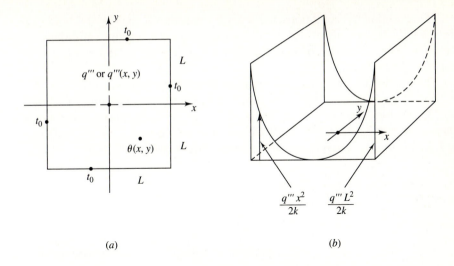

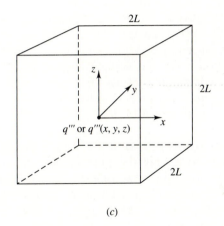

FIGURE 3.6.1
Region geometries: (*a*) a long square rod: (*b*) the boundary conditions for $\theta_2(x, y)$; (*c*) a cubical region, all surfaces at t_0.

and

$$\theta_2(\pm L, y) = \frac{q'''L^2}{2k}$$

The other conditions, at $(x, \pm L)$, are applied:

$$\theta(x, \pm L) = 0 = \theta_1(x, \pm L) + \theta_2(x, \pm L)$$

and

$$\theta_1(x, \pm L) = -q'''x^2/2k$$

$$\theta_2(x, \pm L) = q'''x^2/2k = f(x)$$

(3.6.6)

where $f(x)$ is applied to $\theta_2(x, y)$ at both $y = \pm L$.

These boundary conditions for $\theta_2(x, y)$ are shown in Fig. 3.6.1(b) (see page 135). Thus the solution of $\theta_1(x, y)$ is subtracted from the conditions on $\theta_2(x, y)$, to yield a solution of $\theta(x, y)$, which must be zero at the boundaries, as required by Eq. (3.6.2).

The preceding transformation removed the inhomogeneous term from Eq. (3.6.1), in terms of $\theta_1(x)$. This results in $\nabla^2\theta_2 = 0$ and the boundary conditions, Eqs. (3.6.5) and (3.6.6). The formulation for $\theta_2(x, y)$ is then that for the geometry in Fig. 3.2.1(c). Here, $H = L$ and the simple boundary conditions in Fig. 3.6.1(b) apply. The solution, in the coordinate definitions in Fig. 3.6.1(a), is

$$\theta(x, y) = \frac{q'''L^2}{2k}\left[1 - \left(\frac{x}{L}\right)^2 - 8\sum_{n=0}^{\infty}\frac{(-1)^n}{\beta_n^3}\frac{\cosh(\beta_n y/L)}{\cosh\beta_n}\cos\beta_n\frac{x}{L}\right] \quad (3.6.7)$$

where $\qquad \beta_n = (2n + 1)\pi/2 \quad$ and $\quad \Delta t_{ch} = q'''L^2/2k$

This analysis may be simply modified to a rod of L by H in cross section. The preceding result also includes the circumstance of an anisotropic material, having k_x and k_y, in the two coordinate directions. These are written as k_x and $k_y = k_x/K^2$, for convenience in the result. That is, $k_x = K^2 k_y$. Recall the transformation in Eq. (1.2.15). Then Eq. (3.6.7) is still the solution when k is taken as k_x and β_n is replaced by $\beta_n K$ in the two cosh terms in the summation.

Convection conditions over the surface, instead of $\theta = 0$, may also be accommodated. See the comparable results for $\theta_1(x, y)$ in Özisik (1980). Other more general boundary conditions on $\theta(x, y)$ may be applied, by the method in Sec. 3.2.2, for the determination of $\theta_2(x, y)$.

VARIABLE DISTRIBUTED SOURCE STRENGTH. The technique used previously for q''' uniform, amounts to finding the function $\theta_1(x, y)$ which satisfies the Poisson equation (3.6.4). Then $\theta_2(x, y)$ satisfies the Laplace equation, $\nabla^2\theta_2 = 0$. For a particular form of $q'''(x, y)$, a solution must again be found for the $\theta_1(x, y)$ which satisfies Eq. (3.6.4), given $q'''(x, y)$. Such solutions may be guessed, as in the preceding example.

Moran (1970) gives a systematic procedure for determining $\theta_1(x, y)$, given $q'''(x, y)$. Examples are presented by Özisik (1980), page 67, and are listed in Table 3.6.1, including the one given previously. The same kind of information is given for cylindrical coordinates applied to cylinders of length L by Özisik (1980), page 134.

The procedure is again to determine the boundary conditions which result for $\theta_2(x, y)$, as shown in Fig. 3.6.1(b), for q''' uniform. The next section outlines the analysis for a cubical geometry, with q''' uniform.

GENERATION IN A CUBICAL REGION. This formulation is similar to that given previously for the rod. Now $\theta(x, y, z)$ is

$$\nabla^2\theta(x, y, z) + \frac{q'''}{k} = 0 = \nabla^2\theta_1(x, y, z) + \nabla^2\theta_2(x, y, z) + \frac{q'''}{k} \quad (3.6.8)$$

TABLE 3.6.1
Solution θ_1 for various q''', in two and three dimensions†

q'''	θ_1
Uniform, q'''	$-q'''x^2/2k$
$q_0'''x$	$-q_0'''x^3/2k$
$q_0'''x^2$	$-q_0'''x^4/12k$
$q_0'''x^n$	$-q_0'''x^{n+2}/(n+1)(n+2)k$
$q_0'''x/y^2$	$-q_0'''x \ln y/k$
$q_0'''y/x^2$	$-q_0'''y \ln x/k$

†Where q_0''' is a constant.

Again, θ_1 and θ_2 are solutions of the Poisson and Laplace equations in terms of $\theta = t(x, y, z) - t_0$:

$$\nabla^2\theta_1 + q'''/k = 0 \quad \text{and} \quad \nabla^2\theta_2 = 0$$

For the coordinates shown in Fig. 3.6.1(c) (see page 135), the boundary conditions are

$$\theta(\pm L, \pm L, \pm L) = 0 \tag{3.6.9}$$

Either of $-q'''x^2/2k$, $-q'''y^2/2k$, or $-q''z^2/2k$ satisfy the Poisson equation. Any one of these forms is sufficient.

Again, θ_1 is taken as $-q'''x^2/2k$. This choice determines the boundary conditions to be applied to $\theta_2(x, y, z)$, as in Fig. 3.6.1(b) for a square rod. As before, the values of the function $\theta_1(x, y, z)$ on the boundaries, given as follows, are subtracted from Eq. (3.6.8) to give the boundary conditions for $\theta_2(x, y, z)$:

$$\theta_2(\pm L, y, z) = q'''L^2/2k \tag{3.6.10a}$$

$$\theta_2(x, \pm L, z) = q'''x^2/2k \tag{3.6.10b}$$

$$\theta_2(x, y, \pm L) = q'''x^2/2k \tag{3.6.10c}$$

The solution $\theta_2(x, y, z)$ is then generated as in Sec. 3.2.3.

For variable generation $q'''(x, y, z)$, the procedure given previously applies for the imposed distribution of θ_1 in Table 3.6.1. Boundary conditions on $\theta_2(x, y, z)$ are again then modified as for Eq. (3.6.8), using the appropriate solution of θ_1.

SUMMARY. The preceding method permits a simple solution when the distributed source varies spatially in the restricted ways which give a simple solution to $\nabla^2\theta_1 + q'''/k = 0$. The first three distributions given in Table 3.6.1 have $q''' = 0$ and uniform at $x = 0$. The last two show unbounded source strength in the region. Nevertheless, some of these variations may be useful in some applications and additional ones may be found which are more realistic. Another alternative is numerical calculation, as discussed in Sec. 3.9 for steady-state conduction.

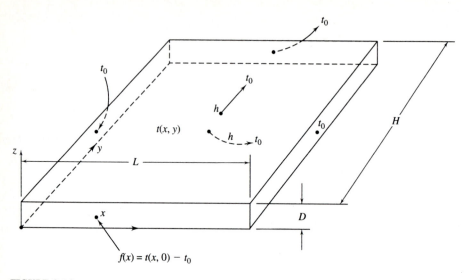

FIGURE 3.6.2

A plate L by H with three edge surfaces at t_0 and one at $t - t_0 = f(x)$, with convection on both sides to a region also at t_0.

3.6.2   Flat Plate with Surface Convection

This simulates a fin or extended surface, as discussed in Sec. 8.2. The conditions are seen in Fig. 3.6.2. The base of the fin, at $y = 0$, is at a temperature $t(x, 0) - t_0 = f(x)$. The fin thickness D is assumed much less than L and H. Then the temperature gradients in the z direction may be ignored and $\theta(x, y, z) = \theta(x, y)$. The temperature distribution is $\theta(x, y) = t(x, y) - t_0$. The plate edges, at $x = 0, L$ and at $y = H$, are at t_0.

There is convection over the two surfaces, at $z = 0$ and D, to an ambient temperature of t_0. The local convective loss $q_c''(x, y)$, per unit area of surface, is $h\theta(x, y)$. This amounts to a variable distributed energy sink of $2h\theta$, per unit area of the plate. The volume, per unit of plate area, is D. Therefore, $q''' = -2h\theta/D$ is a distributed heat sink. The equation and boundary conditions are

$$\nabla^2\theta - \frac{2h\theta}{kD} = \nabla^2\theta - K\theta = 0 \tag{3.6.11}$$

$$\theta(0, y) = \theta(L, y) = \theta(x, H) = 0 \quad \text{and} \quad \theta(0, x) = f(x) \tag{3.6.12}$$

The first three boundary conditions in Eq. (3.6.12) satisfy each term in

$$\theta(x, y) = \sum_{n=1}^{\infty} A_n \sin n\pi \frac{x}{L} \sinh\left[(H - y)\left[K^2 + \frac{(n\pi)^2}{L^2} \right]^{1/2} \right] \tag{3.6.13}$$

If $f(x)$ may be expanded in the series

$$f(x) = \frac{2}{L} \sum_{n=1}^{\infty} \sin \frac{n\pi x}{L} \int_0^L f(x) \sin \frac{n\pi x}{L} \, dx$$

the solution is

$$\theta(x, y) = \frac{2}{L} \sum_{n=1}^{\infty} \frac{\sin(n\pi x/L) \sinh\left[(H - y)(K + n^2\pi^2/L^2)^{1/2}\right]}{\sinh H(K + n^2\pi^2/L^2)^{1/2}}$$

$$\times \int_0^L f(x) \sin \frac{n\pi x}{L} \, dx \tag{3.6.14}$$

For $f(x) = t_1 - t_0$ uniform, the integral becomes $2L(t_1 - t_0)/n\pi$ for n odd, and the solution is

$$\frac{\theta(x, y)}{(t_1 - t_0)} = \phi(x, y)$$

$$= \frac{4}{\pi} \sum_{n=1,3}^{\infty} \frac{1}{n} \frac{\sin(n\pi x/L) \sinh\left[(H - y)(K + n^2\pi^2/L^2)^{1/2}\right]}{\sinh H(K + n^2\pi^2/L^2)^{1/2}}$$

$$\tag{3.6.15}$$

SUMMARY. This result appears to be analogous to that for the strip shown in Fig. 3.1.1, which resulted in Eq. (3.1.8). That solution had an infinite heat flow rate across the boundary at $y = 0$, due to the temperature discontinuities at $x = 0$ and L. The preceding solution, with a convection condition at $z = 0$ and D, to a temperature level t_0, does not have that property. The sine term in Eq. (3.3.15) results in $\theta(0, 0) = \theta(L, 0) = 0$. Alternatively, the evaluation of $q_y'''(0, 0)$ from Eq. (3.6.15) also contains the same term and this flux is zero.

3.7 VARIABLE AND DIRECTIONAL CONDUCTIVITY AND DIFFUSIVITY

In many applications such variations arise because of temperature gradients, directional material properties, and material inhomogeneity. They may also result from material changes which occur with time, due to the changing concentration of a diffusing material or from material curing or aging.

Equations (1.2.8) and (1.3.5) describe the temperature and concentration fields, including effects such as those given previously, as well as many others. Sections 2.2 and 2.3 analyze location- and temperature-dependent effects in one-dimensional steady-state processes.

This section gives two examples of conductivity variation in two dimensions. The first concerns both temperature-dependent conductivity $k(t)$ and concentration-dependent diffusivity $D(C)$. The second example here is conduction in an orthotropic region. Numerical analysis, for $k = k(t)$, is given in Sec. 9.5.

3.7.1 Temperature-Dependent Conductivity

The Kirchhoff transformation in Eq. (1.2.14a) replaces the general result, Eq. (1.2.13), for a purely temperature-dependent isotropic conductivity, $k(t)$. Considering steady state and $q''' = 0$, the transformation is

$$\nabla \cdot k(t)\,\nabla t = \nabla^2 V = 0 \qquad (3.7.1)$$

The effect of variable conductivity will be presented for the long rectangular rod, shown in Fig. 3.2.1(c) and analyzed in Sec. 3.1.1.

The region, along with equivalent boundary conditions in V, are shown in Fig. 3.7.1. Analogous to Eq. (3.1.11), the solution for $V(x, y)$ becomes

$$\frac{V(x, y) - V_0}{V_1 - V_0} = \frac{4}{\pi} \sum_{n=1,3}^{\infty} \frac{1}{n} \frac{\sinh[n\pi(y/L)]}{\sinh[n\pi(H/L)]} \sin n\pi \frac{x}{L} \qquad (3.7.2)$$

where V is here related to t as

$$V(x, y) - V_0 = \int_{t_0}^{t} \frac{k(t)}{k_0}\,dt = \int_{t_0}^{t(x, y)} \frac{k[t(x, y)]}{k_0}\,dt \qquad (3.7.3)$$

As a simple example of the use of Eqs. (3.7.2) and (3.7.3), $k(t)$ is taken linear in absolute temperature T as $k(t) = Kt$. Then k_0 becomes Kt_0 and, from

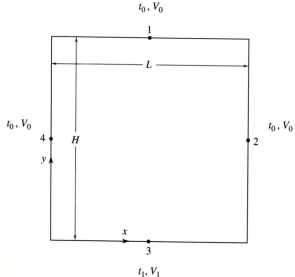

FIGURE 3.7.1

A region of variable conductivity.

Eq. (3.7.3),

$$V(x, y) - V_0 = \frac{\left[t^2(x, y) - t_0^2\right]}{2t_0} \tag{3.7.4}$$

Therefore, $V_1 - V_0 = (t_1^2 - t_0^2)/2t_0$, in terms of imposed temperature conditions t_1 and t_0. As a result, the solution of Eq. (3.7.2), in terms of $t(x, y)$, t_1, and t_0, is

$$\frac{V(x, y) - V_0}{V_1 - V_0} = \frac{t^2(x, y) - t_0^2}{t_1^2 - t_0^2} \tag{3.7.5}$$

Therefore, $t(x, y)$ may be calculated from Eq. (3.7.2) to determine the temperature distribution and local heat flux. A numerical formulation for calculating the effect of temperature-dependent conductivity directly in terms of V_2 is given in Sec. 3.9.3.

3.7.2 Conduction in an Orthotropic Region

The transformation which resulted in Eq. (1.2.15) applies to the long rectangular rod geometry in Fig. 3.2.1(c). Here k_x and k_y apply instead, as seen in Fig. 3.7.2. The new coordinates and the resulting differential equation are

$$X = x\left(\frac{k_a}{k_x}\right)^{1/2} \quad \text{and} \quad Y = y\left(\frac{k_a}{k_y}\right)^{1/2} \tag{3.7.6}$$

$$\frac{\partial^2 t}{dX^2} + \frac{\partial^2 t}{dY^2} = 0 \tag{3.7.7}$$

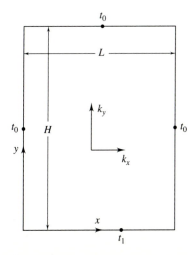

FIGURE 3.7.2
An orthotropic region.

where $k_a = \sqrt[2]{k_x k_y}$. The boundary conditions, t_1 and t_0, in terms of X and Y, are

$$t = t_0 \quad \text{at } X = 0 \text{ and at } L\left(\frac{k_a}{k_x}\right)^{1/2}$$

$$t = t_1 \quad \text{at } Y = 0 \qquad\qquad\qquad (3.7.8)$$

$$t = t_0 \quad \text{at } Y = H\left(\frac{k_a}{k_y}\right)^{1/2}$$

The solution, from Eq. (3.1.11), is obtained by replacing x and y by X and Y in Eq. (3.7.6) and the L and H by the values in Eq. (3.7.8), as

$$\phi(X,Y) = \frac{t(x,y) - t_0}{t_1 - t_0} = \frac{4}{\pi} \sum_{n=1,3}^{\infty} \frac{1}{n} \frac{\sinh[n\pi(Y/L)](k_x/k_y)^{1/2}}{\sinh[n\pi(H/L)](k_x/k_y)^{1/2}} \sin n\pi \frac{x}{L}$$

$$(3.7.9)$$

Therefore, the values of X and Y are known from Eq. (3.7.6), to calculate $t(x, y)$. Heat flux may be found directly by also using the relation between derivatives in x and y and X and Y. The numerical solution for this effect is discussed in Sec. 3.9.3.

3.8   CONTACT RESISTANCE IN MULTIDIMENSIONAL REGIONS

This effect is discussed in Sec. 2.1.4. The local contact resistance between two adjacent conduction regions, R_c, is defined in terms of the heat flux across the contact region and the resulting temperature decrease Δt_c. This decrease is taken as that between the trends of the temperature variations in the two conduction regions a and b. See the idealization in Fig. 2.1.4. Therefore,

$$R_c = \Delta t_c / q_n'' \qquad\qquad (2.1.35)$$

where q_n'' is the heat flux component locally normal to the contact region and R_c is regarded as a property of the contact region.

If the contact region is not flat, as it is between two plane regions, the preceding formulation applies as shown in Fig. 2.1.5. The resulting relation between the two gradients in regions a and b is

$$q_n'' = -k_a \left(\frac{dt}{dn}\right)_a = -k_b \left(\frac{dt}{dn}\right)_b = \frac{\Delta t_c}{R_c} \qquad (2.1.36)$$

This evaluation assumes that the average thickness of the contact region, as Y

in Fig. 2.1.5, is small relative to the local radius of curvature, r_c, shown in Fig. 2.1.5.

Equation (2.1.36) amounts to the conservation of the normal component of the heat flux across the contact region. This restriction applies generally, in steady state or when the contact region contents have negligible local thermal capacity, c'', per unit of contact region area. If c'' is not negligible, then it must be taken into account in unsteady processes, in terms of contact region temperature, $t_c(\tau)$, as

$$-k_a\left(\frac{dt}{dn}\right)_a = -k_b\left(\frac{dt}{dn}\right)_b + c''\frac{dt_c}{d\tau} \qquad (3.8.1)$$

In the preceding equation the contact region contents at each location are assumed to have uniform temperature $t_c(\tau)$ at any time. The magnitude of this effect depends on the magnitude of $c''(dt_c/d\tau)$, compared to the local normal flux component.

For contact regions filled with an appreciable layer of another material, the changing conditions must also account for a transient temperature field in this material. Then the interpretation of the contact region property, in terms of R_c in Eq. (2.1.35), is in terms of the detailed nature of the contact region material.

The previous considerations apply to the normal component, q_n'', of the local heat flux vector $\bar{q}''$. Any components, in regions a and b, parallel to the contact region, are each locally conserved in those regions. The conservation considerations include the local steady or unsteady conduction in each region. These are governed by the Laplace or Fourier equation, when $q''' = 0$. These temperature fields must also correspond to the instantaneous temperature distribution, over the extent of the contact area between regions a and b. These considerations require the matching of the temperature fields in regions a and b over the intervening contact region.

This will be done, as an example, in a two-dimensional region in rectangular coordinates, in Sec. 3.9.4. Section 3.9 outlines the simplest numerical formulation of the basic processes in multidimensional steady-state conduction. These also apply to analogous mass diffusion.

3.9 NUMERICAL MODELING OF STEADY-STATE PROCESSES

Chapter 2 and the preceding sections of this chapter consider analytical and other methods of determining steady-state temperature and concentration distributions and heat and mass transfer rates. The material relates to one-dimensional and to multidimensional regions, for various of the boundary conditions set forth in Sec. 1.4.

Solutions are obtained for many different highly regular geometries, including the effects of distributed sources. Several of the many anomalous characteristic properties of such analytical solutions are shown. Such analysis also does not extend to regions of more complicated kinds of conductivity or mass diffusivity variation. Also, even the simplest of geometric irregularities of a conduction region often make direct analysis impractical.

These and other such considerations led to the early development of numerical formulations, by Emmons (1943), Crank and Nicolson (1947), Dusinberre (1949), and Allen (1954). Subsequently, there has been much continuing further development and improvement of both the methodology and calculation procedures.

The basis of the numerical procedure is to replace the differential equations, developed in Chap. 1, with algebraic approximations of the temperature and concentration fields, in both conduction and mass diffusion processes. The differential equations apply at every location in the region. However, their approximations, in numerical form, apply only at points of finite spacing over the region.

These finite-difference, and comparable finite-element equations, internal to the region, are developed from the differential equations. Similar equations are also developed to apply at the region boundaries, in terms of the kinds of boundary conditions classified in Secs. 1.4.1 and 1.4.2. Emphasis here will be on the finite-difference methodology applied to steady-state processes.

There are several distinct aspects of the development and use of finite-difference methods in conduction and mass diffusion. These parallel the distinctions which arise in the sets of equations developed in Chap. 1. They arise in the differences in the formulations of steady and unsteady processes, and in the formulations for plane, cylindrical, and spherical descriptions of the temperature and concentration fields.

The relations are developed here for steady state. Sections 4.9 and 5.3 concern plane and multidimensional transients. Chapter 9 treats more general and additional aspects of numerical methods. The results in this section are for steady-state processes in Cartesian coordinates, as a simple first example of the methodology. The developments in Sec. 4.9, for plane-region transients, show the role of the time dependence and the resulting kinds of requirements for the stability of the numerical calculations through time. Section 5.3 considers multidimensional transients. Chapter 9 considers the cylindrical and spherical geometries, approximation errors, and many additional aspects and kinds of numerical procedures.

3.9.1 The Finite-Difference Equations

The equations which govern numerical solutions are more easily understood if the analytical and numerical methods are contrasted. For an analytical solution, the differential equation, which includes every point in the conduction region, is solved for the temperature distribution throughout the region. The numerical method uses, instead, a set of algebraic equations, which apply to only a

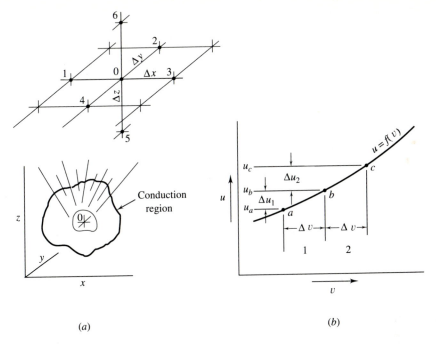

FIGURE 3.9.1
Network for numerical-method analysis.

network of points in the conduction region and at the surfaces. These equations are used to estimate the temperature at these points. These two are different types of results and may be used differently.

Since most one-dimensional steady-conduction circumstances are readily solved analytically, the techniques of numerical methods are developed here for multidimensional conduction, for isotropic materials having uniform thermal conductivity. The Poisson equation, the general equation, in steady state, is replaced by its equivalent "finite-difference" form. This in turn governs the numerical calculations. This finite-difference form is developed for the network point 0 in Fig. 3.9.1(a). The transient term is also included here for later reference:

$$\frac{\partial^2 t}{\partial x^2} + \frac{\partial^2 t}{\partial y^2} + \frac{\partial^2 t}{\partial z^2} + \frac{q'''}{k} = \frac{1}{\alpha}\frac{\partial t}{\partial \tau} \tag{3.9.1}$$

The three spatial second derivatives are estimated in terms of the six temperatures to which t_0 is directly related, $t_1, \ldots, t_6$.

Consider, as shown in Fig. 3.9.1(b), a function $u = f(v)$, which is continuous around v. The first derivative at point b is estimated as the change in u

divided by the change in v as

$$\frac{du}{dv} \approx \frac{\Delta u_1 + \Delta u_2}{2\,\Delta v} = \frac{u_c - u_a}{2\,\Delta v} \tag{3.9.2}$$

This estimate of the first derivative is based upon both the right and left intervals, 1 and 2, around the specific point of interest b. This is called the central-difference estimate. Two "one-sided" estimates are possible. The one based on the backward interval in v, denoted as 1, is called the backward difference. The one based on the forward interval, 2, is called the forward difference. These two estimates are

$$\frac{du}{dv} \approx \frac{\Delta u_1}{\Delta v} = \frac{u_b - u_a}{\Delta v} \tag{3.9.3}$$

$$\frac{du}{dv} \approx \frac{\Delta u_2}{\Delta v} = \frac{u_c - u_b}{\Delta v} \tag{3.9.4}$$

All three of the preceding estimates are only approximations to the value of the slope at point b. Clearly, as Δv becomes smaller, all three estimates become better. The error incurred may be evaluated by expanding u in a Taylor series in v, around point b. The approximation is then seen, for both the backward- and forward-difference estimates, to be due to taking only the first term of the series. That is, the series is truncated. This error is called "truncation error." The first neglected terms in the estimate of the derivative are of the order of Δv times d^2u/dv^2. For the central difference, a cancelation of these terms occurs and the first neglected terms are of the order of Δv^2 times d^3u/dv^3. Error analysis is considered in much more detail in Secs. 9.2 and 9.3.

An estimate of the second derivative at b is obtained by estimating the first derivative in each of the two intervals, 1 and 2, and dividing their difference by the distance between the two points where these two estimates apply, that is, at the centers of the two intervals 1 and 2:

$$\frac{d^2u}{dv^2} \approx \frac{\Delta u_2/\Delta v - \Delta u_1/\Delta v}{\Delta v} = \frac{\Delta u_2 - \Delta u_1}{(\Delta v)^2}$$

$$= \frac{(u_c - u_b) - (u_b - u_a)}{(\Delta v)^2} = \frac{u_a + u_c - 2u_b}{(\Delta v)^2} \tag{3.9.5}$$

The first term of the truncation error for this estimate is of the order of Δv^2 times d^4u/dv^4.

For the previous estimates of both first and second derivatives, the truncation error may be decreased by making Δv smaller. Other possibilities also exist. Recall that the way these estimates will be used is in equations such as the transient conduction equation, where Δv in Eqs. (3.9.2)–(3.9.4) is $\Delta \tau$ and in Eq. (3.9.5) is Δx. The true error is the net error in this equation in terms of

$\Delta\tau$ and Δx together. In some cases, the two truncation errors tend to cancel each other. Also, in a transient, the second derivative may be estimated at different time levels, that would be at point a, b, or c in Fig. 3.9.1(b), for various advantages. For additional discussion of these matters, see Chap. 9, Richtmyer and Morton (1967), and Özisik (1980). Stability requirements and round-off errors are very important considerations in transients.

Considering here steady-state conduction, the preceding estimate of a second derivative, in Eq. (3.9.5), is used to replace those derivatives in Eq. (3.9.1). The curve in Fig. 3.9.1(b) may be looked upon, for example, as the trace of t versus x at constant y and z. From Fig. 3.9.1(a), values of u_a, u_b, and u_c then become t_1, t_0, and t_3, and Δv becomes Δx. Therefore,

$$\frac{\partial^2 t}{\partial x^2} \approx \frac{(t_3 - t_0)/\Delta x - (t_0 - t_1)/\Delta x}{\Delta x} = \frac{t_1 + t_3 - 2t_0}{(\Delta x)^2} \qquad (3.9.6)$$

This estimate is applied to directions y and z in turn. Equation (3.9.1), neglecting the transient term, may then be replaced by the following finite-difference estimate, where q_0''' is the generation rate at point 0:

$$\frac{t_1 + t_3 - 2t_0}{(\Delta x)^2} + \frac{t_2 + t_4 - 2t_0}{(\Delta y)^2} + \frac{t_5 + t_6 - 2t_0}{(\Delta z)^2} + \frac{q_0'''}{k} = 0 \qquad (3.9.7)$$

If $\Delta x = \Delta y = \Delta z$,

$$t_1 + t_2 + t_3 + t_4 + t_5 + t_6 - 6t_0 + \frac{q_0'''(\Delta x)^2}{k} = R_0 \approx 0 \qquad (3.9.8)$$

In most calculations, a large number of equations like Eq. (3.9.8) must be solved simultaneously, to give all of the unknown grid-point temperatures in the region of interest. Many calculation procedures are, in some measure, approximate. The set of equations of the form in Eq. (3.9.8) are then not satisfied exactly. Therefore, the remainder R_0 is not zero, as projected in Eq. (3.9.8). It remains at some small residual value at each initially unknown temperature location, as written in Eq. (3.9.8). For two-dimensional conduction regions, the equation becomes

$$t_1 + t_2 + t_3 + t_4 - 4t_0 + \frac{q_0'''(\Delta x)^2}{k} = R_0 \qquad (3.9.9)$$

The scheme of point numbering shown in Fig. 3.9.1 and used for the subscripts in the preceding three equations may be applied to each of the network points (or node or grid points) of a conduction region, as they are considered in turn. Point 0 is simply a typical node point. However, in the general expression of a numerical approach to a problem, it is convenient to adopt an indexing system which attaches a unique subscript to each point. This is done in terms of a subscript on t, three indices being required for a three-dimensional region. That is, t_0 is written as $t_{i,j,k}$, where the integers i, j, and k each number in one of the three coordinate directions. In this form, Eqs. (3.9.8) and (3.9.9) become

$$t_{i+1,j,k} + t_{i-1,j,k} + t_{i,j+1,k} + t_{i,j-1,k} + t_{i,j,k+1} + t_{i,j,k-1}$$

$$- 6t_{i,j,k} + \frac{q'''_{i,j,k}(\Delta x)^2}{k} = R_{i,j,k} \tag{3.9.10}$$

In two dimensions:

$$t_{i+1,j} + t_{i-1,j} + t_{i,j+1} + t_{i,j-1} - 4t_{i,j} + \frac{q'''_{i,j}(\Delta x)^2}{k} = R_{i,j} \tag{3.9.11}$$

Estimates of the first derivative, $\partial t/\partial \tau$, for conduction transients, lead to a different temperature $t_{i,j,k}$ at each node point at successive times, at intervals of $\Delta \tau$. A time indexing system is used, in the form of a single superscript n. That is, we write $t^n_{i,j,k}$ for the temperature at node point i, j, k at time level n. In this form, the forward-difference first derivative estimate, Eq. (3.9.4), becomes

$$\frac{\partial t}{\partial \tau} \approx \frac{t^{n+1}_{i,j,k} - t^n_{i,j,k}}{\Delta \tau} \tag{3.9.12}$$

where this is, in effect, an estimate of $\partial t/\partial \tau$ at time level $n + 1/2$.

OTHER COORDINATE SYSTEMS AND GEOMETRIES. The foregoing finite-difference techniques may be applied in cylindrical (r, θ, z) and spherical polar coordinates (r, θ, ϕ) as well. The various finite-difference estimates of derivatives are developed in Sec. 9.2.2. Here, the network or grid systems will be indicated, whereby the relations for steady-state conduction in cylindrical and spherical polar coordinates are converted to finite-difference form. The systems of subdivision are shown in Figs. 3.9.2 and 3.9.3.

Consider first the two-dimensional cylindrical coordinates, r and θ. The space network is made by generating a group of symmetric cylindrical surfaces, each at constant r, spaced at Δr. Then radial planes of constant θ, angularly spaced by $\Delta \theta$, are added. The value of Δr between each pair of adjacent surfaces is chosen as $r \Delta \theta$. The value of $\Delta \theta$ does not vary with r. Therefore, these surfaces are more widely spaced at larger r. The resulting network, locally, has elements with sides of approximately equal length Δr. In this arrangement, the appropriate finite-difference equation is the same as that which applies in a two-dimensional Cartesian coordinate geometry, as Eqs. (3.9.9) and (3.9.11), for $\Delta x = \Delta r$. In three-dimensional cylindrical conduction

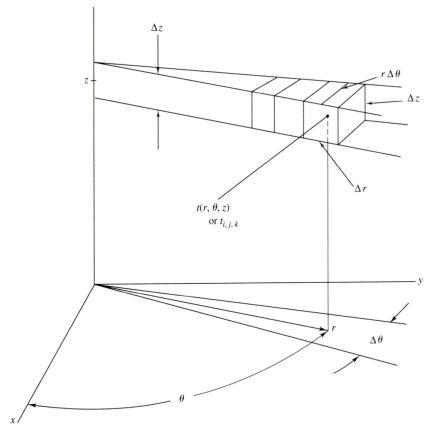

FIGURE 3.9.2
Numerical subdivision of a region described in cylindrical coordinates.

regions, Δz must necessarily be constant for all r. Therefore, the typical resulting three-dimensional network elements do not have equal-length sides at different r. Then the finite-difference equations are similar to those in Cartesian coordinates for nonsquare or noncubical networks, as will be shown in the following discussion.

The network subdivision is seen in Fig. 3.9.3 for spherical coordinates. Now Δr is the distance between concentric spherical surfaces and $\Delta\theta$ and $\Delta\phi$ are subdivisions of θ and ϕ. If $\Delta\theta$ and $\Delta\phi$ are chosen as $r\,\Delta\theta = \Delta r = r\,\Delta\phi$, the edges of each element are approximately equal. Therefore, the node points i, j, k, at the center of each element, lie in an approximately cubical array, locally. However, this and the similar method of subdivision in a cylindrical region, may not be used very near the origin of r.

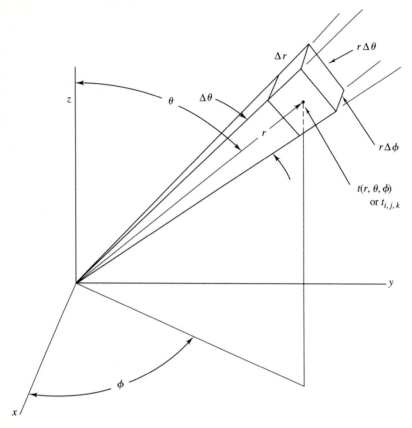

FIGURE 3.9.3
Numerical subdivision of a region described in spherical coordinates.

RECTANGULAR NETWORK ELEMENTS. Many such geometries are not simply represented with square elements. For example, a long rod having an L-shaped cross section may have dimensions which make a square network inconveniently fine. Then a coarser network of rectangular form may be used instead. With $\Delta x \neq \Delta y$, the finite-difference equation, analogous to Eq. (3.9.9), becomes

$$t_1 + t_3 + \left(\frac{\Delta x}{\Delta y}\right)^2 (t_2 + t_4) - 2\left[1 + \left(\frac{\Delta x}{\Delta y}\right)^2\right] t_0 + \frac{q_0'''(\Delta x)^2}{k} = R_0 \quad (3.9.13)$$

Nonuniform networks are also commonly used, as discussed in Chap. 9.

A similar circumstance arises for long rods of triangular, hexagonal, or other polygonal shape. Special relations may be written for each type of geometry. These questions, as well as those concerning irregular shapes, are discussed in some detail in Schneider (1955), Arpaci (1966), Özisik (1980), and Chap. 9.

3.9.2 The Relaxation Method

The finite-difference equation applies to every network point of initially un-known temperature and expresses a relation between the temperature at any point and the temperatures at the surrounding points. That is, the equation expresses the temperature at every point in terms of its surrounding tempera-tures. Since, in any given geometry, one such equation may be written for each unknown network point, there are as many equations as unknown temperatures. The solution of this system of equations is then the finite-difference estimate of the temperature distribution.

A common feature in the use of the numerical method is the simultaneous solution of a large system of linear equations. The number of grid points may be too large for an exact solution of the full set of equations. The number of operations in obtaining a solution to m equations of this type is proportional to the order of m^3. Therefore, approximate answers are obtained by various numerical procedures, carried out in different ways. The relaxation method discussed here, in simplest terms, is an iterative numerical procedure. It is demonstrated in the following discussion for a simple example. Many methods are available for computer calculations, for m large, as briefly discussed in Sec. 9.4.1.

In the relaxation procedure, the unknown temperatures at the grid points may be initially guessed. These guesses do not, in general, satisfy the finite-dif-ference equations for the network points. That is, the left-hand side in Eq. (3.9.8), or Eq. (3.9.9), does not sum to zero. The amount of this sum at any point is termed the "residual," R_0, at that point. For example, in three dimensions,

$$t_1 + \cdots + t_6 - 6t_0 + \frac{q_0'''(\Delta x)^2}{k} = R_0 \tag{3.9.14}$$

Each point has its residual, and the solution is approached by reducing all residuals toward zero by altering the temperatures in the network. The smaller and more randomly distributed the residuals are, the more accurate the esti-mate of the solution.

The numerical method and the heat conduction results are first demon-strated by a simple example. Consider a 2-by-2-ft metal duct insulated ($k = 0.1$ Btu/hr ft °F) as shown in Fig. 3.9.4(a). The temperature distribution and heat loss per unit length will be found for inside and outside insulation surface temperatures of 1100 and 100°F. The accuracy of the results is higher with smaller grid sizes, and this effect will be demonstrated by comparing the results for grid sizes of $\sqrt{2}$, $\sqrt{2}/2$, and $\sqrt{2}/4$ ft.

The solution need not be carried out for the whole region. The smallest representative portion is the part between adjacent planes of symmetry. That is, one-eighth of the cross section for this geometry. For $\Delta x = \Delta y$ and $q_0''' = 0$, the residual equation is

$$t_1 + t_2 + t_3 + t_4 - 4t_0 = R_0$$

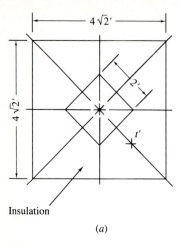

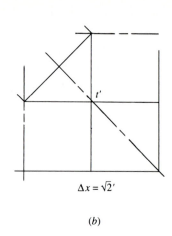

Insulation

(a)   (b)

FIGURE 3.9.4
An insulated duct.

With $\Delta x = \sqrt{2}$ ft, in Fig. 3.9.4(b), there is only one unknown temperature, t'. This may be found directly as

$$1100 + 1100 + 100 + 100 - 4t' = 0$$

$$t' = 600°\mathrm{F}$$

In order to simplify the calculation, $100°\mathrm{F}$ has been subtracted from all temperatures. The result of halving the grid size, in Fig. 3.9.5(a), is that six internal points of unknown temperature arise. The accuracy of the initial guesses was improved by first sketching the three isotherms shown. The values in Fig. 3.9.5(a) are the estimates of the internal temperature distribution. These are used to estimate the heat flow rate through the conduction region. Rods of cross section Δx by Δx are placed along the outside surface, as shown in Fig. 3.9.5(b). The temperatures at the network points are assumed to be the average temperatures on the inside surfaces of the rods. Assuming one-dimensional conduction through the rod, the heat flow per unit length for a complete rod is

$$\frac{k \, \Delta x \, \Delta t}{\Delta x} = k \, \Delta t$$

The total heat flow per unit length is the sum through all the rods, times 8 for the whole region. Note that a $1/2$ appears for the half-rod at the plane of symmetry.

$$\frac{q}{L} = 8k[(418 - 0)/2 + (336 - 0) + (228 - 0) + (114 - 0)]$$

$$= 710 \text{ Btu/hr ft}$$

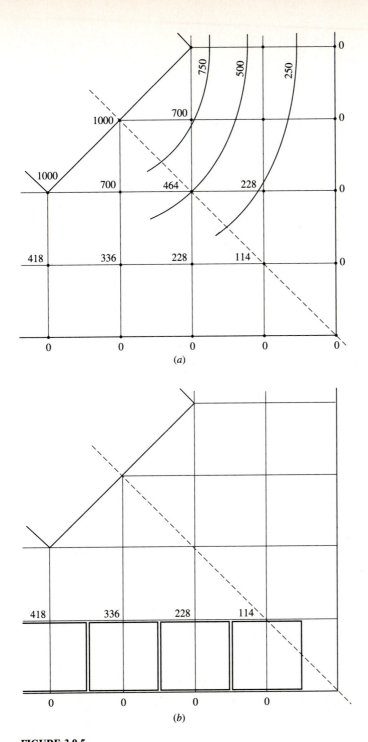

FIGURE 3.9.5
The results for a finer grid, $\Delta x = \sqrt{2}/2$: (a) internal temperatures; (b) rods for heat flow calculation.

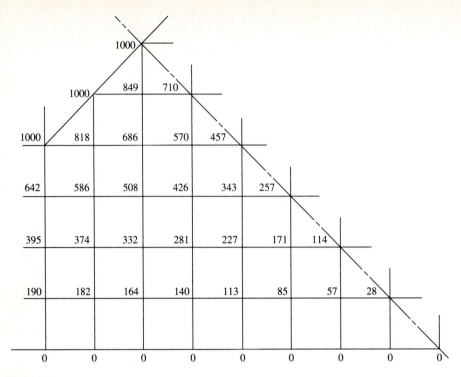

FIGURE 3.9.6
Results for $\Delta x = \sqrt{2}/4$.

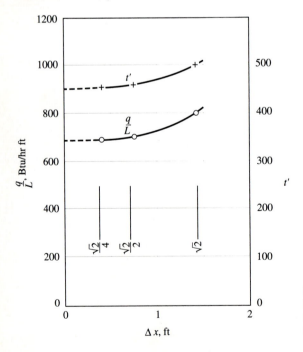

FIGURE 3.9.7
Dependence of the temperature distribution and conduction heat transfer rate on network size.

The heat flow estimate for the $\sqrt{2}$-ft grid, Fig. 3.9.4(b), is

$$\frac{q}{L} = 8k[(1000 - 0)/2 + (500 - 0)]$$

$$= 800 \text{ Btu/hr ft}$$

Better estimates of the temperature distribution and heat flow may be obtained by using even smaller grids. The resulting distribution for $\Delta x = \sqrt{2}/4$ ft is shown in Fig. 3.9.6. The heat flow estimate is 690 Btu/hr ft. A graphical method called "curvilinear squares," see Schneider (1955) for this method (page 138) and a fluid-flow analog, have given values of 686 and 660, respectively.

Comparison of t' and the heat flow rate for the various solutions shows a large change in the first halving of the grid size and a small change in the second. These trends are clearer in Fig. 3.9.7. The extrapolation to a Δx of zero is an estimate of the exact solution.

3.9.3 Temperature-Dependent and Orthotropic Conductivity

Both of these effects are simply accommodated in terms of the transformations in Eqs. (1.2.14) and (1.2.15). The first, the Kirchhoff transformation, exchanges V for t. The other transformation scales the coordinates x, y, and z in terms of the directional components of k, k_x, k_y, and k_z. Each procedure is shown in the following discussion.

TEMPERATURE-DEPENDENT CONDUCTIVITY. If the thermal conductivity in an isotropic material is temperature dependent, the conduction term $\overline{\nabla} \cdot k \nabla t$ in the differential equation is nonlinear. This difficulty may be removed in steady-state problems by the method discussed in Sec. 1.2.3. The procedure is to replace the temperature t with a new variable V defined and calculated as follows:

$$dV = \frac{k(t)}{k_r} \, dt$$

$$V - V_r = \frac{1}{k_r} \int_{t_r}^{t} k(t) \, dt$$

(3.9.15)

where V_r is a value assigned to V at some convenient reference temperature t_r. The quantity k_r is taken as $k(t_r)$. With this substitution, the differential equation in steady state becomes

$$\nabla^2 V + \frac{q'''}{k_r} = 0$$

(3.9.16)

This equation is identical to Eq. (3.9.1) in steady state, in terms of temperature. Therefore, the finite-difference form is written in exactly the same way, replacing t in Eq. (3.9.7) by V. The result for a square network is Eq. (3.9.10) in terms of V:

$$V_{i+1,j,k} + V_{i-1,j,k} + V_{i,j+1,k} + V_{i,j-1,k} + V_{i,j,k+1} + V_{i,j,k-1}$$

$$- 6V_{i,j,k} + \frac{q'''_{i,j,k}(\Delta x)^2}{k_r} = R_{i,j,k} \tag{3.9.17}$$

Therefore, the solution is found, for a particular known $k(t)$, by determining V as a function of t from Eq. (3.9.15), converting t boundary conditions to V, and then proceeding as before with numerical calculations, but in terms of V. After the V distribution has been determined, it is converted back to a t distribution through Eq. (3.9.15).

AN ORTHOTROPIC REGION. The three components of heat flux are shown in Fig. 3.9.8. The differential equation, in steady state for $q''' = 0$, is

$$\frac{\partial}{\partial x}\left(k_x \frac{\partial t}{\partial x}\right) + \frac{\partial}{\partial y}\left(k_y \frac{\partial t}{\partial y}\right) + \frac{\partial}{\partial z}\left(k_z \frac{\partial t}{\partial z}\right) = 0 \tag{3.9.18}$$

The finite-difference estimates of each of the preceding terms is determined for the component heat fluxes, for example, $k_x(\partial t/\partial x)$. This is done in terms of the location indices in Fig. 3.9.1. The fluxes in the second and first intervals in Fig. 3.9.1(*b*) are differenced as follows:

$$q''_2 - q''_1 \approx -\frac{k_{x,2}(u_c - u_b)}{\Delta v} + \frac{k_{x,1}(u_b - u_a)}{\Delta v}$$

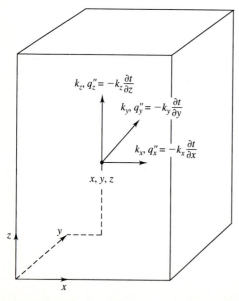

FIGURE 3.9.8
The conduction components in a three-dimensional orthotropic region.

These two flux estimates are at mid-interval. Therefore, the flux gradient at point b is

$$\frac{q_2'' - q_1''}{\Delta v} \approx -\frac{k_{x,2}(u_c - u_b)}{(\Delta v)^2} + \frac{k_{x,1}(u_b - u_a)}{(\Delta v)^2} \qquad (3.9.19)$$

If k_x is taken as uniform over the region, and u is x in Fig. 3.9.1.(a), the estimate of the first term in Eq. (3.9.18) is $-(q_2'' - q_1'')/\Delta v$, as

$$\frac{\partial}{\partial x} k_x \frac{\partial t}{\partial x} \approx k_x \frac{t_1 + t_3 - 2t_0}{(\Delta x)^2} = k_x \frac{t_{i+1,j,k} + t_{i-1,j,k} - 2t_{i,j,k}}{(\Delta x)^2}$$

Then the finite-difference form of Eq. (3.9.18) for $\Delta x = \Delta y = \Delta z$ is

$$k_x \left(t_{i+1,j,k} + t_{i-1,j,k} - 2t_{i,j,k} \right) + k_y \left(t_{i,j+1,k} + t_{i,j-1,k} - 2t_{i,j,k} \right)$$

$$+ k_z \left(t_{i,j,k+1} + t_{i,j,k-1} - 2t_{i,j,k} \right) = R_{i,j,k} \qquad (3.9.20)$$

The proper form for nonuniform conductivities k_x, k_y, and k_z is obtained directly from Eq. (3.9.19). See also Sec. 9.5.

3.9.4 Numerical Modeling of Contact Resistance

Contact resistance, R_c, between adjacent conduction regions of different conductivity is considered in Secs. 2.1.4 and 3.8. Information is given concerning the magnitude of R_c, in terms of materials and contact pressure. A general contact region is shown in Fig. 2.1.5. Equation (2.1.36) indicates the local conservation of the normal component of the heat flux across these regions, in steady-state processes.

Figure 3.9.9(a) shows a duct measuring $2L$ by $2L$. The wall and covering both have thicknesses of L. The conductivities are k_a and k_b. The inside and outside surface temperatures are t_1 and t_0. Four planes of symmetry arise in this simple example. The finite-difference subdivision is shown for $\Delta x = \Delta y = L/2$. Figure 3.9.9($b$) shows a typical segment of the cross section of the geometry.

The points representing the contact region are numbered 1, 2, 3, 4, and 5. Two temperatures are associated with each point: $t_{1,a}, t_{1,b}; \ldots; t_{5,a}, t_{5,b}$. The adjacent points in regions a and b are numbered 6 through 15. Each of these latter temperatures lie inside either region a or b. Each may be expressed in terms of the surrounding temperatures, for $q''' = 0$, by the general relation Eq. (3.9.9), where R_0 is the residual, at any of these points.

$$t_1 + t_2 + t_3 + t_4 - 4t_0 = R_0 \qquad (3.9.21)$$

The subscripts in Eq. (3.9.21) relate to the numbering scheme in Fig. 3.9.1(a), interpreted in two dimensions.

158

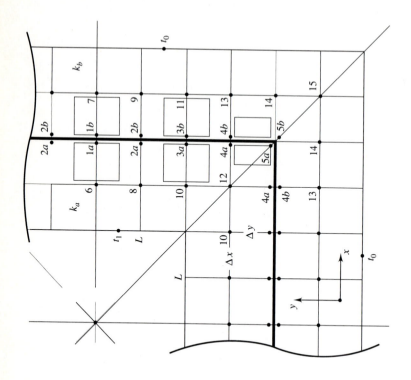

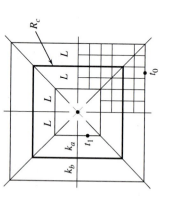

(a)

(b)

FIGURE 3.9.9

A covered duct: (a) the dimensions; (b) the network in a typical subregion.

However, two unknown temperatures arise at points 1, 2, 3, 4, and 5. Therefore, an additional equation is required for each such point, for all the unknown temperatures in the region to be found. For location 1, this additional relation arises from the requirement that the heat flux must be conserved across the contact region, as follows:

$$\frac{(t_{1,a} - t_{1,b})}{R_c} = q_1'' = \frac{k_a}{\Delta x}(t_6 - t_{1,a}) \text{ or } \frac{k_b}{\Delta x}(t_{1,b} - t_7) \qquad (3.9.22)$$

where q_1'' may be evaluated in either of the two ways written previously and R_c is the contact resistance per unit area.

At locations 2, 3, and 4 the heat flux vector $\bar{q}''$ has two components q_x'' and q_y''. Then Eq. (3.9.22) applies for the normal component q_x''. The tangential component q_y'' need not be conserved. For point 3, for example, the additional equation is

$$\frac{(t_{3,a} - t_{3,b})}{R_c} = q_x'' = \frac{k_a}{\Delta x}(t_{10} - t_{3,a}) \text{ or } \frac{k_b}{\Delta x}(t_{3,b} - t_{11}) \qquad (3.9.23)$$

Point 5, at a symmetric corner, is a special circumstance. The heat flux vector has equal components, $q_x'' = q_y''$, to be conserved. Two equations are also necessary, since $t_{5,a}$ does not appear in any other equation, as $t_{5,b}$ does, in the relation for t_{14}. There are several consistent ways to represent this condition. It will be done here in terms of the two tangential components above location 5, one in region a, the other in region b. The balance in region a is

$$\frac{(t_{5,a} - t_{5,b})}{R_c} = q_y = \frac{k_a}{\Delta x}(t_{4,a} - t_{5,a}) \text{ or } \frac{k_b}{\Delta x}(t_{5,b} - t_{14}) \quad (3.9.24a)$$

In region b it is

$$(t_{4,b} - t_{5,b}) = (t_{5,b} - t_{14}) \qquad (3.9.24b)$$

where $t_{5,b}$ appears in the residual equation for point 14 and there are two equations, for $t_{5,a}$ and $t_{5,b}$.

The procedure then is to combine the previous kinds of relations, for points 1 through 5, with the residual equations for points 6 through 15, to determine the temperature field for any values of t_1 and t_0. Then heat conduction may be determined as before. Of course, the grid in Fig. 3.9.9 is very coarse. The approximations made previously would have less severe consequences for $\Delta x = \Delta y$ much smaller.

3.9.5 Numerical Formulation of Surface Conditions

In Sec. 3.9.1 the finite-difference equations which apply at grid locations inside a conduction region are developed. The calculations are formulated, for example, in Eqs. (3.9.8), (3.9.10), (3.9.13), (3.9.14), (3.9.17), and (3.9.20) for the several applications in Secs. 3.9.1 through 3.9.3. These imply that the bounding surface temperatures are specified.

The finite-difference equivalents of the kinds of boundary conditions in Sec. 1.4.1 are developed in the following discussion, for steady-state processes. The very simple result for one dimension is given first, followed by the results for two and three dimensions. Comparable results for one-dimensional and multidimensional transients are given in Secs. 4.9 and 5.3. These formulations are all in Cartesian coordinates. Other coordinate systems and conditions are considered in Chap. 9.

SURFACE CONDITIONS FOR A PLANE REGION. Figure 3.9.10 indicates a plane region divided into layers Δx thick. The material associated with each internal location, at the example locations shown as t_1 and t_2, are the layers Δx thick and centered at each location. This leaves a layer $\Delta x/2$ thick to be associated with the surface location s. An energy balance is written for this half-layer, in terms of the boundary conditions shown and the distributed source $q'''(x) \approx q'''_s$ in this region. The energy balance includes the convection and surface flux loadings, the conduction from plane 1 to plane s, and the distributed energy effect q'''_s associated with location s. Therefore,

$$h(t_e - t_s) + \frac{k}{\Delta x}(t_1 - t_s) + q''_s + q'''_s \, \Delta x/2 = 0$$

Multiplying through by $\Delta x/k$, this equation becomes the following residual equation, for location s:

$$Bt_e + t_1 + \frac{q''_s \, \Delta x}{k} + \frac{q'''_s (\Delta x)^2}{2k} - (B + 1)t_s = R_s \qquad (3.9.25)$$

where $B = h \, \Delta x/k$ is a grid or network Biot number. This simple result

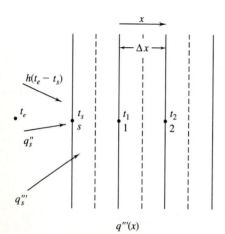

FIGURE 3.9.10
Surface conditions in one-dimensional conduction.

demonstrates the method developed in the following sections for multidimensional regions.

REGULAR TWO-DIMENSIONAL REGIONS. The three typical distinctions which arise are shown in Fig. 3.9.11, all as points s_1, surrounded by points s_2 and s_3. These are on a flat surface, on an outside edge and on an inside edge, respectively. The surface area associated with each location is Δx, per unit depth normal to the figure. However, the volumes associated with the three locations are $(\Delta x)^2/2$, $(\Delta x)^2/4$, and $3(\Delta x)^2/4$, respectively, when Δx and Δy are taken as equal, for convenience. The energy balance for the flat region point is

$$h\,\Delta x(t_e - t_s) + \frac{k\,\Delta x}{\Delta x}(t_1 - t_s) + \frac{k\,\Delta x}{2\,\Delta x}(t_2 - t_s) + \frac{k\,\Delta x}{2\,\Delta x}(t_3 - t_s)$$

$$+ q_s''\,\Delta x + \frac{q_s'''\,\Delta x^2}{2} = 0 \qquad (3.9.26a)$$

where the preceding terms, in t_2 and t_3, are the conduction, toward location s_1, through the half-rod on each side. The resulting equation is

$$Bt_e + t_1 + \frac{t_2 + t_3}{2} + \frac{q_s''\,\Delta x}{k} + \frac{q_s'''(\Delta x)^2}{2k} - (B + 2)t_s = R_s \qquad (3.9.26b)$$

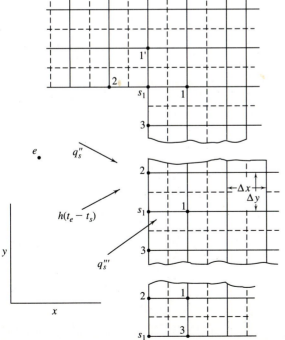

FIGURE 3.9.11
Surface conditions for two-dimensional regions.

For the outside-edge location, the balance is the same as given previously except that t_1 makes no contribution and the source effect is $q_s'''(\Delta x)^2/4$. The result, inferred from Eq. (3.9.26b), is

$$Bt_e + \frac{t_2 + t_3}{2} + \frac{q_s'' \Delta x}{k} + \frac{q_s'''(\Delta x)^2}{4k} - (B + 1)t_s = R_s \qquad (3.9.27)$$

For the inside-edge location, the changes from the flat surface location are that both regions 1 and 1' contribute and the source effect is $3q_s''' \Delta x/4$. Therefore,

$$Bt_e + t_1 + t_{1'} + \frac{t_2 + t_3}{2} + \frac{q_s'' \Delta x}{k} + \frac{3q_s'''(\Delta x)^2}{4k} - (B + 3)t_s = R_s \qquad (3.9.28)$$

These relations then supplement the set of finite-difference relations which apply to the interior network points, which are also of unknown temperature.

The preceding energy balance method also applies if the imposed surface convective and flux loading conditions, $h(t_e - t_s)$ and q_s'', are different on different segments of a particular surface element. For example, consider point s_1 on the flat surface in Fig. 3.9.11. The conditions from point s_1 to point s_2 may be different than those from s_1 to s_3. Then the convection and flux terms in Eq. (3.9.26a) each become two terms, each applying to a surface edge length of $\Delta x/2$. A similar procedure applies to a location where the condition becomes that which applies for an insulating condition. Then only one of these terms remains in Eq. (3.9.26a). Thereafter, the usual insulating condition on the internal temperature difference is used.

REGULAR THREE-DIMENSIONAL REGIONS. The middle portion of Fig. 3.9.11 may also be interpreted as a section through the wall of a thick-walled box. The interior wall region is shown. The additional direction, z, normal to the figure, must be included in the surface point energy balances. The region is subdivided as $\Delta z = \Delta x = \Delta y$. For example, for s_1 on the flat surface, there are four surrounding surface points. They are points 2 and 3 shown, plus points 4 and 5 at Δz and $-\Delta z$, respectively. The surface area and volume associated with s_1 are $(\Delta x)^2$ and $(\Delta x)^3/2$. The energy balance is

$$h(\Delta x)^2(t_e - t_s) + \frac{k(\Delta x)^2}{\Delta x}(t_1 - t_s) + \frac{k(\Delta x)^2}{2\Delta x}(t_2 + t_3 + t_4 + t_5 - 4t_s)$$

$$+ q_s''(\Delta x)^2 + \frac{q_s'''(\Delta x)^3}{2} = 0 \qquad (3.9.29a)$$

$$\text{or} \quad Bt_e + t_1 + \frac{t_2 + t_3 + t_4 + t_5}{2} + \frac{q_s'' \Delta x}{k} + \frac{q_s'''(\Delta x)^2}{2k} - (B + 3)t_s = R_s$$

$$(3.9.29b)$$

Considering the point s_1, along the outside edge at the bottom, the volume is $(\Delta x)^3/4$, the surface area is $(\Delta x)^2$, the conduction area from points 2 and 3 is

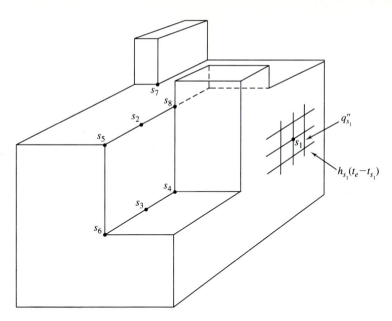

FIGURE 3.9.12
Additional kinds of surface points on three-dimensional regions.

$(\Delta x)^2/2$, and the conduction area from points 4 and 5 is $(\Delta x)^2/4$. For the point s_1, along the inside edge at the top, the volume is $3(\Delta x)^3/4$, the surface area is $(\Delta x)^2$, the two conduction areas from points 1 and 1' are $(\Delta x)^2$, from points 2 and 3 they are $(\Delta x)^2/2$, and from points 4 and 5 they are $3(\Delta x)^2/4$. This result is not given here.

In three-dimensional geometries, there are many additional kinds of surface points. The solid geometry shown in Fig. 3.9.12 shows six additional distinct kinds of conditions which require individual treatment. The three considered previously, as s_1, are points s_1, s_2, and s_3 in Fig. 3.9.12. Surface point relations for transients are considered in Secs. 4.9 and 5.3.

SUMMARY. The preceding developments show the method of determining the effects of surface conditions in square grids in Cartesian coordinates. Extension to rectangular grids is simply done. The parallel consideration of cylindrical and spherical geometries is considered in Chap. 9. Also given in Sec. 9.5.1 are methods of analysis for regions of variable conductivity.

The developments presented immediately before, and in Sec. 3.9 throughout, consider mainly regions bounded at fixed values of the space coordinates. The boundary grid points then also lie on points corresponding to the regular internal array. Irregular bounding surfaces very frequently arise. Their treatment is discussed in Sec. 9.4.2. The finite-element method, which is usually more convenient for such circumstances, is discussed in Sec. 9.6.

PROBLEMS

3.1.1. For a strip of infinite length and subject to the temperature conditions of Fig. 3.1.1, prove that the slope of the isothermals is zero at $x = L/2$, for all θ/θ_1, and zero and infinite at $x = 0$ and L, for $\theta/\theta_1 = 0$ and $\theta/\theta_1 = 1$, respectively.

3.1.2. For the temperature distribution referred to in the preceding problem, find the equation for the adiabatic surfaces in the strip.

3.1.3. A long square rod is at t_1 on the bottom surface and at $t_0 < t_1$ on the other three surfaces.
   (a) Calculate the heat transfer rate q across the surface opposite the one at t_1, in nondimensional form.
   (b) Determine if the heat flow is bounded across any of the other surfaces.

3.1.4. Make the same calculations as in Prob. 3.1.3, but for a uniformly imposed heat flux on the bottom surface instead, and the other surfaces at t_0. Also determine the characteristic temperature difference for this process.

3.2.1. For a long square rod L by L with three surfaces maintained at t_0 and the other subject to an imposed uniform heat flux, determine the equation for the isotherms in the region.

3.2.2. Consider a long metal rod of square cross section (L by L). The upper and lower faces are at t_0. The left face is induction heated; assume that the effect is a uniform flux at the surface. The right face is at t_1.
   (a) Find the temperature distribution in the rod material.
   (b) Obtain a solution for the left-face adiabatic.

3.2.3. Consider the half-infinite strip, as shown in Fig. 3.2.1(a), with the two infinite-length edges at t_0. Find several specific or limiting temperature conditions along the other boundary, $t - t_0 = f(x)$, for which the heat flow across this edge, per unit thickness, would be bounded.

3.2.4. Consider the infinite strip geometry in Fig. 3.2.1(b). Determine if the heat conduction across the region $-\infty \le y \le \infty$, per unit depth, is finite, for $f(y)$ being an exponential decay $e^{-K|y|/L}$ from $y = 0$ in both directions along the surface at $x = 0$. In the preceding exponent, $|y|$ is the absolute value of y.

3.2.5. A long square rod L by L is insulated on two adjacent faces, one other face is subjected to convection, h, to an ambient region at t_e. The remaining face is at a uniform temperature $t_1 > t_e$.
   (a) Derive an expression for the heat flow rate across the two uninsulated surfaces, in terms of t_1 and t_e.
   (b) Determine the temperature at the location where the insulated edge is in contact with the surface at t_1.

3.2.6. A long square rod has convection on two opposite faces, to an ambient fluid at t_e. The other two faces are at t_1. Develop the simplest expression which gives the temperature level at the center of the rod.

3.2.7. Find the steady temperature distribution for a rectangular bar, W by H, having the following boundary conditions:
   at $x = 0$, $t = f(y)$; at $x = W$, $t = t_0$
   at $y = 0$, $q_0'' = $ constant; at $y = H$, convection to an environment at t_e

3.3.1. Calculate the total heat flow rate across the long half-cylindrical region in Fig. 3.3.1(b), per unit length. Estimate the location of the isotherm $\phi = 1/2$.

3.3.2. Consider an infinitely long cylindrical shell of angular section ϕ_0. The inner and outer radii of the shell are R_i and R_0, respectively. The outer surface receives the net radiant heat flux $q''(\phi)$, while the inner surface is maintained at a uniform temperature, t_0. The ends of the shell at $\phi = 0$ and $\phi = \phi_0$ are insulated. Find the steady-state temperature distribution in the shell.

3.3.3. Consider a semiinfinite solid cylinder of radius R whose base is at temperature t_0 and whose periphery is exposed to a fluid at temperature t_e through a heat transfer coefficient h. Determine the steady two-dimensional temperature distribution in the cylinder.

3.3.4. Two infinite regions remotely at t_1 and t_2, of conductivity k_1 and k_2, are in perfect contact only over a small circular region of radius R. The two regions are assumed otherwise perfectly insulated from each other at their interface.
  (*a*) Determine the resistance from a remote location in each region to the assumed isothermal area of direct contact.
  (*b*) Plot the total resistance between the two distant regions as a function of $K = k_1/k_2$, over the range from $K = 0$ to ∞, in nondimensional form.
  (*c*) Plot $(t_c - t_1)/(t_1 - t_2)$ over the same range, where t_c is the uniform temperature over the direct contact region.

3.5.1. A half-infinite region of remote temperature t_∞ is insulated on its surface, except for a small circular area of radius R, which is at a temperature of t_1.
  (*a*) Calculate the shape factor for the region.
  (*b*) Determine the form of the isothermal surfaces at large distances z and r away from the surface temperature condition, compared to the radius R.

3.5.2. Consider a long tube and a spherical shell, each of inside and outside radius a and b, with surface temperatures of t_a and t_b. Calculate the shape factor for each, as S_c and S_s, and give the dimensions for each quantity.

3.5.3. For the contact conduction conditions in Prob. 3.3.4, calculate the shape factor S of each region and give its units.

3.5.4. A radioactive sample is to be stored in a protective box with 7.5-cm-thick walls and inside dimensions of 1 by 1 by 5 cm. The radiation is primarily gamma radiation which is completely absorbed at the inside surface of the box wall. This energy is conducted through the walls of the box. The walls are made of regranulated cork ($k = 0.7$ W/m °C) encased in thin metal. The outside surface of the box will be at 25°C. If the inside surface temperature may not exceed 60°C, what is the maximum permissible sample radiation rate in joules per hour?

3.5.5. For a rectangular box of wall thickness L and inside and outside surface temperatures of t_1 and t_2, calculate the shape factor, for inside edge lengths L_i of:
  (*a*) L, L, and $L/2$.
  (*b*) L, L, and $L/10$.
  (*c*) $L/10$, $L/10$, and $L/10$.

3.5.6. A rectangular box made of cork insulation has inside dimensions of 30 by 20 and 20 and a wall thickness of 10 cm. The inside and outside surface temperatures are 40 and 20°C, respectively.

(a) Calculate the total heat loss through the walls, based on the inside area, A_1.

(b) Repeat part (a), using the outside area A_2.

(c) Calculate the heat loss and shape factor from the appropriate Langmuir equation.

3.5.7. Consider conduction through the walls of thickness L of a rectangular box, in terms of S/L_1, where L_1 is the least of the three inside-edge lengths L_1, L_2, and L_3. Take $L_2/L_1 = 2$ and $L_3/L_1 = 4$. The inside and outside surface areas of the box are A_1 and A_2.

(a) Calculate S/L_1 in terms of the "average" conduction area $(A_1 + A_2)/2$.

(b) Calculate S/L_1 for $L_1/L = L/5$, from the appropriate Langmuir equation.

(c) Repeat part (b) for $L_1 = L/8$.

3.5.8. For the configuration in the preceding problem:

(a) Determine S/L_1 for each condition: $L_1 = L/8$, $L_1 = L/12$, and $L_1 = L/24$.

(b) Calculate the heat conduction rate for each configuration for an inside area A_i of 2400 cm^2, for $t_1 = 60°C$, $t_2 = 10°C$, and $k = 80$ W/m^2 °C. Compare the results.

3.5.9. Calculate the ratio of the transfer rate from the lower sphere in Fig. 3.5.3(a) to that from the single sphere in Fig. 3.5.3(b).

(a) For geometry (a) take the depth to be $D/2$ and the motive temperature difference to be $(t_1 - t_2)/2$, for a more direct comparison with D and $t_1 - t_0$ in (b).

(b) Plot this ratio, vs. D/R, and note any indicated conclusions.

3.5.10. Repeat the calculation in the preceding problem for the ratio of heat transfer for (b) and (c) in Fig. 3.5.3, using D for both, and differences $t_1 - t_0$ and $t_1 - t_\infty$, respectively.

3.5.11. A long cylinder of radius R at t_1 is buried to a center distance D in dry soil whose surface temperature is t_0.

(a) Calculate the cylinder heat loss, in ordinary soil, for $R = 1$ m and for $D = 2$ and 10 m.

(b) In this steady-state model, what is the temperature in the soil far below the cylinder?

3.5.12. Calculate the ratio of the heat flow to the surface, for the geometry in Fig. 3.5.3(d), to that of a sphere buried at a depth $D = L/2$ having an area equal to the submerged part of the cylinder.

3.5.13. Consider an off-centered passage in a steel tube as shown in Fig. 3.5.1(c).

(a) For $R_2/R_1 = 2$, plot S as a function of ϵ/R_1, over the permissible range of ϵ.

(b) Find the ratio of the heat transfer, for $\epsilon = 0$ and $R_2 - R_1 = D$, to that for a cylinder at t_1 of radius R_1, buried at a depth D below a surface at t_2.

3.5.14. Consider the buried cylinder condition in Fig. 3.5.3(e), of uniform surface flux dissipation q'' and $t_0 = 20°C$.

(a) For $D/R = 2$, calculate the shape factor, using the first few terms in the appropriate equation.

(b) For burial in soil, $R = 1$ m and $q'' = 100$ W/m^2, calculate the average surface temperature of the cylinder, t_a.

(c) Find the ratio of the heat conduction rate calculated in part (b) to that for an isothermal cylinder at a uniform surface temperature t_a, for $R = 1$ m and $D/R = 2$.

3.5.15. Consider a buried sphere with a uniformly distributed internal energy source of strength q''', as in a wet slurry of radioactive waste in a storage chamber. The internal conductivity of the sphere may be taken as that of water. The sphere is 10 m in diameter and buried 100 m deep in ordinary soil. The soil surface temperature is 20°C.

(a) For $q''' = 5$ W/m³ calculate the average temperature over the spherical region.

(b) For the same geometry find the temperatures of the surface of the spherical surface if all of the given energy dissipation is to occur on the outside surface at a uniform temperature and at a uniform flux condition.

3.6.1. A long square fresh wooden beam of 30 by 30 cm is heated by microwave radiation for drying. Assume that this heating amounts to a uniformly distributed source q'''. The surface will be kept uniformly at a temperature of 40°C by blowing air over the surfaces. Neglect any mass diffusion effects.

(a) Calculate the maximum value of q''' in W/m² so that the maximum internal temperature will not exceed 90°C.

(b) If the cooling air is at 30°C, estimate the required convection coefficient, h, necessary to keep the surface at 40°C.

3.6.2. A long rectangular rod of L by H is subject to a uniform internal distributed source of strength q'''. The surface temperature is uniformly t_0.

(a) Determine the temperature distribution in the rod.

(b) Calculate the highest temperature in the region, in terms of t_0 and q'''.

(c) Determine the fraction of the total energy input to the region which is conducted across each of the four faces.

3.6.3. Consider a semiinfinite solid cylinder of radius R in which there is a uniform heat generation q'''. The base of the cylinder is maintained at a temperature θ_0 and the periphery is exposed to a fluid at $\theta = 0$ with a resulting heat transfer coefficient h. Determine the steady two-dimensional temperature distribution in the cylinder.

3.6.4. A long square rod is subjected to a distributed internal energy source. The strength varies across the rod in one direction, say x, as $q'''(x) = q_0'''x$, where $x = 0$ at the center of the rod. The surface is maintained at t_0.

(a) Calculate the temperature distribution in the material as $\theta(x, y) = t(x, y) - t_0$.

(b) Identify an appropriate characteristic temperature difference.

(c) Determine the fraction of the total energy input which is conducted across each of the four surfaces.

3.6.5. A long square pure copper conductor, 4 by 4 cm, carries a very high and uniform electric current. It may be cooled to a uniform surface temperature of 40°C by an electrically insulating liquid circulated over its surface. Calculate the maximum current the conductor may carry if the maximum temperature in the material may not exceed 90°C.

3.6.6. A large copper plate fin is to be used to transfer heat from an experiment, at a rate of 10 W, for dissipation to air, at t_0. The arrangement is as shown in Fig. 3.6.2. The process is idealized as a uniform temperature base surface, at $y = 0$, at t_1. All other edge surfaces are uniformly at t_0.

(a) Calculate, in general terms, the rate of conduction across the base of the fin.

(b) If the temperature difference $t_1 - t_0$ may not exceed 30°C and $H/L = 2$, determine the lengths H and L, for a thickness $D = L/10$, for $h = 60$ W/m² °C.

3.7.1. Consider $k(t) = Kt$ in a long rectangular rod with one surface at t_1 and the other three at t_0, as in Fig. 3.7.1.

(a) Calculate the heat flow rate q across the surface opposite the one at t_1.

(b) Compare this result with the value of q determined for k uniform over the region at the value of k at t_0, that is, $k = Kt_0$.

(c) Consider the temperature at the center of the rod, at $x = L/2$ and $y = H/2$. Which of the preceding two postulates of k would result in the highest temperature level, assuming $t_1 > t_0$?

3.7.2. For the same geometry and conditions as in the preceding problem, take $k(t) = k_0[1 + \beta(t - t_0)]$.

(a) Determine the relation between $t(x, y)$ and the transformed variable $V(x, y)$.

(b) Write the solution in terms of $t(x, y)$.

3.7.3. For the same geometry as the preceding two problems, determine if there is any form of the temperature variation of k, that is, of $k(t)$, for which the heat flow across the lower boundary is bounded.

3.7.4. Consider the orthotropic region shown in Fig. 3.7.2. Determine if there are any values of the uniform conductivities k_x and k_y for which the heat flow across the boundary at t_1 would be bounded.

3.7.5. For the orthotropic region in Fig. 3.7.2, and for $k_y/k_x = 2$, would the central temperature be higher or lower than in an isotropic region, for $t_1 > t_0$?

3.7.6. A very long square rod, L by L, has the top face, at $y = L$, maintained at t_1. The other three faces are at t_0. The conduction region in the rod is not isotropic. It is orthotropic, with $2k_x = k_y$, where each value is uniform.

(a) Write the general formulation to determine the temperature distribution, $t(x, y)$, in the rod.

(b) Reduce the equation and boundary conditions to their simplest form.

(c) Write the solution which applies for $t(x, y)$, in the absence of a distributed energy source, in terms of $\phi(x, y)$.

(d) On a diagram, sketch the approximate isotherms $\phi = 0.75, 0.5$, and 0.25, for $k_x = k_y$, and as dashed curves for $2k_x = k_y$.

3.8.1. A long steel pipe, of inner and outer radii r_1 and r_2, and conductivity k_i is separated from a covering layer of 85% magnesia insulation by a fibrous layer of thickness δ. This layer amounts to a contact resistance of R_c, per unit area. The magnesia insulation has a conductivity of k_0 and an outside radius of r_0. The temperatures at r_1 and r_0 are t_1 and t_0, respectively.

(a) Calculate the resistance of each barrier, per unit of tube length.

(b) Determine the expression for the heat transfer rate per unit of tube length.

(c) Calculate q/L for: $r_1 = 4$ cm $= r_2/1.2 = r_3/3$, $\delta = 3$ mm, $t_1 = 200$°C, and $t_0 = 30$°C. The fibrous material in the contact region has a conductivity of six times that of air.

(d) Determine the contact region temperature difference.

3.8.2. Consider the geometry in the preceding problem to be instead spherical. Carry out the same kinds of calculations, with the same particular conditions.

3.8.3. If the geometry in Prob. 3.8.1 is instead two plane layers of thickness of 4 and 6 cm, having the same properties, make the same calculations as in parts (a), (b), (c), and (d) per unit of barrier area, and determine the heat flux.

3.8.4. A long tube of diameter D is insulated by two concentric layers of thickness D, having conductivities of k_1 and k_2. The inside and outside surface temperatures are t_1 and t_2. The contact area resistance between these layers is R_c, per unit area.

    (a) Calculate the heat transfer rate per unit of tube length.

    (b) Repeat part (a), taking each of the insulator layers as plane layers of thickness D, each having an area equal to the average of their inside and outside surface areas, respectively.

    (c) Calculate the shape factor from each result and comment on the causes of the difference.

3.8.5. Consider a long square duct of dimensions L by L covered by two layers of insulation, k_1 and k_2, with inside and outside temperatures of t_1 and t_2, as in Prob. 3.8.4. Assume a contact resistance R_c. Using the same procedure as in part (b) in Prob. 3.8.4, determine the heat flow rate per unit length.

3.9.1. Consider a long tube of inner and outer radii of R_i and R_o, subject to circumferential temperature boundary conditions $t(R_i, \theta) = f(\theta)$ and $t(R_o, \theta) = F(\theta)$. From the relevant differential equation for this conduction circumstance, develop the comparable residual equation for the temperature at an arbitrary internal grid point t_{ij}. Take i and j to be in the radial and circumferential directions, respectively.

3.9.2. A spherical shell, of inner and outer radii R_i and R_o, has surface temperature boundary conditions of $t(R_i, \phi) = f(\phi)$ and $t(R_o, \phi) = F(\phi)$. As in the preceding problem, use the relevant differential equation for this geometry and boundary conditions, to develop the residual equation for the temperature at an arbitrary internal grid point in the region, $t_{i,j}$. Take i and j in the radial and azimuthal directions, respectively.

3.9.3. For the geometry in Fig. 3.9.4 and the results given in Sec. 3.9.2, calculate the shape factor of this geometry for each of the three numerical estimates given of the total heat transfer rate per unit length.

3.9.4. Consider a long square rod at t_1 on one surface and at t_0 on the other three. The conductivity is approximately temperature dependent as $k(t) = Kt$, where t is the absolute temperature.

    (a) Formulate the numerical calculation in terms of the transformation to V, including specific values for the boundary conditions.

    (b) Indicate how the heat transfer rate across the surface at t_1 is to be calculated.

3.9.5. A wall 5 cm thick is strengthened on the outside surface by ribs 5 cm thick and 7.5 cm high, as shown in Fig. P3.9.5. The distance between ribs is 5 cm. The inside and outside temperatures are 250 and 50°C, respectively, $k = 0.9$ W/m °C.

    (a) Find the heat flow through the wall per square centimeter of flat wall area. Use a 2.5-cm-by-2.5-cm network.

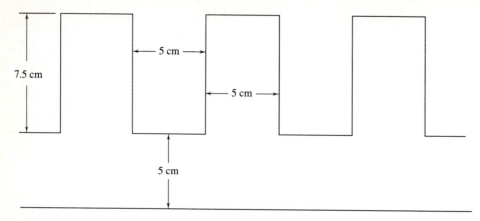

FIGURE P3.9.5

(*b*) Compare the conductance per unit area of flat area of this wall with that of similar wall without stiffeners.

(*c*) Calculate the shape factor, per unit of inside surface area.

3.9.6. A tall chimney made of brick ($k = 0.5$ W/m °C) has inside and outside cross-sectional dimensions of 2 by 1 m and 4 by 4 m. The passage is centered. If the inside and outside surfaces are at 300 and 100°C, respectively, find the rate of heat loss per foot of chimney height and calculate the shape factor of this geometry. Use a 0.5-by-0.5-m network.

3.9.7. For the passage shown in Fig. P3.9.7, compute the rate of heat transfer per unit length and the temperature distribution for surface temperatures of 500 and 100°C. Use a 15-cm square grid. $k = 0.8$ W/m °C.

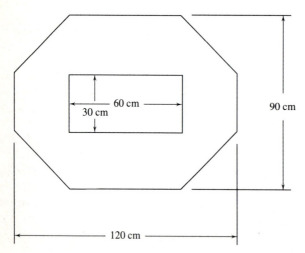

FIGURE P3.9.7

3.9.8. A 60-by-60-cm conduit carrying heated gases is insulated with 85% magnesia, as shown in Fig. P3.9.8. The inside and outside surfaces of the insulation are at 120 and 20°C.

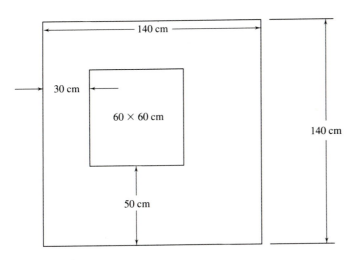

FIGURE P3.9.8

(a) Using a 10-by-10-cm grid, find the heat loss per foot of conduit length.

(b) Calculate the shape factor of this geometry.

3.9.9. Repeat Prob. 3.9.5 with $k = 0.9\{1 + 10^{-4}[t(°C) - 20]\} = 0.01(1 + 10^{-4}\theta)$, in W/m °C, and with the same temperature boundary conditions.

3.9.10. For the duct geometry in Fig. 3.9.9, the heat loss per unit length is to be found for $t_1 - t_0 = 100°C$, $k_a/k_b = 2$, and $\Delta x = L/2 = \Delta y$ and $R_c \Delta x/k_a = 1$.

(a) First do an approximate calculation by modeling each of the two layers of thickness L as a flat surface of thickness L having an area equal to its mean area A_m. That is, $A_{m,a} = 12L$ per unit length. Find q/L, summing three resistances as though they are in series.

(b) Calculate q/L numerically, without the preceding approximation, and compare the result with that in part (a).

(c) Calculate the shape factor, based on the volume average of k in the conduction region, for the results in both parts (a) and (b).

3.9.11. Consider the ribbed wall in Prob. 3.9.5, with a convection process on the outside, $h = 36$ W/m² °C, to a region at 50°C. The inner surface temperature is again taken as 250°C and $k = 0.9$ W/m °C.

(a) Write the surface point equation for each distinctive kind of point in this geometry.

(b) Using a grid of $\Delta x = \Delta y = 2.50$ cm, calculate all unknown grid point temperatures and determine the heat flux across this barrier, per unit of typical inside area.

(c) Calculate the overall conductance, U_o, per unit of inside area.

(d) Calculate U_o in the absence of the external ribs, but including convection, and explain the difference between this value and the one found in part (c).

(*e*) Calculate the ratio of the conductance in part (*c*) to that in part (*d*). What is the value of this ratio as the outside surface convection coefficient approaches zero?

3.9.12. Consider the chimney in Prob. 3.9.6, with a convection process at the inside surface, $h_i = 50$ W/m^2 °C, and at the outside surface, $h_o = 100$ W/m^2 °C.
(*a*) Calculate the heat loss per unit of chimney height.
(*b*) Determine the average temperature of the outside surface of the chimney.

3.9.13. Develop the two surface point equations which relate to inside and outside corner points, respectively, in Fig. 3.9.12.

3.9.14. For the geometry in Prob. 3.8.5, determine the heat flow, per unit length, using a numerical analysis, with $\Delta x = \Delta y = L/2$. Calculate the resulting shape factor.

REFERENCES

Allen, D. N. deG (1954) *Relaxation Methods*, McGraw-Hill, New York.

Arpaci, V. S. (1966) *Conduction Heat Transfer*, Addison-Wesley, Reading, MA.

Attwood, S. S. (1949) *Electric and Magnetic Fields*, John Wiley, New York.

Bassani, J. L., M. W. Nansteel, and M. November (1987) *Int. J. Heat Mass Transfer*, **30**, 903.

Bau, H. H. (1982) *Int. J. Heat Mass Transfer*, **25**, 1701.

Carslaw, H. S., and J. C. Jaeger (1959) *Conduction of Heat in Solids*, 2nd ed., Oxford Univ. Press.

Crank, J. (1975) *The Mathematics of Diffusion*, Clarendon Press, Oxford.

Crank, J., and P. Nicolson (1947) *Proc. Cambridge Philos. Soc.*, **43**, 50.

Dusinberre, G. M. (1949) *Numerical Analysis of Heat Flow*, McGraw-Hill, New York.

Eckert, E. R. G., and R. M. Drake, Jr. (1972) *Analysis of Heat and Mass Transfer*, McGraw-Hill, New York.

Emmons, H. W. (1943) *Trans. ASME*, **65**, 607.

Hahne, E., and U. Grigull (1974) *Int. J. Heat and Mass Transfer*, **17**, 267.

Hahne, E., and U. Grigull (1975) *Int. J. Heat and Mass Transfer*, **18**, 751.

Jakob, M. (1949) *Heat Transfer*, John Wiley and Sons, New York.

Jakob, M., and G. A. Hawkins (1957) *Elements of Heat Transfer*, 3rd ed., John Wiley and Sons, New York.

Langmuir, I., E. Q. Adams, and F. S. Miekle (1913) *Trans. Amer. Electrochem. Soc.* **24**, 53.

Moran, M. J. (1970) *4th Int. Heat Transfer Conf.*, Paris, Vol. 1, Cu 1.4.

Nansteel, M. W., S. S. Sadhal, and P. S. Ayyaswamy (1986) *ASME Winter Annual Meeting*, Anaheim, CA, HTD-Vol. 60.

Özisik, M. N. (1980) *Heat Conduction*, John Wiley and Sons, New York.

Richtmeyer, R. D., and K. W. Morton (1967) *Difference Methods for Initial Value Problems*, 2nd ed., Wiley Interscience, New York.

Rohsenow, W. M., J. P. Hartnett, and E. Ganic, Eds. (1985) *Handbook of Heat Transfer Fundamentals*, McGraw-Hill, New York.

Schneider, P. J. (1955) *Conduction Heat Transfer*, Addison-Wesley, Reading, MA.

Thiyagarajan, R., and M. M. Yovanovich (1974) *J. Heat Transfer*, **96**, 249.

Weihs, D., and R. D. Small (1978) *6th Int. Heat Transfer Conf.*, **3**, 285.

Yovanovich, M. M. (1973) AIAA 11th Aerospace Sciences Meeting, Washington, DC.

Yovanovich, M. M. (1977) AIAA 12th Thermophysics Conference, Albuquerque, NM.

CHAPTER
4

UNSTEADY PROCESSES IN ONE DIMENSION

4.1 METHODS OF ANALYSIS

Chapter 1 sets forth the basic equations and typical bounding conditions for conduction heat transfer and for mass diffusion. The simplest kinds of mass transfer processes are analogs of such conduction heat transfer processes.

Chapter 2 concerns one-dimensional steady-state processes in plane, cylindrical, and spherical regions. In addition to simple conductive transport, the effects of composite barriers and interfacial resistance are analyzed. Also considered in detail are the effects of a variation of thermal conductivity over one-dimensional regions, due to temperature and also to spatial inhomogeneities in the region. The effects of distributed heat and diffusing chemical species sources, q''' and C''', are then analyzed.

Chapter 3 then gives results from the analysis of steady-state conduction processes in two and three dimensions. The effects of distributed sources, variable conductivity, and contact resistance are analyzed. The method of shape factors is also discussed. Formulations of numerical-method representations are also given. This includes the approximating measures used internal to the conduction region and the generalized surface conditions, including temperature conditions, convection, heat flux loading, and radiative exchange.

Therefore, Chaps. 1 through 3 cover the basic mechanisms and geometries of steady-state heat conduction, with some parallel consideration of comparable mass diffusion processes.

This chapter introduces the additional mechanisms which arise when the temperature field in the conduction region is time dependent. One-dimensional processes are considered here. Multidimensional processes are treated in Chap. 5.

The rest of this section presents characteristic general methods commonly used in calculating unsteady temperature fields in conduction regions. Unsteady-state conduction, both transient and periodic, is very important in many applications of heat transfer. Designers in technological areas are often faced with start-up, operating, and instability transients. These must be understood sufficiently well to guide material selection, for example, in solid-fuel rocket nozzles, in reentry heat shields, in reactor components, and in combustion devices. The consideration may relate to the temperature limitations of materials, to heat transfer characteristics, or to the thermal stressing of materials, which may accompany changing temperature distributions.

Unsteady-conduction mechanisms are also very important in the many earth sciences due to the ever-changing effects of solar radiation and atmospheric conditions. For example, both daily and seasonal temperature changes cause complicated time-dependent temperature variations in the soil. Important questions in geophysics are analyzed on the basis of conduction mechanisms in the steady state as well as in terms of transients. The growth characteristics of ice in soil and on the surfaces of bodies of water are also considered in terms of conduction and mass diffusion mechanisms.

There are two basically different general kinds of unsteady processes. One is a transient, wherein the temperature field changes with time, from an initial condition, toward an eventual steady state. An example arises when an object at an initial temperature of t_i throughout is immersed in a surrounding at a different temperature t_e. The temperature difference decays with time.

Another common process is a periodic, in which the temperature at each location in the region continues to vary periodically with time. This arises approximately on the surface layer downward into soil, due both to annual and daily variations of atmospheric conditions. The annual periodic component has a time scale of 365 days, whereas the daily period is 24 hr. Another example is the cylinder wall temperature distribution during the cyclic operation of an internal combustion engine. Then the periodic time is 10^{-3} min for a frequency of 1000 rpm.

There are also additional more complicated and important processes. The imposed surface temperature conditions on the conduction region may be periodic. However, the average temperature level of the imposed condition on the region may also change with time. A very common example is the initial warm-up period of an internal combustion engine. Then the frequency may also increase with time. A similar simpler process is a periodic condition at the boundary of a region, which has a constant average internal temperature gradient.

This chapter analyzes both transients and the simpler steady periodics. Steady periodics are the simplest condition, in which the same imposed periodic driving conditions have continued over a long period of time. Then the average temperature level at each location in the region is constant through time.

Sections 4.2 and 4.3 consider one-dimensional transients in plane, cylindrical, and spherical regions. Periodics are then analyzed in Sec. 4.4. Section 4.5 concerns transients in plane composite and interacting barriers. Section 4.6 analyzes the propagation of a phase change through a plane region, commonly as a melting or freezing front. The similar mechanism of a moving energy input source, as in welding, is formulated in Sec. 4.8. Section 4.9 then develops the simplest procedures for the numerical analysis of time-dependent processes. These are in terms of both region internal temperature changes and for the surface conditions to which the region is subject. This formulation includes internal distributed sources and the requirements for the stability of continuing numerical calculations through time. The remainder of this section considers the application of the methods of separation of variables and Laplace transforms to the solution of one-dimensional unsteady-conduction processes.

The solutions given in Secs. 4.2 through 8 are idealized in the forms of bounding conditions and region geometries. Nevertheless, many of the solutions given provide both detailed information and much valuable insight into comparable practical applications. For example, no plane solid is actually semiinfinite. However, for a short time after the surface temperature of one side of a plate of thickness L is changed, the propagating temperature field into material is largely independent of the presence of the other surface, at $x = L$. There are many other such examples.

4.1.1 Separation of Variables

This method was applied to the Laplace equation in Secs. 3.1.1 and 3.1.2, to determine the forms of admissible steady-state temperature fields, $t(x,y,z)$, $t(r,\theta,z)$, and $t(r,\theta,\phi)$, in rectangular, cylindrical, and spherical coordinates. In unsteady conduction, the Fourier equation applies:

$$\nabla^2 t = \frac{1}{\alpha}\frac{\partial t}{\partial \tau} \qquad (1.2.10)$$

In one-dimensional processes, the forms of this equation for the three geometries are

$$\frac{\partial^2 t}{\partial x^2} = \frac{1}{\alpha}\frac{\partial t}{\partial \tau} \quad \text{for } t(x,\tau) = X(x)F(\tau) \qquad (4.1.1)$$

$$\frac{\partial^2 t}{\partial r^2} + \frac{1}{r}\frac{\partial t}{\partial r} = \frac{1}{\alpha}\frac{\partial t}{\partial \tau} \quad \text{for } t(r,\tau) = \bar{R}_c(r)F(\tau) \qquad (4.1.2)$$

$$\frac{\partial^2 t}{\partial r^2} + \frac{2}{r}\frac{\partial t}{\partial r} = \frac{1}{\alpha}\frac{\partial t}{\partial \tau} \quad \text{for } t(r,\tau) = \bar{R}_s(r)F(\tau) \qquad (4.1.3)$$

These relations are used to determine the forms of $X(x)$, $R_c(r)$, and $R_s(r)$. It will be seen that the form of $F(\tau)$ is the same in all three geometries.

A PLANE REGION. The substitution of $t(x, \tau) = X(x)F(\tau)$ into Eq. (4.1.1) results in the following relation:

$$\frac{\partial^2 X}{\partial x^2} = \frac{1}{\alpha}\frac{\partial F}{\partial \tau} \qquad (= \pm\beta^2) \qquad (4.1.4)$$

As a result of the definitions of X and F, the left-hand side may be a function only of x and the right-hand side a function only of F. Since x and τ are independent, neither side is a function of x or τ and, therefore, both sides are equal to a constant β^2, or $-\beta^2$, as

$$X'' = \pm\beta^2 X$$

$$F' = \pm\alpha\beta^2 F$$

The solutions of these equations are

$$\text{For } +\beta^2: \quad X(x) = C_1 e^{\beta x} + C_2 e^{-\beta x} \qquad (4.1.5)$$

$$\text{For } -\beta^2: \quad X(x) = C_3 e^{i\beta x} + C_4 e^{-i\beta x}$$

$$= C_3 \cos \beta x + C_4 \sin \beta x \qquad (4.1.6)$$

$$\text{For } \pm\beta^2: \quad F(\tau) = C_5 e^{\pm\alpha\beta^2\tau} \qquad (4.1.7)$$

The preceding forms anticipate the various types of processes which may be encountered. The separation constant, β^2, may be real, imaginary, or complex. The choice is based upon the conditions to be satisfied in a given circumstance. For β^2 real, the x dependence is either exponential $(+\beta^2)$ or periodic $(-\beta^2)$, the latter permitting an infinite-series expansion of an initial condition. The accompanying τ dependence is an exponential growth or decay of the temperature field, to an imposed boundary condition. For β^2 imaginary, the time dependence is purely periodic, and the x dependence for this choice results in combined periodic and exponential behavior in x. For β^2 complex, the τ- and x-dependent behaviors are each a combination of exponential and periodic effects. Thus the form of $\pm\beta^2$ is chosen in terms of the kind of process of interest, as seen, for example, in Secs. 4.2 and 4.4.

A CYLINDRICAL REGION. The use of Eq. (4.1.2) with $t(r, \tau) = \overline{R}_c(r)F(\tau)$ results in

$$\frac{1}{\overline{R}_c}\left(\frac{d^2\overline{R}_c}{dr^2} + \frac{1}{r}\frac{d\overline{R}_c}{dr}\right) = \frac{1}{\alpha F}\frac{dF}{d\tau} = \pm\beta^2 \qquad (4.1.8)$$

The equations for $\overline{R}_c(r)$ and $F(\tau)$ are

$$\frac{1}{\overline{R}_c}\left(\frac{d^2\overline{R}_c}{dr^2} + \frac{1}{r}\frac{d\overline{R}_c}{dr}\right) = \pm\beta^2 \qquad (4.1.9)$$

$$F(\tau) = C_c e^{\pm\alpha\beta^2\tau} \qquad (4.1.10)$$

The solution for $F(\tau)$ is of the same form as in Eq. (4.1.7) for the plane region. Similar comments arise for the forms of time dependence which Eq. (4.1.10) may satisfy. The solutions for $\overline{R}_c$ also admit many forms of r dependence.

A SPHERICAL REGION. Equation (4.1.3), in terms of $t(r, \tau) = \overline{R}_s(r)F(\tau)$, becomes

$$\frac{1}{\overline{R}_s}\left(\frac{d^2\overline{R}_s}{dr^2} + \frac{2}{r}\frac{d\overline{R}_s}{dr}\right) = \frac{1}{\alpha F}\frac{dF}{d\tau} = \pm\beta^2 \tag{4.1.11}$$

The equations for $\overline{R}_s(r)$ and $F(\tau)$ are

$$\frac{1}{\overline{R}_s}\left(\frac{d^2\overline{R}_s}{dr^2} + \frac{2}{r}\frac{d\overline{R}_s}{dr}\right) = \pm\beta^2 \tag{4.1.12}$$

$$F(\tau) = C_s e^{\pm\alpha\beta^2\tau} \tag{4.1.13}$$

Again $F(\tau)$ has the same form as before. The particular forms of $\overline{R}_s(r)$ are discussed in terms of particular processes in Sec. 4.3.2.

4.1.2 Laplace Transformations

This method has been widely used in the analysis of many kinds of transients. It is discussed here in terms of solutions of the following Fourier equation, where k and ρc are taken as uniform and constant over the region:

$$\nabla^2 t = \frac{1}{\alpha}\frac{\partial t}{\partial \tau} \tag{4.1.14}$$

The discussion here is in terms of Cartesian coordinates. Then $t = t(x, y, z, \tau)$. This method often gives very simple analysis for many physical mechanisms which are much more difficult to analyze starting from the forms arising from the separation of variables. This transform technique is the basis of many of the solutions given in this and in the following chapters.

The initial advantage of a Laplace transform in any particular circumstance is that the transform removes the time derivative, $\partial t/\partial \tau$, in the preceding equation. The result is an ordinary differential equation, in terms of $\bar{t}(x, y, z)$, called the Laplace transform of $t(x, y, z, \tau)$. The initial and boundary conditions are then applied to the solution of the resulting differential equation in terms of the function $\bar{t}$. The solution of the original formulation is then recovered by inversion of the transform solution $\bar{t}(x, y, z)$, back to $t(x, y, z, \tau)$.

In many conduction circumstances, this inversion is very simply accomplished. A table of transforms, between t and $\bar{t}$, is given in Appendix G. See also Carslaw and Jaeger (1959), Arpaci (1966), Özisik (1980), and Abramowitz and Stegun (1964), for other results. However, circumstances frequently arise in new applications in which the inversion is not tabulated. Generating such transform inversions is sometimes quite difficult.

THE LAPLACE TRANSFORM. The general method is set forth as follows, in terms of the transformation of $t(x, y, z, \tau)$ in three dimensions. Additional theorems are also given which define the results of transforming the several components of the formulation for the solution $t(x, y, z, \tau)$, including initial and boundary conditions. The Laplace transform $L[t(x, y, z, \tau)]$, of $t(x, y, z, \tau)$, written in four useful notations, is

$$L[t(x, y, z, \tau)] = L(t) = \bar{t}(x, y, z) = \int_0^\infty e^{-p\tau} t(x, y, z, \tau)\, d\tau = \bar{t}(p)$$

(4.1.15)

where p may be complex and the real part of p is positive and large enough to make the integral convergent. The integral, a function of p, is the transformation of t to $\bar{t}(x, y, z)$. For example, if $t(\tau) = 1$ and $\sin \omega\tau$ the transforms are

$$L(t) = \bar{t}(p) = 1/p \quad \text{and} \quad \omega/(p^2 + \omega^2)$$

The definition in Eq. (4.1.15) is applied for various x, y, z, τ functions, in the use of this method. The following group of six properties of the Laplace transform is often needed. They are listed as theorems, where t, t_1, t_2, and so on may be functions of x, y, z, τ, for example. Several of the most commonly used theorems are given as follows:

I
$$L[t_1(x, y, z, \tau) + t_2(x, y, z, \tau)] = L(t_1) + L(t_2)$$
(4.1.16)

That is, the transform of a sum of functions is a sum of the transforms.

II
$$L\left(\frac{\partial t}{\partial \tau}\right) = pL(t) - t_0 = p\bar{t}(x, y, z) - t(x, y, z, 0^+)$$
(4.1.17)

where $t(x, y, z, 0^+) = \lim t(x, y, z, \tau)$ as τ goes to 0^+. For a multidimensional function $t(x, y, z, \tau)$, t_0 may be a function of x, y, z.

III
$$L\left(\frac{\partial^n t}{\partial x^n}\right) = \frac{\partial^n \bar{t}}{\partial x^n}$$
(4.1.18)

This applies also for y and z. The result from Eq. (4.1.15) is

$$\int_0^\infty e^{-p\tau} \frac{\partial^n t}{\partial x^n}\, d\tau = \frac{\partial^n}{\partial x^n} \int_0^\infty e^{-p\tau} t\, d\tau$$
(4.1.19)

for t such that the order of integration and differentiation may be exchanged.

IV
$$L\left[\int_0^\tau t(x, y, z, \tau')\, d\tau'\right] = \frac{1}{p} L(t) = \frac{\bar{t}}{p}$$
(4.1.20)

That is, the transform of an integral of t over the time interval 0 to τ is $\bar{t}/p$.

Given a function $t(x, y, z, \tau)$, where τ is replaced by $K\tau$ and K is a positive constant multiple of time τ,

$$V \qquad L[t_K(x, y, z, K\tau)] = \frac{1}{K}\bar{\imath}_K = \frac{1}{K}\bar{\imath}(p/K) \qquad (4.1.21a)$$

where

$$\bar{\imath}_K = \frac{1}{K}\int_0^\infty c^{-(p/K)\tau'}t(x, y, z, \tau')\,d\tau' \qquad (4.1.21b)$$

For b in the following equation being any constant:

$$VI \qquad L[e^{-b\tau}t] = \bar{\imath}(p + b) = \int_0^\infty e^{-(p+b)\tau}t(x, y, z, \tau)\,d\tau \qquad (4.1.22)$$

The following subsection gives the transformation of Eq. (4.1.14) and the formal inversion procedure. Then the method is applied to two transients in semiinfinite plane regions.

TRANSFORMATION AND INVERSION. For $t(x, \tau)$, Eq. (4.1.14) is transformed by the rules in III and II above as follows:

$$L\left(\frac{\partial^2 t}{\partial x^2}\right) = \frac{\partial^2}{\partial x^2}\int_0^\infty e^{-p\tau}t(x, \tau)\,d\tau = \frac{\partial^2\bar{\imath}}{\partial x^2} \qquad (4.1.23)$$

$$L\left(\frac{\partial t}{\partial \tau}\right) = \int_0^\infty e^{-p\tau}\frac{\partial t}{\partial \tau}\,d\tau = pL(\tau) - t(x, 0^+) = p\bar{\imath}(x) - t(x, 0^+) \qquad (4.1.24)$$

Therefore, the subsidiary equation, in $\bar{\imath}(x)$, becomes

$$\frac{d^2\bar{\imath}(x)}{dx^2} - \frac{p}{\alpha}\bar{\imath}(x) = -\frac{t(x, 0^+)}{\alpha} \qquad (4.1.25)$$

where $t(x, 0^+)$ is the initial condition for the transient of interest, specified in theorem II.

The boundary conditions are then also transformed to yield the complete formulation of the solution of the transformation function $\bar{\imath}(x)$. This transform is then inverted to give the solution $t(x, \tau)$. The inversion relation, in terms of $\bar{\imath}(x)$, is

$$t(x, \tau) = \frac{1}{2i}\int_{\gamma-i\infty}^{\gamma+i\infty} e^{\lambda\tau}\bar{\imath}(\lambda)\,d\lambda \qquad (4.1.26)$$

where λ is the complex variable of integration and γ is to be sufficiently large that all singularities of $\bar{\imath}(\lambda)$ lie to the left of the line $(\gamma - i\infty, \gamma + i\infty)$.

SOLUTIONS IN A SEMIINFINITE REGION. The preceding formulation is demonstrated for a semiinfinite region, $x \geq 0$, initially uniformly at t_i. The analysis applies to the two different surface conditions, at $x = 0$, shown in Fig. 4.1.1. In terms of $t(x, \tau) - t_i = \theta(x, \tau)$, the conditions are $\theta(0, \tau) = t_1 - t_i$ constant for $\tau > 0$ and $\theta(0, \tau) = A[t_1(\tau) - t_i]^{n/2}$, where n is any positive integer. The

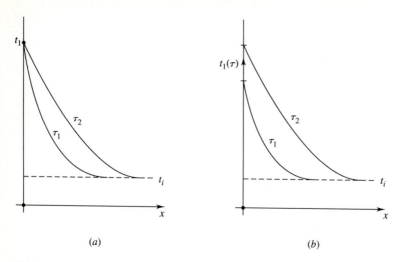

(a)   (b)

FIGURE 4.1.1
Boundary conditions at $x = 0$ for $\tau > 0$: (a) a constant surface temperature $t_1 > t_i$; (b) an increasing surface temperature $t_1(\tau) > t_i$.

subsidiary equation is Eq. (4.1.25), for both circumstances in Fig. 4.1.1. However, the initial condition in both examples in Fig. 4.1.1 is zero, in terms of $\theta(x, \tau)$. Therefore, Eq. (4.1.25) for $x > 0$, in terms of $L(\theta) = \bar{\theta}$, becomes

$$\frac{d^2\bar{\theta}}{\partial x^2} - \frac{p}{\alpha}\bar{\theta} = 0 \tag{4.1.27}$$

The solution of this equation is

$$\bar{\theta}(x) = C_1 e^{\sqrt{p/\alpha}x} + C_2 e^{-\sqrt{p/\alpha}x} = C_2 e^{-\sqrt{p/\alpha}x} \tag{4.1.28}$$

since both $\theta(x, y)$ and $\bar{\theta}(x)$ must be bounded as $x \to \infty$. Therefore, only $e^{-\sqrt{p/\alpha}x}$ may be retained. Next, the solution of $\bar{\theta}(x)$ is determined from the boundary condition at $x = 0$ for $\tau > 0$.

For the simpler circumstance in Fig. 4.1.1(a), the condition $t_1 - t_i$ transforms to

$$L(t_1 - t_i) = \frac{t_1 - t_i}{p} = \bar{\theta}(0)$$

Therefore, from Eq. (4.1.28) the Laplace transform, $\theta(\bar{x})$, of $\theta(x, \tau)$ is

$$\bar{\theta}(x) = \frac{(t_1 - t_i)}{p} e^{-px/\alpha} \tag{4.1.29}$$

The inverse of $[e^{-px/\alpha}]/p$ is erfc $x/2\sqrt{\alpha\tau}$, from Carslaw and Jaeger (1959),

Appendix V, and the solution is

$$\frac{\theta(x,\tau)}{t_1 - t_i} = \text{erfc } x/2\sqrt{\alpha\tau} = \text{erfc } \eta = 1 - \text{erf } \eta \qquad (4.1.30)$$

where erf η is given in Appendix F and is plotted in Fig. 4.2.10, along with erfc η.

The time-dependent surface condition in Fig. 4.1.1(b) is $\theta(0,\tau) = A[t_1(\tau) - t_i]^{n/2}$, where A is the proportionality constant, with the dimension of temperature to the $-n/2$ power. The transform of this condition is

$$L\left[A[t_1(\tau) - t_i]^{n/2}\right] = A\frac{\Gamma(1 + n/2)}{p^1 + n/2} \qquad (4.1.31)$$

where Γ is the gamma function. The inversion is

$$\theta(x,\tau) = A(4\tau)^{n/2}\Gamma\left(1 + \frac{n}{2}\right)i^n \text{ erfc } \eta \qquad (4.1.32)$$

This solution applies for surface temperature variations of the form $\sqrt{\tau}, \tau, \tau\sqrt{\tau}, \dots$.

SUMMARY. The simple first solution, Eq. (4.1.30), is constructed from both the separation of variables and a similarity solution, in Sec. 4.2.3. The Laplace transform has also been very convenient for many of the other kinds of one-dimensional time-dependent processes considered in this chapter, including cylindrical and spherical geometries. Convenient and detailed summaries of a wide range of applications are given by Carslaw and Jaeger (1959), Arpaci (1966), and Özisik (1980). Abramowitz and Stegun (1964) have compiled many transforms.

4.2 TRANSIENTS IN PLANE REGIONS

In such regions, of finite extent, or idealized as infinite, the instantaneous local temperature is $t(x,\tau)$. This temperature level results from the initial temperature level, t_i, or its distribution over the region, at $\tau = 0$, as subsequently changed by other imposed temperature or energy effects. A very common idealization is that the temperature on the surface is suddenly changed to t_0. A simple example is seen in Fig. 4.2.1(a), for a plane layer of thickness L, with a uniform initial temperature t_i. For $t_i(x) = F(x)$, Fig. 4.2.1(b) applies.

The many common thermal boundary conditions are T.1 through T.6 in Sec. 1.4.1. Several analogous mass diffusion conditions are C.1 and C.2 in Sec. 1.4.2. Here, Sec. 4.2.1 gives solutions for regions of finite thickness, as in Fig. 4.2.1(a) and (b), for temperature boundary conditions. Section 4.2.2 analyzes plane regions with a surface convection process, as shown in Fig. 4.2.2. Section 4.2.3 considers idealized semiinfinite and infinite regions, as sketched in Fig. 4.2.9. Solutions are given for several kinds of boundary and initial conditions, in

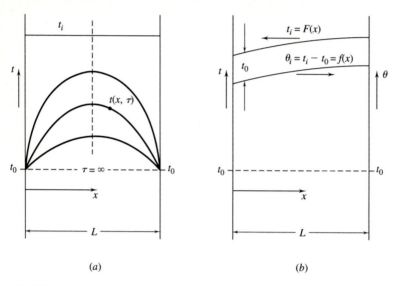

(a)   (b)

FIGURE 4.2.1
A plane layer of thickness L: (a) a uniform initial temperature t_i, where $\theta_i = t_i - t_0$; (b) a variable initial temperature $t_i(x)$, as at some time in a previous transient in the region, where $\theta_i(x) = t_i(x) - t_0 = f(x)$.

each configuration. Also discussed is another conduction formulation which removes the unrealistic result of an infinite propagation rate of transient temperature effects. Then Sec. 4.2.4 considers the effects of distributed energy sources in plane layers.

4.2.1 Transients in Plane Layers

Solutions given here are for the two conditions shown in Fig. 4.2.1, a temperature boundary condition at $x = 0$ and L, and for a uniform and a nonuniform initial temperature distribution. The general solution is determined from Eqs. (4.1.5)–(4.1.7) and the following boundary and initial conditions, in terms of $\theta(x, \tau) = t(x, \tau) - t_0$:

$$\text{At } \tau = 0: \quad \theta_i(x) = t_i(x) - t_0 \quad \text{for } 0 < x < L \qquad (4.2.1a)$$

$$\text{For } \tau > 0: \quad \theta = 0 \quad \text{at } x = 0, L \qquad (4.2.1b)$$

The temperature field decays from θ_i or $\theta_i(x)$ to zero. Therefore, $-\beta^2$ is chosen in Eq. (4.1.7). Also β^2 is taken as real, since no periodics arise for the previously imposed conditions. Therefore, Eq. (4.1.6) is used for $X(x)$ and

$$\theta(x, \tau) = C_5 e^{-\alpha\beta^2\tau}(C_3 \cos \beta x + C_4 \sin \beta x) \qquad (4.2.2)$$

Since $\theta = 0$ at $x = 0$ for all $\tau > 0$, C_3 must be zero. Similarly, $C_4 \sin \beta x$ must be zero at $x = L$ for all $\tau > 0$. Since C_4 may not be zero, $\sin \beta L$ must be zero. This occurs for $\beta L = n\pi$, where n is any integer. The solution, then, is the sum of all such particular solutions:

$$\theta(x, \tau) = \sum_{n=1}^{\infty} C_n e^{-(n\pi/L)^2 \alpha \tau} \sin \frac{n\pi}{L} x \tag{4.2.3}$$

The initial condition, at $\tau = 0$, determines the C_n.

$$\theta_i(x) = f(x) = \sum_{n=1}^{\infty} C_n \sin \frac{n\pi}{L} x$$

This is a Fourier sine series expansion of $f(x)$. The coefficients are

$$C_n = \frac{2}{L} \int_0^L f(x) \sin \frac{n\pi}{L} dx \tag{4.2.4}$$

This result, in conjunction with Eq. (4.2.3), is the solution for a given $f(x)$.

If the initial temperature excess distribution is uniform and equal to θ_i, the constants are evaluated as

$$C_n = \frac{2\theta_i}{L} \int_0^L \sin \frac{n\pi}{L} x \, dx = 0 \qquad \text{for } n = 0, 2, 4, \ldots$$

$$= \frac{4\theta_i}{n\pi} \qquad \text{for } n = 1, 3, 5, \ldots$$

The solution is

$$\phi = \frac{\theta}{\theta_i} = \frac{4}{\pi} \sum_{n=1,3}^{\infty} \frac{e^{-(n\pi)^2 \alpha \tau / L^2}}{n} \sin \frac{n\pi x}{L} \tag{4.2.5}$$

where $\alpha \tau / L^2 = \text{Fo}$, the Fourier number. The local heat flux is calculated as follows:

$$q''(x, \tau) = -k \frac{\partial \theta}{\partial x} = -\frac{4k\theta_i}{L} \sum_{n=1,3}^{\infty} e^{-(n\pi/L)^2 \alpha \tau} \cos \frac{n\pi}{L} x \tag{4.2.6}$$

This is seen to be zero for all τ at $x = L/2$ and infinite at $\tau = 0$ for $x = 0$.

This latter consequence is similar to that shown as an example in Sec. 3.1.1, Eq. (3.1.10), in two-dimensional steady-state conduction. Here, the physical unreality is in assuming that the temperature at $x = 0$ may be instantly increased from $\theta = 0$ to θ_1 at $\tau = 0$.

An additional kind of unrealistic behavior also commonly arises. It is an infinite propagation velocity of a changed surface temperature level. This aspect is more conveniently shown in conjunction with a semiinfinite solid solution in Sec. 4.2.3, and is discussed there in detail.

4.2.2   Convection at Both Surfaces of a Plane Layer

This formulation is shown in Fig. 4.2.2, where $x = 0$ is taken at the midplane of the region of total thickness $2s$. This is more convenient in determining and using the results. The region, initially at t_i throughout, is suddenly, for $\tau > 0$, placed in contact with a fluid at $t_e < t_i$. Convection occurs between the surface and the fluid. The rate of convective heat transfer per unit of surface area is equal to the surface coefficient h times the difference between the surface and fluid temperatures $t_s - t_e$. The midplane is a plane of symmetry. The equation and bounding conditions, in terms of $\theta = t - t_e$, are

$$\frac{\partial^2 \theta}{\partial x^2} = \frac{1}{\alpha}\frac{\partial \theta}{\partial \tau}$$

At $\tau = 0$: $\theta = t_i - t_e = \theta_i$ for $-s \leq x \leq s$

At $\tau > 0$: $\dfrac{\partial \theta}{\partial x} = 0$ at $x = 0$ because of symmetry

At $\tau > 0$: $-k\dfrac{\partial \theta}{\partial x} = -h\theta_s$ at $x = s$

where $\theta_s = t_s(\tau) - t_e$.

The reasoning used in reducing the general solutions in Eqs. (4.1.5)–(4.1.7) again leads to Eq. (4.2.2) as a solution here. The condition at $x = 0$, of a zero derivative, is first applied:

$$\frac{\partial \theta}{\partial x} = C_5 e^{-\alpha\beta^2\tau}(-C_3\beta \sin \beta x + C_4\beta \cos \beta x).$$

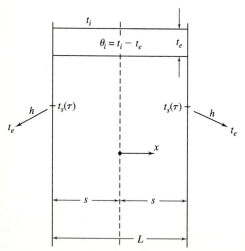

FIGURE 4.2.2
Transient conduction in a plane layer of thickness $2s$.

This is zero at $x = 0$ if $C_4 = 0$ for $\beta \neq 0$. Therefore,

$$\theta = Ce^{-\alpha\beta^2\tau} \cos \beta x \qquad (4.2.7)$$

The last boundary condition, at $x = s$, is applied to determine the permissible values of β:

$$\frac{\partial \theta}{\partial x} = -Ce^{-\alpha\beta^2\tau}\beta \sin \beta x = -\frac{\theta}{\cos \beta x}\beta \sin \beta x$$

at $x = s$:

$$\left(\frac{\partial \theta}{\partial x}\right)_s = -\beta\theta_s \tan \beta s = -\frac{h}{k}\theta_s$$

Therefore,

$$\cot \beta s = \frac{k}{h}\beta = \frac{k}{hs}p \qquad (4.2.8)$$

or

$$\cot p = \frac{p}{hs/k} = \frac{p}{\mathrm{Bi}}$$

where $p = \beta s$. The dimensionless quantity hs/k is the Biot number. There are an infinite number of values of p, called p_j, which satisfy Eq. (4.2.8). These values depend upon the Biot number. That is, a different set of p_j is found for each value of the Biot number. The roots of Eq. (4.2.8) may be estimated by plotting $\cot p$ and p/Bi versus p and finding the intersections, as shown in Fig. 4.2.3.

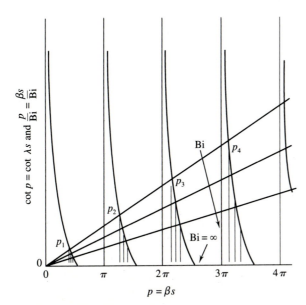

FIGURE 4.2.3
Determination of the roots of Eq. (4.2.8), that is, the eigenvalues for transient conduction in an infinite plate of thickness $2s$.

The first four roots are indicated for several different Bi. We note that for Bi = ∞, $h = \infty$ and $\theta_s = 0$. The values of p_j are then $\pi/2, 3\pi/2, 5\pi/2, \ldots,$. This is the first condition used in this section. The solution is Eq. (4.2.5). The present result, therefore, includes the previous one.

The roots of Eq. (4.2.8) for ranges of Bi are given by Jakob (1949), as calculated by Groeber. Schneider (1955) and Carslaw and Jaeger (1959) list the same information determined with higher accuracy.

Since each root of Eq. (4.2.8) satisfies Eq. (4.2.7), the complete solution for the temperature distribution is obtained by summing over all roots for any given value of Bi:

$$\theta(x, \tau) = \sum_{\beta} C_{\beta} e^{-\alpha\beta^2\tau} \cos \beta x = \sum_{j=1}^{\infty} C_j e^{-p_j^2(\alpha\tau/s^2)} \cos\left(p_j \frac{x}{s}\right)$$

The constants C_j are determined from the initial condition, that is,

$$\theta_i = \sum_{j=1}^{\infty} C_j \cos\left(p_j \frac{x}{s}\right) = C_1 \cos\left(p_1 \frac{x}{s}\right) + C_2 \cos\left(p_2 \frac{x}{s}\right) + \cdots$$

This is an expansion of θ_i in an infinite series. The coefficients may be evaluated by multiplying both sides by $\cos[p_n(x, s)]$ and integrating from $x = 0$ to $x = s$. It can be shown that, for $n \neq j$,

$$\int_0^s C_j \cos\left(p_n \frac{x}{s}\right) \cos\left(p_j \frac{x}{s}\right) dx = 0$$

The term for which $n = j$ is not equal to zero, and the equation for C_j is therefore

$$\int_0^s \theta_i \cos\left(p_j \frac{x}{s}\right) dx = \int_0^s C_j \cos^2\left(p_j \frac{x}{s}\right) dx$$

Successive values of m are taken, and each C_j is evaluated. The general result, valid for all j, is

$$C_j = \theta_i \frac{\int_0^s \cos\left(p_j \frac{x}{s}\right) dx}{\int_0^s \cos^2\left(p_j \frac{x}{s}\right) dx} = \frac{2 \sin p_j}{p_j + \sin p_j \cos p_j} \qquad (4.2.9)$$

The solution is

$$\phi = \frac{\theta}{\theta_i} = \sum_{j=1}^{\infty} \frac{2 \sin p_j \cos[p_j(x/s)]}{p_j + \sin p_j \cos p_j} e^{-p_j^2(\alpha\tau/s^2)} \qquad (4.2.10)$$

where $\alpha\tau/s^2$ is the Fourier number.

This result depends on x as x/s, time as $\alpha\tau/s^2$, a Fourier number, and on the p_j which are determined from the Biot number Bi $= hs/k$. That is, the functional dependence of ϕ is

$$\phi = \phi(x/s, \alpha\tau/s^2, hs/k) = \phi(x/s, \text{Fo}, \text{Bi}) \qquad (4.2.11)$$

Evaluation of this result requires the p_j, discussed previously.

A GRAPHICAL REPRESENTATION. The importance of the preceding result, Eq. (4.2.10), in practical applications, has led to the presentation of this equation in graphical form.

The form of Heisler (1947) is given here, in Figs. 4.2.4 through 4.2.8. The central temperature response information in Fig. 4.2.4 is θ_0/θ_i where $\theta_i = (t_i - t_e)$ and $\theta_0 = \theta(0, \text{Fo}, \text{Bi}) = t(0, \tau) - t_e$, for Bi < 0.01 and Fo > 0.2. Figure 4.2.5 shows $\theta(x/s)/\theta_0$ where $x/s = 0.2, 0.4, 0.6, 0.8, 0.9$, and 1.0, for Fo > 0.2 and Bi < 10. Figure 4.2.6 is a plot of θ_0/θ_i versus m Fo Bi, for Fo > 0.2 and Bi < 0.01. It includes plane regions, long cylinders, and spheres, for $m = 1, 2$, and 3, respectively. In Fig. 4.2.4, $\phi_0 = \theta_0/\theta_i$ depends only on Fo and Bi. In Fig. 4.2.5, $\theta(x/s, \text{Bi})/\theta_0$ is given for a range of Bi over which the form of the temperature distribution in the region does not vary greatly, when normalized by $\theta_0 = \theta(0, \text{Fo}, \text{Bi})$. Figure 4.2.6 applies when the Biot number is very small, that is, for $(k/s) \gg h$. Then, at each time as Fo $= \alpha\tau/s^2$, the temperature distribution remains uniform across the region, as the internal temperature changes from t_i to t_e.

Although Fig. 4.2.4 permits reasonable accuracy for estimates of $\theta_0/\theta_i = \phi_0$, at larger values of Fo, it gives no information for short times, when ϕ_0 is still near 1. Figure 4.2.7 is a short-time response plot, for Fo < 0.2. Contours are values of $1/\text{Bi}$. They are given for $x/s = 0$, that is, $1 - \phi_0$, and for $x/s = 1/2$, or $1 - \phi_{s/2}$. A similar short-time plot, in Fig. 4.2.8, gives the surface temperature response, t_s, as $1 - \phi_s$, as a function of $1/\text{Bi}$. All of these figures are very valuable, at least in making preliminary estimates of transient response for design purposes. Similar information is given in Sec. 4.3.3 for cylinders and spheres.

GRAPHICAL RESULTS WITH AN IMPOSED SURFACE FLUX q_s''. Heisler (1947) has shown that Figs. 4.2.4, 4.2.5, 4.2.7, and 4.2.8 also give the temperature response for a flux condition at $x = s$. This would arise with high-frequency surface induction heating, at a constant rate per unit area of q_s''. For sufficiently high frequencies of heating, in good absorbers such as metals, the absorption of q_s'' occurs entirely in a thin surface layer. The midplane temperature response θ_0/θ_i in Fig. 4.2.4 then becomes instead $[1 - h\theta_0/q_s'']$. Also, θ/θ_0 in Fig. 4.2.5 becomes $[1 - h\theta/q_s'']/[1 - h\theta_0/q_s'']$. In both circumstances, h is the parallel surface convection process. The ordinates in Figs. 4.2.7 and 4.2.8 then become $h\theta/q_s''$, instead of $1 - \theta/\theta_i$ and $1 - \phi_s$, respectively. These same modifications also apply for the cylindrical geometry, in Figs. 4.3.5, 4.3.7, 4.3.9, and 4.3.10, and for spheres in Figs. 4.3.6, 4.3.8, 4.3.11, and 4.3.12.

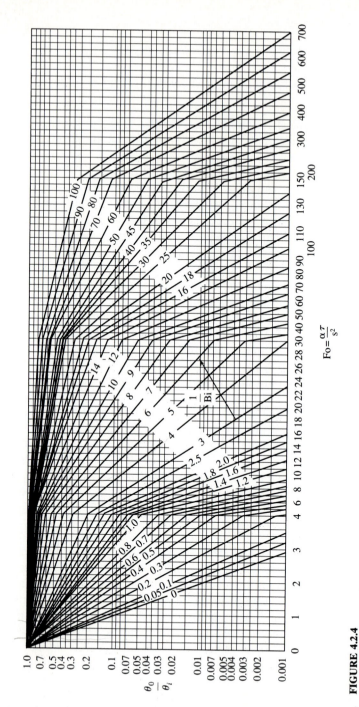

FIGURE 4.2.4
Midplane temperature for a plane region. [From Heisler (1947).]

188

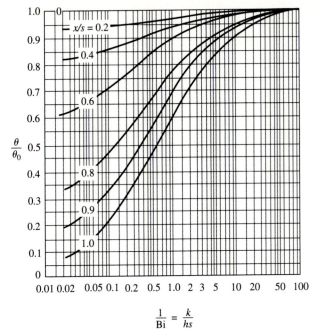

FIGURE 4.2.5
The dependence of the temperature distribution on Bi, for a plane region. [From Heisler (1947).]

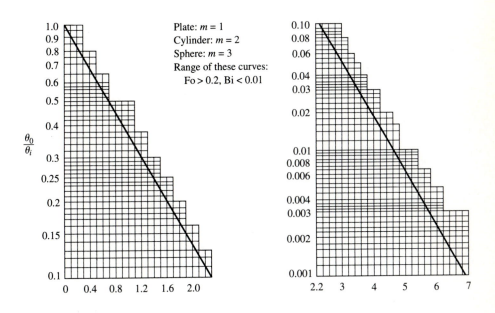

FIGURE 4.2.6
Mid-location temperature for plane regions, long cylinders, and spheres of relatively high internal conductivity. [From Heisler (1947).]

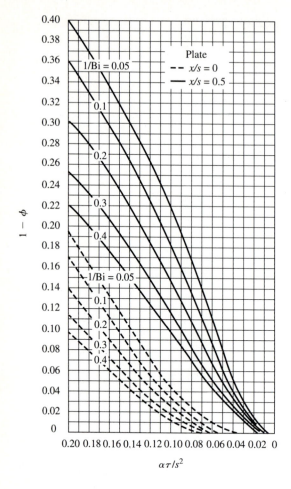

FIGURE 4.2.7
Temperatures at $x = 0$ and $s/2$ as $1 - \phi_0 = 1 - \theta_0/\theta_i$ and $1 - \theta_{s/2} = 1 - \theta_{s/2}/\theta_i$, at short times, vs. Fo $= \alpha\tau/s^2$. [From Heisler (1947).]

4.2.3   Transients in Half-Infinite and Infinite Regions

These geometries are an idealization which permits relatively simple solutions for a wide range of the transient and periodic effects which are important in applications. The effects include flux, convection, radiation, and temperature, and, by analogy, concentration conditions, applied at surfaces. Also included are temperature or concentration conditions which arise internal to such regions. These idealized solutions have many practical and realistic uses.

There is a common requirement for the use of transient solutions in infinite regions, in regions which are actually of limited extent. These solutions apply accurately only for the period of time in which appreciable temperature effects, due to changed surface conditions, have not reached the boundary of the limited region of interest. This consideration is sketched in Fig. 4.2.9(a) where a region initially at t_i has the surface temperature suddenly raised to t_1, and held constant thereafter. Temperature distributions are sketched at later

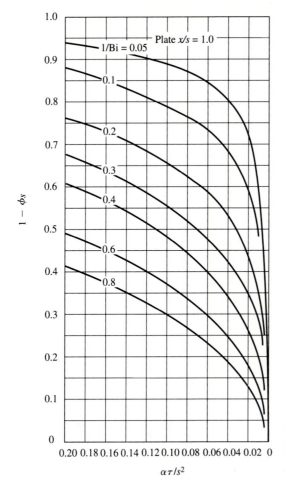

FIGURE 4.2.8
Surface temperature, as $1 - \phi_s = 1 - \theta_s/\theta_i$, for short times, vs. Fo = $\alpha\tau/s^2 < 0.20$. [From Heisler (1947).]

times, τ_1, τ_2, and τ_3. A plane region of thickness L is also shown, superimposed over the semiinfinite region. Clearly the semiinfinite region behavior appropriately models the temperature response in the plane region for times less than $\tau = \tau_2$. Similar simple considerations may also be applied in many other circumstances.

This section first considers conduction in semiinfinite regions, with constant temperature t_1 and also with constant flux q'', each imposed at $x = 0$. The solution is also given for a sudden change in environment temperature from t_i to t_e, with a convection coefficient h.

The infinite solid, see Fig. 4.2.9(b), is then analyzed for the spreading temperature field resulting from a sudden internal energy input temperature effect. The first effect is the sudden appearance of a varying temperature field, as $t_i - t_\infty = f(x)$. The second is an instantaneous finite-plane heat pulse, Q'', per unit area, at $x = 0$, at $\tau = 0$. Another example is shown later, in Fig. 4.2.12. It is a sudden uniform temperature increase, from t_i to t_1, in an internal plane

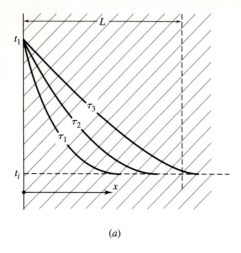

(a)

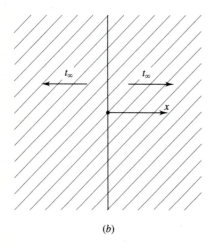

(b)

FIGURE 4.2.9
Transients in regions of infinite extent: (a) semi-infinite; (b) infinite.

region of thickness $2s$. The last example, seen in Fig. 4.2.13, is two half-infinite regions, initially at t_L and t_R, respectively, suddenly placed together. Assuming that the resulting contact resistance R_c is negligible, the temperature of the common interface, t_{if}, is determined.

SEMIINFINITE SOLID SOLUTIONS. The first solution is for the conditions in Fig. 4.2.9(a) as follows, where $\theta = t - t_i$. This solution was determined in Sec. 4.1.2 as an example of the use of Laplace transforms.

$$\frac{\partial^2 \theta}{\partial x^2} = \frac{1}{\alpha} \frac{\partial \theta}{\partial \tau} \qquad \theta = t(x, \tau) - t_i \qquad (4.2.12a)$$

At $\tau = 0$ $\theta = 0$ for $x \geq 0$

For $\tau > 0$ $t_1 - t_i = \theta_1$ at $x = 0$ $\qquad (4.2.12b)$

A real value of β^2 is necessary, since no periodic temperature changes arise. A negative exponent is required in Eq. (4.1.7) for the decay of the imposed difference. Therefore, Eq. (4.1.6) is used:

$$\theta = C_5 e^{-\alpha\beta^2\tau}(C_3 \cos \beta x + C_4 \sin \beta x) \tag{4.2.13}$$

The first condition yields $C_3 = 0$, but since no condition at L is specified here, all values of β are permissible. Summing all possibilities, we have

$$\theta = \int_0^\infty C_\beta e^{-\alpha\beta^2\tau} \sin \beta x\, d\beta \tag{4.2.14}$$

The methods of determining C_β from the initial condition are involved. The analysis is given by Schneider (1955). That analysis, and the Laplace transform solution in Sec. 4.1.2, lead to the following result:

$$\phi = 1 - \operatorname{erf}\frac{x}{2\sqrt{\alpha\tau}} = \operatorname{erfc}\frac{x}{2\sqrt{\alpha\tau}} = \operatorname{erfc}\eta \tag{4.2.15}$$

The temperature excess ratio ϕ is a function of x and τ only in the combination $x/2\sqrt{\alpha\tau}$. The function erf, in Eq. (4.2.15), is known as the error function or as the probability integral and is plotted in Fig. 4.2.10 and tabulated in Appendix F.

The preceding solution, Eq. (4.2.15), may be obtained in a very simple form by initially assuming that the x and τ dependence of $\theta(x, \tau)$ may be

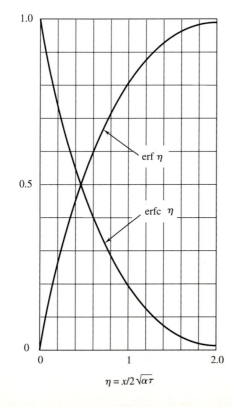

FIGURE 4.2.10
The error function, Eq. (4.2.15), as listed in Appendix F, and the conjugate error function.

combined into a new dependent variable η as follows:

$$\eta = \frac{x}{2\sqrt{\alpha\tau}}$$

For this variable, the differential equation (4.2.12a) and boundary conditions become

$$\frac{d^2\theta}{d\eta^2} + 2\eta\frac{d\theta}{d\eta} = 0 \tag{4.2.16a}$$

$$\eta = 0 \ (\text{i.e., } x = 0) \qquad \theta = t_1 - t_i = \theta_i$$
$$\eta \to \infty \ (\text{i.e., } \tau \to 0) \qquad \theta = 0 \tag{4.2.16b}$$

The solution of Eq. (4.2.16), with the relevant conditions, is again

$$\theta = A \, \text{erf} \, \eta + B = \theta_1(1 - \text{erf} \, \eta) \tag{4.2.16c}$$

This kind of formulation shows the role of the variable $\eta \propto x/\sqrt{\tau}$. This form will arise as part of the solution for many processes assumed to occur in infinite regions.

OTHER SOLUTIONS. The transient considered previously, of a temperature step at the surface of a semiinfinite solid, is important in many applications. It applies approximately to materials of relatively low thermal diffusivity, in which an external temperature change results essentially in an immediate and equal surface temperature change.

However, in many practical circumstances the transient is actually caused by a change in the environment temperature to t_e, causing a convection exchange with the surface, controlled by a convection coefficient, h. Sometimes it may be due to a sudden thermal flux loading q'' at the surface, caused perhaps by radiation or by induction heating absorbed very near the surface. The boundary conditions for these two different circumstances, in terms of $\theta = t(x, \tau) - t_i$, are

$$\text{At } x = 0: \quad q'' = -k\frac{\partial\theta}{\partial x} = h[t_e - t(0,\tau)] = h[\theta_e - \theta(0,\tau)]$$

$$\text{At } x = 0: \quad q'' = -k\frac{\partial\theta}{\partial x} \quad \text{or} \quad \frac{\partial\theta}{\partial x} = -\frac{q''}{k}$$

The solutions, obtained very simply by the method of Laplace transforms, are

$$\frac{t(x,\tau) - t_i}{t_e - t_i}$$

$$= \frac{\theta}{\theta_e} = \text{erfc} \, \eta - \exp\left(\frac{hx}{k} + \frac{h^2}{k^2}\alpha\tau\right)\text{erfc}\left(\eta + \frac{h}{k}\sqrt{\alpha\tau}\right) \tag{4.2.17}$$

$$\theta = \frac{2q''\sqrt{\alpha\tau}}{k}\left(\frac{e^{-\eta^2}}{\sqrt{\pi}} - \eta \, \text{erfc} \, \eta\right) = \frac{q''}{k}\sqrt{\frac{4\alpha\tau}{\pi}}\left(e^{-\eta^2} - \sqrt{\pi} \, \eta \, \text{erfc} \, \eta\right) \tag{4.2.18}$$

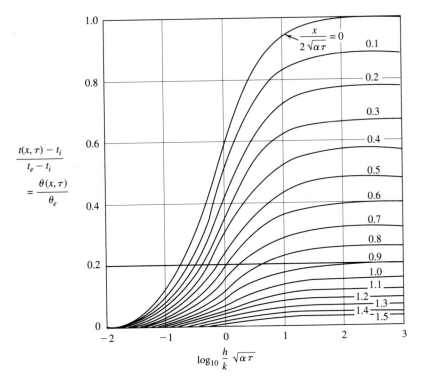

FIGURE 4.2.11
Temperature response $\theta(x,\tau)/\theta_e$, in Eq. (4.2.17). [From Eckert and Drake (1972).]

where $\eta = x/2\sqrt{\alpha\tau}$, k is the thermal conductivity of the conduction region, and erfc is the conjugate error function defined in Eq. (4.2.15). The characteristic temperature in Eq. (4.2.18) is time dependent as $\sqrt{\tau}$. Equation (4.2.17) is a very useful result and is plotted in Fig. 4.2.11.

These three solutions, Eqs. (4.2.15), (4.2.17), and (4.2.18), include many circumstances commonly encountered. They also apply for an initial transient period in a plane region during which the temperature disturbance has not reached the far side in an appreciable way. For solutions with the imposed temperature condition at $x = 0$ being time dependent, see Carslaw and Jaeger (1959), Sec. 2.5. Many other solutions are also given by Carslaw and Jaeger (1959), Chap. 12.

TRANSIENT SOLUTIONS FOR AN INFINITE REGION. Many solutions may be obtained. Their practical use is for transients generated by various effects internal to finite conduction regions, for the time periods in which these effects have not yet reached any bounding surfaces to an appreciable extent.

Three transient results in an infinite solid, $-\infty < x < \infty$, are given. The first is called the Laplace solution and is the transient temperature field which follows the conduction–diffusion of an isolated initial temperature distribution

$\theta_i = t_i - t_\infty = f(x)$. The following relation of x and τ satisfies Eq. (4.2.12a), for any value of $x - x'$:

$$\frac{1}{2\sqrt{\pi\alpha\tau}}e^{-(x-x')^2/4\alpha\tau}$$

where x' is any value. The following result also satisfies the initial condition and, therefore, is a solution:

$$\theta(x,\tau) = \frac{1}{2\sqrt{\pi\alpha\tau}}\int_{-\infty}^{\infty} f(x')e^{-(x-x')^2/4\alpha\tau}\,dx' \qquad (4.2.19)$$

where x' is the variable of integration.

An example of this formulation in an infinite region, initially at t_i, is seen in Fig. 4.2.12(a). An internal energy source suddenly raises the temperature to t_1, in the region $-s < x < s$. The resulting transient temperature field, for $-\infty < x < \infty$, is

$$\phi = \frac{t - t_i}{t_1 - t_i} = \frac{1}{2}\left[\text{erf}\frac{s - x}{2\sqrt{\alpha\tau}} + \text{erf}\frac{s + x}{2\sqrt{\alpha\tau}}\right] \qquad (4.2.20)$$

The temperature effect is conducted away from the region $-s < x < s$, at the expense of the remaining temperature level there. The effect is shown in Fig. 4.2.12(b), where the contours shown are at times τ such that $\alpha\tau/s^2 = 0$ to 5. See Carslaw and Jaeger (1959), Sec. 2.2, for other similar solutions, including those for cylindrical and spherical temperature fields.

The third process is the infinite solid, at an initially uniform temperature of t_i throughout, subjected to an instantaneous heat pulse, per unit area, of Q'', having finite magnitude at $x = 0$ and at $\tau = 0$. This is also the problem of a semiinfinite solid subject to an instantaneous surface heat pulse of $Q''/2$, the exposed surface being thereafter insulated. The solution in terms of $\theta = t - t_i$ is

$$\theta(\eta,\tau) = \frac{Q''}{2\rho c\sqrt{\alpha\tau}}e^{-\eta^2} \qquad (4.2.21)$$

where $\eta = x/2\sqrt{\alpha\tau}$.

THE SUDDEN CONTACT OF TWO HALF-INFINITE REGIONS. This is seen in Fig. 4.2.13 where the two regions L and R are initially at t_L and t_R, respectively, and have conductivities and diffusivities of k_L, α_L and k_R, α_R. After contact at $\tau = 0$, they have a common interface temperature t_{if}, if there is no contact resistance, R_c, between the two surfaces. A temperature distribution at some later time is sketched in Fig. 4.2.13(b). The unequal slopes at $x = 0$, where $q_L''(0, \tau) = q_R''(0, \tau)$, indicate that $k_L < k_R$. The equation and boundary condi-

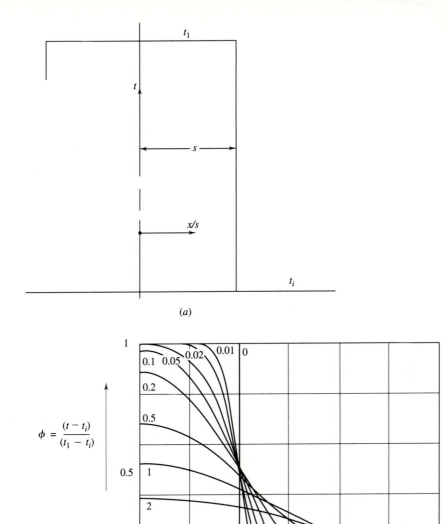

$$\phi = \frac{(t - t_i)}{(t_1 - t_i)}$$

FIGURE 4.2.12
A sudden increase of temperature from t_i to t_1 in a layer $2s$ thick in an infinite region: (a) the imposed temperature increase; (b) the temperature distribution $\phi = (t - t_i)/(t_1 - t_i)$ at later times, as $\alpha \tau / s^2$.

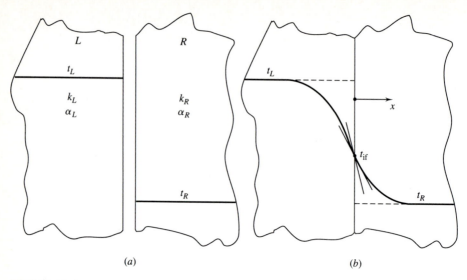

FIGURE 4.2.13
Sudden contact between two extensive regions: (a) the regions; (b) the temperature field at some time τ after contact.

tions in each region are written in terms of $\theta_L(x, \tau) = t_L(x, \tau) - t_R$ and $\theta_R(x, \tau) = t_R(x, \tau) - t_R$:

$$\frac{\partial^2 \theta_R}{\partial x^2} = \frac{1}{\alpha_R} \frac{\partial \theta_R}{\partial \tau} \quad \text{and} \quad \frac{\partial^2 \theta_L}{\partial x^2} = \frac{1}{\alpha_L} \frac{\partial \theta_L}{\partial \tau} \tag{4.2.22}$$

$$\theta_L(-\infty, \tau) = t_L - t_R = \theta_1 \qquad \theta_R(+\infty, \tau) = 0 \tag{4.2.23a}$$

$$\theta_L(x, 0) = t_L - t_R \quad \text{and} \quad \theta_R(x, 0) = 0 \tag{4.2.23b}$$

The interface conditions which connect the two temperature distributions θ_L and θ_R are

$$\theta_L(0, \tau) = \theta_R(0, \tau) \tag{4.2.24a}$$

$$-k_L \left(\frac{\partial \theta_L}{\partial x} \right)_0 = -k_R \left(\frac{\partial \theta_R}{\partial x} \right)_0 \tag{4.2.24b}$$

Equation (4.2.15) indicates that the error function is a solution of both relations in Eqs. (4.2.22) as

$$t_L(x, \tau) = A_L + B_L \operatorname{erf} \frac{x}{2\sqrt{\alpha_L \tau}} \tag{4.2.25a}$$

$$t_R(x, \tau) = A_R + B_R \operatorname{erf} \frac{x}{2\sqrt{\alpha_R \tau}} \tag{4.2.25b}$$

Using the conditions in Eqs. (4.2.23) and (4.2.24) determines the four constants and gives the two solutions $\theta_L(x, \tau)$ and $\theta_R(x, \tau)$. Equating these in accordance with Eq. (4.2.24a) gives the following relationship of the dependence of t_{if}. See Eckert and Drake (1972) for the details.

$$\frac{t_{if} - t_R}{t_L - t_R} = \frac{k_L\sqrt{\alpha_R}}{k_L\sqrt{\alpha_R} + k_R\sqrt{\alpha_L}} \qquad (4.2.26)$$

The interface heat flux is

$$q''(\tau) = \frac{k_R\sqrt{\alpha_L}}{k_L\sqrt{\alpha_R} + k_R\sqrt{\alpha_L}} \frac{k_L(t_L - t_R)}{\sqrt{\pi \alpha_L \tau}} \qquad (4.2.27)$$

This result has many important applications. Consider pouring a liquid metal at a temperature $t_m = t_L$, above the melting point, into a mold at $t_M = t_R$, below the melting point. Equation (4.2.26) indicates whether or not the solid phase is immediately formed at the interface. This effect is also important in sensors, to determine surface temperatures of a region, for example, as t_R given previously. The region L is the thermometric element material. For $t_{if} - t_R$ to be very small, for a small error, $k_R\sqrt{\alpha_L}$ should be much greater than $k_L\sqrt{\alpha_R}$ in Eq. (4.2.26).

A similar solution is given by Crank (1957), page 40, with an interposed mass diffusive contact resistance between the two regions after contact. The initial condition was $C = 0$ on one side and $C = C_i$ on the other side. Transient response is graphed, for several values of the interposed contact resistance.

PROPAGATION VELOCITY OF TEMPERATURE EFFECTS. Many transient solutions of the Fourier equation show the anomaly of an infinite propagation rate of an imposed temperature condition. An example is the semiinfinite solid solution for the surface temperature step, Eq. (4.2.15) given later,

$$\phi = \text{erfc} \frac{x}{2\sqrt{\alpha\tau}} \qquad (4.2.15)$$

The conjugate error function decays only asymptotically as $x/2\sqrt{\alpha\tau}$ goes to infinity. Thus, at any small value of time τ, ϕ is not zero at all increasing values of x. Therefore, the propagation velocity of the effect of the temperature increase at $x = 0$ is infinite. Another aspect of this solution is that an infinite initial surface flux $q''(0, 0)$ arises at $x = 0$ at $\tau = 0$. These characteristics result from the approximations inherent in the Fourier law of conduction, Eq. (1.1.3), being used in the kind of energy balance which resulted, for example, in Eq. (1.2.6).

On the other hand, the five basic microscopic conduction mechanisms discussed in Sec. 1.1.3 are modeled as proceeding successively along steps of length λ, in Fig. 1.1.2. There, λ is, for example, the mean free path in a gas or the spacing of adjacent elements of the structure of a solid. These microscopic models, of a step-by-step propagation, are inconsistent with an infinite velocity of the diffusing thermal effect.

This matter continues to be studied. However, it has not been analytically resolved in the form of a new widely applicable formulation. One apparently reasonable postulate, for example, by Vernotte (1958), modifies the Fourier law, with an additional term. This is written by Frankel et al. (1987) as

$$q'' + \tau_T \frac{\partial q''}{\partial \tau} = -k \frac{\partial t}{\partial x} \qquad (4.2.28)$$

where the Fourier law is seen as a first approximation and τ_T is introduced as a finite transport or relaxation time. It is sometimes estimated to lie in the range from 10^{-12} to 10^{-14} s. The preceding measure changes the Fourier equation from parabolic to hyperbolic form.

An investigation of this matter by Baumeister and Hamill (1969, 1971) expresses τ_T as α/C^2, where C is the speed of heat propagation. The conduction equation is then

$$\frac{\partial^2 t}{\partial x^2} = \frac{1}{\alpha} \frac{\partial t}{\partial \tau} + \frac{1}{C^2} \frac{\partial^2 t}{\partial \tau^2} \qquad (4.2.29)$$

This is called the telegraph or damped wave equation. The presence of the thermal inertia term $\partial^2 t/\partial \tau^2$ results in the maximum flux, at $x = 0$, being finite. It also results in a thermal wave that travels at a finite speed.

In general, C is a large number. Therefore, appreciable effects from the last term in Eq. (4.2.29) arise only in regions of very rapidly changing temperature gradient. This characteristic would be evaluated by comparing the magnitude of the new term in Eq. (4.2.29) to either of the other two terms, as

$$\frac{1}{C^2} \frac{\partial^2 t}{\partial \tau^2} \bigg/ \frac{\partial^2 t}{\partial x^2} \quad \text{or} \quad \frac{1}{C^2} \frac{\partial^2 t}{\partial \tau^2} \bigg/ \frac{1}{\alpha} \frac{\partial t}{\partial \tau}$$

A short summary of the history of the study of these phenomena is given by Gembarovic and Majernik (1988). Specific calculations were also made concerning the propagation of surface heat pulses into a finite medium of thickness L. Non-Fourier effects, at $x = L$, were found to be negligible for $Ve = \sqrt{\alpha \tau_T}/L \leq 0.04$, where Ve is called the Vernotte number. See also the study of Kaminski (1990) concerning the value of τ_T in Eq. (4.2.28), as determined by measured propagation rates. A detailed summary is given.

SUMMARY. The collection of results in this section has many practical applications in conduction. Although the formulations are idealized, the solutions retain most of the principal features of the process. This permits general evaluations in design and in performance. An example of such utility is the Heisler chart results. Most of the preceding solutions may also be applied in multidimensional conduction geometries. At sufficiently short times, the effect of a changed surface temperature condition remains very thin and is not strongly dependent on local surface curvature.

The previous solutions, through Eq. (4.2.27), also may be used for mass diffusion when Fick's law applies in the form of Eq. (1.3.1), which resulted in Eq. (1.3.3), with $C''' = 0$. The results in the following section, for a distributed source q''', also apply in mass diffusion with $C''' \neq 0$.

4.2.4 A Distributed Source in a Plane Layer

In many transients, a distributed energy input is a principal effect in determining subsequent temperature response in the region. Sources arise through many different kinds of effects. Two different kinds of inputs are analyzed in the following discussion. In both, the initial condition is $t(x,0) = t_i$. For $\tau > 0$, fixed temperatures, and an insulating condition, are applied at the surfaces. Also, at $\tau = 0$, a thereafter constant internal energy generation rate begins.

Two forms of distributed sources are considered. The first is uniform across the region, as q'''. This might arise from a sudden application of an electrical current in a resistive region. The steady-state result is in Sec. 2.4.1. The other kind of source also analyzed is an exponential decay in strength, $q'''(x)$, across the region. This is a formulation for the absorption of high-energy

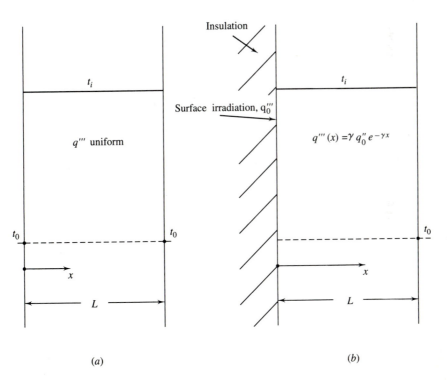

(a)　　　　　　　　　　　　　(b)

FIGURE 4.2.14
Transients subject to a distributed energy source q''': (a) a uniform source; (b) an exponentially decaying effect.

particles or of microwave irradiation across the region, as considered in Sec. 2.4.2 in steady state. The regions and conditions for the two circumstances are shown in Fig. 4.2.14.

UNIFORM DISTRIBUTED SOURCE. The equation, initial and boundary conditions, for Fig. 4.2.14(a), are written in terms of $\theta(x, \tau) = t(x, 0) - t_0$ and $\theta_i = t_i - t_0$ as

$$\frac{\partial^2 \theta}{\partial x^2} + \frac{q'''}{k} = \frac{1}{\alpha} \frac{\partial \theta}{\partial \tau} \tag{4.2.30}$$

and
$$\theta(0, \tau) = \theta(L, \tau) = 0 \qquad \theta(x, 0) = \theta_i \tag{4.2.31}$$

The inhomogeneous term in Eq. (4.2.30) is removed in solving for the transient temperature response $\theta(x, \tau)$. This is done in Sec. 3.6, for multidimensional steady-state processes. The procedure here is as follows:

$$\theta(x, \tau) = \theta_1(x) + \theta_2(x, \tau) \tag{4.2.32}$$

where $\theta_1(x)$ is a steady-state solution, with generation.

The equations and boundary conditions for $\theta_1(x)$ and $\theta_2(x, \tau)$, the Poisson and Fourier equations, respectively, are

$$\theta_1(x): \quad = \frac{d^2\theta_1}{dx^2} + \frac{q'''}{k} = 0 \qquad \theta_1(0) = \theta_1(L) = 0 \tag{4.2.33}$$

$$\theta_2(x, \tau): \quad \frac{\partial^2\theta_2}{\partial x^2} = \frac{1}{\alpha} \frac{\partial \theta_2}{\partial \tau} \qquad \theta_2(0, \tau) = \theta_2(L, \tau) = 0 \tag{4.2.34a}$$

The initial condition for θ_2 is written to satisfy Eq. (4.2.31) as

$$\theta_i(x, 0) = \theta_1(x) + \theta_2(x, 0)$$

$$\tag{4.2.34b}$$

or
$$\theta_2(x, 0) = \theta_i - \theta_1(x) = \theta_{2,i}$$

The solution for $\theta_1(x)$ is determined as in Sec. 2.4.1. However, for the conditions in Eq. (4.2.33) it is

$$\theta_1(x) = \frac{q'''L^2}{2k}\left(1 - \frac{x}{L}\right)\frac{x}{L} \tag{4.2.35}$$

Therefore,

$$\theta_{2,i} = \theta_i - \frac{q'''L^2}{2k}\left(1 - \frac{x}{L}\right)\frac{x}{L} = f(x) \tag{4.2.36}$$

where $f(x)$ is now the initial condition to be satisfied by the formulation in Eq. (4.2.34b). The general solution for $\theta_2(x, \tau)$ is Eq. (4.2.2). For the boundary conditions in Eq. (4.2.34a), the solution becomes

$$\theta_2(x, \tau) = \sum_{n=1}^{\infty} C_n e^{-(n\pi/L)^2 \alpha\tau} \sin\frac{n\pi}{L}x \tag{4.2.3}$$

The initial condition, Eq. (4.2.36), determines the C_n as follows; see Schneider

(1955):

$$C_n = \frac{2}{L} \int_0^L f(x) \sin\frac{n\pi}{L} dx = \frac{4\theta_i}{n\pi} - \frac{4q'''L^2}{k(n\pi)^3} \qquad (4.2.37)$$

where $n = 1, 3, 5$ and $C_n = 0$ for n even.

The first component in the previous result for C_n is the initial condition θ_i, without the source, q''', as in Eq. (4.2.5). The second is the correction arising from the source. For the solution $\theta(x, \tau)$, Eq. (4.2.37) is applied to Eq. (4.2.3). Then $\theta_2(x, \tau)$ is added to Eq. (4.2.35), for $\theta_1(x)$, to give the solution as follows:

$$\frac{\theta(x,\tau)}{Q_1} = \frac{1}{2}\left(1 - \frac{x}{L}\right)\frac{x}{L} + \frac{4}{\pi}\sum_{n=1,3}^{\infty}\frac{e^{-(n/L)^2\alpha\tau}}{n}\left[\frac{\theta_i}{Q_1} - \frac{1}{(n\pi)^2}\right]\sin\frac{n\pi}{L}x$$

$$(4.2.38)$$

where $Q_1 = q'''L^2/k$ is a characteristic temperature difference Δt_{ch} resulting from the distributed source.

AN EXPONENTIAL DECAY IN SOURCE STRENGTH. In Sec. 2.4.2 a linear absorption process indicated that local distributed source strength $q'''(x)$ varies across the region as

$$q'''(x) = \mu q_0'' e^{-\mu x} \qquad (4.2.39)$$

where μ is the attenuation coefficient in the region for the irradiation and q_0'' is the irradiation intensity at $x = 0$. Equation (4.2.30) then becomes

$$\frac{\partial^2\theta}{\partial x^2} + \frac{\mu q_0''}{k}e^{-\mu x} = \frac{1}{\alpha}\frac{\partial\theta}{\partial\tau} \qquad (4.2.40)$$

The initial and boundary conditions shown in Fig. 4.2.14(b) (see page 201), in terms of $\theta(x, \tau) = t(x, 0) - t_0$ and $\theta_i = t_i - t_0$, with the left boundary insulated, are

$$\theta(x, 0) = \theta_1(x) + \theta_2(x, 0) = \theta_i$$

$$\theta(L, \tau) = 0 \quad \text{and} \quad (\partial\theta/\partial x) = 0 \quad \text{at } x = 0 \qquad (4.2.41)$$

The form of the solution is again taken as Eq. (4.2.32) and the equations and conditions in Eq. (4.2.41) become

$$\theta_1(x): \quad \frac{d^2\theta_1}{dx^2} + Q'e^{-\mu x} = 0 \qquad \theta_1(L) = 0 = (d\theta_1/dx) \quad \text{at } x = 0$$

$$(4.2.42)$$

$$\theta_2(x, \tau): \quad \frac{\partial^2\theta_2}{\partial x^2} = \frac{1}{\alpha}\frac{\partial\theta_2}{\partial\tau} \qquad \theta_2(L, \tau) = 0 = (\partial\theta_2/\partial x) \quad \text{at } x = 0$$

$$(4.2.43)$$

$$\theta_{2,i}(x, 0) = \theta_i - \theta_1(x) = f(x) \qquad (4.2.44)$$

where Q' in Eq. (4.2.42) is $\mu q_0''/k$. The second conditions in Eqs. (4.2.42) and (4.2.43) satisfy the insulating condition at $x = 0$.

Referring to the general solution in Sec. 2.4.2, the result for $\theta_1(x)$ becomes

$$\theta_1(x) = \frac{Q'}{\mu^2}\left[\mu\left(1 - \frac{x}{L}\right) + e^{\mu x} - e^{-\mu x}\right] \tag{4.2.45}$$

The conditions in Eq. (4.2.43) indicate the following form of $\theta_2(x, \tau)$:

$$\theta_2(x, \tau) = \sum_{n=1}^{\infty} C_n e^{-(2n-1/2)^2(\pi/L)^2\alpha\tau} \cos\frac{(2n-1)\pi}{2L}x \tag{4.2.46}$$

At $\tau = 0$, this must satisfy the initial condition, in Eq. (4.2.44), using Eq. (4.2.45). Thereby, the C_n are determined and the solution is known.

SUMMARY. The first solution given previously shows the effects of Q_1 and θ_i on the temperature distribution. For Q_1 large, the effect of the transient is small. For θ_i large compared to Q_1, the effect of internal generation is small. In the second solution, large magnitudes of Q' and θ_i play the same roles. In the two examples, Q_1 and Q'/μ^2 are the characteristic temperatures. Their magnitude, compared to θ_i, indicates the importance of the effect of generation on the temperature distribution.

4.3 TRANSIENTS IN CYLINDRICAL AND SPHERICAL REGIONS

The foregoing section gives results for many kinds of transient responses in bounded and infinite plane regions. Analysis was given in detail for characteristic examples. Peculiarities of some aspects of the results are discussed. These examples are useful in many applications. In addition, the results are often of simple form and the detailed nature of the solution is apparent.

This section concerns one-dimensional transients, in long cylinders, and in spheres, with uniform initial and boundary conditions. Then, $t = t(r, \tau)$. Results for hollow spheres are also given. Sections 4.3.1 and 4.3.2 both consider a temperature boundary condition and also a distributed energy source q''', for each geometry, respectively. Section 4.3.3 then considers both geometries for a convection boundary condition, $h(t_s - t_e)$. The presentations of these convection results are very similar and are in the same form as for the plane layer in Sec. 4.2.2, which resulted in Eqs. (4.2.10) and (4.2.11).

4.3.1 Cylindrical Regions with Temperature Conditions

The geometry is seen in Fig. 4.3.1, along with the two conditions considered. Solutions of the following equation are in terms of Bessel functions:

$$\frac{\partial^2 t}{\partial r^2} + \frac{1}{r}\frac{\partial t}{\partial r} + \frac{q'''}{k} = \frac{1}{\alpha}\frac{\partial t}{\partial \tau} \tag{4.3.1}$$

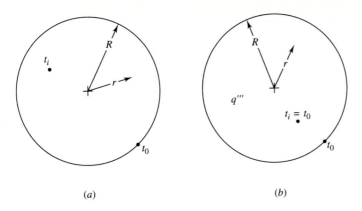

(a)   (b)

FIGURE 4.3.1
The cylindrical geometry, with the two conditions used.

The solutions are given in the following discussion in terms of $\phi(r,\tau) = (t - t_0)/(t_i - t_0)$, where applicable.

INITIAL TEMPERATURE t_i. The solution for $t(R,\tau) = t_0$ for $\tau > 0$ and $q''' = 0$ is

$$\phi = \frac{t(r,\tau) - t_0}{t_i - t_0} = \frac{2}{R} \sum_{n=1}^{\infty} e^{-\alpha\beta_n^2\tau} \frac{J_0(r\beta_n)}{\beta_n J_1(R\beta_n)} \tag{4.3.2}$$

where $J_0(r\alpha_n)$ and J_1 are Bessel functions of the first kind of order zero and one. The β_n are determined as the positive roots of $J_0(\beta_n R) = 0$ in

$$1 = A_1 J_0(\beta_1 r) + A_2 J_0(\beta_2 r) + \cdots \tag{4.3.3}$$

and

$$A_n = \frac{2}{R^2 J_1^2(R\beta_n)} \int_0^R r J_0(r\beta_n)\, dr \tag{4.3.4}$$

The average of the temperature excess ϕ at any time is

$$\bar{\phi} = \frac{4}{R^2} \sum_{n=1}^{\infty} \frac{1}{\beta_n^2} e^{-\beta_n^2\tau} \tag{4.3.5}$$

INITIAL TEMPERATURE t_0, SUDDEN UNIFORM INTERNAL GENERATION. The solution is similar to that given previously. However, there is the usual added term for the eventual steady state. Also a characteristic temperature $\Delta t_{\text{ch}} = q'''R^2/4k$ arises in $\phi(r,\tau)$:

$$\phi = \frac{t(r,\tau) - t_0}{q'''R^2/4k} = 1 - \left(\frac{r}{R}\right)^2 - 8 \sum_{n=1}^{\infty} e^{-\alpha\beta_n^2\tau} \frac{J_0(r\beta_n)}{R^3\beta_n^3 J_1(R\beta_n)} \tag{4.3.6}$$

where the α_n are determined as the positive roots of $J_0(R\beta_n) = 0$. The result is plotted in Fig. 4.3.2, where the response is in terms of $\alpha\tau/R^2 = \text{Fo}$.

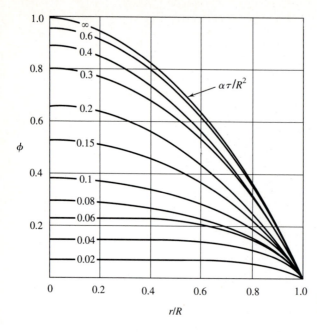

FIGURE 4.3.2
Temperature response from a uniformly distributed input q''', where the contours are $\alpha\tau/R^2$. [From Carslaw and Jaeger (1959).]

SUMMARY. These kinds of solutions are much more complicated than for plane regions. The evaluations of β_n are more complicated, although some values are tabulated. For further results see Carslaw and Jaeger (1959) and Özisik (1980).

4.3.2 Spherical Regions with Temperature Conditions

The geometry is shown in Fig. 4.3.3 with several kinds of boundary conditions. The solutions are determined from

$$\frac{\partial^2 t}{\partial r^2} + \frac{2}{r}\frac{\partial t}{\partial r} + \frac{q'''}{k} = \frac{1}{\alpha}\frac{\partial t}{\partial \tau} \tag{4.3.7}$$

These solutions are again given in terms of ϕ.

INITIAL TEMPERATURE t_i. For $t(R,\tau) = t_0$ for $\tau > 0$ the result, for $q''' = 0$, in terms of Fig. 4.3.3(a), is

$$\phi = \frac{t - t_0}{t_i - t_0} = \frac{R}{r}\sum_{n=0}^{\infty}\left[\text{erfc}\frac{(2n+1)R - r}{2\sqrt{\alpha\tau}} - \text{erfc}\frac{(2n+1)R + r}{2\sqrt{\alpha\tau}}\right] \tag{4.3.8}$$

The decay characteristic of this solution at large $\sqrt{\tau}$ is in terms of the conjugate error functions.

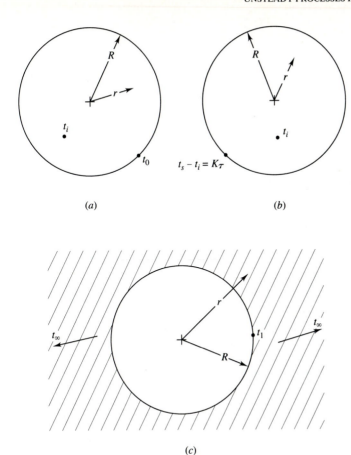

FIGURE 4.3.3
The spherical geometry with the three conditions used.

LINEAR SURFACE TEMPERATURE INCREASE. The initial temperature is t_i and $t_s - t_i = t(R, \tau)$ increases as $t_s - t_i = K\tau$. See Fig. 4.3.3(b). The temperature in the spherical region increases as

$$\frac{t(x, \tau) - t_i}{K\tau} = \left[1 - \frac{1}{6\,\mathrm{Fo}}\left(1 - \frac{r^2}{R^2}\right)\right]\left[-\frac{2}{\pi^3\,\mathrm{Fo}}\frac{R}{r}\right.$$

$$\left. \times \sum_{n=1}^{\infty} \frac{(-1)^n}{n^3} e^{-\alpha n^2 \pi^2 \tau / R^2} \sin\frac{n\pi r}{R}\right] \qquad (4.3.9)$$

where, for large $\mathrm{Fo} = \alpha\tau/R^2$, the two subtractive terms Fo^{-1} decay.

A SPHERICAL CAVITY IN A REGION AT t_∞. For the surface temperature $t(R, 0) = t_\infty$ and $t(R, \tau) = t_1$ for $r > 0$, the temperature response, for $r \geq R$, see Fig. 4.3.3(c), is

$$\phi = \frac{t(r,\tau) - t_\infty}{t_1 - t_\infty} = \frac{R}{r}\,\text{erfc}\,\frac{r - R}{2\sqrt{\alpha\tau}} \tag{4.3.10}$$

This decays toward R/r as $\sqrt{\tau}$ increases.

SUMMARY. These three solutions show behavior characteristics of spherical regions. The solutions are relatively simple. The following section gives results for convection surface conditions for both the cylindrical and spherical geometries.

4.3.3 Surface Convection Conditions

These are very important results, in terms of applications and design. The solution, ϕ, for a plane layer of thickness $2s$, is given in Eqs. (4.2.10) and (4.2.11). The same form results for both the long cylinder and the sphere. The notation is seen in Fig. 4.3.4. Solutions for both geometries are given by Carslaw and Jaeger (1959) and by Jakob (1949). However, given the importance of these solutions in use, they have often been represented graphically. The form is the same as for the plane region, as shown in Figs. 4.2.4, 4.2.5, 4.2.7, and 4.2.8.

The functional form of the solutions is as given in Eq. (4.2.11), for plane regions, with x replaced by r, as

$$\phi = \phi(r/s, \alpha\tau/s^2, hs/k) = \phi(r/s, \text{Fo}, \text{Bi}) \tag{4.3.11}$$

Figures 4.3.5 and 4.3.6 give θ_0/θ_i versus $\text{Fo} = \alpha\tau/s^2$, with contours of $\text{Bi}^{-1} = (hs/k)^{-1}$, for the long cylinder and sphere, respectively. Again, $\theta_0 = \theta(0, 0, \text{Bi})$ and $\theta_i = t_i - t_e$. These results give the midpoint temperature decay with increasing Fo.

In Figs. 4.3.7 and 4.3.8, $\theta(r/s)/\theta_0$, where $r/s = 0.2, 0.4, 0.6, 0.8, 0.9$, and 1.0, is given as a function of Bi^{-1}. The results, normalized by θ_0, apply accurately for $\text{Bi} < 100$ and for times $\text{Fo} = \alpha\tau/s^2 > 0.2$. These curves determine the instantaneous temperature distributions, in terms of θ.

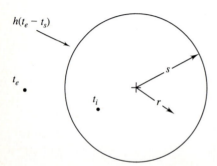

$h(t_e - t_s)$

t_e

s

t_i

r

FIGURE 4.3.4
The cylindrical and spherical regions of radius s, initially at $t(r, 0) = t_i$.

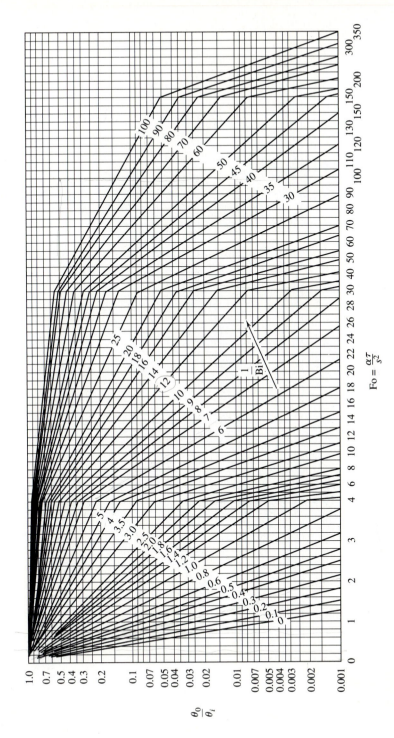

FIGURE 4.3.5
Axis temperature of a long cylinder. [From Heisler (1947).]

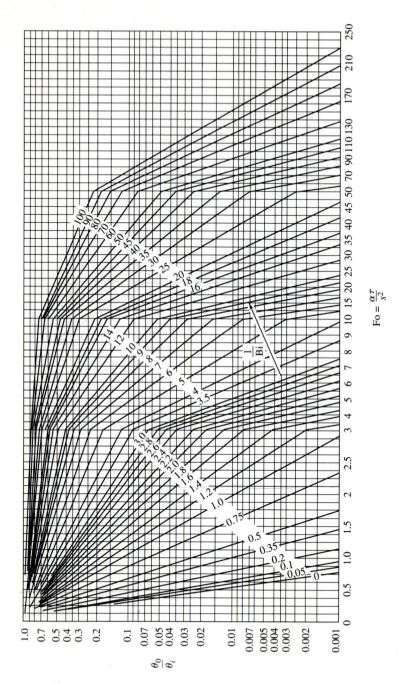

FIGURE 4.3.6
Center temperature of a sphere. [From Heisler (1947).]

210

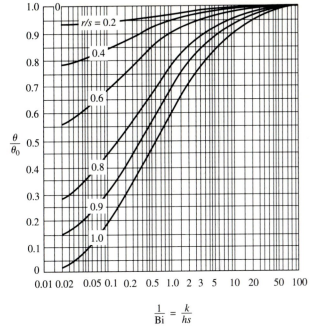

FIGURE 4.3.7
The dependence of the temperature distribution on Bi, for a long cylinder. [From Heisler (1947).]

$$\frac{1}{\text{Bi}} = \frac{k}{hs}$$

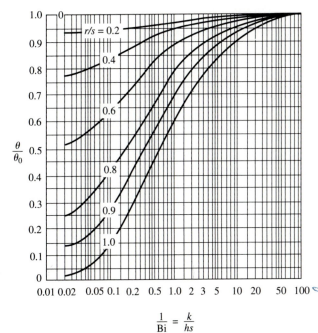

FIGURE 4.3.8
The dependence of the temperature distribution on Bi for a sphere. [From Heisler (1947).]

$$\frac{1}{\text{Bi}} = \frac{k}{hs}$$

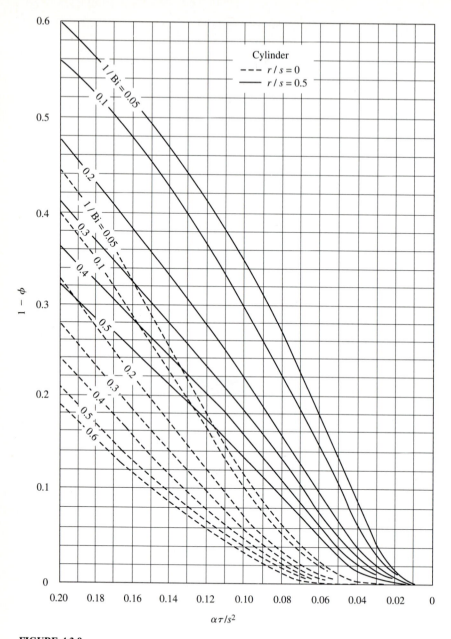

FIGURE 4.3.9
Temperature in long cylinders at $x = 0$ and $s/2$, as $1 - \phi_0$ and $1 - \phi_{s/2}$, at short times. [From Heisler (1947).]

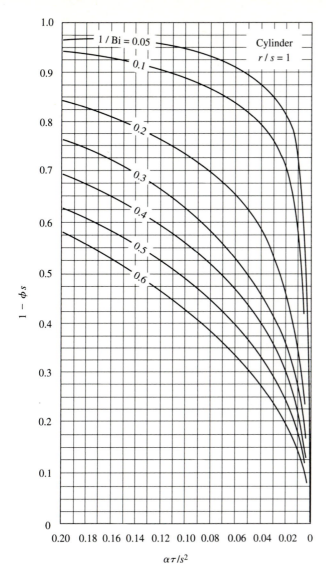

FIGURE 4.3.10

Temperature in long cylinders at $r = s$, as $1 - \phi_s$, at short times. [From Heisler (1947).]

Finally, midpoint temperature decay θ_0/θ_i is plotted for all three geometries in Fig. 4.2.6. The value of $m = 1, 2, 3$ selects the geometry. These results apply for Fo $= \alpha\tau/s^2 > 0.2$ and Bi < 0.01. A small value of Bi results from $h \ll k/s$, that is, h is small compared to internal conductance k/s. All of these solutions are very valuable in obtaining convenient estimates of temperature response.

As for the plane layer in Sec. 4.2.2, Figs. 4.3.5 and 4.3.6 do not give sufficient information at small values of Fo. Short-time response plots are shown

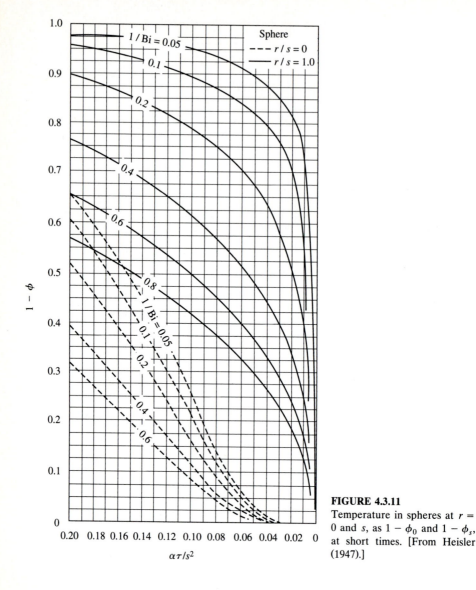

FIGURE 4.3.11
Temperature in spheres at $r = 0$ and s, as $1 - \phi_0$ and $1 - \phi_s$, at short times. [From Heisler (1947).]

in Figs. 4.3.9 through 4.3.12. They give high resolution of the early part of the transient.

Recall from Sec. 4.2.2 that the Heisler charts in Figs. 4.2.4, 4.2.5, 4.2.7, and 4.2.8, for a plane region of thickness $2s$, also apply when an induction surface-heating process, q_s'', parallels surface convection. This same situation also applies for the cylindrical and spherical region results in Figs. 4.3.5 through 4.3.12. The ordinates in these figures are then modified as discussed in Sec. 4.2.2 for the plane region.

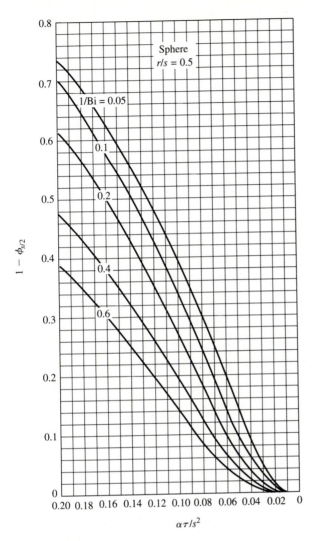

FIGURE 4.3.12
Temperatures in spheres at $r = s/2$, as $1 - \phi_{s/2}$, at short times. [From Heisler (1947).]

4.4 PERIODIC PROCESSES

The foregoing processes described in this chapter concern transient conduction. Most concerned an initial state in a region, t_i, suddenly subject to changed bounding or to internal conditions, such as a distributed energy source. The conduction process which follows is then toward a new eventual steady-state distribution, throughout the region.

However, in many circumstances the imposed boundary or internal conditions are approximated as simply periodic in time, at some frequency f. Common examples are the daily variation of air temperature or of solar loading at a bounding surface. However, such an effect may also be combined with another periodic component, for example, due to the annual variation of such effects. Then there are two periodic components. These might be modeled as two superimposed purely steady periodic effects, of frequencies f_1 and f_2. There are also many periodic processes which are combined with a transient effect, arising from an initial condition. An example is the cylinder wall of a reciprocating combustion engine, heating up after being started.

Section 4.4.1 gives solutions of several kinds of steady periodics in a half-infinite region. The solution for the starting periodic transient is also discussed. Then steady periodics are discussed in which the input is a combination of harmonics. Section 4.4.2 concerns steady periodics in a plane region of thickness $2s$.

4.4.1   Periodics in a Half-Infinite Region

For the geometry in Fig. 4.4.1(a), the surface temperature $\theta(0, \tau)$ will be taken as a steady periodic. The first example is an imposed surface temperature condition of the following form:

$$\theta(0, \tau) = \theta_a \cos \omega \tau \quad \text{at } x = 0 \tag{4.4.1}$$

where τ is measured from a time level after which all transient effects of the beginning of this periodic surface variation have died out. Thereafter, at each

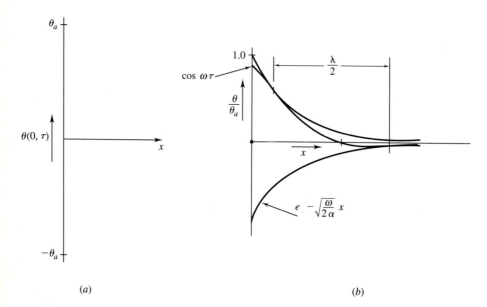

(a)   (b)

FIGURE 4.4.1

A periodic in a half-infinite region: (a) the boundary condition; (b) an instantaneous temperature distribution $\theta(x, \tau)$ and the envelopes of local maxima and minima.

location, the temperature will vary periodically with time at a fixed amplitude. In Eq. (4.4.1) θ_a is the amplitude of the imposed periodic effect. The local amplitude, $\theta_a(x)$, damps into the region. The long time average temperature at all locations, t_m, is zero. The net conduction at all locations in the region, over a cycle, is also zero.

The following relevant equation has solutions of the form given in Eqs. (4.1.5)–(4.1.7):

$$\frac{\partial^2 \theta}{\partial x^2} = \frac{1}{\alpha}\frac{\partial \theta}{\partial \tau}$$

A purely time-periodic solution is obtained only with an imaginary exponent of e in Eq. (4.1.7). Since the "angular velocity" of the periodic variation must be ω, in radians per unit time, the quantity β^2 may be

$$\beta^2 = \frac{i\omega}{\alpha} \quad \text{and} \quad f = \frac{\omega}{2\pi}$$

Then β is complex as

$$\beta = \pm\sqrt{\frac{i\omega}{\alpha}} = \pm(1+i)\sqrt{\frac{\omega}{2\alpha}}$$

Equation (4.1.5) is used for $X(x)$ in this circumstance, and the following is a solution:

$$\theta = Ce^{-i\omega\tau}\left\{ C_1 \exp\left[\pm i(1+i)\sqrt{\frac{\omega}{2\alpha}}\,x\right] + C_2 \exp\left[\mp i(1+i)\sqrt{\frac{\omega}{2\alpha}}\,x\right]\right\}$$

$$= A\exp\left[-i\left(\omega\tau \mp \sqrt{\frac{\omega}{2\alpha}}\,x\right)\right]\exp\left(\mp\sqrt{\frac{\omega}{2\alpha}}\,x\right)$$

$$+ B\exp\left[-i\left(\omega\tau \pm \sqrt{\frac{\omega}{2\alpha}}\,x\right)\right]\exp\left(\pm\sqrt{\frac{\omega}{2\alpha}}\,x\right)$$

Neither the plus nor the minus sign is consistent with the requirement that temperature must be finite, unless either A or B is zero. Setting $B = 0$,

$$\theta = \exp\left(\mp\sqrt{\frac{\omega}{2\alpha}}\,x\right)\left[a_1\cos\left(\omega\tau \mp \sqrt{\frac{\omega}{2\alpha}}\,x\right) - a_2\sin\left(\omega\tau \mp \sqrt{\frac{\omega}{2\alpha}}\,x\right)\right]$$

The plus sign in the exponent is inadmissible, and the boundary condition, at $x = 0$, requires that $a_2 = 0$ and $a_1 = \theta_a$. Therefore, the solution is

$$\theta(x,\tau) = \theta_a e^{-\sqrt{\omega/2\alpha}\,x}\cos\left(\omega\tau - \sqrt{\frac{\omega}{2\alpha}}\,x\right) = \theta_a(x)\cos\left(\omega\tau - \sqrt{\frac{\omega}{2\alpha}}\,x\right)$$

$$(4.4.2)$$

where the local amplitude is $\theta_a(x)$.

The kinds of conduction processes for which this is a solution have many interesting characteristics. The coefficient of the cosine term in Eq. (4.4.2) is the local amplitude. This amplitude decays very fast, exponentially with x. This characteristic is shown in Fig. 4.4.1(b) (see page 216). As a result of this rapid decay of periodic disturbances, the equations for semiinfinite solids may often be applied to thick plane regions.

The temperature varies periodically everywhere. The distance between adjacent maxima at any instant in time is called the wavelength λ. This may be found as the difference in x which will cause a difference of 2π in the argument of the cosine, at any given value of τ:

$$\Delta\left(\omega\tau - \sqrt{\frac{\omega}{2\alpha}}\,x\right) = \sqrt{\frac{\omega}{2\alpha}}\,\Delta x = \sqrt{\frac{\omega}{2\alpha}}\,\lambda = 2\pi$$

$$\lambda = 2\pi\sqrt{\frac{2\alpha}{\omega}} = \sqrt{4\pi\alpha/f}$$

(4.4.3)

Thus λ varies with frequency $f = \omega/2\alpha$ as $1/\sqrt{f}$. For soil, $\lambda = 2.4$ cm, 93 cm, and 17.8 m, for f of one cycle per minute, day, and year, respectively. In steel, $\lambda = 0.14$ and 1.1 cm for $f = 60$ and 1 Hz, respectively.

The velocity of the waves may be found as the distance between local maxima, divided by the period, which is the time necessary for a crest to travel this distance:

$$v = 2\pi\frac{\sqrt{2\alpha/\omega}}{2\pi/\omega} = \sqrt{2\alpha\omega} = \sqrt{4\pi\alpha f}$$

(4.4.4)

This result shows that these thermal waves are propagated at a velocity which depends upon the characteristic of the material, α, and upon the frequency of the disturbance. This is in contrast with the propagation of sound in a gas. The velocity of small pressure waves is independent of frequency. The heat flux at $x = 0$ is also periodic as

$$q''(0, \tau) = -k\left(\frac{\partial\theta}{\partial x}\right)_0 = k\theta_a\frac{\sqrt{\omega}}{\alpha}\cos\left(\omega\tau + \frac{\pi}{4}\right)$$

(4.4.5)

The flux and temperature variations are 45° out of phase with each other. The flux leads. Recall that the net heat input per cycle is zero. Also, from the temperature distribution function, Eq. (4.4.2), the time average temperature excess θ is zero at all x. The steady-periodic solution for an imposed periodic surface flux, $q''(0, \tau) = q''_{0,a}\cos\omega\tau$, is indicated in Eq. (4.4.5), compared with Eq. (4.4.2). The time variation of θ is the same. The temperature response now lags the flux by $\pi/4$.

The preceding solution is the steady periodic. It applies only after any starting transient has decayed. The result, over the whole region, is a uniform average temperature t_m. Carslaw and Jaeger (1959), page 64, give the solution for the whole process, including the starting transient. The initial temperature is

t_m and the eventual solution is Eq. (4.4.2). The added transient term is

$$-\frac{\theta_a}{\sqrt{\pi}} \int_0^{x/2\sqrt{\alpha\tau}} \cos\left[\omega(\tau - x^2/4\alpha\mu^2)\right]e^{-\mu}\,d\mu$$

where μ is the variable of integration. This term decays to zero as τ increases.

CONVECTION AT THE SURFACE. If the exposed face of a semiinfinite solid is in contact with a fluid whose temperature varies sinusoidally and the constant convection coefficient is h, the formulation in terms of $\theta(x, \tau) = t(x, \tau) - t_m$ is

$$\frac{\partial^2\theta}{\partial x^2} = \frac{1}{\alpha}\frac{\partial\theta}{\partial\tau}$$

At $x = 0$: $q''(0, \tau) = -k\dfrac{\partial\theta}{\partial x} = h\left[t_e(\tau) - t(0,\tau)\right] = h\left[\theta_e - \theta(0,\tau)\right]$

$$(4.4.6a)$$

$$\theta_e = t_e(\tau) - t_m = \theta_{a,e}\cos\omega\tau \qquad (4.4.6b)$$

where $t_e(\tau)$ and t_m are, respectively, the instantaneous and average temperatures of the fluid. The solution is

$$\theta(x, \tau) = a\theta_{a,e}e^{-\sqrt{\omega/2\alpha}\,x}\cos\left(\omega\tau - \frac{\sqrt{\omega}}{2\alpha}x - b\right) \qquad (4.4.7a)$$

where

$$a = \sqrt{\frac{1}{2c^2 + 2c + 1}} \qquad (4.4.7b)$$

$$b = \arctan\frac{c}{c + 1} \qquad (4.4.7c)$$

$$c = \sqrt{\frac{k\omega\rho c}{2h^2}} \qquad (4.4.7d)$$

The factor a in Eq. (4.4.7b) indicates the damping amplitude due to the surface resistance of the convection process. A large surface resistance, that is, a small h, leads to large values of c and small values of a. There is an additional lag angle b for the temperature variation. It applies even at $x = 0$.

SUMMARY. The preceding results are the conduction responses to the single periodic component surface conditions in Eqs. (4.4.1) and (4.4.6b). These results may be very simply extended to imposed oscillations which contain many individual frequency components, as $\omega, 2\omega, \ldots$, formulated as follows.

$$\theta(0, \tau) = \theta_{a,1}\cos\omega\tau + \theta_{a,2}\cos 2\omega\tau + \cdots \qquad (4.4.8)$$

and

$$\theta(0, \tau) = \theta_{e,1}\cos\omega\tau + \theta_{e,2}\cos 2\omega\tau + \cdots \qquad (4.4.9)$$

Since the differential equation is linear, these effects may be simply superimposed. That is, $\theta(x, \tau)$ becomes a sum of solutions like Eqs. (4.4.2) and (4.4.7a)

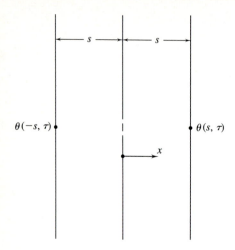

$\theta(-s, \tau)$

$\theta(s, \tau)$

x

FIGURE 4.4.2
A surface of thickness $2s$, subject to an imposed periodic surface temperature variation on each face, in phase.

in terms of the amplitudes of the harmonic components $\theta_{a,i}$ and $\theta_{e,i}$. Carslaw and Jaeger (1959), page 69, apply the surface temperature formulation in Eq. (4.4.8) as a Fourier series representation of a "square-wave" form of surface input $t(0, \tau) - t_m$.

4.4.2 A Periodic in a Plane Layer

The total thickness, as shown in Fig. 4.4.2, is $2s$. Both surfaces are subject to the same periodic temperature variation, in phase, as

$$\theta(s, \tau) = \theta(-s, \tau) = \theta_{s,a} \cos \omega \tau \qquad (4.4.10)$$

Only the solution, from Jakob (1949), Chap. 14, is given here, as

$$\theta(x, \tau) = \theta_{s,a} F\left(\frac{x}{s}, \sqrt{\frac{\omega s^2}{2\alpha}}\right) \cos\left[\omega \tau + T\left(\frac{x}{s}, \sqrt{\frac{\omega s^2}{2\alpha}}\right)\right] \qquad (4.4.11)$$

The function T is as follows:

$$T = \tan^{-1} \frac{f_1(\sigma) f_2(\sigma x/s) - f_2(\sigma) f_1(\sigma x/s)}{f_1(\sigma) f_1(\sigma x/s) + f_2(\sigma) f_2(\sigma x/s)} \qquad (4.4.12a)$$

where σ and $\sigma x/s$ are denoted by z, and $f_1(z)$ and $f_2(z)$ are

$$f_1(z) = \cos z \cosh z \qquad f_2(z) = \sin z \sinh z \qquad (4.4.12b)$$

and

$$\sigma = \sqrt{\frac{\omega s^2}{2\alpha}} \qquad (4.4.12c)$$

Values for F are given in Table 4.4.1.

TABLE 4.4.1
The values of the amplitude decay function F in Eq. (4.4.11),
as it depends on $\sigma = \sqrt{\omega s^2 / 2\alpha}$ and x / s

σ	0	$\frac{1}{8}$	$\frac{2}{8}$	$\frac{3}{8}$	$\frac{4}{8}$	$\frac{5}{8}$	$\frac{6}{8}$	$\frac{7}{8}$	1
0	1	1	1	1	1	1	1	1	1
0.5	0.98	0.98	0.98	0.98	0.98	0.98	0.98	0.99	1
1.0	0.77	0.77	0.77	0.78	0.79	0.81	0.85	0.91	1
1.5	0.47	0.47	0.47	0.48	0.52	0.58	0.68	0.83	1
2.0	0.27	0.27	0.28	0.30	0.36	0.45	0.58	0.77	1
4.0	0.04	0.04	0.05	0.08	0.13	0.22	0.37	0.64	1
8.0	0.00	0.00	0.01	0.01	0.02	0.05	0.14	0.36	1
∞	0	0	0	0	0	0	0	0	...

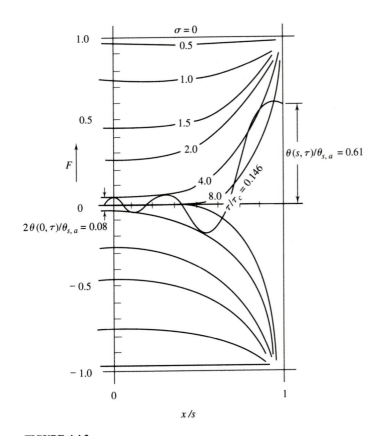

FIGURE 4.4.3
A temperature distribution in a plane region of thickness $2s$ for $\sigma = 4.0$, at time $\tau/\tau_c = 0.146$, where $\tau_c = 2/\omega$, the period of the oscillation. [From Jakob (1949).]

For a given circumstance, $\sigma = \sqrt{\omega s^2/2\alpha}$ is known. Then T is evaluated from Eq. (4.4.12) for various values of x/s. The attenuation factor, or decay function, F, is found in Table 4.4.1. Then Eq. (4.4.11) is the solution.

This temperature distribution is symmetric about $x = 0$, as shown in Fig. 4.4.3 for $\sigma = 4$, at a time when a local maximum appears at $x/s = 0$. Then $\theta(0, \tau)/\theta_{s,a} = 0.61$ and is decreasing. The other curves shown, for $\sigma = \sqrt{\omega s^2/2\alpha} = 0$ to 8, are the amplitude envelopes F. At higher σ, that is, higher ω, damping is much more rapid into the region.

An example of the practical value of such results is in assessing the effectiveness of regenerative heaters. These operate as a fixed thermal capacity element, being periodically subjected to fluid streams at different temperature levels. Then σ indicates the amount of penetration of the periodic temperature effect into the thermal storage region. High effectiveness is calculated for σ small, as implied by the amplitude distributions seen in Fig. 4.4.3.

4.5   TRANSIENTS IN INTERACTING CONDUCTION REGIONS

There are many processes in which transient temperature fields in adjacent materials interact at their interfaces. Steady-state conduction through composite barriers is discussed in Sec. 2.1.3. Interfacial contact resistance is described in Secs. 2.1.4 and 3.8, including the effect of contact region thermal capacity in transients.

At the interface between any two regions in steady state, the temperature levels and normal components of steady-state heat flux q_n'' are matched. With contact resistance, R_c, a contact region temperature difference, $\Delta t_c = q_n'' R_c$, arises. In transients, the instantaneous interface normal heat flux components q_n'' again are matched, but in terms of the local transient temperature distribution in each adjacent region. The Δt_c, as defined previously, may vary with time. However, if the contact region contents have appreciable thermal capacity, any energy storage in this contact region also changes with time. Then the normal component q_n'' is not conserved across the region. For a contact region material having a thermal capacity of c'' per unit of contact area, and a high conductivity compared to those of the adjacent regions, for example L and R in Fig. 4.5.1, the relation is

$$q_{n,L}'' = c'' \frac{dt_{c''}}{d\tau} + q_{n,R}'' \tag{4.5.1}$$

or

$$-k_L\left(\frac{\partial t}{\partial n}\right)_L = c'' \frac{dt_{c''}}{d\tau} - k_R\left(\frac{\partial t}{\partial n}\right)_R \tag{4.5.2}$$

These fluxes are evaluated on the two sides and $t_{c''}(\tau)$ is the instantaneous temperature of the contact region material.

The immediate interaction region is sketched in Fig. 4.5.1. Since, in this example, $\partial^2 t/\partial x^2$ is taken as positive in both regions, heating is imposed from

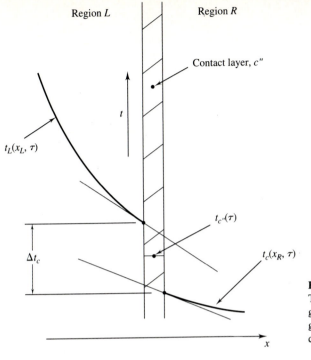

FIGURE 4.5.1
The interface between two regions L and R, with contact region resistance and thermal capacity.

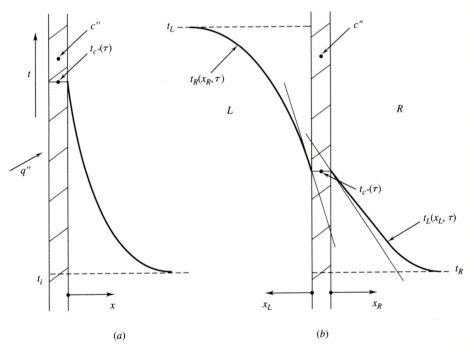

(a)

(b)

FIGURE 4.5.2
Boundary thermal capacity effects: (*a*) at the surface of a region; (*b*) between two regions brought into contact.

the left. The interface slopes of $t_L(x_L, \tau)$ and $t_R(x_R, \tau)$ may result from two different effects; different k_L and k_R and increasing thermal storage in capacity c''.

The following sections concern two aspects of region interaction in one-dimensional transients. The first aspect concerns the effect of a surface layer of thermal capacity as shown in Fig. 4.5.2(a). The surface condition is a flux loading, q'', into a semiinfinite region, with a surface layer having thermal capacity. Figure 4.5.1(b) is an example of two regions brought together, with a layer between. This latter circumstance is similar to the analysis in Sec. 4.2, Fig. 4.2.13, which gave the results in Eqs. (4.2.26) and (4.2.27). However, no intervening layer was included.

The second aspect is the transient response in plane composite barriers, as in Fig. 2.1.2, analyzed there for steady-state heat transfer. The first example which follows, shown in Fig. 4.5.3, is a surface layer of thickness L on a semiinfinite region. The second example is more general, the formulation for transient response in a barrier of M layers, as shown in Fig. 4.5.4.

4.5.1 The Effect of a Surface Layer

A half-infinite region, with a high conductivity surface layer, is shown in Fig. 4.5.2(a), subject to a constant flux loading at the surface. The following equation applies to the region $x \geq 0$:

$$\frac{\partial^2 t}{\partial x^2} = \frac{1}{\alpha}\frac{\partial t}{\partial \tau} \qquad (4.5.3)$$

The conditions written as follows, in terms of $\theta(x, t) = t(x, t) - t_i$, are $\theta(\infty, \tau) = 0$ and

$$q'' = c''\frac{d\theta_{c''}(\tau)}{d\tau} - k\left(\frac{\partial \theta}{\partial x}\right)_{x=0} \qquad (4.5.4)$$

where $\theta_{c''}(\tau) = \theta(0, \tau)$ in the conduction region and the imposed flux q'' is taken at a constant input for $\tau > 0$. The following solution is written in terms of the thermal capacity ratio $\rho c/c'' = b$ and $H = x/2\sqrt{\alpha\tau}$ as

$$\theta(x, \tau) = \frac{q''}{bk}\left[2b\frac{\sqrt{\alpha\tau}}{\pi}e^{-\eta^2} - (1 + bx)\,\text{erfc}\,\eta + e^{b(x+\alpha\tau b)}\,\text{erfc}\left(n + b\sqrt{\alpha\tau}\right)\right]$$

$$(4.5.5)$$

For $c'' = 0$ or $1/b = 0$, this reduces to Eq. (4.2.18), which applies in the absence of a surface thermal capacity layer. A two-region semiinfinite contact is shown in Fig. 4.5.2(b).

4.5.2 Composite Barrier Transients

The results in Sec. 4.5.1 assume one region to be a thin layer of relatively very high diffusivity material. A general consequence of this condition is that the

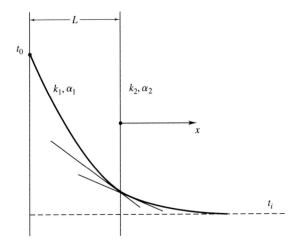

FIGURE 4.5.3
A thin region of thickness L in contact with an infinite region, both initially t_i.

material temperature is uniform at any time, at $t_{c''}(\tau)$, in general. This section considers multiple regions of comparable diffusivity. The first example is the two regions 1 and 2, initially at t_i, in Fig. 4.5.3. The temperature response is given for each region.

The analysis of a more general circumstance of a composite region is then considered. The barrier seen in Fig. 4.5.4 is composed of M conduction layers. The particular transient taken as an example begins from a uniform t_i throughout. For $\tau > 0$ the barrier is subject to convection processes, h_0 to t_0 and h_e to t_e, at the two surfaces. These are at $x = 0$ and at $x = \Sigma_j^M \delta_j = L$. The component layers have thicknesses δ_j and the properties shown. All material interfaces are assumed to be in perfect thermal contact.

A DOUBLE-LAYER REGION. The origin is taken at the interface, in Fig. 4.5.3, for convenience. The properties are k_1, α_1 and k_2, α_2. The local transient temperature is expressed in terms of $\theta_1(x, \tau) = t_1(x, \tau) - t_i$ and $\theta_2(x, \tau) = t_2(x, \tau) - t_i$. The equations and conditions are

$$\frac{\partial^2 \theta_1}{\partial x^2} = \frac{1}{\alpha_1} \frac{\partial \theta_1}{\partial \tau} \quad \text{and} \quad \frac{\partial^2 \theta_2}{\partial x^2} = \frac{1}{\alpha_2} \frac{\partial \theta_2}{\partial \tau} \tag{4.5.6a}$$

$$\theta(x, 0) = 0 \qquad \theta_1(-L, \tau) = t_0 - t_i \qquad \theta_2(\infty, \tau) = 0 \tag{4.5.6b}$$

$$\theta_1(0, \tau) = \theta_2(0, \tau); \quad \text{at } x = 0 \quad -k_1 \frac{\partial \theta_1}{\partial x} = -k_2 \frac{\partial \theta_2}{\partial x} \tag{4.5.6c}$$

The relations in Eq. (4.5.6b) are initial and boundary conditions. In Eq. (4.5.6c) they are interface conditions.

Solutions of the preceding formulation, and of the one to follow for the more general example, are usually found using Laplace transforms, when practical. Carslaw and Jaeger (1959), Sec. 12.8, give the following results for the formulation in Eq. (4.5.6), of the circumstance in Fig. 4.5.3:

$$\frac{t_1(x,\tau) - t_i}{t_0 - t_i} = \phi_1(x,\tau)$$

$$= \sum_{n=0}^{\infty} c^n \left[\text{erfc} \, \frac{(2n+1)L + x}{2\sqrt{\alpha_1 \tau}} - c \, \text{erfc} \, \frac{(2n+1)L - x}{2\sqrt{\alpha_1 \tau}} \right]$$

$$(4.5.7a)$$

$$\frac{t_2(x,\tau) - t_i}{t_0 - t_i} = \phi_2(x,\tau) = \frac{2}{1+b} \sum_{n=0}^{\infty} c^n \, \text{erfc} \, \frac{(2n+1)L + ax}{2\sqrt{\alpha_2 \tau}} \qquad (4.5.7b)$$

where $\quad a = \sqrt{\alpha_1/\alpha_2} \qquad b = ak_2/k_1 \qquad c = (b-1)/(b+1) \qquad (4.5.7c)$

The instantaneous heat flux at $x = -L$ is given as

$$q''(-L,\tau) = -k_1 \frac{\partial \theta_1}{\partial x} = \frac{k_1(t_0 - t_i)}{\sqrt{\pi \alpha_1 \tau}} \left[1 + 2 \sum_{n=1}^{\infty} c^n e^{-n^2 L^2 / \alpha_1 \tau} \right] \quad (4.5.8)$$

For very large times, the quantity in brackets becomes $(k\rho c)_2/(k\rho c)_1$.

A COMPOSITE BARRIER. As shown in Fig. 4.5.4, the origin of x is at the left face. The temperatures of the bounding media are t_0 and t_e, with convection as

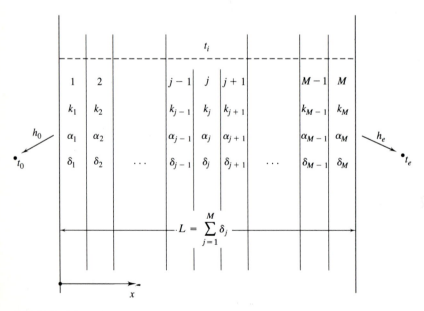

FIGURE 4.5.4
A barrier composed of M layers, with surface convection at both $x = 0$ and $x = \sum_j^M \delta_j$.

h_0 and h_e. Any solution for the transient temperature distribution across the region, $t(x, \tau)$, amounts to finding the temperature distributions in each layer $t_j(x, \tau)$. These result from the imposed ambient temperatures t_0 and $t_e \neq t_i$. Each solution $t_j(x, \tau)$ must be matched to the two adjacent ones, $t_{j-1}(x, \tau)$ and $t_{j+1}(x, \tau)$. The interface temperatures and heat fluxes must be the same on each side of the interface. Consider the jth layer, $j \neq 1$ or M. The equation and bounding conditions are

$$\frac{\partial^2 t_j}{\partial x^2} = \frac{1}{\alpha_j} \frac{\partial t_j}{\partial \tau} \qquad t_j(x, 0) = t_i \qquad (4.5.9)$$

and

$$-k_j \frac{\partial t_j}{\partial x} = -k_{j+1} \frac{\partial t_{j+1}}{\partial x} \quad \text{and} \quad t_j(x, \tau) = t_{j+1}(x, \tau) \qquad (4.5.10)$$

at the interface between layers j and $j + 1$. On the two sides of the barrier, the convection processes result in the necessary interface conditions missing in Eq. (4.5.10), at $x = 0$ and at $x = \sum_j^M \delta_j = L$.

$$\text{at } x = 0: \quad h_0[t_0 - t_1(0, \tau)] = -k_1 \left(\frac{\partial t_1}{\partial x} \right)_{x=0} \qquad (4.5.11)$$

$$\text{at } x = L: \quad h_e[t(L, \tau) - t_e] = -k_M \left(\frac{\partial t_M}{\partial x} \right)_L \qquad (4.5.12)$$

The general method for the formulation of the solution $t(x, \tau)$ is straight-forward, as discussed in Carslaw and Jaeger (1959), Sec. 12.8, and in Özisik (1980), Sec. 8-1 and 8-2. However, obtaining solutions for a number of layers requires very intricate algebraic operations, to match all interface temperature and flux conditions. Therefore, the preceding formulation is often more important as a guide to the numerical modeling of such transient response in composite barriers.

The foregoing procedure amounts to simultaneously matching both the interface temperature level and the heat flux at each material interface. An alternative method, in a numerical analysis, is to directly match the heat flux across each internal interface of the composite region. This kind of numerical model is developed in Sec. 9.5.1, including the effects of variable properties.

The complicating aspects of the formulation in Eqs. (4.5.9)–(4.5.12) are that all calculations must be made simultaneously. Also, the interface temperatures and conductivities are not known, as between layers j and $j + 1$ in Fig. 4.5.4. This deficiency may be overcome, in a numerical method, by postulating a single effective conductivity k_s at each material interface location across the composite region. For the same grid spacing Δx across each interface, $k_s = 2k_L k_R/(k_L + k_R)$. Here, k_L and k_R are the instantaneous values of k one grid spacing away from the interface, on each side. Recall the definition of k_s in Eq. (3.3.12). Also see the analysis leading to Eq. (9.5.9).

The advantage of this numerical procedure is that the condition of flux continuity across each interface is automatically satisfied. Only the interface temperatures then need be calculated. For a discussion concerning nonuniform grids and other aspects of this procedure, see Patankar (1980).

4.6   FREEZING AND MELTING FRONTS

In many processes the initial and changed temperature conditions span the phase equilibrium temperature t_{PE} of the material in the conduction region. The propagating temperature field then results in solid melting, or in freezing into a body of liquid. A common example is a region of ordinary ice at a temperature of $t_i \leq 0°C$, when the surface is raised to $t_0 > 0°C$. Then a melting front, at $t = 0°C$, propagates into the conduction region. An inverse process occurs when a liquid metal is poured into a mold to solidify. Then a freezing front propagates into the liquid metal, from the colder surfaces of the mold. These two kinds of processes are idealized in Fig. 4.6.1(a) and (b).

The instantaneous phase interface is at $X(\tau)$. The interface velocity, $u(\tau) = dX(\tau)/d\tau$, decreases with time, for fixed t_0 and t_∞, as the conduction resistance between the surface and the phase interface increases. The process in Fig. 4.6.1(a) is a transient conduction of heat inward to $x = X(\tau)$. The heat flux, in the liquid at the interface, is $q_L''(X, \tau)$. The latent heat absorption rate, per unit area at the front, is $q_{PC}'' = \rho u H$, where H is the latent heat per unit mass and ρ is the material density. Here it is assumed that $\rho_L = \rho_S$, for simplicity. The instantaneous temperature distributions in the liquid and solid are written as $t_L(x, \tau)$ and $t_S(x, \tau)$, as shown in Fig. 4.6.1(a).

Since the phase equilibrium temperature is greater than the remote temperature t_∞ in this example, heat is also conducted away from the moving interface, into the remaining solid. The flux into the solid at the interface is $q_S''(X, \tau)$. The energy balance at the interface is then

$$q_L''(X, \tau) = q_{PC}'' + q_s''(X, \tau) = \rho u H + q_s''(X, \tau) \qquad (4.6.1a)$$

That is, the conduction inward from the left supplies the latent heat, as well as the heat conducted into the solid beyond. Since $q_L''(X, \tau)$ is greater than $q_S''(X, \tau)$, by the amount $\rho u H$, the slope of the temperature gradient in the liquid is greater than in the solid at $x = X(\tau)$, if $k_L = k_S$. This is assumed in the sketch in Fig. 4.6.1(a), merely to emphasize this effect. For water, $k_S/k_L = 2.7$. Equation (4.6.1a) is written in terms of the gradients, at $x = X(\tau)$, as

$$-k_L \left(\frac{\partial t}{\partial x} \right)_L = \rho u H - k_S \left(\frac{\partial t}{\partial x} \right)_S \qquad (4.6.1b)$$

where the two derivatives are evaluated at the moving interface at $x = X(\tau)$, from $t_L(X, \tau)$ and $t_S(X, \tau)$ at time τ.

This analysis has not been formulated in terms of a moving coordinate system, of velocity u, with $x = 0$ located at the phase interface $X(\tau)$. This is not convenient here because velocity, u, is a function of time in these processes. The other procedure is followed in Sec. 4.7, for energy fronts in which u is constant.

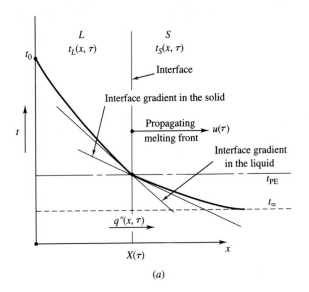

(a)

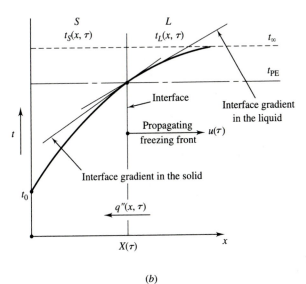

(b)

FIGURE 4.6.1
Phase change fronts in a conduction region: (a) a melting front; (b) a freezing front.

In the freezing example in Fig. 4.6.1(b), the heat flux is toward the left, that is, negative. The latent heat liberation rate is still $\rho u H$. The gradients at $x = X(\tau)$ are related as before. The energy balance at $x = X(\tau)$ is now

$$-k_S\left(\frac{\partial t}{\partial x}\right)_S = -\rho u H - k_L\left(\frac{\partial t}{\partial x}\right)_L \tag{4.6.2}$$

where $\rho u H$ is now the latent heat liberation, which is conducted toward the left.

In Eq. (4.6.2) the derivatives are also evaluated from $t_S(x, \tau)$ and $t_L(x, \tau)$. Since u is still taken as positive, the resulting flux is written as $-\rho u H$. The slope in the left region is still of greater magnitude, for $k_L = k_S$, because it is still the sum of the sensible heat conduction and the phase change effect.

As a consequence of moving interfaces, with conditions such as Eqs. (4.6.1) and (4.6.2) applied there, the formulation is nonlinear. This is discussed in Carslaw and Jaeger (1959) and demonstrated by Lunardini (1981). As a result, solutions for particular imposed conditions may not be constructed by superposition, as in linear processes. The solutions given in Secs. 4.6.1 and 4.6.2 are constructed of $\mathrm{erf}[x/2\sqrt{\alpha \tau}\,]$, in terms of α_L and α_S, which are a particular solution of the Fourier equation in each region. The available solutions then are those of that form, which may also be made to satisfy the relevant interface and bounding conditions, as for example at $X(\tau)$ in Fig. 4.6.1.

THE INTERFACE TEMPERATURE CONDITION. The other condition which also applies at $x = X(\tau)$ is a phase change relationship between the temperature levels on the two sides of the moving front. If the phase change is at least approximately in thermodynamic equilibrium, between pure phases of the same material, the temperatures in the two phases at the interface are approximated as equal, as $t_L(X, \tau) = t_S(X, \tau) = t_{\mathrm{PE}}$, a constant. The distributions are drawn in Fig. 4.6.1, assuming this equality.

However, if the material contains a dissolved solute which has different solubilities in the two phases, the process may be much more complicated. As an example, consider a freezing front advancing into quiescent saline water. The equilibrium solubility of the chemical components of salinity in ice is small. Therefore, these components are in part excluded back into the water at the advancing freezing interface. Referring to Fig. 4.6.1(b), this means that the salinity in the water at $x \geq X(\tau)$ is above that of the more distant water. This results in a lowered interface freezing temperature. However, there is also saline-component diffusion toward the right, into the liquid region. Then the heat conduction mechanism is also linked directly to this parallel mass diffusion process, in the water. Such effects are considered further in Chap. 6. In this section it is assumed that the interface temperature t_{PE} is applied as the constant temperature boundary condition, at $x = X(\tau)$, for both $t_L(x, \tau)$ and $t_S(x, \tau)$.

ANALYSIS. The preceding formulation is supplemented by the Fourier equation in each region, L and S, as follows:

$$\frac{\partial^2 t}{\partial x^2} = \frac{1}{\alpha_L}\frac{\partial t}{\partial \tau} \quad \text{and} \quad \frac{\partial^2 t}{\partial x^2} = \frac{1}{\alpha_S}\frac{\partial t}{\partial \tau} \tag{4.6.3}$$

The solutions of these equations are matched at the interface. The following relatively simple results assume that the solid and liquid phases have equal density. Section 4.6.1 considers both the melting of a solid at its phase equilibrium temperature t_{PE} and the freezing of a supercooled liquid, initially at $t_\infty < t_{PE}$. These two processes are shown in Fig. 4.6.2. In Sec. 4.6.2, the two processes in Fig. 4.6.1 are considered. Section 4.6.3 then considers additional aspects.

4.6.1 One Region of Temperature Gradient during Phase Change

These two simpler circumstances are shown in Fig. 4.6.2. They are a melting and a freezing front. The first is a solid at the phase equilibrium temperature $t_{PE} = t_\infty$, with the surface temperature maintained at $t_0 > t_{PE}$ for $\tau > 0$. The melting front propagates into the region as shown. The remaining temperature gradient shown at $x - X(\tau)$ provides the heat flux q''_m for melt propagation, at the velocity $u - dX(\tau)/d\tau$, as

$$q''_m = -k_L \frac{\partial t}{\partial x} = \rho u H \quad \text{at } x = X(\tau) \tag{4.6.4}$$

Supercooling very commonly occurs in relatively pure liquids in clean environments. For example, both pure and saline water may be routinely supercooled by -5 to $-10°C$. Much greater supercooling is also possible. Sudden dendritic freezing structures and fronts may arise and propagate out into the liquid, changing its phase and raising its temperature to t_{PE}, as idealized in Fig. 4.6.2(b). The temperature in the region $x \le X(\tau)$ remains at t_{PE}, if the imposed temperature t_0 is the same as t_{PE}, as assumed in this example. The solidification at location $X(\tau)$ amounts to a plane source of strength $q'' = \rho u H$. All of this energy is conducted into the supercooled liquid, at the right, and has raised its temperature from $t_\infty < t_{PE}$, as the freezing front $X(\tau)$ arrives.

MELTING A SOLID AT THE PHASE EQUILIBRIUM TEMPERATURE. In Fig. 4.6.2(a) the propagating distribution $t_L(x, \tau)$ is seen. It was shown in Sec. 4.2.3, Eq. (4.2.16c), that the following relation is a solution of the Laplace equation:

$$t_L(x, \tau) = A \operatorname{erf} \eta + B \quad \text{where } \eta = x/2\sqrt{\alpha_L \tau} \tag{4.6.5}$$

Since $t(0, \tau) = t_0$ at $\eta = 0$:

$$t_L(x, \tau) = A \operatorname{erf} \eta + t_0 \tag{4.6.6}$$

This function must always be t_{PE} at the phase interface X, and

$$t_L(X, \tau) = t_{PE} = A \operatorname{erf}\left[X/\sqrt{\alpha_L \tau}\right] + t_0$$

or
$$\operatorname{erf}\left[X/2\sqrt{\alpha_L \tau}\right] = \frac{t_{PE} - t_0}{A} \tag{4.6.7}$$

The right side in Eq. (4.6.7) is a constant. Therefore, $X/2\sqrt{\alpha_L \tau}$ is a constant

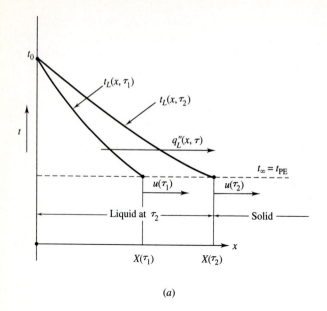

(a)

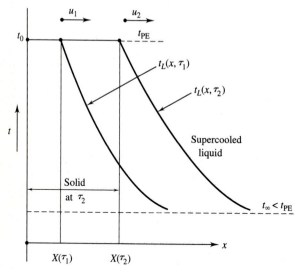

(b)

FIGURE 4.6.2
One region of temperature gradient: (a) melting of a solid at the phase equilibrium temperature; (b) freezing into a supercooled liquid.

through time, say λ, and

$$A = \frac{t_{PE} - t_0}{\mathrm{erf}\,\lambda} \tag{4.6.8}$$

where $X(\tau)$ is

$$\lambda = X/2\sqrt{\alpha_L \tau} \quad \text{or} \quad X(\tau) = 2\lambda\sqrt{\alpha_L \tau} \tag{4.6.9}$$

Thus $X \propto \sqrt{\tau}$ and $u \propto 1/\sqrt{\tau}$. Eliminating A between Eqs. (4.6.6) and (4.6.8) results in

$$\frac{t_L(x,\tau) - t_0}{t_{PE} - t_0} = \frac{\mathrm{erf}\left[x/2\sqrt{\alpha_L \tau}\right]}{\mathrm{erf}\,\lambda} \tag{4.6.10}$$

At the phase interface the conduction heat flux in the liquid, $q_L''[X(\tau), \tau]$, must equal the heat required to cause a melting rate of $u = \partial X/\partial \tau$. This condition is written as

$$q_L''[X(\tau), \tau] = -k_L\left(\frac{\partial t}{\partial x}\right)_X = \rho u H \tag{4.6.11}$$

Equations (4.6.9) and (4.6.10) are used to evaluate the terms in Eq. (4.6.11). The following relation is obtained between the value of λ, the imposed temperature difference $t_0 - t_{PE}$, and properties c_L and H, where c_L is the liquid specific heat:

$$\lambda e^{\lambda^2} \mathrm{erf}\,\lambda = \frac{c_L(t_0 - t_{PE})}{\sqrt{\pi}\,H} = N_1(\lambda) \tag{4.6.12}$$

where N_1 and λ are related as shown by curve (a) in Fig. 4.6.3. The value of N_1 is determined for each particular application. Then λ and, for example, $X(\tau)$, may be found.

FREEZING INTO A SUPERCOOLED LIQUID. This is the process in Fig. 4.6.2(b). The freezing front propagates as a plane energy source, due to the sudden release of the latent heat H, on freezing. This energy is conducted ahead into the liquid to heat it toward $t_0 - t_{PE}$. The conduction solution in the liquid phase at $x > X(\tau)$ is now required and the resulting temperature distribution is taken as

$$t(x, \tau) = t_\infty + A\,\mathrm{erfc}\left[x/2\sqrt{\alpha_L \tau}\right] \tag{4.6.13}$$

As before $t(x, \tau) = t_0 = t_{PE}$ at the phase interface and

$$t_{PE} = t_\infty + A\,\mathrm{erfc}\left[X/2\sqrt{\alpha_L \tau}\right] \tag{4.6.14}$$

Therefore, $X/2\sqrt{\alpha_L \tau} = \lambda$, a constant, and

$$X(\tau) = 2\lambda\sqrt{\alpha_L \tau} \tag{4.6.15}$$

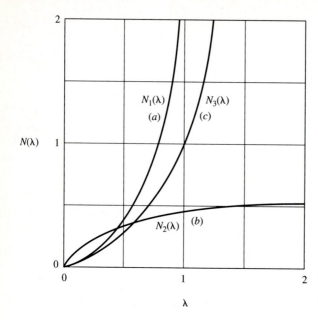

FIGURE 4.6.3
The dependence of the parameter N on λ: (a) N_1 in Eqs. (4.6.12) and (4.6.24); (b) N_2 in Eq. (4.6.18); (c) N_3 in Eq. (4.6.25).

The temperature distribution in the liquid, at $x \geq X(\tau)$, is

$$\frac{t(x, \tau) - t_\infty}{t_{PE} - t_\infty} = \frac{\mathrm{erfc}\left[x/2\sqrt{\alpha_L \tau}\right]}{\mathrm{erfc}\,\lambda} \qquad (4.6.16)$$

At the moving phase interface, in the liquid, the condition is again

$$-k_L\left(\frac{\partial t}{\partial x}\right)_X = \rho u H \qquad (4.6.17)$$

The condition on λ is now

$$\lambda e^{\lambda^2} \mathrm{erfc}\,\lambda = \frac{c_L(t_{PE} - t_\infty)}{\sqrt{\pi}\,H} = N_2(\lambda) \qquad (4.6.18)$$

where this relationship between N_2 and λ is plotted as curve (b) in Fig. 4.6.3.

SUMMARY. Again a value of N depends on the circumstance. Then λ is determined. The freezing-front velocity may then be found. For water subcooling of 5°C, the value of N_2 in Eq. (4.6.18) is about 0.036. Then e^{λ^2} and erf λ are near 1.0 and $\lambda \approx 0.037$. Note that the velocity of the freezing front decreases as $1/\sqrt{\tau}$.

4.6.2   Temperature Gradients in Both Phases

Two examples of this more general circumstance are considered, as shown in Fig. 4.6.1. The first is for the melting of a solid phase initially at a lower temperature t_∞ than t_{PE} in Fig. 4.6.1(a). The second, in Fig. 4.6.1(b), is a propagating freezing front into a liquid at $t_\infty > t_{PE}$. The latter circumstance was analyzed by Franz Neumann more than a century ago.

The melting and freezing fronts propagate into the regions. In both circumstances, conduction temperature fields arise in both phases, since $t_\infty \neq t_{PE}$. The differential equations are Eq. (4.6.3). The phase interface conditions are Eqs. (4.6.1) and (4.6.2), for melting and freezing, respectively.

The solution will be given here for the circumstance of Neumann's analysis, the advancing freezing interface in Fig. 4.6.1(b). Solutions are constructed as before. Each temperature field in the differential equations (4.6.3) is satisfied by $\mathrm{erf}[x/2\sqrt{\alpha\tau}]$ and $\mathrm{erfc}[x/2\sqrt{\alpha\tau}]$ as before. The forms in the solid and liquid regions are taken as

$$t_S(x,\tau) = t_0 + B_S\,\mathrm{erf}\left[x/2\sqrt{\alpha_S\tau}\right] \tag{4.6.19}$$

$$t_L(x,\tau) = t_\infty + B_L\,\mathrm{erfc}\left[x/2\sqrt{\alpha_L\tau}\right] \tag{4.6.20}$$

At the interface $X(\tau)$, $t_S = t_L = t_{PE}$ and

$$t_0 + B_S\,\mathrm{erf}\left[X/2\sqrt{\alpha_S\tau}\right] = t_\infty + B_L\,\mathrm{erfc}\left[X/2\sqrt{\alpha_L\tau}\right] = t_{PE} \tag{2.6.21}$$

Since this relation applies at all times τ, X must be proportional to $\sqrt{\tau}$. This condition may be written as before, in terms of a constant λ:

$$X = 2\lambda\sqrt{\alpha_S\tau} \tag{4.6.22}$$

The constants B_S and B_L may again be found as in Sec. 4.6.1, as they were for the conditions in Fig. 4.6.2. Then $t_S(x,\tau)$ and $t_L(x,\tau)$ are known. See Carslaw and Jaeger (1959) or Özisik (1980). The interface heat conduction–phase change relation, Eq. (4.6.2), gives the relation between L, t_{PE}, the properties, and the imposed temperature conditions t_0 and t_∞ as

$$\frac{e^{-\lambda^2}}{\mathrm{erf}\,\lambda} + \frac{k_L}{k_S}\left(\frac{\alpha_S}{\alpha_L}\right)^{1/2}\frac{(t_{PE} - t_\infty)}{(t_{PE} - t_0)}\frac{e^{-\lambda^2(\alpha_S/\alpha_L)}}{\mathrm{erfc}\left[\lambda\sqrt{\alpha_S/\alpha_L}\right]} = \frac{\lambda H\sqrt{\pi}}{c_S(t_{PE} - t_0)} \tag{4.6.23}$$

This general result includes the circumstance of $t_\infty = t_{PE}$. Then Eq. (4.6.23) becomes

$$\lambda e^{\lambda^2}\,\mathrm{erf}\,\lambda = \frac{c_S(t_{PE} - t_0)}{\sqrt{\pi}\,H} = N_1(\lambda) \tag{4.6.24}$$

This is the same relation, as Eq. (4.6.12), for the melting of a solid at $t_\infty = t_{PE}$, the Stefan model, except that there c_L appears, instead of c_S, as before.

SUMMARY. The preceding solutions are useful for many applications in solidification. The principal parameter in all of these solutions is $c\,\Delta t/\sqrt{\pi}\,H$. This is small for water–ice processes at low temperature differences. Then Eq. (4.6.23) may be approximated, as by Stefan (1891), see Carslaw and Jaeger (1959), as follows and plotted in Fig. 4.6.3 as curve (c):

$$\lambda^2 = c_S(t_{PE} - t_0)/2H = N_3(\lambda) \tag{4.6.25}$$

For large temperature differences this parameter is near unity. Cho and Sunderland (1974) have reanalyzed the preceding Neumann formulation for a linear variation of thermal conductivity in each phase. Extensive graphical results are given in terms of the rate of property variation with temperature. This effect is larger with smaller latent heat H.

4.6.3   Additional Considerations

The foregoing solutions cover some very important kinds of processes. Approximations are made to give exact solutions. A number of other such solutions have been developed. These are very useful in indicating the kind of parameters which arise in general. They also give direct estimates of rates of conduction and phase change. However, assumptions, such as $\rho_L = \rho_S$ and uniform constant properties are often inaccurate. Also, many configurations arise which require approximate methods. The following subsections give further information in these matters.

OTHER SIMILAR SOLUTIONS. In addition to the three examples in Secs. 4.6.1 and 4.6.2, Carslaw and Jaeger (1959) give results for a half-infinite solid region, placed in contact with a liquid region. Also, many substances do not have fixed melting points, but fuse over a range from t_1 to t_2. An equivalent specific heat is used for this layer, to simulate the heat of fusion. Multiple transformation temperature phase change is also analyzed. Özisik (1980) also gives the analysis for the cylindrical solid formed around a line heat sink in a liquid at $t_\infty > t_{PE}$.

The requirement that $\rho_L = \rho_S$ may be removed in analyses of a similar form to those given here. A so-called convection term arises in the formulation. The eventual result is approximately the same, in a calculation given by Carslaw and Jaeger (1959) for water, $\rho_L/\rho_S = 1.091$. The analysis of Cho and Sunderland (1974), referred to previously, includes this effect, along with that of variable conductivity in each phase. Although the effect of relative motion between the phases, arising from $\rho_L \neq \rho_S$, is often called "convection"; it does not imply any buoyancy-induced flows arising in a gravitational field. It refers to the flow which arises to accommodate the density change due to phase change.

APPROXIMATE METHODS. Reliance on such techniques is very widespread. The range of exact solution applicability is small compared to the wide diversity of important phase change mechanisms. This diversity in applications includes many kinds of bounding conditions and convection effects in both solid–liquid and liquid–gaseous combinations. Variable conductivity commonly arises in one or both of the phases. Very commonly, mass diffusion processes occur in parallel with conduction and convection effects. The role of parallel mass diffusion is discussed briefly earlier in this section. Further consideration is deferred to Chap. 6.

Pedroso and Domoto (1973) consider the special circumstance of $t_\infty = t_{PE} > t_0$, sometimes called the Stefan circumstance, for variable conductivity and specific heat of the solid phase. The perturbation analysis of the temperature was in terms of $c(t_{PE} - t_0)/H$, which was assumed small. General estimates of the importance of these effects are given.

Many other approximate formulations have been used, both to simplify analysis and to analyze phase fronts under other important circumstances. A common measure is to assume a quasi-static conduction process in each of the single-phase regions of variable temperature. One method neglects the motion of the phase interface in evaluating the transient conduction temperature field in any region of variable temperature. This field is then used to calculate the phase change rate at the phase interface. A further simplification calculates the temperature field, without any transient effect, as a steady-state distribution. This applies more accurately when thermal diffusion is rapid, compared to the energy effect of the moving phase change interface. These methods and additional approximation techniques are discussed in detail by Lunardini (1981) and Eckert and Drake (1972). See also Prud'homme and Nguyen (1989) for results in plane, cylindrical, and spherical geometries, using numerical and singular perturbation methods. Other related results are cited. The extensive review by Yao and Prusa (1989) discusses relevant methods and results for conduction- and convection-dominated melting and freezing mechanisms, as well as in nonhomogeneous materials. Lecomte and Batsale (1991) have developed an approximate method for removing phase fronts. It is based on a polynomial approximation of the temperature profiles, quite simple calculation procedures are shown to give high accuracy, when applied to the Neumann problem.

A finite-difference analysis for phase change processes, also with multiple moving boundaries, was developed by Kim and Kaviany (1990). Results are given for one-dimensional processes, including density differences between phases, heat generation, and multiple boundaries. Two-dimensional processes are analyzed for phases of equal density. The realism of the results is a assessed by comparison with analytical and other solutions. Excellent accuracy is reported. Many references are given to other related results.

4.7   MOVING SOURCES

The moving phase front considered in Sec. 4.6 amounts to a plane energy source, or sink, as solidification or melting occurs. A common characteristic of those mechanisms is that the phase front, at $X(\tau)$, moves with changing velocity, $u = dX/d\tau$, through time. For example, this arises in the mechanisms in Fig. 4.6.2(a) and (b). As the melting or freezing front propagates into the region, its velocity decreases. This is seen from the decreasing slope of $\partial t/\partial x$, at the interface, in the liquid phase, with time. Recall that $-k\,\partial t/\partial x$ is the latent heat release ρuH; see Eq. (4.6.4). Also $X(\tau) = K/\sqrt{\tau}$, from Eqs. (4.6.9) and (4.6.15). However, at very long times, the distribution of $t_L(x, \tau_2)$ in Fig. 4.6.2(b) eventually changes relatively slowly during any given interval in time.

Many other applications arise in which the concentrated source of thermal energy or of a chemical species moves at a constant velocity u, through the region. An example is laser or arc welding of a seam between two adjacent edges of surfaces. Either the welder or the plates may be moving. Another example is a rod being reduced in diameter by being pulled through a die. The mechanical deformation in the immediate region of area reduction releases thermal energy in the material. Then the residual internal stress reduction may be achieved downstream of the die by a concentrated induction coil around the rod. This is a second energy source. A similar mass diffusion process arises when the energy source is a moving source of the production of species C. An example is the motion of a catalytic effect due to concentrated irradiation. Another is the propagation of a strong shock or detonation wave, which produces new chemical species.

The last two examples include plane sources, such as shock waves. Others are concentrated moving sources in a multidimensional region, as in arc or laser welding. All appear to be transients to a stationary observer. However, they appear to be independent of time to an observer moving at the velocity of the propagating process. This kind of process is conventionally called a quasi-steady or quasi-stationary state. The diffusing distribution appears static to an observer moving at the velocity of an energy source of constant strength.

To a stationary observer, the passing temperature effect is a transient. In coordinates fixed to the region material, the temperature field would be a solution of the following differential equation. Here, any energy source is assumed to be a concentrated effect, not like q''', as a distributed source. The general equation in fixed coordinates is

$$\frac{\partial^2 t}{\partial x^2} + \frac{\partial^2 t}{\partial y^2} + \frac{\partial^2 t}{\partial z^2} = \frac{1}{\alpha} \frac{\partial t}{\partial \tau} \qquad (4.7.1)$$

This equation is first modified into the form which applies to the temperature field seen in a quasi-steady reference system. The following analysis then gives results for plane moving sources, of strength q_0''. Multidimensional quasi-steady and concentrated processes, which model applications in welding, are considered in Sec. 8.3.

THE QUASI-STEADY-STATE FORMULATION.   A source of energy arising in some particular geometric configuration is considered to move at constant and linear velocity u. The region is idealized as of infinite extent. See Fig. 4.7.1, which assumes a particular moving source element, of strength dq, concentrated at a point x, y, z, at time τ. The quasi-steady temperature distribution arising from the source is to be written in terms of a new x-direction coordinate ξ whose value remains zero at the chosen source effect location. That is, the origin of ξ moves with the source, in Fig. 4.7.1.

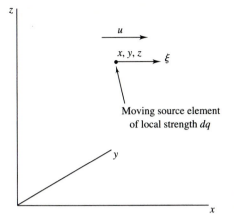

FIGURE 4.7.1
Moving coordinate ξ, with an origin remaining fixed at the location of the moving sphere source element dq.

Rosenthal (1946) postulated the following transformation, which has these properties:

$$\xi = x - u\tau \qquad \tau' = \tau, \qquad y' = y, \qquad z' = z \qquad (4.7.2)$$

At any value of ξ, say ξ_1, $x_1 = \xi_1 + u\tau$. The resulting location of ξ_1, in terms of x_1, then propagates at velocity u. Therefore, the quasi-steady temperature distribution is now in the form $t(\xi, y, z)$. Equation (4.7.1) is revised to this form as follows:

$$\frac{\partial t}{\partial x} = \frac{\partial t}{\partial \xi}\frac{\partial \xi}{\partial x} + \frac{\partial t}{\partial \tau'}\frac{\partial \tau'}{\partial x} = \frac{\partial t}{\partial \xi} \quad \text{and} \quad \frac{\partial^2 t}{\partial x^2} = \frac{\partial^2 t}{\partial \xi^2}$$

where, from Eq. (4.7.2), $\partial \xi/\partial x = 1$ and τ does not depend on x. Also

$$\frac{\partial t}{\partial \tau} = \frac{\partial t}{\partial \xi}\frac{\partial \xi}{\partial \tau} + \frac{\partial t}{\partial \tau'}\frac{\partial \tau'}{\partial \tau} = -u\frac{\partial t}{\partial \xi} + \frac{\partial t}{\partial \theta'}$$

Since $\dfrac{\partial t}{\partial \tau'} = 0$, for a quasi-static process.

$$\frac{\partial t}{\partial \tau} = \frac{\partial t}{\partial \xi}\frac{\partial \xi}{\partial \tau} = -u\frac{\partial t}{\partial \xi}$$

where $\partial \xi/\partial \tau = -u$ from Eq. (4.7.2). Therefore, Eq. (4.7.1) becomes

$$\frac{\partial^2 t}{\partial \xi^2} + \frac{\partial^2 t}{\partial y^2} + \frac{\partial^2 t}{\partial z^2} = \frac{-u}{\alpha}\frac{\partial t}{\partial \xi} \qquad (4.7.3)$$

This formulation was applied to several configurations by Rosenthal (1946) and has since been used in many additional applications.

A PLANE SOURCE. The uniform concentrated moving source, of strength q_0'' per unit area, is at $\xi = 0$ in Fig. 4.7.2. In the moving coordinate system the observer sees the region material moving uniformly toward the left, shown as the direction of u. The remote temperature, at large $\xi > 0$, is t_i. The equation,

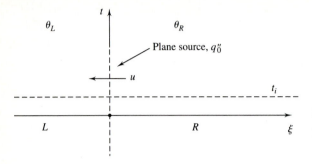

FIGURE 4.7.2
A plane source of strength q_0'' propagating toward the right in an infinite region.

from (4.7.3), and the solution for $t(\xi)$, as $\theta(\xi) = t(\xi) - t_i$, are

$$\frac{d^2\theta}{d\xi^2} + \frac{u}{\alpha}\frac{d\theta}{d\xi} = 0 \qquad (4.7.4)$$

$$\theta(\xi) = C_1 e^{-(u/\alpha)\xi} + C_2 \qquad (4.7.5)$$

The apparent boundary conditions are

$$\theta \to 0 \quad \text{as } \xi \to +\infty$$

and
$$\frac{d\theta}{d\xi} \to 0 \quad \text{as } \xi \to \pm\infty \qquad (4.7.6)$$

Consider the right region $\xi \geq 0$. There, C_2 must be zero by the first condition in Eq. (4.7.6) and

$$\theta_R(\xi) = C_{1,R} e^{-(u/\alpha)\xi} \qquad (4.7.7)$$

This is an exponential decay, which also satisfies the second condition in Eq. (4.7.6), as $\xi \to +\infty$. To satisfy the other condition at $\xi \to -\infty$, in Eq. (4.7.5), requires that

$$\theta_L = C_{2,L}$$

Therefore, the solution is a constant for $\xi < 0$ and an exponential decay for $\xi > 0$. Equality of $\theta_R(0)$ and $\theta_L(0)$ results for $C_{2,L} = C_{1,R} = C$.

The physical meaning of this result is that the energy from the plane source, q_0'', is conducted entirely ahead into the region to the right, which has the exponentially decaying temperature distribution in Eq. (4.7.7). This condition is next applied at $\xi = 0$.

$$q_0'' = -k\left(\frac{d\theta_R}{d\xi}\right)_{\xi=0} = kC_{1,R}(u/\alpha)$$

and
$$C_{1,R} = \alpha q_0''/uk = q_0''/\rho cu \qquad (4.7.8)$$

This evaluates $C_{1,R}$ and $C_{2,L}$ and the solutions in the two regions are

$$\theta(\xi) = q_0''/\rho cu \qquad \text{for } \xi \leq 0 \qquad (4.7.9a)$$

$$\theta(\xi) = (q_0''/\rho cu)e^{-(u/\alpha)\xi} \quad \text{for } \xi \geq 0 \qquad (4.7.9b)$$

SUMMARY. This solution has an interesting consequence. The plane source is merely preheating the material, as it approaches $\xi = 0$, up to a temperature level $q_0''/\rho cu$, at $\xi = 0$. No conduction occurs in the material after it passes $\xi = 0$. This behavior is similar to the freezing-front propagation into a supercooled liquid, in Sec. 4.6.1. There, the plane energy source was the release of the latent heat, to preheat the supercooled liquid to the phase equilibrium temperature t_{PE}. Moving line and point sources are analyzed in Sec. 8.3.

4.8 THE INTEGRAL METHOD APPROXIMATION

The foregoing sections of this chapter have principally concerned formal and complete solutions, for many geometries and kinds of boundary and initial conditions. However, there are many circumstances, even in regular geometries, in which such methodology does not apply or is very cumbersome. Section 4.6 indicates some common simplifying approximations often made in analyzing phase front propagation. Sections 3.9 and 4.9 indicate numerical alternatives of approximation.

This section concerns an approximate integral method of calculation which is often simple, convenient, and sufficiently accurate. It is applied to transients in one-dimensional plane regions, using an integral form developed from the following Fourier equation:

$$\frac{\partial^2 t}{\partial x^2} = \frac{1}{\alpha}\frac{\partial t}{\partial \tau} \quad \text{or} \quad \frac{\partial^2 \theta}{\partial x^2} = \frac{1}{\alpha}\frac{\partial \theta}{\partial \tau} \tag{4.8.1}$$

where $\theta = t - t_i$. The method is similar to the flow boundary layer integral method of von Karman, as conveniently available in Schlichting (1968). A summary of the early applications of this method to conduction was given by Goodman (1964).

4.8.1 The Formulation

The method and its approximations are first discussed for the application in Fig. 4.8.1, a semiinfinite solid initially at t_i with the surface suddenly changed to t_1 at $\tau = 0$. The analytical solution is given in Sec. 4.2, in terms of $\theta = t - t_i$ and $\theta_1 = t_1 - t_i$, as

$$\frac{t - t_i}{t_1 - t_i} = \frac{\theta}{\theta_1} = \phi = \text{erfc } \eta \qquad \eta = x/2\sqrt{\alpha\tau} \tag{4.8.2a}$$

The resulting transient heat flux at the surface is

$$q''(0, \tau) = k\theta_1/\sqrt{\pi\alpha\tau} \tag{4.8.2b}$$

The integral method approximates this process as a growing thermal layer of instantaneous thickness, or penetration depth, $\delta(\tau)$, as seen in Fig. 4.8.1. The assumed approximate solution ϕ_A is taken as zero at $x = \delta(\tau)$. The integral

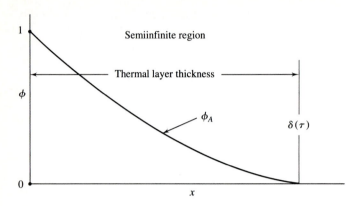

FIGURE 4.8.1
The integral formulation of a transient in a semiinfinite region, where ϕ_A is an assumed normalized temperature distribution.

method calculates this growing thickness, using additional approximations. The heat flux at $x = 0$, $q''(0, \tau)$, may then be determined.

Taking $\phi_A = 0$ at $\delta(\tau)$ is the first approximation. Recall that the erfc η is not zero, at any time $\tau > 0$, at all increasing values of x. Refer to Sec. 4.2.3. A frequent additional approximation is that ϕ_A is also taken to have zero slope at $x = \delta(\tau)$. The formulation of the following integral method also requires another major approximation. The distribution ϕ_A must be assumed. This is usually done in the form of a polynomial in $x/\delta(\tau) = X$, up to some degree. In general, the constants of the polynomial are determined from what are called compatibility conditions. These are known and assumed characteristics of the function $\phi_A(x/\delta_t) = \phi_A(X)$. Several such conditions are

$$\phi_A(0) = 1 \qquad \phi_A(1) = 0 \qquad \phi'_A(1) = 0$$

where the prime indicates the differentiation of ϕ_A with respect to X. Another condition arises from Eq. (4.8.1), applied at $x = 0$. There, $\partial t/\partial \tau = 0 = \alpha \, \partial^2 t/\partial x^2$. Therefore, $\phi''_A(0) = 0$. As an example, these four conditions are applied to the following third-degree polynomial, to determine a, b, c, and d:

$$\phi_A = a + bX + cX^2 + dX^3 \tag{4.8.3}$$

The result, from the previous four conditions, is

$$\phi_A = 1 - \frac{3}{2}\left(\frac{x}{\delta}\right) + \frac{1}{2}\left(\frac{x}{\delta}\right)^3 = 1 - \frac{3}{2}X + \frac{1}{2}X^3 \tag{4.8.4}$$

For a second-degree polynomial, neglecting the condition $\phi''_A(0) = 0$, the result is instead

$$\phi_A = 1 - 2X + X^2 \tag{4.8.5}$$

Other forms are also used.

A basic limitation of such results concerns the penetration of the thermal layer, of thickness $\delta(\tau)$, as shown, for example, in Fig. 4.8.1. This approximate formulation applies only during the time interval, 0 to τ, in which the distribution ϕ_A has not yet reached a region in which other physical effects have arisen. These effects might be a distant surface or another propagating temperature field. In general, the total conduction region must be considerably thicker than $\delta(\tau)$.

The method is developed in the following discussion. It is then applied in the following section for several kinds of processes, in plane regions. Section 4.8.2 gives results for three characteristic boundary conditions at the surface of a semiinfinite region. Results are also given, in the same geometry, for a uniform and time-dependent distributed source $q'''(\tau)$. Additional solutions and methods, including those with variable thermal properties, are given in Carslaw and Jaeger (1959), Eckert and Drake (1972), and Özisik (1980). The integral form of Eq. (4.8.1) will be developed first.

THE INTEGRAL EQUATION. This equation is determined by integrating Eq. (4.8.1), over the region from $x = 0$ to $x = \delta(\tau)$, as

$$\int_0^{\delta(\tau)} \frac{\partial^2 t}{\partial x^2} \, dx = \frac{1}{\alpha} \int_0^{\delta(\tau)} \frac{\partial t}{\partial \tau} \, dx$$

This is evaluated in terms of $\theta = t - t_i$, where t_i is uniform, as

$$\left(\frac{\partial \theta}{\partial x} \right)_{\delta(\tau)} - \left(\frac{\partial \theta}{\partial x} \right)_{x=0} = \frac{1}{\alpha} \int_0^{\delta(\tau)} \frac{\partial \theta}{\partial \tau} \, dx \qquad (4.8.6)$$

The right side may be written in terms of the time derivative of an integral of $\theta(x, \tau)$ as

$$\left(\frac{\partial \theta}{\partial x} \right)_{\delta(\tau)} - \left(\frac{\partial \theta}{\partial x} \right)_{x=0} = \frac{1}{\alpha} \left[\frac{d}{d\tau} \int_0^{\delta(\tau)} \theta(x, \tau) \, dx \right] \qquad (4.8.7)$$

The first term on the left is zero if the slope of ϕ_A is taken as zero at $x = \delta(\tau)$. The integral on the right side is $\theta(x, \tau) = t(x, \tau) - t_i$, summed across the thermal layer, at time τ. This quantity is called $Q(\tau)$. Equation (4.8.7) then becomes the integral energy equation, written as

$$-\alpha \left(\frac{\partial \theta}{\partial x} \right)_{x=0} = \frac{d}{d\tau} \int_0^{\delta(\tau)} \theta(x, \tau) \, dx = \frac{d}{d\tau} Q(\tau) \qquad (4.8.8a)$$

Then $\rho c Q(\tau)$, where ρc is the volumetric specific heat, is the instantaneous increased level of energy storage in the region $0 \le x < \delta(\tau)$, at time τ. Then,

$$q''(0, \tau) = -k \left(\frac{\partial \theta}{\partial x} \right)_{x=0} = \frac{d}{d\tau} \rho c Q(\tau) \qquad (4.8.8b)$$

4.8.2 Results in Semiinfinite Regions

Three surface conditions are shown in Fig. 4.8.2. They are an imposed constant surface temperature t_1, a time-varying surface temperature $t(0, \tau)$, including also a flux condition, and an adjacent ambient convection region at t_e. A solution is also given for a region at t_i suddenly subjected to a uniformly distributed heat source $q'''(\tau)$, as shown in Fig. 4.8.3. The region is initially

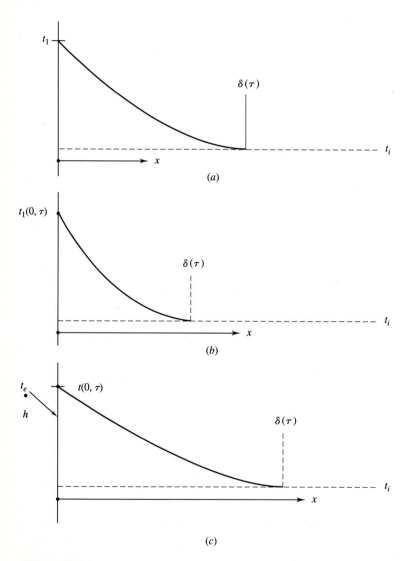

FIGURE 4.8.2

Three surface conditions: (*a*) constant value t_1 at the surface; (*b*) a time-dependent surface temperature; (*c*) convection with an ambient medium at t_e.

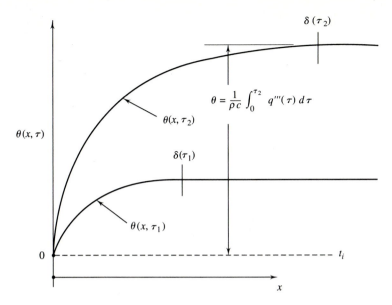

FIGURE 4.8.3
A distributed heat source $q'''(\tau)$ throughout the region, the surface maintained at t_i.

uniformly at t_i, in all four examples. The integral method results are given in the following discussion.

A SUDDEN INCREASE IN SURFACE TEMPERATURE LEVEL. This is the mechanism which leads to the solution $\phi = \operatorname{erfc} x/2\sqrt{\alpha\tau}$ in Eq. (4.2.15). The resulting surface flux was

$$q''(0, \tau) = k(t_1 - t_i)/\sqrt{\pi\alpha\tau} \qquad (4.8.2a)$$

The result from the integral method will be determined using the cubic for ϕ_A, in Eq. (4.8.4). The right side in Eq. (4.8.8a) is evaluated as

$$\frac{d}{d\tau}\int_0^{\delta(\tau)}\theta(x, \tau)\,dx = \theta_1\frac{d\delta(\tau)}{d\tau}\int_0^{\delta(\tau)}\phi_A(x, \tau)\frac{dx}{\delta(\tau)}$$

$$= \theta_1\frac{d\delta(\tau)}{d\tau}\int_0^1\phi_A(X)\,dX \qquad (4.8.9)$$

The integral is 3/8. The left side of Eq. (4.8.8a) is evaluated from ϕ_A, as $\phi_A'(0) = -3/2$, and

$$-\alpha\left(\frac{\partial\theta}{\partial x}\right)_{x=0} = -\frac{\alpha\theta_1}{\delta(\tau)}\left(\frac{\partial\phi}{\partial X}\right)_0 = \frac{3\alpha\theta_1}{2\delta(\tau)} \qquad (4.8.10a)$$

The result is

$$\delta(\tau)[d\delta(\tau)/d\tau] = 4\alpha$$

and for $\delta(0) = 0$, $\delta(\tau)$ is

$$\delta(\tau) = \sqrt{8\alpha\tau} \qquad (4.8.10b)$$

The surface heat flux, from Eq. (4.8.10), in terms of $\theta_1 = (t_1 - t_i)$, is

$$q''(0, \tau) = -k\left(\frac{d\theta}{\partial x}\right)_0 = \frac{3k\theta_1}{2\sqrt{8\alpha\tau}} = \frac{k\theta_1}{\sqrt{(32/9)\alpha\tau}} \qquad (4.8.11)$$

This value of $q''(0, \tau)$ is 13% less than the exact solution, given in Eq. (4.8.2b). Using a fourth-degree polynomial, with the additional compatibility condition that $\phi''(\delta) = 0$, gives a result 3% less than the exact solution.

A TIME-DEPENDENT SURFACE TEMPERATURE. This is shown in Fig. 4.8.2(b), where $\theta = t - t_i$. This may arise from an imposed surface flux condition, as $q''(0, \tau)$, in general. A quadratic is used, with the listed conditions as follows:

$$\theta(\tau) = a + bx + cx^2$$

$$-k\left(\frac{\partial\theta}{\partial x}\right)_0 = q''(0, \tau) = q_0''(\tau) \qquad (4.8.12)$$

$$\theta[\delta(\tau), \tau] = 0 = \left(\frac{\partial\theta}{\partial x}\right)_{x = \delta(\tau)}$$

This formulation results in the following quadratic distribution:

$$\theta(x, \tau) = \frac{q_0''(\tau)\delta(\tau)}{2k}\left[1 - 2\left(\frac{x}{\delta}\right) + \left(\frac{x}{\delta}\right)^2\right]$$

$$= \frac{q_0''(\tau)\delta(\tau)}{2k}[1 - 2X + X^2] \qquad (4.8.13)$$

This result is applied to the general formulation, Eq. (4.8.8a), and the resulting relation between $q_0''(\tau)$ and $\delta(\tau)$ is

$$\frac{d}{d\tau}\left[q_0''(\tau)\delta^2(\tau)\right] = 6\alpha q_0''(\tau)$$

The integration results in

$$\delta(\tau) = \left[\frac{6\alpha}{q_0''(\tau)}\int_0^\tau q_0''(\tau)\,d\tau\right]^{1/2} \qquad (4.8.14)$$

Substituting into the preceding relation $q_0''(\tau) = q''(0, \tau)$ from the exact solution for a surface temperature step, Eq. (4.8.2b), results in $\delta(\tau) = \sqrt{3\alpha\tau}$. This suggests good accuracy for the quadratic form in Eq. (4.8.12).

Applying then a constant flux q_0'' at $x = 0$ results in $\delta(\tau) = \sqrt{6\alpha\tau}$ and the following temperature distribution:

$$\theta(x, \tau) = \frac{q_0''}{k}\frac{\sqrt{3\alpha\tau}}{2}\left(1 - \frac{2x}{\sqrt{6\alpha\tau}} + \frac{x^2}{6\alpha\tau}\right) \qquad (4.8.15)$$

The preceding characteristic temperature difference, Δt_{ch}, is 9% above that for the exact solution, Eq. (4.2.18), when the $\sqrt{\pi}$ in that relation is included in the coefficient. For $\phi_A = (1 - X)^3$, the agreement is within 2% and the temperature distribution is

$$\theta(x,\tau) = \frac{q_0''(\tau)\delta(\tau)}{3k}[1 - X]^3 \qquad (4.8.16)$$

The integral equation then gives the following differential equation relating $\delta(\tau)$ and $q_0''(\tau)$:

$$\frac{d}{d\tau}[q_0''(\tau)\delta(\tau)] = 12\alpha q_0''(\tau) \qquad (4.8.17)$$

CONVECTIVE HEATING AT THE SURFACE. This is shown in Fig. 4.8.2(c). The boundary condition in terms of $\theta = (t - t_i)$ is

$$h[t_e - t(0,\tau)] = h(\theta_e - \theta_0) = -k\left(\frac{\partial \theta}{\partial x}\right)_{x=0} = q_0''(\tau) \qquad (4.8.18)$$

The variable surface temperature distribution in Eq. (4.8.16) is evaluated at $x = 0$. The resulting value of $\delta(\tau)$, substituted into Eq. (4.8.17), results in

$$\frac{d}{d\tau}\left[\frac{\theta^2(0,\tau)}{q_0''(\tau)}\right] = \frac{4\alpha}{3k^2}q_0''(\tau) \qquad (4.8.19)$$

This is one equation for two quantities, $\theta^2(0,\tau)$ and $q_0''(\tau)$. It has a solution if one of these is a function F only of the other, as $q_0''(\tau) = F[\theta(0,\tau)]$. The convective condition in Eq. (4.8.18) is of this form. Then Eq. (4.8.19) becomes

$$\frac{3}{4F}d[\theta(0,\tau)/F] = \alpha\,d\tau \qquad (4.8.20)$$

The following solution was given by Goodman (1964), in terms of $\theta_e = t_e - t_i$:

$$\frac{4}{3}\left(\frac{h}{k}\right)^2\alpha\tau = \frac{1}{2}\left[\frac{1}{[1 - \theta(0,\tau)/\theta_e]} - 1\right] + \ln\{[1 - \theta(0,\tau)]/\theta_e\} \qquad (4.8.21)$$

The comparable analytical solution is Eq. (4.2.17), plotted in Fig. 4.2.11. The preceding integral result is in very close agreement.

A UNIFORMLY DISTRIBUTED ENERGY SOURCE. Figure 4.8.3 indicates the transient response at times τ_1 and τ_2. The equation and conditions in terms of $\theta(x, y) = t(x, \tau) - t_i$ are

$$\frac{\partial^2 \theta}{\partial x^2} + \frac{q'''(\tau)}{k} = \frac{1}{\alpha}\frac{\partial \theta}{\partial \tau} \qquad (4.8.22)$$

$$\text{At } x = 0 \quad \tau \geq 0 \qquad \theta = 0$$
$$\text{At } x > 0 \quad \tau = 0 \qquad \theta = 0$$

Another condition, at $x > \delta(\tau)$, is apparent from Eq. (4.8.22). Since $\partial^2\theta/\partial x^2 = 0$, at $x > \delta(\tau)$,

$$\frac{q'''(\tau)}{k} = \frac{1}{\alpha}\frac{\partial\theta}{\partial\tau} \quad \text{or} \quad \theta(x,\tau) = \frac{1}{\rho c}\int_0^\tau q'''(\tau)\,d\tau \qquad (4.8.23)$$

Integration of Eq. (4.8.22) then results in

$$-\alpha\left(\frac{\partial\theta(x,\tau)}{\partial x}\right)_{x=0} = \frac{d}{d\tau}\left[\int_0^\delta \theta(x,\tau)\,dx + \delta(\tau)\int_0^\pi \frac{q'''(\tau)}{\rho c}\,d\tau\right] \qquad (4.8.24)$$

The following relation satisfies all conditions:

$$\theta(x,\tau) = \frac{\tau q'''(\tau)}{\rho c}\left[1 - (1-X)^3\right] \qquad (4.8.25)$$

The resulting differential equation relating $\delta(\tau)$ and $q'''(\tau)$ is

$$\frac{d}{d\tau}\left[\tau q'''(\tau)\delta(\tau)\right] = \frac{12\alpha\tau q'''(\tau)}{\delta(\tau)} \qquad (4.8.26)$$

Thus $\delta(\tau)$ is found when $q'''(\tau)$ is specified. For $q''' = $ constant, the calculated surface heat flux $q''(0,\tau)$ is about 6% below that from the exact solution, $q''(0,\tau) = q'''\sqrt{4\alpha\tau/\pi}$.

SUMMARY. The previous calculations apply the integral method to four different kinds of conditions. The results are generally in good agreement with exact analysis. This is the indirect support for the use of this approximate method for a wide range of transient response mechanisms which are not amenable to full analysis. This method also sometimes has an advantage over numerical formulations, in more clearly indicating the relevant parameters of a process and their functional dependence. It also often simply includes an allowance for variable spatial and thermophysical properties in the conduction region. Additional such analyses are given in Özisik (1980), Sec. 9.2. These are for a cylindrical region, for a time-dependent surface heat flux, for variable properties, and for two types of bounding conditions for a plane layer.

4.9   NUMERICAL METHODS IN UNSTEADY PROCESSES

Numerical methods are discussed in detail in Sec. 3.9 for multidimensional steady-state processes described in Cartesian coordinates. Here, numerical methods are developed for one-dimensional unsteady processes, in their simplest form. Plane regions are considered. These results are extended to multidimensional unsteady processes formulated in Cartesian coordinates, at the end of Chap. 5, which treats unsteady multidimensional processes. The additional numerical method considerations relevant to cylindrical and spherical coordinates are included in Chap. 9, along with the methodologies of additional

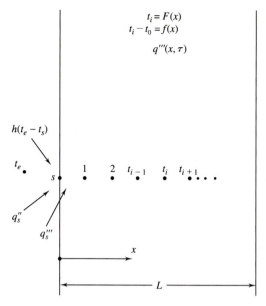

$$t_i = F(x)$$
$$t_i - t_0 = f(x)$$
$$q'''(x, \tau)$$

FIGURE 4.9.1
Subdivisions and initial surface conditions
for a plane conduction region.

approximations, accuracy, other formulations, and general numerical procedures.

This section first gives the procedures for transforming time-dependent
differential equations, and the simplest boundary conditions, into their finite-
difference equivalents. Considerations concerning the stability of the numerical
calculation, through time, immediately arise. A simple numerical analysis of
transient response is given to demonstrate procedures.

General surface processes, such as those in Fig. 4.9.1, are then formulated
in numerical terms. Different stability requirements arise. The last development
here is an example of an alternative finite-difference formulation, the implicit
method, which alters stability requirements.

4.9.1 Numerical Formulations

The numerical procedure for steady processes, in Sec. 3.9, is the development of
finite-difference relations. These permit the calculation of the temperatures
over the grid point representation, which replaces the continuous region. In one
common calculation procedure, initial guesses are adjusted to reduce the
residuals, by iteration, while satisfying definite boundary conditions.

For an unsteady temperature field, on the other hand, the procedure
calculates later temperature distributions, after a known initial temperature
distribution, or equivalent condition. Each solution through time must continue
to satisfy the boundary conditions. The numerical method determines the
changing internal distribution, for a network of locations which cover the
conduction region. That is, starting from the given initial distribution of temper-

ature, subsequent distributions are determined, step by step through time. Therefore, the equations which guide the calculation at any given value of time τ are those which permit the determination of the distribution at time $\tau + \Delta\tau$, from the distribution at time τ.

THE FINITE-DIFFERENCE EQUATION. The calculation equations are determined by replacing the derivatives of temperature in the differential equation by estimates of their value. The result, a finite-difference equations, is developed in the following discussion for an isotropic region, having a uniform and constant thermal conductivity. The relevant differential equation is

$$\frac{\partial^2 t}{\partial x^2} + \frac{q'''}{k} = \frac{1}{\alpha}\frac{\partial t}{\partial \tau} \tag{4.9.1}$$

The simplest estimate of second derivatives is given by Eq. (3.9.5), with the index notation given in Fig. 3.9.1(b). The first derivative in Eq. (4.9.1), in terms of time, may be estimated in terms of the central, backward, or forward difference, as discussed in Sec. 3.9.1. The finite difference for these estimates is a difference in time τ, an interval $\Delta\tau$, at a given point in space. Three temperature levels may be involved in such estimates: t_b, which is the present temperature at time level n, called t_i^n; t_a, which is the previous temperature at the same point, called t_i^{n-1}; and t_c, which is the later temperature at the same point, called t_i^{n+1}. Using the notation t_i^n, the forward, backward, and central estimates of $\partial t/\partial\tau$ become

$$\frac{\partial t}{\partial \tau} \approx \frac{t_i^{n+1} - t_i^n}{\Delta\tau} \tag{4.9.2a}$$

$$\frac{\partial t}{\partial \tau} \approx \frac{t_i^n - t_i^{n-1}}{\Delta\tau} \tag{4.9.2b}$$

$$\frac{\partial t}{\partial \tau} \approx \frac{t_i^{n+1} - t_i^{n-1}}{2\Delta\tau} \tag{4.9.2c}$$

Each of the preceding relations, in terms of t_i^n, is in the form of the indexing scheme discussed in Sec. 3.9.1. At location x_i, at any time level n, the temperature is written as t_i^n.

The three estimates have different error levels, as discussed briefly in Sec. 3.9.1. They also have different stability properties, as will be shown in the following discussion. However, of principal interest here are the different characteristics in the numerical procedures which result from their use. The forward-difference estimate is in terms of the present and subsequent temperatures at any given point. This is combined with an estimate of the second derivative in Eq. (4.9.1), in terms of the present temperatures, Eq. (3.9.6). This procedure results in a finite-difference equation containing only one unknown, t_i^{n+1}. This equation, for a one-dimensional region, is written as follows. Note that q_i''' is the local energy generation rate at location i, per unit volume and time, at time τ:

$$\frac{t_{i+1}^n + t_{i-1}^n - 2t_i^n}{(\Delta x)^2} + \frac{q_i'''^n}{k} = \frac{t_i^{n+1} - t_i^n}{\alpha \, \Delta \tau} \qquad (4.9.3a)$$

This forward-time, central-space estimate becomes

$$t_{i+1}^n + t_{i-1}^n - 2t_i^n + \frac{q_i'''^n(\Delta x)^2}{k} = \frac{1}{F}\left(t_i^{n+1} - t_i^n\right) \qquad (4.9.3b)$$

where F is the network Fourier number as follows:

$$F = \frac{\alpha \, \Delta \tau}{(\Delta x)^2} = \text{Fo}_\Delta \qquad (4.9.4)$$

Solving for t_i^{n+1},

$$t_i^{n+1} = F(t_{i+1}^n + t_{i-1}^n) + (1 - 2F)t_i^n + \frac{q_i'''^n(\Delta x)^2 F}{k} \qquad (4.9.5)$$

where n is the present time level in the calculation.

The preceding relation, based on the forward-difference estimate of $\partial t / \partial \tau$, is seen to permit the explicit calculation of the unknown t_i^{n+1}, at time $\tau + \Delta \tau$, from known adjacent values in the network at time τ. This formulation is called the "explicit" method of transient numerical analysis. In this method a known distribution throughout the network is used to calculate the distribution after a time interval $\Delta \tau$. This procedure is then repeated with these results to give the distribution after $2 \, \Delta \tau$, and so, until the desired information is obtained.

The use of the backward difference, Eq. (4.9.2b), results in a finite-difference equation which is similar to the preceding relation. It has both relative advantages and disadvantages in actual use. A disadvantage is that t_i^{n-1} is involved. Then t_i^n must be calculated in terms of unknown surrounding temperatures. That is, t_i^n is given implicitly in terms of other unknowns. The method, therefore, is called "implicit" and requires the simultaneous solution of the whole temperature distribution, at time τ, from a system of linear equations. This disadvantage is offset to some extent by the more advantageous stability characteristics of the method. The implicit method is discussed in Sec. 4.9.3, after the detailed consideration of the explicit method in Sec. 4.9.2.

The size of the network chosen affects the accuracy of the estimate of the derivatives and, therefore, the accuracy of the result. This matter is discussed in Sec. 3.9.1. In general, the smaller the intervals Δx and $\Delta \tau$, the more accurate is the result. However, in the transient explicit method, these intervals may not be chosen in a completely arbitrary manner, no matter how small Δx and $\Delta \tau$ are chosen. Noting that the coefficients of the temperatures in Eq. (4.9.5) indicate the magnitude of the effect of the various temperatures on t_i^{n+1}, it must be concluded that none of these coefficients may be negative. A negative coefficient would mean that a high temperature at any point, or in its vicinity, would tend to produce a lower subsequent temperature at that point. This often leads to instability in the numerical procedure. It may cause temperatures to oscillate, giving an unreasonable physical behavior. See the discussion in Smith (1965). Therefore, a coefficient of zero is the minimum acceptable value. In order for the coefficient of t_i^n to be equal to or greater than zero,

$$F \leq 1/2 \qquad (4.9.6)$$

Therefore, a choice of network spacing Δx fixes a maximum permissible value for the time interval $\Delta \tau$. In Sec. 5.3.1 the stability limits for locations internal to two- and three-dimensional regions are shown to be more stringent, as $1/4$ and $1/6$, respectively.

It is appropriate to note here that the choice of F for one-dimensional transients may be adjusted to reduce the combined effect of truncation errors in the finite-difference approximation of time and space derivatives in the differential equation.

4.9.2   Explicit-Method Results

Equation (4.9.5) is the basis of the explicit numerical method for transient analysis and it has been adapted for hand or computer calculation and even for graphical procedures. The first application here will be for a one-dimensional transient for $q''' = 0$, and for the value of F chosen to permit the simplest procedure. The maximum value of $1/2$ is chosen for F, and Eq. (4.9.5) becomes

$$t_i^{n+1} = \frac{t_{i+1}^n + t_{i-1}^n}{2} \qquad (4.9.7)$$

This very simple result amounts to the condition that the subsequent temperature at point i is the average of the preceding temperatures at the adjacent points. This result is the basis of what is often called the Schmidt–Binder method. It is applied in the following discussion to a particular transient circumstance.

A 32-cm-thick wall, initially at 20°C throughout, suddenly has one face raised to 52°C. The other face is perfectly insulated. The thermal diffusivity of the material is 256 cm²/hr. We shall estimate the temperature distribution after 45 min. A Δx of 8 cm results in the network shown in Fig. 4.9.2(a). The adiabatic condition on the right face may be met by placing another wall of equal thickness on the right side, which is subjected to the same surface temperature condition of 52°C. See Fig. 4.9.2(b). For $\Delta x = 8$ cm, the time interval for $F = 1/2$ is

$$F = \frac{1}{2} = \frac{\alpha \, \Delta \tau}{(\Delta x)^2} \qquad \Delta \tau = \frac{(\Delta x)^2}{2\alpha} = \frac{8^2}{2 \times 256} = \frac{1}{8} \text{ hr}$$

Therefore, six time intervals, or repetitions, will be required to reach 45 min. The calculation is carried out graphically in Fig. 4.9.2(c), in terms of $\theta = t - t_i$. Each temperature is the average of the previous ones at adjacent points. Parallel calculations are given in Table 4.9.1.

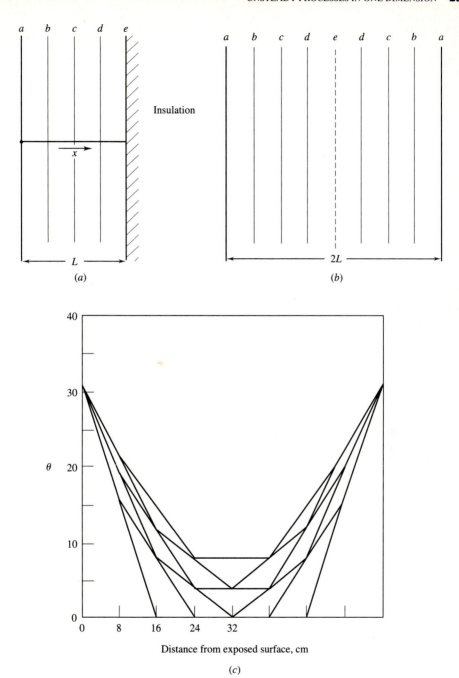

FIGURE 4.9.2
Numerical results for a transient in a wall, insulated on one side: (a) the geometry; (b) the equivalent geometry; (c) the developing temperature distribution, θ, for $F = 1/2$.

TABLE 4.9.1
**Numerical calculation of the temperature response
for the geometry of Fig. 4.9.2(a)**

n	a	b	c	d	e	d'	c'	b'	a'
0	32	0	0	0	0	0	0	0	32
1	32	16	0	0	0	0	0	16	32
2	32	16	8	0	0	0	8	16	32
3	32	20	8	4	0	4	8	20	32
4	32	20	12	4	4	4	12	20	32
5	32	22	12	8	4	8	12	22	32
6	32	22	15	8	8	8	15	22	32

In the list of values in Table 4.9.1, certain of the unreasonable results of having chosen the upper limit $F = 1/2$ are apparent. The temperature at any given point increases and then remains constant over the next interval, before increasing again. This is seen in Fig. 4.9.3(a), where θ_b and θ_e are plotted against time. The points are joined by lines. The heat flux would show the same characteristic. The temperature distributions at various times, in Fig. 4.9.2(c), are also unrealistic.

Greatly improved results may be obtained by an almost equally simple calculation technique, if $F = 1/3$ is used. Equation (4.9.5) then indicates that the subsequent temperature at each interior location is the average of the preceding temperatures at all three points:

$$t_i^{n+1} = \frac{t_{i+1}^n + t_i^n + t_{i-1}^n}{3} \tag{4.9.8}$$

The time interval is now

$$\Delta\tau = \frac{(\Delta x)^2}{3\alpha} = \frac{1}{12} \text{ hr}$$

Nine steps are required. The results are also plotted in Fig. 4.9.3 along with the results of the exact solution from Eq. (4.2.5). This solution applies for boundary conditions on two surfaces, a distance L apart, as seen in Fig. 4.2.1(a), and for a surface temperature decrease. It was converted to the circumstance in Fig. 4.9.2(a) as seen in Fig. 4.9.2(b). The change from $F = 1/2$ to $F = 1/3$ made a great improvement in the form of the temperature response and the distribution at $\tau = 30$ min. The results are in closer agreement with the exact solution.

The preceding results are a specific example of the effect of time step, $\Delta\tau$, on accuracy. The forward-difference estimate of $\partial t/\partial\tau$, in Eq. (4.9.2a), resulted in the requirement $F \le 1/2$. This is an upper limit on $\Delta\tau$, given any Δx. As Δx is chosen smaller, as for better spatial representation, the maximum permissible value of $\Delta\tau$ also decreases. These considerations are further discussed in Sec. 5.3, for multidimensional transients, and in Chap. 9, in a more general way. As relevant here, the following section shows a formulation of the implicit method,

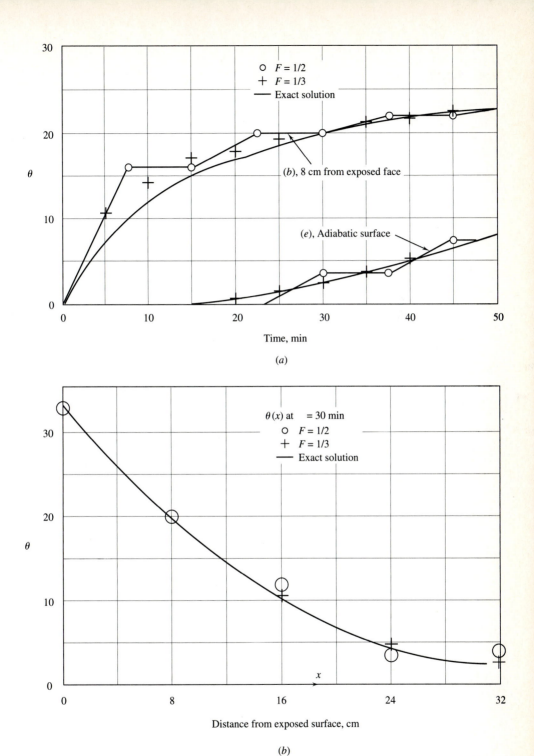

FIGURE 4.9.3

Comparison of solutions for $F = 1/2$ and $1/3$ for a transient in a wall: (a) temperature change with time at $x = L/4$ and L; (b) temperature distribution at $\tau = 30$ min.

255

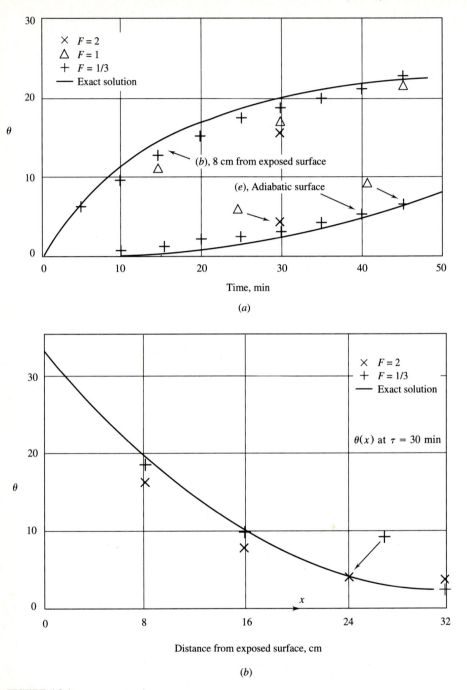

FIGURE 4.9.4
An implicit-method solution for the same transient as in Fig. 4.9.3: (*a*) temperature changes with time; (*b*) temperature distribution at 30 min.

which removes the preceding limit on stability. The results are compared with those given previously.

4.9.3 The Implicit Method

This technique uses the backward-difference estimate of the time derivative, as written in Eq. (4.9.2b). Combining this with the second derivative estimate, given in Eq. (3.9.6), and as used in Eq. (4.9.3a), results in

$$\frac{t_{i+1}^n + t_{i-1}^n - 2t_i^n}{(\Delta x)^2} + \frac{q_i'''^n}{k} = \frac{t_i^n - t_i^{n-1}}{\alpha \Delta \tau} \qquad (4.9.9)$$

where the quantities and t_{i+1}^n, t_{i-1}^n, t_i^n, and $q_i'''^n$ are at the same time level τ and t_i^{n-1} is at the previous time. This equation again applies to each location in the region of unknown temperature level during the transient.

In starting a calculation of transient response, the values of the t_i^{n-1} over the region are the initial condition. Then the calculation of the temperature distribution, at the end of the first time step $\Delta \tau$, requires the determination of t_{i+1}^n, t_{i-1}^n, t_i^n, and $q_i'''^n$, at each location simultaneously. This procedure is implicit, in terms of the calculation of the temperature t_i^n. It must be calculated in conjunction with t_{i+1}^n, t_{i-1}^n, and $q_i'''^n$, simultaneously with all the other temperatures and values of q''' inside the region.

For calculation, Eq. (4.9.9) is written as follows, where $F = \alpha \Delta \tau / (\Delta x)^2$, as before:

$$t_i^n(1 + 2F) = F(t_{i+1}^n + t_{i-1}^n) + \frac{q_i'''^n(\Delta x)^2 F}{k} + t_i^{n-1} \qquad (4.9.10)$$

Denoting time level n, as $n + 1$ instead, this result, centered at location i, becomes

$$t_i^{n+1}(1 + 2F) = F(t_{i+1}^{n+1} + t_{i-1}^{n+1}) + \frac{q_i'''^{n+1}(\Delta x)^2 F}{k} + t_i^n \qquad (4.9.11)$$

where, for example, t_i^n may be the initial condition.

Although this implicit method requires that all of the unknown temperatures at each subsequent level of time must be solved for simultaneously, it has one clear advantage over the explicit formulation. Equation (4.9.10) indicates that considerations of reasonableness, or of stability, place no upper limit on the value of F. Recall that the explicit-method limit was $F \le 1/2$. The absence of a limit means that a larger value of $\Delta \tau$ may be used for a given Δx and α, resulting in a reduction in a number of necessary time steps. Values as large as $F = 2$ may produce reasonable results, with one-fourth of the number of steps required with $F = 1/2$.

These characteristics of the implicit method are shown in Fig. 4.9.4. Solutions are compared for the transient which gave the results in Fig. 4.9.3, by the explicit method. It is seen that very large values of F result in quite reasonable predictions, compared with the exact solution.

4.9.4 Surface Conditions in Numerical Form

The numerical calculation procedures for locations in the conduction region are to be supplemented by comparable relations at surfaces. These numerically model both steady and unsteady boundary conditions. This is done here in terms of the forward difference, or explicit method, applied to the surface region in Fig. 4.9.5.

An energy balance is written in terms of the surface layer of thickness $\Delta x/2$, with a distributed source in that region of local strength $q_s'''^n$. The surface processes are a flux loading, q_s''', and convection with an adjacent ambient fluid at t_e. Conduction from point 1, generation in the surface region $\Delta x/2$ thick, the flux loading and the convection gain are equated to the increased thermal storage in the region at s of thickness $\Delta x/2$. All effects are written per unit area of surface

$$\frac{k}{\Delta x}(t_1^n - t_s^n) + \frac{q_s'''^n \Delta x}{2} + q_s'''^n + h(t_e^n - t_s^n) = \frac{\rho c \, \Delta x}{2 \, \Delta \tau}(t_s^{n+1} - t_s^n) \quad (4.9.12)$$

This result is solved for t_s^{n+1} to yield the relation which is used, in conjunction with Eq. (4.9.5) for the interior points:

$$t_s^{n+1} = 2BFt_e^n + 2Ft_1^n + \frac{2F \Delta x q_s'''^n}{k} + \frac{Fq_s'''^n(\Delta x)^2}{k} + [1 - 2F(B + 1)]t_s^n$$

$$(4.9.13)$$

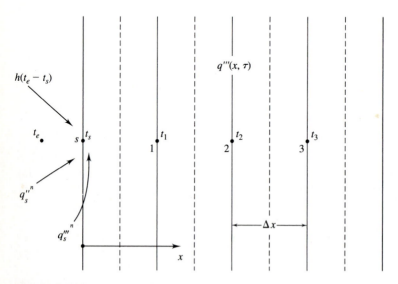

FIGURE 4.9.5
Surface conditions for a plane conduction layer.

This form is consistent with the interior location result in Eq. (4.9.5), where, as before, in Eq. (4.9.4):

$$F = \frac{\alpha \, \Delta \tau}{(\Delta x)^2}$$

and, as in Eq. (3.9.25), the network Biot number B is

$$B = \frac{h \, \Delta x}{k}$$

STABILITY LIMITS. Equation (4.9.13) indicates a different stability limit than that in Eq. (4.9.5). For the interior locations, the limit is still $F \leq 1/2$, Eq. (4.9.6). For the surface location equation, Eq. (4.9.13), the requirement is still that the coefficients of all of the temperature terms must be positive, or zero at the least value. In Eq. (4.9.13) the coefficient of t_e is positive, or zero in the absence of convection. The coefficient of t_i^n is positive. However, the coefficient of t_s^n is zero or positive only if

$$F = \frac{\alpha \, \Delta \tau}{(\Delta x)^2} \leq \frac{1}{2(B+1)} \qquad (4.9.14)$$

If there is a convection process, that is, if $B > 0$, this stability condition is more conservative than that which applies at interior locations, $F \leq 1/2$. That is, the upper limit of $\Delta \tau$, for given α and $(\Delta x)^2$, is less. Since the time step $\Delta \tau$ must be the same throughout the region, the smaller limit at the surface is the maximum value. It applies throughout the region.

The surface condition equations in two and three dimensions are developed in Sec. 5.3. As seen, the stability limits associated with surface points may be more restrictive than those which apply to interior points. A transient surface point energy balance, like that in Eq. (4.9.12), also correctly models changing convection, as $h(\tau)$, surface flux, as $q_s'''^n(\tau)$, and generation, as $q_s'''^n(\tau)$. There are also circumstances in which these quantities may depend on the local temperature level, as $q_s'''^n(t)$. Then additional considerations concerning stability may arise. The only effect considered in this book is a temperature effect on the strength of the distributed source, as $q_s'''^n(t)$.

PROBLEMS

4.2.1. A plate of thickness δ initially has a sinusoidal temperature distribution varying from t_0 at $x = 0$ to t_m at $x = \delta/2$ and to t_0 at $x = \delta$. If the surfaces of the plate are held at t_0 for subsequent times, find the temperature distribution as a function of time.

4.2.2. An extensive plane region is initially uniformly at t_i throughout. Suddenly the temperature of its two surfaces is changed to $t_0 > t_i$ and held constant thereafter. We are interested in several aspects of the temperature response, and distribution, in the plate material at later times τ.

(a) Find the location and time at which the greatest local temperature gradient occurs.

(b) Calculate the local heat flux distribution in the material, $q''(x,\tau)$, as a function of location and time.

(c) At time τ_1, the region is suddenly insulated on its two surfaces. Calculate the temperature t_f the plate material will attain eventually, in terms of material properties ρ and c.

4.2.3. Consider an extensive slab of thickness $2s$, initially at a uniform temperature t_0 throughout. It is subjected to a thereafter constant surface flux q'' on each face. Assume $q''' = 0$ and k constant and uniform. Take the original at the midplane.

(a) Write the complete problem statement in terms of $\theta = t - t_0$.

(b) Calculate the Laplace transform $\bar{\theta}$ of θ, and evaluate all constants.

4.2.4. A plate initially has a linear temperature distribution, from t_a at the left face to t_b at the right face. The two surface temperatures are suddenly changed to a thereafter constant value t_0. Find the temperature as a function of location within the plate material, and time.

4.2.5. Wide oak boards 1 in. thick are to be heated from an initial temperature of 60°F by hanging then in a steam-filled enclosure.

(a) If the steam condenses at 212°F on both surfaces and if the surface resistance is neglected, plot the board midplane temperature vs. time in minutes.

(b) Estimate the time necessary for the midplane temperature to reach 180°F.

$$\rho = 42 \text{ lb/ft}^3 \qquad c = 0.33 \text{ Btu/lb°F} \qquad k = 0.105 \text{ Btu/hr ft°F}$$

4.2.6. (a) For a plate of thickness L, initially at t_0 throughout, with surface temperatures suddenly changed to t_b and thereafter maintained constant, derive an expression relating average plate temperature to time.

(b) For the circumstance of Prob. 4.2.5, estimate the time necessary for the average temperature of the oak boards to reach 180°F.

4.2.7. A very thick low-carbon-steel plate, initially at 60°F, is placed in a heat treatment bath at 860°F. The surface process is characterized by a convection coefficient of 150 Btu/ft^2 °F hr. Find the surface temperature variation with time, for short times.

4.2.8. Repeat the preceding problem if the source of the heating is radio frequency induction heating, which results approximately in the equivalent of a surface heat flux of 50,000 Btu/ft^2 hr. The plate is in air at 150°F, and the heat losses from the surface to the air amount to a total coefficient h of 10 Btu/hr ft^2 °F.

4.2.9. Steel plates being thinned by rolling must be periodically reheated. How long must plates 3 in. thick, which are at 1000°F, be kept in a furnance surrounding at 1600°F, $h = 30$ Btu/hr ft^2°F, in order to reach a minimum metal temperature of 1300°F?

$$k = 20 \text{ Btu/hr ft °F} \qquad \alpha = 0.37 \text{ ft}^2/\text{hr}$$

4.2.10. For the condition of Prob. 4.2.9, how long must the plate be kept in the furnace in order to reach an average metal temperature of 1300°F?

4.2.11. The nozzle of a rocket engine is made of a ceramic material 1/2 in. thick. The combustive gases passing through the nozzle are at a temperature of 3400°F, and a convection coefficient of 1000 Btu/hr ft^2 °F is expected. If the maximum operating temperature the ceramic material will stand is 2700°F, find the permissible engine operating period if the initial material temperature is 100°F. The nozzle wall may be approximated by a flat plate with the unexposed surface being considered adiabatic.

$$k = 10.3 \text{ Btu/hr ft °F} \qquad \rho = 150 \text{ lb/ft}^3 \qquad c = 0.2 \text{ Btu/lb °F}$$

4.2.12. For the temperature response of the ceramic material, in Prob. 4.2.11, determine the behavior, including very short times after the beginning of operation.
(a) Plot the surface temperature, up to the time when it reaches 2700°F.
(b) Also plot both the temperature variation at the other face and the increasing total energy content of the material, per unit area.

4.2.13. Large ingots of steel may be approximated as flat plates 2 ft thick. The temperature of the ingot material is uniform at 600°F when the molds are removed. In surroundings at 60°F, how long will it take for the maximum metal temperature to reach 150°F if the combined convection and radiation coefficient is 12 Btu/hr ft^2 °F? What is the surface temperature at this time?

$$k = 20 \text{ Btu/hr ft °F} \qquad \alpha = 0.37 \text{ ft}^2/\text{hr}$$

4.2.14. A large steel plate of 10 cm in thickness is to be induction heated on both sides. It is at a uniform initial internal temperature of 100°C. It is to be heated to a midplane temperature of 800°C, for further rolling, down to a thickness of 4 cm. The surface convective losses to an environment of 100°C is 50 W/m °C, $k = 60.5$ W/m K, $\alpha = 17.7 \times 10^{-6}$ m^2/s.
(a) Find the heating rate necessary for the midplane temperature to reach 800°C in 30 min.
(b) Plot the midplane temperature increase with time.

4.2.15. For the conditions in Prob. 4.2.14:
(a) Estimate the time at which the surface temperature is 700°C, for the induction heating rate given in Prob. 4.2.14.
(b) Plot the surface temperature increase with time.

4.2.16. A 4-in. concrete building wall is very well insulated on its inside surface. The wall is in equilibrium with the outside air at 60°F when the outside-air temperature changes to 40°F in a very short period. Find the temperatures of the outside and inside surfaces of the wall after 2 hr and after 6 hr. Estimate the amount of heat transferred to the air per square foot of wall area during the first 2 hr. The convection coefficient is 2.0 Btu/hr ft^2 °F.

$$k = 1.1 \text{ Btu/hr ft °F} \qquad \alpha = 0.017 \text{ ft}^2/\text{hr}$$

4.2.17. A concrete wall 2 ft thick is initially at 60°F throughout. Its surface temperatures are suddenly changed to 100°F. From the appropriate solution and from the solution for a semiinfinite solid, plot the temperature vs. ln τ for a point 3 in.

from one surface and for the midpoint. If a 1% error in temperature change is permissible, find the conditions under which the proper solution may be replaced by the simpler semiinfinite solid solution.

$$k = 1.00 \text{ Btu/hr ft }^\circ\text{F} \qquad \alpha = 0.016 \text{ ft}^2/\text{hr}$$

4.2.18. A semiinfinite solid initially at a uniform temperature t_0 is suddenly subjected to a thereafter constant surface temperature t_b.

(a) Find the velocity at which the temperature disturbance is propagated through the solid.

(b) If the velocity of propagation is defined as the propagation rate of a temperature increase of 1% of $(t_b - t_a)$, determine the velocity of propagation.

4.2.19. A semiinfinite solid initially at t_0 throughout is in contact at its face with an environment at t_e. A surface coefficient h may be used for a linear process at the surface.

(a) If t_e may be a function of time, determine whether there is any such function (or functions) which will permit a "similarity" type of solution, that is, in terms of $x/2\sqrt{\alpha\tau}$.

(b) For a surface flux condition $q''(\tau)$ instead of a convection condition, find the function $q''(\tau)$ for which a similarity solution exists. Find the solution for $\theta(x,\tau)$ and calculate θ at $x = 0$.

4.2.20. For the conditions in Prob. 4.2.14, and at short times:

(a) Calculate the temperature gradient in the steel, at the surface.

(b) Plot this gradient for the short times during which it is a good approximation in the geometry of Prob. 4.2.14.

4.2.21. A large oak board of thickness L, initially at 20°C throughout, is suddenly placed in an environment at 120°C. The surface coefficient is 10 W/m² °C, to an environment at 20°C.

(a) Calculate the time for which the process on each side may be treated as though it occurred in a semiinfinite solid.

(b) Plot the surface flux and temperature as a function of time.

4.2.22. In an infinite region of material at a uniform temperature t_∞, a sudden concentrated energy input results in a plane region of thickness L of uniform temperature $t_1 > t_\infty$.

(a) Calculate the resulting heat flux variation with time, at the edges of the region L.

(b) Compare the temperature response at the middle of this region with that following from the same temperature step occurring at the face of a semiinfinite region.

4.2.23. For the input of a plane heat pulse in an infinite region at $x = 0$:

(a) Determine Δt_{ch} and plot $\theta(0, \tau)/\Delta t_{ch} = \phi(0, \tau)$.

(b) Find the velocity of propagation of a temperature rise of $\phi(0, \tau) = 10^{-3}$.

4.2.24. Extensive steel and copper regions, at 100 and 0°C, are suddenly put in perfect contact at their plane surfaces. Consider the early time period in which each region might be approximated as semiinfinite.

(a) Calculate the time variation of temperature and heat flux at the interface.

(b) Determine the ratio of the temperature propagation velocity in the copper and steel regions, in terms of a temperature change of $10^{-3}°C$ in each region.

4.2.25. A semiinfinite slab of material is initially at a uniform temperature $\theta = 0$.

(a) Determine how the surface temperature may vary with time such that a similarity solution can be obtained. That is, how may $\theta(0, \tau)$ vary so as to allow solutions to be found of the form: $\theta(x, \tau) = \phi(\tau)f(\eta)$ where $\eta = x/g(\tau)$?

(b) For $\theta(0, \tau) = \theta_0$ constant, determine $\theta(x, \tau)$ by completing the similarity solution.

4.2.26. A plate, initially at t_i throughout, is suddenly subjected to surface temperatures of t_0 and to a uniform distributed source of strength q'''.

(a) Derive an expression for the heat flux at the surface for all times and determine the value of the heat flux at $\tau \to \infty$.

(b) Determine the form of the transient temperature distribution and locate any maxima or minima in the distribution.

4.2.27. An extensive vertical wall of thickness $2L$ has its two surfaces maintained at the same temperature t_0. At a certain instant a distributed dissipation process begins in the left half of the wall. The source remains of constant and uniform strength q''' thereafter. Consider both the transient and eventual steady-state temperature fields in the wall.

(a) Sketch the initial, several transient, and the final steady-state distributions of $t - t_0$ through the wall.

(b) Formulate the equations and boundary conditions which will yield the general solution for the transient temperature distribution.

(c) Calculate the steady-state distribution $t - t_0$ from the results of part (b) or by any other method.

4.2.28. A plate of thickness L is subject to a distributed source of strength q''' and at $\tau = 0$ has a temperature distribution given by

$$\theta_1(x) = \theta_0 \sin \pi \frac{x}{L}$$

Find the temperature distribution at time τ if $\theta = 0$ at $x = 0, L$, for $\tau \geq 0$.

4.2.29. An aluminum plate of 2 cm thickness is initially uniformly at 25°C. It is then suddenly subjected to a uniform and continuing electrical current which dissipates energy at a rate of 6 W/cm³. The surfaces are maintained at 25°C by cooling. Calculate Δt_{ch} and plot the approximate central temperature variation with time.

4.2.30. A large board of $L = 3$ cm thick pine initially at $t_i = 25°C$, is heated by microwave from its otherwise insulated left face. The irradiation rate at the surface is 5 W/cm². Assume the extinction coefficient to be 0.2 cm⁻¹. The right face is maintained at 25°C.

(a) Calculate the distributed energy source strength distribution.

(b) Determine the radiated heat flux at $x = L$.

(c) Plot $\theta(0, \tau)$.

4.2.31. For the configuration and thermal conditions in Prob. 4.2.30, determine $\theta(x, \tau)$ as a function of x/L.

4.3.1. A long copper rod of 1 cm radius, initially at 20°C, is suddenly subject to an electric current, I, whose energy dissipation rate amounts to a uniformly distributed source of 300 W, per cm length. The rod surface is maintained at 20°C.
(a) Determine Δt_{ch} and the eventual axis temperature in steady-state.
(b) Plot the transient temperature change at $r = 0$ and $R/2$.

4.3.2. The limiting axial temperature in Prob. 4.3.1 is to be 40°C.
(a) Calculate the limiting current I_L as I_L/I. Assume constant electrical conductivity.
(b) Calculate I_L from the resistivity of copper.

4.3.3. A spherical region is initially at t_i, with the surface maintained at t_0 for $\tau > 0$. Calculate the central temperature and the surface heat loss rate as functions of time.

4.3.4. The surface temperature of a spherical region, initially at 0°C throughout, increases as $t_s = K\tau$. Estimate the temperature response, $t(0, \tau)$, at the center of the sphere.

4.3.5. The temperature of the wall of a spherical cavity, buried deeply in rock strata at t_∞, changes to t_0.
(a) For $t_\infty = 20°C$, $t_0 = 100°C$, and $R = 10$ m, calculate the heat loss rate $q(\tau)$.
(b) Determine the transient penetration distance of this thermal effect, defined as the distance from $r = 0$ where $\phi = 10^{-4}$.
(c) What is this penetration at steady state?

4.3.6. Consider a small spherical solid, $k = k_1$, of diameter D embedded in a very large solid region at temperature t_∞, with conductivity k_∞. There is a contact resistance between the sphere and its environment, of R_c, per unit area.
(a) If the sphere material has an energy source of strength Q per unit time, write down the complete formulation for determining the temperature fields in the whole region, in steady state.
(b) Calculate the temperature field.
(c) Determine the central sphere temperature and the temperature drop across the interface.

4.3.7. In one frequently used method of measuring convection coefficients for fluid flow normal to cylindrical elements, long heated cylinders at a high temperature are placed in the flow. The temperature variation with time is measured at the cylinder axis. A long, solid copper cylinder of 1 in. diameter, with an initially uniform temperature of 160°F, is to be placed in streams of fluids with temperatures of 60°F. Find the convection coefficient for two circumstances, the axis temperature reduced to 110°F in 2.5 min and in 4.0 s. Discuss the differences between these two transient processes.

$$\rho = 588 \text{ lb/ft}^3 \qquad c = 0.092 \text{ Btu/lb}°F \qquad k = 224 \text{ Btu/hr ft}°F$$

4.3.8. For the circumstance in Prob. 4.3.7, plot the axial and surface temperature response, for both processes, from the beginning of the transient to $\phi = 0.01$. Use logarithmic coordinates.

4.3.9. Steel wire is being reduced to a diameter of 0.10 in. by being drawn through a die at 3 fps. Because the cold working in the die, the wire temperature is raised to 300°F. Neglecting axial conduction inside the wire, find the distance along the wire from the die at which the wire temperature has been reduced to 180°F if the surface loses heat to the surroundings at 60°F according to a convection coefficient of 7.0 Btu/hr ft^2°F. Assume one-dimensional radial conduction in the wire.

4.3.10. For the process in Prob. 4.3.9, calculate the axial conduction divided by the radial loss, both calculated per unit area of the appropriate conduction area.

4.3.11. Long cylindrical billets of brass 4 in. in diameter are heated to 1100°F preparatory to an extrusion process. If the minimum metal temperature permissible for extrusion is 900°F, how long may these billets stand in 80°F surroundings if the surface coefficient is 15 Btu/hr ft^2 °F? What is the temperature at the cylinder axis at this time?

$$\alpha = 1.1 \text{ ft}^2/\text{hr} \qquad k = 65 \text{ Btu/hr ft}°\text{F}$$

4.3.12. Steel spheres of 3 in. diameter, heated to 500°F, are to be cooled by immersion in an oil bath which is at 100°F. A convection coefficient of 70 Btu/hr ft^2 °F is expected. Find the surface and midpoint temperatures for times of 10 s and 1 and 6 min after immersion.

$$k = 19 \text{ Btu/hr ft}°\text{F} \qquad \alpha = 0.35 \text{ ft}^2/\text{hr}$$

4.3.13. Steel spheres of 0.3 in. diameter, initially at 300°F, are to be cooled by allowing them to fall by gravity through an oil bath at 100°F. The convection coefficient will be $h = 70$ Btu/hr ft °F. The velocity of fall will be 0.3 fps. Determine how deep, D, the oil pool must be so that the center temperature of the spheres will be 200°F when they reach the bottom. $k = 19$ Btu/hr ft °F, $\alpha = 0.35$ ft^2/hr.

4.3.14. Copper spheres of 0.2 in. diameter, initially at 200°F, are to be cooled by being dropped through a gas at 100°F. The convection coefficient will be 10 Btu/hr ft^2°F. Neglect radiation. Find the time of fall necessary for the central sphere temperature to decrease to 150°F. For copper, $\alpha = 4.31$ ft^2/hr and $k = 222$ Btu/hr ft °F.

4.3.15. Copper spheres of 10 mm diameter are to be cooled from an initial temperature of t_i by dropping them vertically downward in air uniformly at t_∞. As they accelerate downward the convection coefficient increases with time τ as $h = h_0(1 + \tau/b)$, where $h_0 = 20$ W/m^2 K is a constant and $b = 10$ s. For copper, $k = 401$ W/m K, $\rho = 9000$ kg/m^3, and $c = 385$ J/kg K.
(*a*) Determine the kind of conduction process which arises at short times.
(*b*) Write the equation which relates the cooling rate to the convection process, in terms of all the relevant quantities.
(*c*) Determine the relation between the sphere temperature and time, τ, during its fall after being dropped.
(*d*) Determine a relation for the drop time, τ_d, for the temperature to be reduced to $t = (t_i + t_\infty)/2$.

4.3.16. For the spherical circumstance in Prob. 4.3.12, plot the transient temperature variation, from $\phi(r,0) = 1$ to $\phi(r,\tau) = 10^{-2}$, for locations $r/s = 0, 1/2, 1$.

4.3.17. Steel spheres of 6 cm diameter, initially at 40°C, are to be induction heated.
(*a*) Neglecting surface losses, calculate the required surface flux to increase the minimum internal temperature to 500°C in 20 s.
(*b*) What is the surface temperature at this time?
(*c*) Repeat parts (*a*) and (*b*) for long cylinders of 6 cm diameter.

4.3.18. For the required temperature conditions in Prob. 4.3.11, consider surface induction instead of convection heating.

(a) Determine the necessary surface flux level which will result in a central temperature of 900°F in 3 min.

(b) What is the surface temperature at this time?

4.3.19. If the geometry in Prob. 4.3.18 is instead a sphere of 4 in. diameter, determine the necessary induction surface flux.

4.4.1. Estimate the depth to which frost penetrates in the ground at a latitude where the yearly surface temperature variation may be considered sinusoidal, with maximum and minimum surface temperatures of 75 and 25°F. Compute the time lag of the temperature wave at the maximum depth of frost penetration. Assume that the water content of the soil is sufficiently low so that the latent heat of fusion and the change in properties may be ignored. Typical properties of soil are

$$k = 0.6 \text{ Btu/hr ft °F} \qquad c = 0.20 \text{ Btu/lb °F} \qquad \rho = 90 \text{ lb/ft}^3$$

4.4.2. For a semiinfinite solid subjected to a sinusoidal variation in surface temperature:

(a) Find the depth at which the amplitude of the temperature variation is $1/a$ of the surface value.

(b) What are the relative damping rates of two simultaneous disturbances, one being twice the frequency of the other?

(c) Show that the net heat flow per cycle is zero at all locations.

(d) Derive an expression for the heat input at $x = 0$, for the positive half of the heat flow cycle.

(e) Prove that the instantaneous maxima of the temperature distribution through the solid are not the local temperature maxima.

4.4.3. Find the conditions under which the solution for a sinusoidal temperature variation at the surface of a semiinfinite solid may be applied to a plate of thickness L subject to a temperature variation on one of its two surfaces, the other surface being insulated, if temperature fluctuations of the order of 1% in the semiinfinite solid solution may be considered negligible.

4.4.4. At a certain latitude the daily variation in the surface temperature of the earth may be considered sinusoidal with 4°F amplitude. Find the wavelength and the velocity of the wave. At what depth will the amplitude be reduced to 1°F?

$$\alpha = 0.012 \text{ ft}^2/\text{hr}$$

4.4.5. The thermal diffusivity of soil is to be estimated at a given location by noting the propagation characteristics of the daily temperature variation of the soil. This variation is assumed to be due to a sinusoidal variation in surface temperature. Thermocouple junctions are buried at depth of 1 and 4 in.

(a) If the temperature maximum at 1 in. depth occurs at 3 P.M. and the maximum at 4 in. depth at 5 P.M., compute the thermal diffusivity of the soil.

(b) If the maximum and minimum temperatures at 1 in. are 63 and 57°F, find the amplitude of the variation at the surface and the average surface temperature.

4.4.6. A semiinfinite region, initially uniformly at t_i, is subject at its surface to a periodic temperature variation $t(0, \tau) - t_i = \theta_a \cos \omega \tau$, for $\tau > 0$. Estimate the time period after which the nonperiodic component of the temperature field

$\theta(x, \tau) = t(x, \tau) - t_i$ has become small, compared to the eventual purely periodic variation, at some location. Determine, for an x location two wavelengths in from the surface, the time at which the transient is 1% of the local maximum amplitude of the eventual pure periodic.

4.4.7. Repeat Prob. 4.4.3 for a sinusoidal variation of the temperature of a fluid in contact with the surface of the semiinfinite solid, accounting for convective resistance.

4.4.8. A very thick concrete retaining wall is in contact with air on its exposed side. If, during a particular season, the daily temperature variation of the air may be assumed to be sinusoidal over the range 50 to 80°F, find the variation in temperature at the surface of the wall and at a point 2 in. inside the wall. The expected convection coefficient is 2.0 Btu/hr ft^2 °F.

$$\alpha = 0.016 \text{ ft}^2/\text{hr} \qquad k = 1.0 \text{ Btu/hr ft °F}$$

4.4.9. A semiinfinite conduction region is subject to a periodic surface temperature condition. This surface condition amounts to two periodic temperature variations of angular frequencies ω_1 and ω_2. They are in phase at the surface. That is,

$$\theta(0, \tau) = \theta_{0, a_1} \cos \omega_1 \tau + \theta_{0, a_2} \cos \omega_2 \tau$$

Calculate $\theta(x, \tau)$ in the conduction region, subject to this surface condition.
(*a*) Write the equation and bounding conditions.
(*b*) Give the solution for $\theta(x, \tau)$ without detailed derivation if possible.
(*c*) Comment on the phase relation between the two periodic components at any given location within the conduction region, for $\omega_2 = 4\omega_1$.

4.4.10. The surface of a very thick region is subject to two simultaneous periodic components of temperature variation. An example is the surface of soil subject to both daily and annual variations, the periods being 24 hr and 365 days, respectively. Determine the resulting function $t(x, \tau) - t_m$.

4.4.11. A large concrete wall, of thickness $L = 2$ ft and $\alpha = 0.016$ ft^2/hr, is subject to a periodic surface temperature variation, of the same half-amplitude of 10°F, on each surface. For a sufficient thickness, the changing temperature field in the wall may be approximated as that in a semiinfinite solid, in from each side. The condition for which this may be considered sufficiently accurate may be taken as a temperature half-amplitude of 0.1°F, at the location of their interaction.
(*a*) If the two surface temperature variations are of the same magnitude and frequency and in phase, find the range of frequency for which this approximation is met.
(*b*) If the frequency on the right is four times that on the left, indicate how to find the limiting frequency on the left.

4.4.12. The surface temperatures of a concrete wall of 20 cm thickness both vary daily, periodically, and in phase between 25 and 15°C.
(*a*) Determine the amplitude of the periodic at the midplane of the wall.
(*b*) How many local amplitude maxima occur instantaneously across the wall material?

4.4.13. Consider the starting transient for a steady periodic as diagrammed in Fig. 4.4.1. Assume that the whole region was initially at $t(x, 0) = t_\infty$. At $\tau = 0$ the periodic boundary condition is applied. Estimate the time τ_P at which the remaining transient effect at $x = \lambda$ is less than 1% of θ_a.

4.5.1. A thick slab of material, coated by a highly conductive surface layer of thermal capacity c'', is to be heated by subjecting the surface, for $\tau > 0$, to a constant radiative flux q''.

(a) Determine the characteristic temperature difference for this circumstance.

(b) Calculate the rate of temperature rise of the surface layer.

(c) Calculate the fraction of the total heat input, at time τ, which has been stored in the surface layer.

4.5.2. The surface layer in Prob. 4.5.1 is taken as silver of 0.5 mm thickness on a region of glass. The imposed heat flux is 100 W/cm^2.

(a) Calculate the temperature response at the surface and at $x = 2$ cm.

(b) Estimate the actual temperature drop across the layer of silver at $x = 2$ cm at 5 s into the transient.

4.5.3. A layer of plaster and a very thick region of concrete, in contact with each other with relatively negligible contact resistance, are initially uniformly at a temperature of t_i, as in Fig. 4.5.3. Calculate the interface temperature and heat flux, as a function of τ, after the left surface is increased to $t_0 = 30°C$, and $t_i = 10°C$.

4.5.4. Consider two thick and extensive conduction regions, L and R for left and right. At their common flat interface there is a very thin layer of electrically conducting metal. Perfect thermal contact is assumed and the thermal resistance and thermal capacity of the metal layer may be neglected. The initial temperature throughout is t_i. Then a thereafter constant current flows through the metal layer. The total electrical energy dissipation rate, per unit area, is q''.

(a) Define a coordinate system, give the governing equations for the temperature excess in both regions, and write all applicable initial, boundary, and other conditions which apply to the temperature fields.

(b) Give the solutions in both regions, for k_L and k_R and $\alpha_L = \alpha_R$.

(c) For $k_L = 2k_R$ and $\alpha_L = \alpha_R$, sketch the expected temperature distributions in the whole region at several later times.

(d) For the conditions in part (c), determine how $\theta_L(0, \tau)$ and $\theta_R(0, \tau)$ are related to each other.

(e) Suggest any likely solution for the temperature field in the two regions, for the conditions in part (c).

4.5.5. Two plane conduction regions, 1 and 2, each of thickness L, are in perfect thermal contact. They are initially uniform at t_i. The temperature of the two exposed surfaces are reduced to t_0. Take the origin of the x coordinate at the left surface of the wall. For the two materials, $k_1 \neq k_2$ and $(\rho c)_1 = (\rho c)_2$.

(a) Define the temperature excess ratios ϕ_1 and ϕ_2 in each region.

(b) Write the differential equations and all of the sufficient boundary and initial conditions for the regions, in terms of the temperature excess ratios.

(c) Write the appropriate general solutions for ϕ_1 and ϕ_2.

(d) Evaluate the constants in these solutions.

(e) Indicate only the changes in the boundary conditions in part (b) if there is actually a contact resistance, R_c, between the two regions, that is, at $x = L$.

4.6.1. Consider an extensive region of ice, at $0°C$, which is subjected to a sudden constant surface flux loading q'', at $x = 0$. Neglect convection.

(a) Sketch the internal temperature at several specific later times.

(b) Write the complete formulation, for the determination of $t(x, \tau)$, in the resulting processes, taking the diffusivities to be the same in both phases.

(c) Determine whether the given equation may be used to assess the instantaneous temperature field.

(d) Indicate how any such correct equation in the melt is related to the instantaneous location of the phase interface.

4.6.2. An extensive layer of pure ice, initially at its phase equilibrium temperature, suddenly has the exposed horizontal upper surface temperature raised to $t_0 = 5°C$ and held at that value. The latent heat is 79.5 cal/g m.

(a) Calculate the resulting value of λ and the instantaneous location of the melting front.

(b) Determine the temperature distribution throughout the region.

(c) Calculate the velocity of propagation of both the phase interface and the location at which the local temperature is 2°C.

4.6.3. A deep pool of water, supercooled to $-5°C$, suddenly begins to freeze downward at its surface. Neglect the density effect of the phase change. The latent heat is 79.5 cal/g m.

(a) Determine the relevant value of λ.

(b) Calculate the location and velocity of the phase interface as a function of time.

(c) At what velocity does the $-4.5°C$ isotherm propagate?

4.6.4. A deep pool of water is uniformly at 4°C. A cold environment suddenly lowers the surface temperature to $-4°C$, which remains approximately constant thereafter. $H = 79.5$ cal/g m.

(a) Determine the value of λ from both Eqs. (4.6.23) and (4.6.24).

(b) Calculate the velocity of phase front propagation.

(c) Determine the temperature distribution in both phases.

4.6.5. Consider the numerical analysis of a melting front advancing into a solid initially at t_{PE}. The surface of the region is at t_0. Take the densities of the solid and liquid phases, ρ_S and ρ_L, to be the same.

(a) Sketch the geometry and a suitable grid system.

(b) Develop the finite-difference form of the relation which applies at the phase interface at each level of time, τ.

(c) Investigate the condition for the stability of the numerical calculations.

4.7.1. A cold-drawn steel wire of 3 mm diameter is to be heated afterward to 500°C for tempering, while being drawn at a velocity of u. The wire will pass through an insulated heater holder. The heat will be applied by a continuous and concentrated electric current flowing across a very thin plane of the moving wire. Assume that the heating effect amounts to a stationary plane source, $q_0''(W/m^2)$. The wire is at 100°C when it enters the heater.

(a) Indicate the appropriate conduction model.

(b) For a drawing rate of 0.5 m/s, what must the heating rate be, in terms of q_0''?

(c) Consider the conduction temperature field ahead of the location of concentrated energy input. At what distance will a 1% temperature increase be found in the approaching cold wire?

4.8.1. Compare the three different polynomial relations given in Sec. 4.8 for $\phi(X) = \phi/\Delta t_{ch}$, along with the conjugate error function.

(a) Plot the three polynomial relations, and erfc η to erfc $\eta = 0.02$.

(b) Calculate the slope at each at $x = 0$ (and at $\eta = 0$).

(c) Calculate the integral of each of these functions and compare the results.

4.8.2. For the geometry and temperature conditions in Prob. 4.2.7, consider the amount of time, τ, after the beginning of the transient that the integral method and the semiinfinite solid solution may each be accurately used. Do this by finding how long the appropriate results from the integral method and the semiinfinite solid solutions apply at a distance 5 cm in from the surfaces of the cube? Consider a 1% response there for the erfc η solution.

4.8.3. Consider the sudden contact of two semiinfinite regions of different properties and initial temperature as discussed in Sec. 4.2 and diagrammed in Fig. 4.2.13. Assume that a thin layer of material of thermal capacity c'' and high conductivity is present over the surface of region R, at t_R, at the instant of contact, that is, $t_{c''} = t_R$ at $\tau = 0$.
  (a) Develop the integral method formulation which will determine $[t_{c''}(\tau) - t_R)]/(t_L - t_R)$ and the transient temperature distributions in regions L and R.
  (b) Using the second-degree polynomial form, determine the temperature distributions in regions L and R.

4.8.4. Consider a time-dependent source, $q'''(\tau)$, in the semiinfinite solid.
  (a) Determine any forms of $q'''(\tau)$ for which $\delta(\tau)$ may be explicitly calculated.
  (b) For a simple special condition $q'''(\tau)$, determine the form of $\delta(\tau)$.
  (c) Calculate the resulting heat flux at the surface and the function $\delta(\tau)$.

4.8.5. For a 1-ft-thick concrete slab, $\alpha = 0.016$ ft^2/hr, initially at t_i, both surfaces are raised to t_0. Using the distribution in Eq. (4.8.4), find:
  (a) The time interval in which $\delta(\tau) \le L/2$.
  (b) The time at which ϕ_A is 0.01 at $L/2$.

4.9.1. Repeat Prob. 4.2.5 using a numerical method with $F = 1/2$ and $1/3$ and $\Delta x = 1/4$ in.

4.9.2. A steel plate 6 in. thick is taken from a furnace with a uniform metal temperature of 800°F. The plate remains in air at 60°F for 20 min; the convection coefficient is 25 Btu/hr ft^2 °F. The plate is then immersed in water at 60°F, and a negligible surface resistance may be assumed. Find the necessary immersion time to reduce the maximum metal temperature to 140°F. Use $\Delta x = 0.6$ in.

$$\alpha = 0.35 \text{ ft}^2/\text{hr} \qquad k = 20 \text{ Btu/hr ft °F}$$

4.9.3. A concrete wall 16 in. thick initially has a sinusoidal temperature distribution, being 60°F at both surfaces and 90°F at the center. If the left and right surfaces are suddenly changed to 40 and 70°F, respectively, find the time necessary for the maximum temperature in the wall to be reduced to 75°F, using a numerical technique. Use $F = 1/3$ and $\Delta x = 2$ in.

4.9.4. A 12-in.-thick concrete wall is initially at 60°F. At a certain time, heating is applied on both surfaces of the wall at a rate that increases the surface temperature at a rate of 10°F/h. At the same time, a uniform internal distributed source arises, for which $q''' = 288$ Btu/hr ft^3.
  (a) Using a numerical method, determine the time that the minimum temperature in the wall has reached 90°F.
  (b) Where is the minimum temperature found? Take $\Delta x = 2$ in., $\alpha = 0.016$ ft^2/hr, $k = 1.0$ Btu/ft hr °F.

4.9.5. A metal pressure vessel of a reactor may be idealized as a flat plate 2 in. thick. In operation the inside surface of the vessel will be adiabatic, and the outside surface will be maintained at 500°F by circulating a coolant. During normal periods of reactor operation, the vessel wall is subjected to an irradiation which may be approximated by a uniformly distributed source of strength 3.68×10^5 Btu/hr ft^3. We are interested in the transient temperature distribution in the vessel wall as a function of time for an abrupt start-up to full normal operating level. Assume that the material is initially at 500°F throughout.

$$\alpha = 0.5 \text{ ft}^2/\text{hr} \qquad k = 40 \text{ Btu/hr ft °F}$$

(a) Using a 1/2-in. grid and the simplest calculation procedure, find the temperature distribution after 25 s.

(b) Find the location and magnitude of the maximum temperature in the wall after steady state has been achieved.

4.9.6. An extensive metal plate of 1-ft thickness is initially at 100°F throughout. It is thermally insulated on its right surface. Suddenly the plate material is irradiated and a uniform and steady distributed energy source, of $q''' = 2 \times 10^4$ Btu/ft^3 hr, results. The left face of the plate, initially at 100°F, is controlled during the transient so that its temperature increase with time is only 200°F/hr.

$$\alpha = 0.5 \text{ ft}^2/\text{hr} \qquad k = 20 \text{ Btu/hr ft °F}$$

(a) Present the necessary equation for an explicit-method finite-difference calculation.

(b) Using $\Delta x = 0.2$ ft, find the temperature distribution in the plate and the surface heat flux after 0.12 hr.

4.9.7. Determine the solution of Prob. 4.2.14 numerically, using a grid size of $\Delta x = 1$ cm. Compare the result with that obtained from the Heisler charts.

4.9.8. Determine the solution of Prob. 4.2.15 numerically, using $\Delta x = 1$ cm.

(a) Compare the time found in part (a) with that found from the Heisler charts.

(b) In redoing part (c), plot both the midplane temperature, obtained numerically, and the result from the Heisler charts.

4.9.9. Consider the numerical analysis of a melting front advancing into a solid initially at its phase equilibrium temperature. The surface of the region is at t_0. Take the densities of the liquid and solid phases, ρ_L and ρ_S, to be the same.

(a) Sketch the geometry and a suitable grid system.

(b) Develop the finite-difference form of the relation which applies at the phase interface at each level of time, τ.

(c) Find the condition for the stability of the numerical calculations.

4.9.10. We wish to find the time necessary to heat an extensive flat steel plate, of 1.6-ft thickness, from an initial temperature of 240°F to a midplane temperature of 740°F. Heating is to be achieved by passing an electric current through the metal, which will result in a Joulean dissipation rate q''' of 1.5×10^5 Btu/hr ft^3. The surfaces will be in contact with air at 40°F moving at a high enough velocity that $h = 50$ Btu/hr ft^2 °F on each side. The plate is to be in a very large enclosure whose surfaces are at 40°F. For the steel, $\epsilon = 0.9$, $k = 20$ Btu/hr ft °F, $\alpha = 0.333$ ft^2/hr. In a numerical procedure use $\Delta x = 0.2$ ft, and do the following:

(a) Determine the proper internal and surface point equations.

(b) Find the appropriate time step.

(c) Calculate the necessary time to achieve the desired effect.

4.9.11. A masonry wall of 1-ft thickness is initially uniformly at 60°F throughout at sunrise. The solar intensity on the wall suddenly becomes 100 Btu/hr ft^2 and the absorbtivity may be taken as 0.6. The outside air temperature increases, from 60°F at sunrise, by 10°F per hour, for 3 hr. The total surface coefficient on the outside surface may be taken as 4 Btu/hr ft °F. Assume that the inside wall surface is insulated. For the wall material $k = 1.0$ Btu/hr ft °F and $\alpha = 0.039$ ft^2/hr.

(a) Find the wall temperature distribution after 2 hr, using a numerical method. Take $\Delta x = 0.25$ ft.

(b) Calculate the surface heat flux during the first two hours:

4.9.12. An exterior concrete wall of 6-in thickness is initially at 40°F throughout. Assume the inside surface is perfectly insulated. The outside surface is suddenly subjected to a surrounding temperature of 60°F, $h = 4$ Btu/hr ft^2 °F, and a time-dependent net radiation input flux $q'' = 80\tau$ (hr), in Btu/hr ft^2. Use the properties $k = 1$ Btu/hr ft °F, $\rho = 144$ lb/ft^3, and $c = 0.2$ Btu/lb °F.

(a) For a numerical method, and $\Delta x = 3$ in., determine the maximum time step for calculation.

(b) Using this value, write the relations to be used in a numerical computation of the wall temperature response.

(c) Determine the necessary elapsed time for the minimum temperature in the wall to reach 43°F. What is the exposed surface temperature at this time?

4.9.13. An extensive metal plate of thickness $L = 1.0$ ft is initially uniformly at 100°F throughout. It is thermally insulated on its right surface. An irradiation of the material suddenly occurs. It amounts to a location-dependent distributed energy source q''', which does not change with time.

$$q'''(x) = q_0''' e^{-x/L} = 2 \times 10^4 \, e^{-x/L} \, (\text{Btu/ft}^3 \, \text{hr})$$

The left face of the plate is exposed to an ambient medium, which remains at 100°F, $h = 100$ Btu/ft^2 °F hr. For the plate material, $\alpha = 0.5$ ft^2/hr and $k = 20$ Btu/hr ft °F.

(a) Present the necessary equations for an explicit-method finite-difference calculation.

(b) using $\Delta x = 0.2$ ft, find the temperature distribution in the plate and the surface heat flux after 0.10 hr.

4.9.14. The 1-ft-thick stainless-steel wall of a large nuclear reactor pressure vessel is initially at 200°F. After start-up the fluid inside the vessel heats the inside vessel wall surface and the temperature rises at a rate of 40°F per hour. Radioactive by-products penetrate the wall from the inside at an equivalent energy rate of $I_0 = 320$ Btu/hr ft^2. This energy is absorbed as it penetrates the wall, in x. The local intensity through the wall is $I(x) = I_0 e^{-x}$. The other side of the wall is assumed to be insulated. In a numerical calculation, for $\Delta x = 3$ in., the wall material temperature response is to be determined.

(a) Write the equations for an explicit numerical calculation, for each point.

(b) Determine the maximum time step.

(c) Write the equations in terms of the specific inputs which apply for this vessel.

(d) Calculate the time, after start-up, that the mid-wall temperature reaches 215°F. $k = 10$ Btu/hr ft °F and $\alpha = 0.50$ ft^2/hr.

REFERENCES

Abramowitz, M., and I. A. Stegun (1964) *Handbook of Mathematical Functions*, AMS, 55, Natl. Bur. Standards.

Arpaci, V. S. (1966) *Conduction Heat Transfer*, Addison-Wesley, Reading, MA.

Baumeister, K. J., and T. D. Hamill (1969) *J. Heat Transfer*, **91**, 543.

Baumeister, K. J., and T. D. Hamill (1971) *J. Heat Transfer*, **93**, 126.

Carslaw, H. S., and J. C. Jaeger (1959) *Conduction of Heat in Solids*, 2nd ed., Oxford Univ. Press, New York.

Cho, S. H., and Sunderland, J. E. (1974) *J. Heat Transfer*, **96**, 214.

Crank, J. (1957) *The Mathematics of Diffusion*, Oxford Univ. Press, London.

Eckert, E. R. G., and R. M. Drake, Jr. (1972). *Analysis of Heat and Mass Transfer*, McGraw-Hill, New York.

Frankel, J. I., B. Vick, and M. N. Özisik, (1987) *Int. J. Heat Mass Transfer*, **30**, 129.

Gembarovic, J., and V. Majernik (1988) *Int. J. Heat Mass Transfer*, **31**, 1073.

Goodman, T. R. (1964) *Adv. in Heat Transfer*, **1**, 52.

Heisler, M. P. (1947) *Trans. ASME*, **69**, 227.

Ingersoll, L. R., O. J. Zobel, and A. C. Ingersoll (1954) *Heat Conduction* Univ. of Wisconsin Press, Madison.

Jakob, M. (1949) *Heat Transfer*, vol. 1, John Wiley and Sons, New York.

Kaminski, W. (1990). *ASME J. Heat Transfer*, **112**, 555.

Kim, C.-J. and M. Kiviany (1990). *Int. J. Heat Mass Transfer*, **33**, 2721.

Lecomte, D. and J.-C. Batsale, (1991) *Int. J. Heat Mass Transfer*, **34**, 894.

Lunardini, V. J. (1981) *Heat Transfer in Cold Climates*, Van Nostrand Reinhold, New York.

Özisik, M. N. (1980). *Heat Conduction*, John Wiley and Sons, New York.

Patankar, S. V. (1980) *Numerical Heat Transfer and Fluid Flow*, Hemisphere, Washington.

Pedroso, R. I., and G. A. Domoto (1973) *J. Heat Transfer*, **95**, 553.

Prud'homme, M., and T. H. Nguyen (1989) *Int. J. Heat Mass Transfer*, **32**, 1507.

Rosenthal, D. (1946) *Trans. ASME*, **68**, 849.

Schlichting, H. (1968) *Boundary Layer Theory*, 6th ed., McGraw-Hill, New York.

Schneider, P. J. (1955) *Conduction Heat Transfer*, Addison-Wesley, Reading, MA.

Smith, G. D. (1965) *Numerical Solutions of Partial Differential Equations with Exercises and Worked Solutions*, Oxford Univ. Press, London.

Stefan, J. (1891) *Ann. Phys u Chem. NeueFolge* **42**(2), 269.

Vernotte, M. P. (1958) *Compte Rendu*, **246**, 3154.

Yao, L. S., and J. Prusa (1989) *Adv. in Heat Transfer*, **19**, 1.

CHAPTER

5

MULTIDIMENSIONAL TIME-DEPENDENT PROCESSES

5.1 METHODS OF ANALYSIS

Many conduction processes of practical interest occur in multidimensional regions. Chapter 3 concerns those in steady state, as $t(x, y, z)$, $t(r, \theta\, z)$, and $t(r, \phi, \theta)$ in common coordinates. Chapter 4 concerns one-dimensional transients, as $t(x, \tau)$ and $t(r, \tau)$, for many kinds of imposed conditions and constituent processes. Actual transients are often accurately idealized as one dimensional. The material in this chapter concerns some of the techniques and results which are available for processes in which multidimensional effects must be retained for a realistic representation of transient transport.

The general method of separation of variables indicates the forms of solutions which will satisfy the following linear differential equation:

$$\nabla^2 t = \frac{1}{\alpha} \frac{\partial t}{\partial \tau} \tag{5.1.1}$$

For multidimensional steady state, the formulations in Sec. 3.1.2, for the rectangular, cylindrical, and spherical coordinates, are

$$t(x, y, z) = X(x)Y(y)Z(z) \tag{5.1.2}$$

$$t(r, \theta, z) = \bar{R}_c(r)\Theta(\theta)Z(z) \tag{5.1.3}$$

$$t(r, \phi, \theta) = \bar{R}_s(r)\Phi(\phi)M(\mu) \tag{5.1.4}$$

where $\mu = \cos\theta$. These postulates were applied in turn to Eq. (5.1.1) for steady state, $\nabla^2 t = 0$. Thereby, the admissible forms of the nine functions: $X(x)$, $Y(y)$, $Z(z)$; $\overline{R}_c(r)$, $\Theta(\theta)$, $Z(z)$; and $\overline{R}_s(r)$, $\Phi(\phi)$, and $M(\mu)$ were determined. These are Eqs. (3.1.15), (3.1.19)–(3.1.21), and (3.1.26)–(3.1.28). They are the basis of the analyses in Secs. 3.2 through and 3.4.

The separation of variables method was applied in Sec. 4.1.1. to one-dimensional transients, to determine the consequences of the transient effect in each of the three geometries, as

$$\frac{\partial^2 t}{\partial x^2} = \frac{1}{\alpha}\frac{\partial t}{\partial \tau} \qquad t(x,\tau) = X(x)F(\tau) \qquad (5.1.5)$$

$$\frac{\partial^2 t}{\partial r^2} + \frac{1}{r}\frac{\partial t}{\partial r} = \frac{1}{\alpha}\frac{\partial t}{\partial \tau} \qquad t(r,\tau) = \overline{R}_c(r)F(\tau) \qquad (5.1.6)$$

$$\frac{\partial^2 t}{\partial r^2} + \frac{2}{r}\frac{\partial t}{\partial r} = \frac{1}{\alpha}\frac{\partial t}{\partial \tau} \qquad t(r,\tau) = \overline{R}_s(r)F(\tau) \qquad (5.1.7)$$

The results for $X(x)$, $\overline{R}_c(r)$, and $\overline{R}_s(r)$ are similar to those in the steady state. The transient effect in all three geometries was shown to be

$$F(\tau) = Ce^{\pm \alpha\beta^2\tau} \qquad (5.1.8)$$

where β^2 may be real, imaginary, or complex, with the consequences discussed following Eq. (4.1.7). Thereby, many different forms of transient and periodic processes may be analyzed.

The net effect of the separations, postulated in Eqs. (5.1.2)–(5.1.7), is that Eqs. (5.1.2)–(5.1.4) represent $\nabla^2 t$ in Eq. (5.1.1) and Eq. (5.1.8) represents $\partial t/\partial \tau$, in Eq. (5.1.1), in a general formulation of the method. Therefore, solutions of multidimensional transients are constructed in this way, for particular applications.

A convenient and comprehensive analysis of the general consequences of the method of separation of variables is given by Özisik (1980). Sections 2-2, 3-1, and 4-1 apply to the Cartesian, cylindrical and spherical coordinates, respectively. Many different and specific kinds of conduction circumstances are then analyzed. See also Myers (1971), Chap. 3, and Arpaci (1966), Sec. 4-5.

Another consequence of the separation of variables method is that some multidimensional transient solutions may be generated as simple products of one-dimensional solutions. This is done in Sec. 5.1.1 for two- and three-dimensional transients. This method applies to regions which are described in terms of combined plane and cylindrical boundaries. Examples are short cylinders and rectangular regions.

Section 5.1.2 concerns several additional general aspects of solution techniques. A general formulation is also given for generating solutions in regions with time-dependent surface conditions. Section 5.2 gives several multidimensional solutions. Many circumstances arise in which analytical solutions are not available. Section 5.3 formulates transient two- and three-dimensional finite-

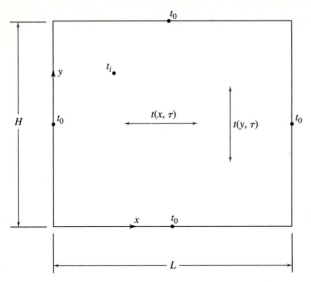

FIGURE 5.1.1
The formulation of a two-dimensional transient $\phi(x, y, \tau)$, as the product of two one-dimensional transients.

difference methods, in Cartesian coordinates, in terms of surface conditions and stability limits.

5.1.1 Multidimensional Product Solutions

The generation of solutions as products of one-dimensional solutions provides very valuable multidimensional transient results. As an example, consider the one-dimensional transient in the plane layer, diagrammed in Fig. 4.2.1(a), initially at t_i, or at $\theta_i = t_i - t_0$. The solution for the transient response is Eq. (4.2.5) as

$$\phi_P = \frac{\theta}{\theta_i} = \frac{4}{\pi} \sum_{n=1,3}^{\infty} \frac{e^{-(n\pi)^2 \alpha \tau / L^2}}{n} \sin n\pi x/L = \phi_P(\text{Fo}, x/L) \quad (5.1.9)$$

where the subscript P denotes the plane layer.

The temperature field $\phi(x, y, \tau)$ in a long rectangular rod of L width and H height is to be generated as the product of the following two solutions $t(x, \tau)$ and $t(y, \tau)$, as shown in Fig. 5.1.1. The two solutions, in the form of Eq. (5.1.9), are

$$\phi(x, y, \tau) = \phi_{P,L}(x, \tau)\phi_{P,H}(y, \tau) = \phi_{P,L}\phi_{P,H} \quad (5.1.10)$$

The preceding three functions are each to be solutions of

$$\frac{\partial^2 \phi}{\partial x^2} + \frac{\partial^2 \phi}{\partial y^2} - \frac{1}{\alpha} \frac{\partial \phi}{\partial \tau} = 0$$

and $\phi_{P,L}$ and $\phi_{P,H}$ satisfy

$$\frac{\partial^2 \phi_{P,L}}{\partial x^2} - \frac{1}{\alpha}\frac{\partial \phi_{P,L}}{\partial \tau} = 0$$

and

$$\frac{\partial^2 \phi_{P,H}}{\partial y^2} - \frac{1}{\alpha}\frac{\partial \phi_{P,H}}{\partial \tau} = 0$$

From Eq. (5.1.10):

$$\nabla^2 \phi = \nabla^2(\phi_{P,L}\phi_{P,H}) = \phi_{P,H}\frac{\partial^2 \phi_{P,L}}{\partial x^2} + \phi_{P,L}\frac{\partial^2 \phi_{P,H}}{\partial y^2}$$

$$= \frac{1}{\alpha}\frac{\partial \phi}{\partial \tau} = \frac{1}{\alpha}\frac{\partial(\phi_{P,L}\phi_{P,H})}{\partial \tau}$$

$$= \frac{1}{\alpha}\left[\phi_{P,H}\frac{\partial \phi_{P,L}}{\partial \tau} + \phi_{P,L}\frac{\partial \phi_{P,H}}{\partial \tau}\right]$$

This is rearranged as

$$\left[\nabla^2 \phi - \frac{1}{\alpha}\frac{\partial \phi}{\partial \tau}\right]$$

$$= \phi_{P,H}\frac{\partial^2 \phi_{P,L}}{\partial x^2} - \frac{1}{\alpha}\phi_{P,H}\frac{\partial \phi_{P,L}}{\partial \tau} + \phi_{P,L}\frac{\partial^2 \phi_{P,H}}{\partial y^2} - \frac{1}{\alpha}\phi_{P,L}\frac{\partial \phi_{P,H}}{\partial \tau}$$

$$= \phi_{P,H}\left[\frac{\partial^2 \phi_{P,L}}{\partial x^2} - \frac{1}{\alpha}\frac{\partial \phi_{P,L}}{\partial \tau}\right] + \phi_{P,L}\left[\frac{\partial^2 \phi_{P,H}}{\partial y^2} - \frac{1}{\alpha}\frac{\partial^2 \phi_{P,H}}{\partial \tau}\right]$$

Therefore, if $\phi_{P,L}$ and $\phi_{P,H}$ each satisfy the one-dimensional Fourier equation, $\phi(x, y, \tau)$ satisfies the two-dimensional form.

The boundary conditions on $\phi_{P,L}$ and $\phi_{P,H}$ are next interpreted to obtain proper boundary conditions on $\phi(x, y, \tau)$, as follows:

$$\phi(0, y, \tau) = \phi(L, y, \tau) = \phi(x, 0, \tau) = \phi(x, H, \tau) = 0$$

where, for example, $\phi(0, y, \tau) = \phi_{P,L}(0, \tau)\phi_{P,H}(y, \tau)$ is zero because $\phi_{P,L}(0, \tau)$ was taken as zero to obtain the solution in Eq. (5.1.9). Thus the four boundary conditions on $\phi_{P,L}$ and $\phi_{P,H}$ collectively result in the same conditions on the four faces on the rectangular rod, L by H. This applies, in general, for product solutions satisfying linear and homogeneous boundary conditions. The initial condition for $\phi(x, y, \tau)$ is satisfied as

$$\frac{\theta(x, y, 0)}{\theta_i} = \left[\frac{\theta(x, 0)}{\theta_i}\right]_{P,L}\left[\frac{\theta(y, 0)}{\theta_i}\right]_{P,H} = 1 \qquad (5.1.11)$$

The two solutions $\phi_{P,L}$ and $\phi_{P,H}$ satisfy the boundary conditions on ϕ as interpreted in the following way. Consider the boundary at $y = 0$ to H, at $x = 0$ or L. The boundary heat flow contributed by $t(y, \tau)$ is zero there, since $\partial t(y, \tau)/\partial y = 0$. Thus $\partial t(y, \tau)/\partial y$ supplies the heat flux along the boundaries at $x = 0$ and L. The solution $\phi(x, \tau)$ brings the product solution $\phi(x, y, \tau)$ down to zero along these boundaries. The same kind of effect arises along the boundaries $x = 0$ to L, at $y = 0$ and H.

The solution for $\phi(x, y, \tau)$ is then written from Eq. (5.1.9) as $\phi_{P,L}(\text{Fo}_L, x/L)\phi_{P,H}(\text{Fo}_H, y/H)$. The result for a square rod is then simply $\phi(x, y, \tau) = \phi_{P,L}(x, \tau)\phi_{P,H}(y, \tau)$, over an equal range of x and y.

An effective way to interpret the preceding product solution is to consider the region in which $\phi(x, y, \tau)$ is found to merely consist of the common region arising from the orthogonal intersection of two one-dimensional regions of thicknesses L and H. The value of ϕ at any point x, y, at time τ, is then simply the product of the two solutions in that region. The previous solutions also indicate that the midpoint temperature excess, $\phi(L/2, H/2, \tau) \leq 1$, decays more rapidly than either $\phi_{P,L}(L/2, \tau) \leq 1$ or $\phi_{P,H}(H/2, \tau) \leq 1$.

The extension of the preceding analysis, for a long L by H rectangular rod, to a L by H by W parallelpiped, with $\phi(x, y, z, \tau)$, is apparent. The product solution then has an additional factor $\phi_{P,W}(z, \tau)$. This component then is of the form $\phi_{P,W}(\text{Fo}_W, z/L)$.

OTHER PRODUCT SOLUTIONS. It was convenient before to place the coordinate origin at the corner, as seen in Fig. 5.1.1, because the origin for the solution for ϕ_P in Eq. (5.1.9) is at the left face of the one-dimensional geometry, in Fig. 4.2.1(a). However, the origin used in the components of a product solution are commonly placed at a plane of symmetry. This would be at $x = L/2$ and $y = H/2$ in the geometry in Fig. 5.1.1. This is the procedure in Fig. 4.2.2 for a plane region with surface convection to an ambient at t_e on both sides. An exception to this procedure is for product solution components ϕ_{SI}, that is, for a semiinfinite solid, for example, as in Fig. 4.1.1. Then $x = 0$ is at the surface, as shown.

Using the plane, cylindrical, and semiinfinite solid one-dimensional transient solutions, a group of interesting multidimensional transient solutions may be assembled. Nine such geometries are shown in Fig. 5.1.2. The one-dimensional solutions are ϕ_P, ϕ_C, and ϕ_{SI}, where ϕ_C and ϕ_{SI} are given in Eqs. (4.3.2) and (4.2.15). The preceding discussion of the way in which the product solution one-dimensional components collectively satisfy the boundary conditions on, for example, $\phi(x, y, z, \tau)$, also applies to the additional geometries in Fig. 5.1.2.

A SURFACE FLUX CONDITION. The preceding kinds of product solutions do not apply for an imposed flux condition at a surface, for example, as $q''(x)$.

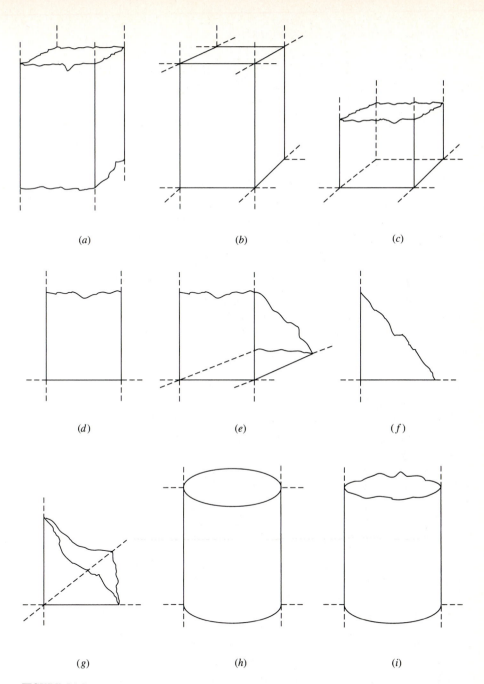

FIGURE 5.1.2

Geometries in which multidimensional transient response may be determined as products of plane, cylindrical, and semiinfinite one-dimensional transient solutions as: $\phi_P(x, \tau)$, $\phi_C((r, \tau)$, and $\phi_{SI}(x, \tau)$. (a) A long, rectangular bar L by H; $\phi_{P,1}\phi_{P,2}$. (b) A rectangular solid L by H and W; $\phi_{P,1}\phi_{P,2}\phi_{P,3}$. (c) End region of a long, rectangular bar; $\phi_{P,1}\phi_{P,2}\phi_{SI}$. (d) Edge region of a large and long plate; $\phi_P\phi_{SI}$. (e) Corner region of a large and long plate; $\phi_{P,1}\phi_{SI,1}\phi_{SI,2}$. (f) Edge region of a quarter-infinite solid; $\phi_{SI,1}\phi_{SI,2}$. (g) Corner region of an eighth-infinite solid; $\phi_{SI,1}\phi_{SI,2}\phi_{SI,3}$. (h) A short cylinder; $\phi_P\phi_C$. (i) End region of a long, cylindrical rod; $\phi_C\phi_{SI}$.

279

Consider such a condition along $x = 0$ to L, $y = 0$, in Fig. 5.1.1:

$$q''(x) = -k\left(\frac{\partial t}{\partial y}\right)_{y=0} = -k\left(\frac{\partial \phi}{\partial y}\right)_{y=0}(t_i - t_0)$$

$$= -k\left[\frac{\partial(\phi_{P,L}\phi_{P,H})}{\partial y}\right]_{y=0}(t_i - t_0)^2$$

$$= -k\phi_{P,L}\left[\frac{\partial \phi_{P,H}}{\partial y}\right]_{y=0}(t_i - t_0)^2 \qquad (5.1.12)$$

or

$$\left(\frac{\partial \phi_{P,H}}{\partial y}\right)_{y=0} = -\frac{q''(x)}{k\phi_{P,L}(t_i - t_0)^2} = F(x,\tau) \qquad (5.1.13)$$

However $\phi_{P,H}(y,\tau)$ cannot satisfy a function of x, either as $q''(x)$ or in $\phi_{P,L}$. Therefore, the only admissible condition is $q''(x) = 0$. This is an insulating condition at $y = 0$. This solution is a rod of height $2H$ and width L, with uniform surface conditions of $\phi = 0$ at $y = H$ and $y = -H$.

SURFACE CONVECTION CONDITIONS. Product solutions also apply for the geometries in Fig. 5.1.2, initially uniformly at t_i, suddenly immersed in a convective environment uniformly at t_e with a uniform convection coefficient h over the surfaces. The one-dimensional solutions ϕ_P and ϕ_{SI} are given in Eqs. (4.2.10) and (4.2.17). They give multidimensional solutions for all but the cylindrical geometries in Fig. 5.1.2. For those, the Heisler charts may be used to estimate ϕ_C. The Heisler charts may also be used to estimate local temperature levels internal to the geometries in Fig. 5.1.2.

MULTIDIMENSIONAL INITIAL CONDITIONS. All of the preceding results apply only for the initial condition, for example, $\phi(x, y, z, 0) = \phi_i = 1$. However, many applications may amount to a nonuniform initial condition over the region, as $t(x, y, z, 0) = t_i(x, y, z)$. Such a distribution may remain from an earlier transient process in the region. However, a solution for any one-dimensional transient may accommodate only one initial temperature distribution, for example, as $t(x, 0) = t_i(x)$. Therefore, the only admissible form of $t(x, y, z, 0) = t_i(x, y, z)$ is a product, as $t_i(x, y, z) = t_i(x), t_i(y), t_i(z)$. That is, $t_i(x, y, z)$ must be separable as a product of independent functions of x, y, and z. Then each of the one-dimensional solutions to be used in a product solution is determined in terms of one particular component of t_i. See Carslaw and Jaeger, (1959), Sec. 6.6, for an alternative method of expressing $t_i(x, y, z)$, when not separable, in terms of a triple sine series, as discussed in Sec. 5.2.1.

SUMMARY. The previous developments indicate how products extend one-dimensional transient solutions to multidimensional temperature fields. This permits many more uses of one-dimensional results in design and other applications. The method applies directly for uniform initial temperatures t_i, for changed surface temperatures, and for suddenly imposed convection conditions. The method does not apply generally for imposed surface heat flux conditions.

5.1.2 Several General Methods

The preceding section indicates the method of products solutions. These are very valuable extensions of relatively simple one-dimensional transient solutions, to represent multidimensional ones. However, the specific results given apply only for uniform initial temperatures, for example, $t(x, y, z, 0) = t_i$, as a single value. They also accommodate only a uniform and constant condition over the bounding surfaces, as a single temperature level, t_0, or convection with both h and t_e uniform and constant. Also, the method does not include the effects of distributed energy sources.

This section considers several among the available alternatives, for use in more complicated processes. The first aspect is a transformation which accommodates a location-dependent distributed energy source $q'''(x, y, z)$ in transients, analogous to the technique in Sec. 3.6.1 for steady state. Thereafter, Duhamel's method is given. It permits the determination of the transient temperature response, in a region, which results from a time-dependent surface temperature, $t_s(x, y, z, \tau)$, for $q''' = 0$.

The general transport equations are then determined for orthotropic mechanisms of heat conduction, k_i, and mass diffusion, D_i, in Secs. 1.2.1 and 1.3.2. Formulations of anisotropic processes, in terms of k_{ij} and D_{ij}, are also given. Consideration is given in the following discussion for reducing the complexity of this general formulation, in terms of principal axes and associated conductivities.

The last matter, briefly discussed in this section, concerns what is called the inverse heat conduction problem (IHCP). This method to be contrasted to the common analytical techniques. Those techniques determine solutions from the differential equations, combined with well-posed ancillary boundary and initial conditions. Instead, the inverse techniques use some collection of measured interior temperatures, for example, $t_1(x, t), t_2(x_2, \tau), \ldots$ in a one-dimensional transient. These temperatures are used to infer the boundary surface condition which would result in this collection of internal responses. Common objectives are to infer the imposed surface flux, or surface temperature condition, which led to this response. It amounts to using remote sensors to determine unknown surface conditions. This method may also be used to determine the thermal diffusivity of a conduction region. Such procedures are ill-posed problems, in that the data are commonly an inadequate basis for general conclusions. Calculation procedures are also commonly unstable.

A LOCATION-DEPENDENT ENERGY SOURCE. Section 3.6.1 concerns multidimensional steady-state processes with internal generation. The Poisson equation was replaced by a sum of homogeneous and inhomogeneous parts. Solutions were given for q''' uniform, for two- and three-dimensional regions. A similar procedure is given in Sec. 4.2.4 for one-dimensional transients, for q''' uniform and constant, for $\tau \geq 0$.

The general procedure is formulated for q''' independent of time, as

$$\nabla^2 t + \frac{q'''(x, y, z)}{k} = \frac{1}{\alpha} \frac{\partial t}{\partial \tau} \qquad (5.1.14)$$

with conditions $t_i(x, y, z)$ at $\tau = 0$ and $t_s(x, y, z)$ over the surface. The inhomogeneous part is separated as follows:

$$t(x, y, z, \tau) = t_1(x, y, z) + t_2(x, y, z, \tau) \qquad (5.1.15)$$

where t_1 is a solution of

$$\nabla^2 t_1 + q'''/k = 0 \qquad (5.1.16)$$

This must satisfy the boundary condition $t_s(x, y, z)$. Then $t_2(x, y, z, \tau)$ over the region is determined from

$$\nabla^2 t_2 = \frac{1}{\alpha} \frac{\partial t_2}{\partial \tau} \qquad (5.1.17)$$

with initial and boundary conditions as follows:

$$t_2(x, y, z, 0) = t_1(x, y, z) - t_1(x, y, z) \quad \text{and} \quad t_{s,2} = 0$$

Thereby, Eq. (5.1.15) satisfies Eq. (5.1.14), and the conditions t_i and t_s, as the sum of two solutions. One is a transient, and the other is steady state. A convective surface condition may be accommodated in a similar analysis.

TIME-DEPENDENT BOUNDARY CONDITIONS. These kinds of processes commonly arise. Imposed surface flux and convection conditions often have this consequence, as seen in many of the results in Chap. 4. The consequences of several purely time-dependent transient surface conditions, $t_s(\tau)$, are given in Secs. 4.3.2 and 4.8.2 and in Sec. 6.3.1, for mass diffusion. The Duhamel theorem, of 1833, provides a general formulation for such analysis. Several aspects of this method are outlined as follows.

Consider the formulation for $t(x, y, z, \tau)$ in a conduction region, for a uniform initial condition t_i, subject to a transient surface temperature condition $t_s(x, y, z, \tau)$. These functions are written as $t(x, y, z, \tau) - t_i = \theta(x, y, z, \tau)$ and $t_s(x, y, z, \tau) - t_i = \theta_s(x, y, z, \tau)$. Consider a transient response $\theta'(x, y, z, \tau)$, in this same conduction region, for $\theta_i = 0$, subject to a constant boundary condition of the same form, as $t_s(x, y, z, \tau)$, which applies at any particular time $\tau = \lambda$. That is, θ' is the solution for $\theta_i = 0$ and $\theta_s(x, y, z, \lambda)$. It may be shown that the solution for $\theta(x, y, z, \tau)$ with $\theta_i = 0$ and $\theta_s(x, y, x, \tau)$ is

$$\theta(x, y, z, \tau) = \int_0^\tau \frac{\partial}{\partial \tau} \theta'(x, y, z, \lambda, \tau - \lambda) \, d\lambda \qquad (5.1.18)$$

That is, θ' is the transient solution for the boundary conditions at time $\tau = \lambda$,

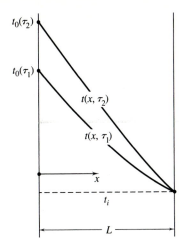

FIGURE 5.1.3
A plane region at t_i subject to a time-dependent surface temperature level, as $t_0(\tau) - t_i = K\tau$.

for the solution $\theta(x, y, z, \tau)$, as

$$\nabla^2 \theta' = \frac{1}{\alpha} \frac{\partial \theta'}{\partial \tau} \qquad \theta_i' = 0 \quad \text{and} \quad \theta'(x, y, z, 0) = \theta_s'(x, y, z, \lambda) \quad (5.1.19)$$

where $\theta_s'(x, y, z, \lambda)$ applies over the surface at time $\tau = \lambda$. This formulation also applies for a convection surface condition in which $\theta_e = t_e(x, y, z, \tau) - t_i$ and $\theta_e(x, y, z, \tau) = 0$.

The preceding formulation is much simpler when the surface, or ambient temperature, that is, θ_s or θ_e, is uniform and only a function of time, as $\theta_s(\tau)$ or $\theta_e(\tau)$. Then, for $\theta_s(\tau)$,

$$\theta(x, y, z, \tau) = \int_0^\tau \theta_s(\lambda) \frac{\partial}{\partial \tau} \theta'(x, y, z, \tau - \lambda) \, d\lambda \qquad (5.1.20)$$

The procedure is to determine $\theta'(x, y, z, \tau - \lambda)$, for $\theta_s(\lambda)$ or $\theta_e(\lambda)$, then evaluate the time derivative of $\theta'(x, y, z, \tau - \lambda)$. The result is then integrated between 0 and τ. Myers (1971) discusses the application of Duhamel's theorem to the plane region transient shown in Fig. 5.1.3, subject to a time-dependent surface temperature condition at the left face, as $t_0(\tau)$. The construction of a solution is given, including the possibility of discontinuities in the time-dependent surface condition $t_0(\tau)$.

A specific result is given as follows for $t(0, \tau) - t_i = K\tau$ on the left face, with t_i maintained at $x = L$. The solution for $\theta'(x, \tau)$ for any given time τ_0, when $t(0, \tau) - t_i = K\tau_0$, is

$$\frac{\theta'(x, \tau)}{K\tau_0} = \frac{t(x, \tau) - t_i}{t(0, \tau_0) - t_i} = \left(1 - \frac{x}{L}\right) - \frac{2}{\pi} \sum_{n=1}^\infty \frac{\sin n\pi x/L}{n} e^{-(n\pi)^2 \alpha \tau / L^2}$$

$$(5.1.21)$$

The function $\theta'(x, \tau - \lambda)$ in Eq. (5.1.20) is then

$$\frac{\theta'(x,\tau)}{K\tau_0} = 1 - \frac{x}{L} - \frac{2}{\pi} \sum_{n=1}^{\infty} \frac{\sin n\pi x/L}{n} e^{-(n\pi)^2(\tau-\lambda)\alpha/L^2} \qquad (5.1.22)$$

Therefore, $\theta'(x, \tau)$ is the function in Eq. (5.1.20) and $\theta_s(\lambda) = k\lambda$. The result, in terms of Fo $= \alpha\tau/L^2$, is

$$\frac{t(x,\tau) - t_i}{K\tau} = \left(1 - \frac{x}{L}\right) - \frac{2}{\pi^3} \sum_{n=1}^{\infty} \frac{\sin n\pi x/L}{n^3} \left[1 - e^{-(n\pi)^2 \text{Fo}}\right] \qquad (5.1.23)$$

The use of Duhamel's method is discussed in generality, and in terms of application, in Carslaw and Jaeger (1959), Sec. 1.14, Özisik (1980), Chap. 5, and in Myers (1971), Sec. 4.2.

ANISOTROPIC REGIONS. The formulations for such regions are Eqs. (1.2.3) and (1.3.2) for the three orthogonal components of heat conduction flux, q''_{x_i}, and diffusing mass flux, m''_{x_i}, respectively. Such heat conduction and mass diffusion processes commonly occur in laminated and crystalline materials. The effect of anisotropic conduction is seen in writing one component of the heat flux, q''_x. This is in terms of k_{ij}, as k_{11}, k_{12}, k_{13}:

$$-q''(x) = k_{11}\frac{\partial t}{\partial x} + k_{12}\frac{\partial t}{\partial y} + k_{13}\frac{\partial t}{\partial z} \qquad (5.1.24)$$

The general anisotropic conductivity postulate of k_{ij} is seen in Eq. (1.2.3a). However, in ordered crystalline materials, not all nine components arise. In some materials, some of the k_{ij} also have the same value. See Carslaw and Jaeger, (1959), Sec. 1.17, and Özisik (1980), Chap. 5, for the several classes of behavior.

The complexity of the kind of formulation of $q''(x)$, $q''(y)$, and $q''(z)$, as implied in Eq. (5.1.24), may be reduced by a suitable transformation, to a new system of rectangular coordinates ξ, η, and ζ. Then there are the three principal axes of conductivity, k_1, k_2, and k_3. In terms of these principal conductivities, the conduction equation becomes

$$k_1\frac{\partial^2 t}{\partial \xi^2} + k_2\frac{\partial^2 t}{\partial \eta^2} + k_3\frac{\partial^2 t}{\partial \zeta^2} = \rho c \frac{\partial t}{\partial \tau} \qquad (5.1.25)$$

The preceding formulation is of the form of the anisotropic formulation given in Eq. (1.2.6). The further transformation in Eq. (1.2.15) removed the orthotropic effect. Equation (5.1.25) is in an orthotropic form, but now in terms of ξ, η, and ζ and k_1, k_2, and k_3. The transformation in Eq. (1.2.15) retrieves the isotropic formulation, with $k_a = \sqrt[3]{k_1 k_2 k_3}$. See Carslaw and Jaeger (1959), Sec. 1.18, and also Eckert and Drake (1972).

There are a number of limitations to the utility of the preceding coordinate transformation, to the principal axes of conductivity. For example, the conduction region geometry of interest may have corresponded to bounding surfaces at particular values of x, y, and z, as for example, for a rectangular region. Under transformation to ξ, η, and ζ coordinates, the boundaries may not retain this feature of simplicity, in terms of fixed values of ξ, η, and ζ. However, this difficulty does not arise for conduction in an infinite region. Then remote boundary conditions are to be matched only at large distances from a concentrated embedded temperature effect. An extensive consideration of these matters, and of many other aspects of anisotropic region conduction, is given by Özisik (1980).

INVERSE HEAT CONDUCTION PROBLEMS (IHCP). There are many practical circumstances in which one or more temperature sensors, internal to the conduction region, might profitably be used to infer the boundary or other conditions that have caused the internal temperature field. One such application occurs at a bounding surface, whose temperature or heat flux level may not be accurately measured directly. The surface condition is then inferred from the internal sensors. An example is an extreme imposed surface condition, which is not consistent with practical temperature or heat flux measurement devices. Then the internal temperature data are used to approximately reconstruct the conditions at the surface.

Such procedures are the inverse of the kinds of direct conduction analysis procedures used, for example, in Secs. 4.2 and 4.3 and in Sec. 5.2. Those solutions use the differential equation and all relevant conditions to determine the temperature distribution at all locations internal to the conduction region, as, for example, $t(x, y, z, \tau)$. The inverse method uses instead a very small sample of direct internal measurement locations, say $t_1, t_2, \ldots$, to infer surface conditions. This is the basic limitation of both the reality and the accuracy of the inverse method. Several specific locations are taken to represent an unknown and continuous conduction region, including the boundary conditions.

Since the data do not completely represent the region, the solution does not depend continuously on the measurement data in the region. Therefore, the procedure is ill-posed, in the sense of the input data not fully representing the region. Another aspect of this is that the inferred surface heat flux values may be very sensitive to the measured temperature values internal to the region. See Beck et al. (1985) and Hensel (1991) for extensive reviews of methods that have been developed for IHCP procedures, both by analysis and by numerical procedures, including finite elements.

Another important kind of inverse heat conduction problem arises in the determination of transport properties, commonly of the internal thermal diffusivity α. This procedure is called parameter estimation (PE). This may be done with a heated sample of the material. A simple example would be a thick layer of a uniform material, initially at t_i. If one exposed surface is suddenly subjected to a surface flux of q'', the temperature response $t(x, \tau)$ is Eq. (4.2.18), in Sec. 4.2.3. This solution applies approximately for the time, from $\tau = 0$, to the time when the temperature effect has become appreciable at the other surface. In this simple circumstance, one internal temperature measurement, at any known time, would give an estimate of α. Multiple sensors and measurements would also permit the inference of α as a function of temperature, as $\alpha(\tau)$. See Minkowycz et al. (1988).

SUMMARY. The foregoing developments indicate the ways in which a distributed source, a time-dependent boundary condition, and anisotropic conduction mechanisms may be formulated for multidimensional transients. The distributed source is easily accommodated. The Duhamel method is also instructive in giving an alternative picture of the construction of a solution of transient response for time-dependent boundary conditions. This method is also a common basis for the interpretation required in inverse heat conduction problems. The anisotropic transformation indicates the relation of such conduction mechanisms to the simpler orthotropic and isotropic formulations.

5.2 OTHER MULTIDIMENSIONAL TRANSIENTS

The preceding section first indicates in Sec. 5.1.1 how multidimensional transient response characteristics may be generated from one-dimensional results, such as those given in Secs. 4.2 and 4.3. The method of separation of variables provides solutions for the components, which may be combined as a product, to represent a multidimensional response. This technique applies for a uniform or separable initial condition t_i in the multidimensional region. Uniform surface conditions are also necessary, as t_s constant and uniform, or with surface convection, with h and t_e constant and uniform. The product solution geometries are seen in Fig. 5.1.2.

Section 5.1.2 concerns a number of general techniques whereby more complicated mechanisms may be included in analysis. A distributed location-dependent energy source $q'''(x, y, z)$ may be simply accommodated as the sum of steady-state Poisson and transient Fourier effects. A transformation is also indicated which may be effective in reducing anisotropic diffusion formulations to much simpler forms. The Duhamel method for time-dependent boundary conditions was also given.

In this section several characteristic kinds of multidimensional transient solutions are given, for geometries described in Cartesian and in cylindrical coordinates, in Secs. 5.2.1 and 5.2.2. These are included to indicate some general aspects of the form of such solutions, for simple and classical imposed conditions. Section 5.3 and Chap. 9 indicate the numerical methodologies applied to the frequently more practical formulations which are required in most applications.

5.2.1 Rectangular Region Transients

Two characteristic results are given here. The first is a three-dimensional region, initially at a variable initial temperature $t_i(x, y, z)$, with a constant and

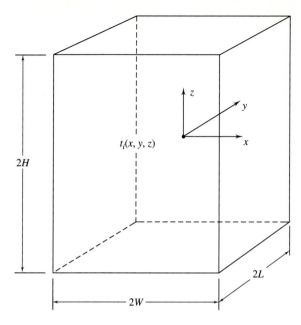

FIGURE 5.2.1
A rectangular region, initially at $t_i(x, y, z)$.

uniform surface temperature t_0, for $\tau > 0$. The second result is for a semiinfinite strip, of infinite length in the z direction, initially at $t_i(x, y)$. The base, at $x = 0$, and one lateral surface, at $y = L$, are at t_0, for $\tau > 0$. The other lateral surface, at $x = L$, is subjected to a convection condition h to an ambient also at t_0, where t_0 is taken as uniform and constant. The general equation is

$$\nabla^2 t = \frac{1}{\alpha} \frac{\partial t}{\partial \tau} \tag{5.2.1}$$

A RECTANGULAR REGION. The surface condition is taken as t_0 on all six surfaces in Fig. 5.2.1. The origin is at the center of the geometry, for convenience. Assume that $t_i(x, y, z)$ is an odd function. It may be expanded in a triple sine series, as follows. For an even function, cosines are used. In terms of $\theta_i = t_i(x, y, z) - t_0$,

$$\theta_i(x, y, z) = \sum_{m=1}^{\infty} \sum_{n=1}^{\infty} \sum_{p=1}^{\infty} A_{m,n,p} \sin \frac{m\pi x}{W} \sin \frac{n\pi y}{L} \sin \frac{p\pi z}{H} \tag{5.2.2}$$

Then $A_{m,n,p}$ is expressed in terms of $\theta_i(x, y, z)$ as

$$A_{m,n,p} = \frac{8}{WLH} \int_0^W \int_0^L \int_0^H \theta_i(x', y', z') \sin \frac{m\pi x'}{W} \sin \frac{n\pi y'}{L} \sin \frac{p\pi z'}{H} \, dx' \, dy' \, dz' \tag{5.2.3}$$

The time dependence is given in Eq. (5.1.8), as an exponential decay. Therefore,

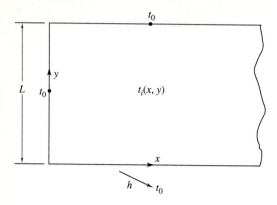

FIGURE 5.2.2
A half-infinite plane layer transient, subject to temperature and convection surface conditions.

the general term in Eq. (5.2.2) is

$$e^{-\alpha\tau\beta^2_{m,n,p}} \sin\frac{m\pi x}{W} \sin\frac{n\pi y}{L} \sin\frac{p\pi z}{H} \tag{5.2.4}$$

where

$$\beta^2_{m,n,p} = \pi^2\left[\left(\frac{m}{W}\right)^2 + \left(\frac{n}{L}\right)^2 + \left(\frac{p}{H}\right)^2\right] \tag{5.2.5}$$

Then the conditions $\theta(\pm W, \pm L, \pm H, \tau)$ and Eq. (5.2.1) are satisfied and the solution is

$$\theta(x, y, z, \tau) = \sum_{m=1}^{\infty}\sum_{n=1}^{\infty}\sum_{p=1}^{\infty} e^{-\alpha\tau\beta^2_{m,n,p}} A_{m,n,p} \sin\frac{m\pi x}{W} \sin\frac{n\pi y}{L} \sin\frac{p\pi z}{H}$$
$$\tag{5.2.6}$$

A HALF-INFINITE PLANE LAYER. The geometry, the boundaries, and the initial condition $t_i(x, y)$ are seen in Fig. 5.2.2. The form of the solution, as $\theta(x, y, \tau) = t(x, y, \tau) - t_0$ in terms of Fo $= \alpha\tau/L^2$, is

$$\theta(x, y, \tau) = \sum_{m=1}^{\infty}\sum_{n=0}^{\infty}\int_0^{\infty} C_m(n) e^{-\alpha\tau(n^2+2\delta/m)} X(n, x) Y(\gamma_m, y)\, dn \tag{5.2.7}$$

where $X(n, x)$ and $Y(\gamma_m, y)$ are the separation of variables functions satisfying

$$\frac{d^2X(x)}{dx^2} + n^2 X(x) = 0$$

$$\frac{d^2Y(y)}{dy^2} + \gamma^2 Y(y) = 0$$

The application of the initial condition $\theta_i(x, y)$ and the boundary conditions results in

$$\theta(x, y, \tau) = AB$$

where

$$A = \theta = \frac{1}{\sqrt{\pi \, \mathrm{Fo}}} \sum_{m=1}^{\infty} e^{-\Delta_m^2 \, \mathrm{Fo}} \frac{(\Delta_m^2 + \mathrm{Bi}^2)}{(\Delta_m^2 + \mathrm{Bi}^2 + \mathrm{Bi})} \sin \Delta_m \left(1 - \frac{y}{L} \right)$$

and

$$B = \int_{x'=0}^{\infty} \int_{y'=0}^{L} \theta_i \left(\frac{x'y'}{L^2} \right) \sin \Delta_m \left(1 - \frac{y'}{L} \right)$$

$$\times \left[\exp \left[-\frac{(x/L - x'/L)^2}{4 \, \mathrm{Fo}} \right] - \exp \left[-\frac{(x/L + x'/L)^2}{4 \, \mathrm{Fo}} \right] \right] \frac{dx'}{L} \frac{dy'}{L} \quad (5.2.8)$$

where the Δ_m are the successive roots of

$$\Delta_m \cot \Delta_m = -\frac{hL}{k} = -\mathrm{Bi} \quad (5.2.9)$$

and the variables of integration are x'/L and y'/L.

SUMMARY. The last result is an example of the complexity which often arises for multidimensional transients. The formulation involves only one generalization, the initial distribution $t_i(x, y)$. However, even for t_i taken as uniform, the calculation of any specific result is demanding. However, exact solutions in any given circumstance do retain various anomalies which are lost and sometimes overlooked in numerical formulations and calculated results. A common such effect is the loss of the singularities which arise in exact solutions. These may appear only as grid-size-dependent effects in numerical methods.

5.2.2 Cylindrical Region Transients

Many transient conductive processes arise in which the region geometry may be taken as cylindrical. One-dimensional transients, as $t(r, \tau)$, are considered in Sec. 4.3.1. Solutions are also given in Sec. 4.3.3 in graphical form. These apply for long solid cylinders, initially at a uniform temperature t_i, suddenly subjected to a convective environment at t_e.

Section 5.1.1 indicates how the method of product solutions may be used to develop two-dimensional solutions for short cylinders and for semiinfinite cylinders. This method is limited to a restricted class of initial internal temperature conditions. This is that $t_i(x, r)$ must be separable, as $t_i(x, r) = t_i(x)t_i(r)$, where each component applies separately to one component of the product solution. The boundary conditions, over the whole region, must also be linear and homogeneous.

Results are given here for three two-dimensional transients, as $t(r, \theta, \tau)$. They are for a solid cylinder of infinite length, and for an angular section of such a geometry. The cylindrical coordinates are seen in Fig. 5.2.3. The general

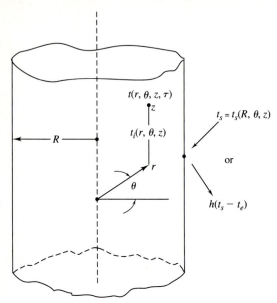

FIGURE 5.2.3
Cylindrical coordinates, with general bounding conditions.

three-dimensional equation is

$$\frac{\partial^2 t}{\partial r^2} + \frac{1}{r}\frac{\partial t}{\partial r} + \frac{1}{r^2}\frac{\partial^2 t}{\partial \theta^2} + \frac{\partial^2 t}{\partial z^2} = \frac{1}{\alpha}\frac{\partial t}{\partial \tau} \qquad (5.2.10)$$

The spherical geometry also very commonly arises in applications. The coordinate system is seen in Fig. 5.2.4 and the general equation is given as follows, for

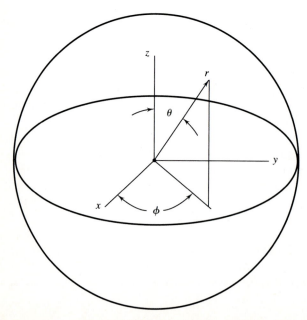

FIGURE 5.2.4
Spherical coordinates.

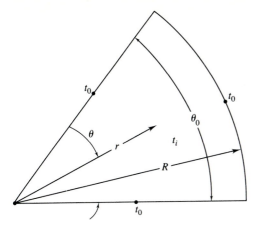

FIGURE 5.2.5
An infinitely long wedge region.

reference:

$$\frac{\partial^2 t}{\partial r^2} + \frac{2}{r}\frac{\partial^2 t}{\partial r} + \frac{1}{r^2 \sin\theta}\frac{\partial}{\partial\theta}\left(\sin\theta\frac{\partial t}{\partial\theta}\right) + \frac{1}{r^2 \sin^2\theta}\frac{\partial^2 t}{\partial\phi^2} + \frac{1}{r^2 \sin^2\theta}\frac{\partial t}{\partial\theta} = \frac{1}{\alpha}\frac{\partial t}{\partial\tau} \quad (5.2.11)$$

Three two-dimensional solutions are given here for the infinite-length cylindrical region. In the first two circumstances, the initial nonuniform internal temperature condition is $t_i(r,\theta)$. The surface is suddenly changed to a uniform temperature t_0 or subjected to a convective process with an environment at t_e, as $h(t_s - t_e)$. A very similar condition is $t_i(r,\theta)$, in a pie-shaped sector of a cylindrical geometry, extending from $\theta = 0$ to $\theta = \theta_0$, as in Fig. 5.2.5. The initial condition is t_i uniform. These are all two-dimensional processes $t(r,\theta,\tau)$ and the term in z is omitted from Eq. (5.2.10).

A TEMPERATURE SURFACE CONDITION, t_0. The following form satisfies Eq. (5.2.10) for $t(r,\theta,\tau)$, or for $t(r,\theta,\tau) - t_0$.

$$e^{-\alpha\beta^2\tau}J_n(\beta r)(A_n \cos n\theta + B_n(\sin n\theta)$$

The initial condition $T(r,\theta,0) = t_i(r,\theta) - t_0$ is expressed as a Fourier series and the β are the positive roots of $J_n(\beta R) = 0$, $\beta_1, \beta_2, \ldots, \beta_s, \ldots$. The solution is in terms of

$$A_{0,s} = \frac{1}{\pi R^2[J_0'(\beta_s R)]^2}\int_0^R\int_{-\pi}^{\pi} T(r,\theta)J_0(\beta_s r)r\,dr\,d\theta \quad (5.2.12)$$

$$A_{n,s} = \frac{2}{\pi R^2 J^2}\int_0^R\int_{-\pi}^{\pi} T(r,\theta)\cos n\theta\, J_n(\beta_s r)r\,dr\,d\theta \quad (5.2.13)$$

$$B_{n,s} = \frac{2}{\pi R^2 J^2}\int_0^R\int_{-\pi}^{\pi} T(r,\theta)\sin n\theta\, J_n(\beta_s r)r\,dr\,d\theta \quad (5.2.14)$$

where $J = [J'_n(\beta_s R)]$. The solution is

$$T(r, \theta) = t_i(r, \theta) - t_i = \sum_{s=1} \sum_{n=0} (A_{n,s} \cos n\theta + B_{n,s} \sin n\theta) J_n(\beta_s r) e^{-\alpha\beta_s^2\tau}$$

(5.2.15)

A CONVECTION SURFACE CONDITION. The same form again applies and $t(r, \theta, \tau) - t_e$. The β now become the positive roots of $\beta J'_n(\beta R) + hJ_n(\beta R)/k = 0$ instead. The solution is still Eq. (5.2.15). However, the three coefficients of the double integrals in Eqs. (5.2.12)–(5.2.14) now also depend on a Biot number as $\mathrm{Bi} = h/k\beta_s$. The additional product term $[1 + (h/k\beta_s)^2]$ arises in the denominator in Eq. (5.2.12). The additional product term in each Eqs. (5.2.13) and (5.2.14) is $[1 + (h/k\beta_s)^2 - (n/\beta_s R)^2]$.

A PIE-SHAPED WEDGE. Refer to Fig. 5.2.5. The wedge region is from $\theta = 0$ to θ_0. The initial temperature is taken as uniform at t_i. For $\tau > 0$ the three surfaces, at $\theta = 0$ and θ_0 and at $r = R$, are at t_0. The solution is

$$\frac{t(r, \theta, \tau) - t_0}{t_i - t_0} = \frac{8}{\pi R^2} \sum_{n=0} \frac{1}{(2n+1)} \sin(s\theta)$$

$$\times \sum_{m=1} e^{-\alpha\beta_m^2\tau} \frac{J_s(r\beta_m)}{[J'_s(R\beta_m)]^2} \int_0^R r J_s(r\beta_m) \, dr \quad (5.2.16)$$

where $s = (2n+1)\pi/\theta_0$ and the $\pm\beta_m$, for $m = 1, 2 \ldots$ are the roots of $J_s(R\beta) = 0$.

SUMMARY. The first two solutions indicate how a nonuniform initial condition changes toward a uniform internal decaying temperature field, for both surface conditions. The last solution is a decay of $t(r, \theta, \tau)$ to t_0. Comparable solutions for the spherical geometry are much more complicated. They commonly require the use of Legendre and Bessel functions, in complicated forms. A range of solutions may be found in Chap. 9 of Carslaw and Jaeger (1959) and in Özisik (1980), Chap. 4. For many practical applications, and for more complicated initial and boundary conditions, direct recourse to numerical methods is commonly favored.

5.3 NUMERICAL METHODS IN MULTIDIMENSIONAL TRANSIENTS

General formulations and many specific aspects of numerical analysis are given in detail in Chap. 9. However, initial consideration of this method is given in Sec. 3.9 for steady-state processes described in several dimensions. Simple central-difference estimates of second derivatives were used. Section 4.9 is the first approach to the numerical modeling of transients, in plane regions. For these transients, stability considerations arise in terms of the spatial and temporal subdivisions, Δx and $\Delta \tau$, used in the numerical formulation. These two earlier considerations also outline the general techniques of finite-difference analysis. Here, those results are augmented to also include multidimensional transients, again using the same kind of derivative estimates.

Two principal matters arise. The first is the resulting altered stability requirements in transient analysis. For one-dimensional transients, using the explicit method of numerical calculation, the limits are in terms of $F = \alpha \, \Delta \tau/(\Delta x)^2$. For a purely conductive process, the limit in a plane layer was found to be $F \leq 1/2$, where $F = \alpha \, \Delta \tau/(\Delta x)^2$. With surface heat transfer processes, postulated in terms of the coefficient h, the result is $F = \alpha \, \Delta \tau/(\Delta x)^2 \leq 1/2(B + 1)$, where F is the grid Fourier number. Here $B = h \, \Delta x/k$ is a grid Biot number. It is positive. Therefore, for any choice of Δx, the maximum value of $\Delta \tau$ in the calculation is reduced, with such a surface condition. Further required reduction arises for the multidimensional surface points shown in Fig. 3.9.12. These conditions are considered here.

The second matter, which interacts with the preceding consideration, is the large multiplicity of different kinds of surface point equations which arise. Recall Fig. 3.9.12. Each different kind of surface point, in general, gives a different surface point equation. These were given in Sec. 3.9.5, for steady state, for several kinds of surface points, as Eqs. (3.9.25)–(3.9.28) and (3.9.29b).

The following analysis first determines the stability limits for internal conduction region grid points in two- and three-dimensional transients. Then several typical multidimensional surface point equations are developed for transients, as examples. Thereafter, the resulting more restrictive stability considerations are discussed.

5.3.1   Numerical Stability in Purely Conductive Processes

The relevant differential equation is the basis of the numerical finite-difference equation. The subdivision scheme in a Cartesian formulation is shown in Fig. 3.9.1(a). A time–temperature response at a given location is shown in Fig. 3.9.1(b). Consider the following relation, to be satisfied by $t(x, y, z, \tau)$:

$$\frac{\partial^2 t}{\partial x^2} + \frac{\partial^2 t}{\partial y^2} + \frac{\partial^2 t}{\partial z^2} + \frac{q'''}{k} = \frac{1}{\alpha} \frac{\partial t}{\partial \tau} \tag{5.3.1}$$

Three first derivative estimates of $\partial t/\partial \tau$ are given in Eqs. (3.9.2)–(3.9.4). The forward difference, or explicit method, is used here. This is Eq. (3.9.4). In indicial notation, this is Eq. (3.9.12) as

$$\frac{\partial t}{\partial \tau} \approx \frac{t_{i,j,k}^{n+1} - t_{i,j,k}^{n}}{\Delta \tau} \tag{5.3.2}$$

A second-derivative estimate at time τ, in one dimension, is Eq. (3.9.5). In the notation of Fig. 3.9.1(a), for the explicit method, this three-consecutive-point central-difference estimate is

$$\frac{\partial^2 t}{\partial x^2} \approx \frac{(t_3 - t_0) - (t_0 - t_1)}{(\Delta x)^2} = \frac{t_1 + t_2 - 2t_0}{(\Delta x)^2}$$

This, written in the notation of Eq. (5.3.2), is

$$\left(\frac{\partial^2 t}{\partial x^2}\right)^n_{i,j,k} \approx \frac{t^n_{i+1,j,k} + t^n_{i-1,j,k} - 2t^n_{i,j,k}}{(\Delta x)^2}$$

where n denotes a particular time level during the transient. The estimate of $\nabla^2 t$ in three dimensions is then

$$\nabla^2 t \approx \left(t^n_{i+1,j,k} + t^n_{i-1,j,k} + t^n_{i,j+1,k} + t^n_{i,j-1,k}\right.$$

$$\left. + t^n_{i,j,k+1} + t^n_{i,j,k-1} - 6t^n_{i,j,k}\right)/(\Delta x)^2 \qquad (5.3.3)$$

Then the forward-difference estimate of Eq. (5.3.1) is

$$t^n_{i+1,j,k} + t^n_{i-1,j,k} + t^n_{i,j+1,k} + t^n_{i,j-1,k} + t^n_{i,j,k+1} + t^n_{i,j,k-1} - 6t^n_{i,j,k}$$

$$+ \frac{q'''^n_{i,j,k}(\Delta x)^2}{k} = \frac{(\Delta x)^2}{\alpha \, \Delta \tau}\left(t^{n+1}_{i,j,k} - t^n_{i,j,k}\right) = \frac{1}{F}\left(t^{n+1}_{i,j,k} - t^n_{i,j,k}\right) \quad (5.3.4)$$

where $F = \alpha \, \Delta \tau/(\Delta x)^2$.

This is the forward-difference estimate of $t^{n+1}_{i,j,k}$ at point i, j, k, at time level n. This is the basis of the calculation of the network temperature distribution at all i, j, k, at time level $n + 1$. Equation (5.3.4) applies at all points i, j, k which lie inside a purely conductive region. This relation is explicit in that the $t^{n+1}_{i,j,k}$ are calculated directly from the $t^n_{i,j,k}$. The known initial temperature conditions, at $n = 0$ and over the region, are $t^0_{i,j,k}$. Solving Eq. (5.3.4) for the unknown value $t^{n+1}_{i,j,k}$,

$$t^{n+1}_{i,j,k} = F\left(t^n_{i+1,j,k} + t^n_{i-1,j,k} + t^n_{i,j+1,k} + t^n_{i,j-1,k} + t^n_{i,j,k+1} + t^n_{i,j,k-1}\right)$$

$$+ (1 - 6F)t^n_{i,j,k} + \frac{q'''^n_{i,j,k}(\Delta x)^2 F}{k} \qquad (5.3.5)$$

The one-dimensional form is given in Eq. (4.9.5). The coefficient of t^n_i is then $(1 - 2F)$. In two dimensions the coefficient of $t^n_{i,j}$ is $(1 - 4F)$. In Eq. (5.3.5), the coefficient is $(1 - 6F)$.

The stability consideration for the continuing calculation $t^{n+1}_{i,j,k}$ at increasingly later time levels n, arises from the coefficient of $t^n_{i,j,k}$, as discussed in Sec. 4.9.1, for one-dimensional transients. The least value of this coefficient must be zero. Otherwise a higher temperature at point i, j, k, at time n has a negative effect on the later temperature at the same location, $t^{n+1}_{i,j,k}$. This is in conflict with the role of local thermal capacity energy storage. Higher earlier storage levels should tend to persist as higher temperatures at a later time. Therefore, $(1 - 2F)$, $(1 - 4F)$, and $(1 - 6F)$ must be a least value zero. A value of zero is often used. Including this value, the three limits are

$$F = \frac{\alpha \, \Delta \tau}{(\Delta x)^2} \leq 1/2, \, 1/4, \text{ and } 1/6 \qquad (5.3.6)$$

These lower limits result in the terms in t_i^n, $t_{i,j}^n$, and $t_{i,j,k}^n$ being omitted from each of the three equations, expressed in the form of Eq. (5.3.5). However, this unrealism has been convenient in past calculations because of the resulting simplicity. For example, for $q_{i,j,k}^{\prime\prime\prime n} = 0$, the temperature at time level $n + 1$ is then simply the average of the surrounding temperatures at time level n. The role of the $q_{i,j,k}^{\prime\prime\prime n}$ term seen in Eq. (5.3.5) is very direct. Positive values increase $t_{i,j,k}^{n+1}$ and negative ones, as for an energy sink, decrease it.

The stringent stability limits which arose previously, for the forward-difference formulation, may be avoided using a backward-difference time derivative estimation. This is Eq. (3.9.3) and the resulting analysis and some results are given in Sec. 4.9.3. There are also other possibilities, in the time levels used in the estimation of spatial second derivatives, in $\nabla^2 t$. In Eq. (5.3.3) the distributions at time level n are used. Many other possibilities have been proposed. One such measure is discussed in Sec. 5.3.3. The general consideration of these matters is in Sec. 9.3.3.

5.3.2   Surface Point Equations

The preceding finite-difference equations apply only to network points within the purely conductive region. At the network points on any bounding surfaces, the energy effects of the bounding surface conditions must also be included. This was done for multidimensional steady-state conduction in Sec. 3.9.5, for one-, two-, and three-dimensional processes. Section 4.9.4 considered surface conditions for one-dimensional transients. In both analyses, the surface region effects of convection h, of surface flux loading q'', and of a distributed energy source q_s''' were included. See the representations in Figs. 3.9.10 through 3.9.12 for one, two, and three dimensions in steady state. For one-dimensional transients the same effects were considered; see Fig. 4.9.5.

In this section surface point equations are given for several typical surface conditions and local geometric surface configurations in two- and three-dimensional transients. As in Secs. 3.9 and 4.9, the numerical forms are determined by energy balances, rather than by converting boundary conditions into finite-difference form.

TWO-DIMENSIONAL SURFACE REGIONS. Consider the region subdivided by a square network, as in Fig. 3.9.11. In general, the exposed surface is not isothermal, t_s^n represents the average temperature only of the surface between the dashed lines surrounding point s_1. The mass associated with the point s_1 on the flat portion of the surface is $\rho(\Delta x)^2/2$, per unit depth in the z direction. This mass receives energy by conduction from the vicinity of points 1, 2, and 3.

Convection and surface flux are also assumed present. A distributed energy source has a local intensity near the surface of q_s'''.

The sum of energy gain rates is set equal to the time rate of change of stored energy, as

$$\frac{k\,\Delta x}{\Delta x}(t_1^n - t_s^n) + \frac{k\,\Delta x}{2\,\Delta x}(t_2^n - t_s^n) + \frac{k\,\Delta x}{2\,\Delta x}(t_3^n - t_s^n) + h\,\Delta x(t_e^n - t_s^n)$$

$$+ q_s'' \,\Delta x + \frac{q_s'''(\Delta x)^2}{2} = \frac{\rho c(\Delta x)^2}{2}\frac{t_s^{n+1} - t_s^n}{\Delta\tau} \tag{5.3.7}$$

This relation is rewritten in terms of F, defined in Eq. (5.3.6), and $B = h\,\Delta x/k$:

$$(t_1^n - t_s^n) + \frac{t_2^n - t_s^n}{2} + \frac{t_3^n - t_s^n}{2} + B(t_e^n - t_s^n) + \frac{q_s''\,\Delta x}{k} + \frac{q_s'''(\Delta x)^2}{2}$$

$$= \frac{1}{2F}(t_s^{n+1} - t_s^n) \tag{5.3.8}$$

The expression for t_s^{n+1} is then

$$t_s^{n+1} = 2FBt_e^n + 2Ft_1^n + F(t_2^n + t_3^n) + \frac{2q_s''\,\Delta x F}{k}$$

$$+ \frac{q_s'''(\Delta x)^2 F}{k} + [1 - 2F(B + 2)]t_s^n \tag{5.3.9}$$

These relations are similar to those for the one-dimensional region, except for the effects of conduction from points 2 and 3.

If point s_1 is on the outside or inside edge, as also shown in Fig. 3.9.11, the preceding results do not apply since the masses associated with these points are $\rho(\Delta x)^2/4$ and $3\rho(\Delta x)^2/4$, instead of $\rho(\Delta x)^2/2$. Also there are zero and two internal (points 1 and 1') conduction paths, respectively, instead of the single path (point 1) for a point on the flat surface. Thereby, the relevant energy balances will be different from Eq. (5.3.7). The results for these points are as follows.

For an outside-edge point:

$$t_s^{n+1} = 4FBt_e^n + 2F(t_2^n + t_3^n) + \frac{4q_s''\,\Delta x F}{k} + \frac{q_s'''(\Delta x)^2 F}{k}$$

$$+ [1 - 4F(B + 1)]t_s^n \tag{5.3.10}$$

For an inside-edge point:

$$t_s^{n+1} = \frac{4}{3}FBt_e^n + \frac{4}{3}F(t_1^n + t_{1'}^n) + \frac{2}{3}F(t_2^n + t_3^n) + \frac{4q_s''\,\Delta x F}{3k}$$

$$+ \frac{q_s'''(\Delta x)^2 F}{k} + \left[1 - \frac{4}{3}F(B + 3)\right]t_s^n \tag{5.3.11}$$

This collection of relations then permits the calculation of two-dimensional transients. The appropriate relation is used at each surface point of unknown temperature.

THREE-DIMENSIONAL SURFACE REGIONS. These often arise in circumstances in which complicated geometry and surface processes occur. The relevant equations for a point on a flat surface, s_1 in Fig. 3.9.12, will be given. The scheme is the same as that shown in Fig. 3.9.11, except that temperature gradients may also occur in the z direction. Therefore, conduction may also occur to the region s_1 from two additional adjacent surface network point directions, say 4 and 5. Choosing $\Delta x = \Delta y = \Delta z$, the mass associated with point s_1 is $\rho(\Delta x)^3/2$. The conduction path from point 1 inside the region has an area of $(\Delta x)^2$ and from points 2, 3, 4, and 5 an area of $(\Delta x)^2/2$. The surface area associated with point s_1 is $(\Delta x)^2$. An energy balance is written, and the transient finite-difference equation for point s_1 becomes

$$t_s^{n+1} = 2FBt_e^n + 2Ft_1^n + F(t_2^n + t_3^n + t_4^n + t_5^n) + \frac{2q_s'''^n(\Delta x)F}{k} + \frac{q_s'''^n(\Delta x)^2 F}{k}$$

$$+ [1 - 2F(B + 3)]t_s^n \tag{5.3.12}$$

Four different additional surface point equations are needed for inside and outside edges and corners. These locations are denoted by s_2, s_3, s_4, and s_5 in Fig. 3.9.12. Additional possibilities are shown as s_6, s_7, and s_8. These are to be analyzed by the same procedure as before.

5.3.3 Stability Limits

The calculated stability limits of $F = \alpha \Delta\tau/(\Delta x)^2 \le 1/2, 1/4$, and $1/6$, in Eq. (5.3.6), represent one-, two-, and three-dimensional regions only when surface temperatures are specified values. Also, these limits were derived considering only grid locations internal to the region. The presence of other kinds of surface processes commonly reduces these limits. Recall the analysis in Sec. 4.9.4 for one-dimensional transients. A surface heat transfer process, characterized in terms of h, reduced the stability limit from $F_{1D} = 1/2$ to $F_{1D} = 1/2(B + 1)$, where $B \ge 0$.

This effect is also apparent in the two-dimensional flat surface point result in Eq. (5.3.9). There $F_{2D} = 1/4$ in Eq. (5.3.6) becomes $F_{2D} \ge 1/(2B + 4)$, from Eq. (5.3.9). The presence of edge and corner points has further reduced the stability criterion. For a two-dimensional region, the only other equations are for outside- and inside-edge points, Eqs. (5.3.10) and (5.3.11). The stability limits there are $F_{2D} \le 1/(4B + 4)$ and $1/[(4B/3) + 4]$, respectively. Therefore, the stability limit is determined by the most restrictive condition, the outside edge, as

$$F_{2D} \le 1/(4B + 4) \tag{5.3.13}$$

This effect arises from the relative thermal capacity of the material associated with each of the distinctive kinds of network or grid points. This is $\rho(\Delta x)^2$, $\rho(\Delta x)^2/2$, $\rho(\Delta x)^2/4$, and $3\rho(\Delta x)^2/4$ for the internal, surface, and outside- and inside-edge points. Thus the lowest level of thermal capacity determines the upper limit of F for stability throughout the two-dimensional region.

This effect may be countered by concentrating more thermal capacity in the layer immediately adjacent to the surface points, for example, between points s and 1 in Fig. 4.9.5. This may be done by associating a thicker layer, $\Delta x_s > \Delta x$ between the points 1 and s, at t_s and t_1, respectively, in a coarser grid there. This would increase the upper limit on F for the surface condition. Then $\Delta x_s/\Delta x$ could be chosen to give the same stability limit as that which applies internally, $F \leq 1/2$.

A similar effect also arises in multidimensional regions. Equation (5.3.12), for a point on the surface, indicates a limit of $F_{3D} \leq 1/(2B + 6)$, compared to the larger value $F_{3D} \leq 1/6$ in Eq. (5.3.6). Edge and corner point characteristics will further reduce the limit. In any circumstance, it is the minimum value of F_{1D}, F_{2D}, or F_{3D}, over the whole network, which indicates the stability limit. In any calculation, this value applies at all internal network points, at all time levels n.

Chapter 9 also considers other estimates of second derivatives than those used in Eq. (5.3.3). For example, the early Crank and Nicolson (1947) modified implicit method employs a forward difference for the time derivative. However, it averages two second-derivative estimates, one at time level n and the other at time $n + 1$. The resulting form for a one-dimensional transient is

$$t_i^{n+1} - t_i^n = \frac{F}{2}\left(t_{i-1}^{n+1} + t_{i+1}^{n+1} - 2t_i^{n+1} + t_{i-1}^n + t_{i+1}^n - 2t_i^n\right) \quad (5.3.14)$$

This method is implicit in that it simultaneously involves several temperatures at time level $n + 1$. However, there is no formal upper limit on F and the truncation error is smaller than for either the explicit method or the simple implicit method. However, Patankar and Baliga (1978) have shown that unrealistic solutions result when the time interval is large. The calculational effort of an implicit method may become large for the large number of points often encountered in multidimensional problems. However, methods of estimate and computational procedures that increase the economy of calculation time may be employed to ameliorate these disadvantages. See Sec. 9.3.3.

The preceding limits arise for uniform grid spacing across the whole region under consideration. Increased values of Δx, Δy, or Δz may be used for the surface layer, normal to the surface of a geometric region. Thereby, higher values of the limiting value of $F = \alpha \Delta\tau/(\Delta x)^2$ result. This gives a larger permissible value of $\Delta\tau$, the time step of the calculation. However, due regard must be given to the accuracy of the finite-difference representation, when surface region gradients may be large.

PROBLEMS

5.1.1. A rectangular bar is initially at uniform temperature $\theta(x, y, 0) = \theta_i = 0$. At time $\tau = 0$, a uniform internal heat generation q''' begins, along with a constant net radiant heat flux q_s'' over the surface $y = b$. The other surfaces are insulated. Indicate, in terms of equations and conditions, a means for determining analytically the two-dimensional, unsteady temperature distribution $\theta(x, y, \tau)$.

5.1.2. Consider the transient cooling of a cubical stack of horizontal steel plates, initially at t_i throughout, after the surface temperature is dropped to t_0 and held at that value thereafter. The conductivity in the plates, in the two horizontal directions, say x and y, is that of steel k_s. That across the plates, in the vertical direction, say z, is less. Take it as k_z. Take the coordinate system at the middle of the cube, of edge length $2L$.
(a) Write out the complete formulation which will lead to the conduction solution temperature response, in terms of the temperature excess $\theta(x, y, z, \tau)$.
(b) Determine whether or not $\theta(x, y, z, \tau)$ may be determined as a product of one-dimensional solutions. If so, write out the formulation of each one-dimensional solution, in specific terms.
(c) Apply the orthotropic transformation for the general analysis of this circumstance, giving the resulting differential equation.
(d) Determine the solution $\theta(x, y, z, \tau)$.

5.1.3. Transient responses determined for semiinfinite conduction regions are often used to estimate transient temperature responses for short times in completely bounded geometries. Consider a cubical solid of size $L \times L \times L$, initially at t_i throughout, whose surface temperature is suddenly changed to a thereafter constant level t_0. If the relevant semiinfinite solid solution is considered sufficiently accurate until a 5% change of $t_i - t_0$ occurs at the center, how long may this approximation be used to describe the response?

5.1.4. A very long square ($2s$ by $2s$) is initially at t_i throughout. Suddenly two opposite faces, say normal to the x direction, are changed to t_0. The other two opposite faces, normal to y, are, at the same time, subjected to a fluid region also at t_0. A convection coefficient h applies there. The temperature field at later times is to be found.
(a) Write the complete equations and all applicable conditions for the temperature response inside the rod material, in terms of $\theta = t - t_0$.
(b) Prove that the solution θ may be determined as a product solution.
(c) Write out any such solution, in terms of the parameters of the transport circumstance.

5.1.5. A steel cylinder with a diameter and length of 8 cm and 12 cm, respectively, is initially at 600°C. It is placed in oil at 20°C for rapid cooling, for case hardening of the surface layer. The expected value of h is 60 W/m² °C.
(a) Determine the time when the maximum internal temperature level has decreased to 500°C.
(b) What is the minimum metal temperature at this time and where does it occur?
(c) Plot the temperature distribution at that time, along the cylinder axis.

5.1.6. Indicate a method of solution which would be useful for each of the following types of conduction.
(*a*) Temperature varying thermal conductivity.
(*b*) Nonhomogeneous linear differential equation and/or nonhomogeneous boundary conditions.
(*c*) Problems which do not possess a characteristic length or time.
(*d*) Multidimensional transients with a uniform initial condition and uniformly homogeneous boundary conditions.
(*e*) Time-varying effects such as in the heat generation rate or boundary temperature level.

5.1.7. A cubical region, L by L by $L = 20$ cm, is initially uniformly at $100°C$ throughout. For $\tau > 0$ the surfaces are maintained at $50°C$. The material conductivity and diffusivity are $k = 90$ cal/cm hr $°C$ and $\alpha = 200$ cm^2/hr. Calculate the time at which the maximum temperature in the region is reduced to $55°C$. What is the minimum value at this time?

5.1.8. For the geometry and temperature conditions given in problem 5.1.7, consider the amount of time, τ, after the beginning of the transient that the integral method and the semiinfinite solid solution may each be accurately used. How long do the appropriate results from the integral method and the semiinfinite solid solutions apply at a distance 5 cm in from the surfaces of the cube?

5.1.9. A long square rod ($2s$ by $2s$) is initially at t_i, uniformly throughout. Suddenly, the two faces normal to the x direction are changed to t_0. The other two faces are then subject to a convection process and to an ambient medium also at t_0.
(*a*) Sketch the geometry and the conditions. Completely formulate the conditions and equations necessary to determine the temperature distribution, in terms of θ and ϕ.
(*b*) Determine any additional requirements which might arise for θ to be determined in terms of a product solution.
(*c*) Write out the solution.
(*d*) Consider the magnitude of the local heat flux over all boundaries of the square rod. At which locations is this flux unbounded in magnitude?

5.1.10. A solid cube of material of side length 6 in. has a uniform temperature of $100°F$. The material properties are $k = 1.0$ Btu/hr ft $°F$, $\rho = 144$ lb/ft^3, and $c_p = 0.2$ Btu/lb $°F$. It is placed in a liquid at $50°F$. The convection coefficient will be 4 Btu/hr ft^2 $°F$.
(*a*) Where will the highest and lowest temperatures be found in the cube material at later times?
(*b*) When will the highest temperature have dropped to $55°F$?
(*c*) When will the lowest temperature be $55°F$?
(*d*) For the time found in part (*c*), what is the temperature at midlength along the edges of the cube?

5.1.11. A ceramic brick ($2 \times 4 \times 6$ in.) is annealed by heating it to $800°F$ uniformly. It is then allowed to cool in air at $80°F$. The surface heat transfer coefficient and material properties are $h = 2$ Btu/hr ft^2 $°F$, $k = 0.8$ Btu/hr ft $°F$, $c = 0.2$ Btu/lb $°F$, and $\rho = 100$ lb/ft^3.
(*a*) Show how the temperature distribution $\theta(x, y, z, \tau)$ is given by the product of three one-dimensional transient solutions.

(b) After cooling for 1 hr, what is the location and temperature of the coldest place in the brick, and the hottest place in the brick.

(c) If the brick material is assumed to have uniform internal temperature during cooling, at all times, what would its temperature be after 1 hr?

5.1.12. A long, square bar ($2s$ by $2s$ in cross section), initially at t_i throughout, is suddenly immersed in a fluid at t_0. Assume that the surface resistance to heat transfer is negligible compared with the internal resistance. Plot from the Heisler charts the midpoint temperature excess (divided by $\theta_i = t_i - t_0$) vs. a dimensionless time variable. Use logarithmic coordinates.

5.1.13. Repeat Prob. 5.1.12 for a cube having sides of length $2s$.

5.1.14. A long, square steel bar (2 by 2 in.), initially at 300°F throughout, is immersed in water at 60°F. Neglecting surface resistance, find the immersion time necessary to reduce the maximum temperature in the bar to 90°F.

5.1.15. Repeat Prob. 5.1.14 for a steel cube of size 4 by 4 by 4 in.

5.1.16. A semiinfinite region is initially at t_i throughout. The surface temperature $t(0, \tau)$ increases from t_i as $t_i + K\tau^n$. Determine $t(x, \tau) - t_i$ in the region at later times, using the Duhamel theorem.

5.1.17. Do as requested in the preceding problem if $t(0, \tau) - t_i$ varies as $\theta_{0, a} \cos \omega\tau$.

5.1.18. A slab of thickness $2s$ is initially at t_i throughout. At $\tau = \tau_1$, a spatially uniform energy dissipation begins. It increases as $q'''(\tau/\tau_1) = q_0'''[1 + (\tau/\tau_1)^n]$. Determine the internal temperature response $\theta(x, \tau)$.

5.2.1. A cubical region, of side length $2L$, is initially at $\theta_i(x, y, z) = \theta_0 \sin \pi x/L$.
(a) Determine the subsequent temperature distribution in the region.
(b) Calculate the central temperature, as a function of time.
(c) Calculate the surface heat flux at each of the surfaces of the volume.

5.2.2. Perform the same calculations as in the preceding problem, for $\theta_i(x, y, z) = \theta_0(\sin \pi x/2L)(\sin \pi y/2L)(\sin \pi z/2L)$.

5.2.3. Consider the half-infinite geometry in Fig. 5.2.2, with a uniform initial temperature $t_i(x, y) = t_i$.
(a) Evaluate the expression for the total heat flow rate $q(\tau)$, across the surface at $x = 0$.
(b) Determine this rate for each of the limits, Bi $\to 0$ and Bi $\to \infty$.
(c) Comment on the implications of these results.

5.2.4. For the circumstance in the preceding problem, and $t_i(x, y)$ again uniform, consider convection at both surfaces instead, that is, at $y = 0$ and $y = L$, to a region at t_0. Take h as the same at both surfaces. Retain the condition $\theta(0, y, \tau) = t(0, y, \tau) - t_0 = 0$. Investigate if the solution for this circumstance may be inferred from the solution given in Sec. 5.2.1.

5.2.5. For the half-infinite plane layer in Fig. 5.2.2:
(a) Determine the solution for t_i uniform.
(b) Calculate $q''(y)$ at $x = 0$.
(c) Determine the total heat flow rate across the surface at $x = 0$.

5.3.1. A long square rod, 4.8×4.8 in., is initially at 100°F throughout, $\alpha = 0.01$ ft^2/hr, $k = 10$ Btu/hr °F. All four lateral surfaces are suddenly reduced to 0°F and held at that temperature thereafter. We wish to determine the resulting center temperature, t_c, 45 min later. Make this determination using each of the two following methods.

(a) For a conduction solution, write down the complete first term. Use only this term to evaluate t_c.

(b) The simplest numerical method, using $\Delta x = \Delta y = 1.2$ in.

(c) Comment on causes for any differences.

5.3.2. For the circumstance in Prob. 5.3.1:

(a) Using $\Delta x = 1.2$ in., calculate the center temperature change with time and the time at which the temperature there reaches 20°F.

(b) Determine the center temperature change with time from an appropriate exact conduction and from a product solution and compare the three results.

5.3.3. A long 1-by-1-ft square metal duct is insulated with a 6-in. layer of 85% magnesia. The initial temperature of the insulation is 60°F throughout. A high-temperature fluid suddenly begins flowing through the duct, changing the temperature of the inside surface of the insulation to a thereafter constant temperature of 460°F. The outside surface temperature remains constant at 60°F. Use a 2-in. grid.

(a) Determine the temperature distribution and heat transfer rate after steady state has been achieved.

(b) For the transient process, plot the heat loss rate from the outside surface vs. time, from the beginning of the process until 80% of the steady-state value has been achieved.

$$\rho = 10 \text{ lb/ft}^3 \qquad c = 0.2 \text{ Btu/lb °F}$$

5.3.4. Repeat Prob. 5.1.14 using a numerical method with $F = 1/4$ and $1/5$ and $\Delta x = 1/4$ in. Compare the two solutions.

5.3.5. Using the backward-difference estimate, generate the finite-difference equivalent of Eq. (5.3.5).

5.3.6. In a metallurgical process, cubical steel blocks $3 \times 3 \times 3$ in. are to be preheated to a minimum metal temperature of 170°F by suspending them in a special salt bath maintained at 550°F. They are initially at 100°F throughout. The properties of steel ($k = 25$ Btu/hr ft °F and $\alpha = 0.6$ ft^2/hr) are assumed independent of temperature. A surface coefficient of 300 Btu/hr ft^2 °F is expected. For the numerical analysis of this circumstance by the explicit method:

(a) Determine the equations to be used in the calculation of subsequent surface point temperatures.

(b) Indicate the stability limit for the calculation, for $\Delta x = \Delta y = 0.5$ in.

5.3.7. Using the forward-difference estimate:

(a) Develop the surface point equations for the inside- and outside-edge points, for a three-dimensional transient.

(b) Compare the resulting stability limit with that for the flat surface portions of the region.

5.3.8. Repeat Prob. 5.3.7 for inside and outside corner points.

5.3.9. A parallelepiped of dimensions H, W, and D is initially at t_i throughout. It is suddenly immersed in a fluid at t_e. It is subject to a convective effect h and also to a surface flux loading q_s'', on all surfaces.

(*a*) Formulate this circumstance in terms of $t(x, y, z, \tau)$, h, q_s'', and so on in appropriate nondimensional forms.

(*b*) Take $H = 4L$, $W = 2L$, and $D = 2L$, where L is some length and determine the stability limit for $\Delta x = \Delta y = \Delta z = L/2$.

5.3.10. For the circumstance in Prob. 5.3.9, take the material to be stone concrete and $L = 40$ cm.

(*a*) For $q_s'' = 0$ and h very large, calculate the time at which the midpoint reaches a temperature of $(t_i + t_e)/2$.

(*b*) Compare this result with that obtained using the method of product solutions.

5.3.11. Consider the surface points for a numerical formulation of a conduction transient in a three-dimensional region like that shown in Fig. 3.9.12.

(*a*) Determine the surface point equations for both inside and outside corners.

(*b*) Compare the resulting stability limits with those for flat surfaces and for inside- and outside-edge points.

(*c*) Where would the most restrictive stability limit arise?

5.3.12. Recall the surface point stability limit found for the one-dimensional numerical formulation in Fig. 4.9.5.

(*a*) What would be the stability limit at the surface location s, choosing $\Delta x_s = 2\Delta x$, where Δx is the internal grid spacing.

(*b*) Find the value of $\Delta x_s / \Delta x$ which would give a stability limit equal to that in the interior of the region.

5.3.13. Perform the analysis requested in Prob. 5.3.14 for point s_1, on the flat surface in Fig. 3.9.11 and indicate how the resulting rectangular elements would be treated.

5.3.14. Consider the problem of the heat loss rate from a 3-by-6-ft (high) heating tunnel 3 ft below the surface of the ground in a cold climate. Conduits and pipes inside the tunnel keep the air at a temperature of 65°F. The "remote" ground temperature is assumed to be 50°F. We wish to estimate the rate of heat loss from the tunnel for a winter condition of no snow cover, day and night air temperatures of 10 and -10°F, respectively, and a completely clear sky on January 1.

(*a*) Consider the calculations for this problem and investigate the validity of assuming a steady-state process at air temperatures of 10 and -10°F despite the 20°F diurnal variation.

(*b*) If steady state is assumed at a seasonal average air temperature, estimate the heat loss from the tunnel. Make any reasonable assumptions.

(*c*) Describe a procedure for estimating the tunnel loss for a winter season, given its length and average air temperature.

REFERENCES

Arpaci, V. S. (1966) *Conduction Heat Transfer*, Addison-Wesley, Reading, MA.

Beck, J. V., B. Blackwell, and C. R. St. Clair, Jr. (1985) *Inverse Heat Conduction: Ill-Posed Problems*, Wiley and Sons, New York.

Carslaw, H. S., and J. C. Jaeger (1959) *Conduction of Heat in Solid*, 2nd ed., Oxford Univ. Press, New York.

Crank, J., and P. Nicolson (1947) *Proc. Cambridge Philos. Soc.*, **43**, 50.

Eckert, E. R. G., and R. M. Drake, Jr. (1972) *Analysis of Heat and Mass Transfer*, McGraw-Hill, New York.

Hensel, E. (1991) *Inverse Theory and Application for Engineers*, Prentice-Hall, Englewood Cliffs, N.J.

Minkowycz, W. J., E. M. Sparrow, G. W. Schneider, and R. H. Pletcher (1988) *Handbook of Numerical Heat Transfer*, Wiley & Sons, New York.

Myers, G. E. (1971) *Analytical Methods in Conduction Heat Transfer*, McGraw-Hill, New York.

Özisik, N. M. (1980) *Heat Conduction*, John Wiley and Sons, New York.

Patankar, S. V., and B. R. Baliga (1978) *Numerical Heat Transfer*, **1**, 27.

CHAPTER
6

OTHER
MASS
DIFFUSION
PROCESSES

6.1 MASS DIFFUSION MECHANISMS

The formulations which govern mass diffusion in a region are developed in Sec.
1.3. These results apply for the local concentration of the substance $C(x, y, z, \tau)$,
diffusing through a region of material which remains stationary at all locations
and later times. The basic relation between the local mass flux m''_{x_i}, the
concentration gradient $(\partial C / \partial x)_j$, and the diffusion coefficient D_{ij} was taken as
Fick's first law of diffusion, Eq. (1.3.2b):

$$m''_{x_i} = -D_{ij}\left(\frac{\partial C}{\partial x}\right)_j \tag{6.1.1}$$

where $j = 1, 2, 3$ for each direction $i = 1, 2, 3$. This law is analogous to Fourier's
law of heat conduction. In isotropic regions this relation becomes

$$m''_{x_i} = -D\left(\frac{\partial C}{\partial x}\right)_i \tag{6.1.2}$$

Each of the preceding two mass flux formulations was used in a local mass
balance, or conservation condition, of the diffusing substance C. Recall the
notation in Fig. 1.3.1. That balance also includes the effects of local adsorption
or desorption of substance C interior to the region. The net rate of any such
processes, producing substance C locally, is written as C'''. This is a local source,
for $C''' > 0$, or sink, for $C''' < 0$. The units of C''' are concentration, in the same

units as used for C, per unit of region volume and per unit time. This distributed source is analogous to q''', which is the local strength of any distributed local heat generation or absorption rate, per unit of volume and time.

The two general equations for mass diffusion, which result from Eqs. (6.1.1) and (6.1.2), are

$$\frac{\partial}{\partial x_i} D_{ij} \left(\frac{\partial C}{\partial x} \right)_j + C''' = \frac{\partial C}{\partial \tau} \tag{6.1.3}$$

or, for $D_{ij} = D_i = D_x, D_y, D_z$, as for an orthotropic region,

$$\frac{\partial}{\partial x} \left(D_x \frac{\partial C}{\partial x} \right) + \frac{\partial}{\partial y} \left(D_y \frac{\partial C}{\partial y} \right) \frac{\partial}{\partial z} \left(D_z \frac{\partial C}{\partial z} \right) + C''' = \frac{\partial C}{\partial \tau} \tag{6.1.4}$$

and, for an isotropic region,

$$\nabla \cdot D \nabla C + C''' = \frac{\partial C}{\partial \tau} \tag{6.1.5}$$

For mass diffusion processes accompanied by simultaneous heat conduction in the region, the analogous equations for the temperature distribution are

$$\frac{\partial}{\partial x_i} \left(k_{ij} \frac{\partial t}{\partial x_j} \right) + q''' = \rho c \frac{\partial t}{\partial \tau} \tag{6.1.6}$$

or, for $k_{ij} = k_i = k_x, k_y, k_z$, again for an orthotropic heat conduction region,

$$\frac{\partial}{\partial x} \left(k_x \frac{\partial t}{\partial x} \right) + \frac{\partial}{\partial y} \left(k_y \frac{\partial t}{\partial y} \right) + \frac{\partial}{\partial z} \left(k_z \frac{\partial t}{\partial z} \right) + q''' = \rho c \frac{\partial t}{\partial \tau} \tag{6.1.7}$$

and, for an isotropic region,

$$\nabla \cdot k \nabla t + q''' = \rho c \frac{\partial t}{\partial \tau} \tag{6.1.8}$$

In all of the preceding equations, D_{ij}, D_i, D, C''', k_{ij}, k_i, k, q''', and the specific heat c each may be functions of location (x, y, z) and of time τ. Also, the diffusion coefficient D and conductivity k may each depend on concentration C and on temperature t.

However, the density ρ of the structure, in Eqs. (6.1.6)–(6.1.8) and in the use of Eqs. (6.1.3)–(6.1.5), may only depend on location, for example, as $\rho(x, y, z)$. That is, the diffusing region density may be inhomogeneous, but the material is assumed to remain at constant strain for the balances which lead to Eqs. (6.1.3)–(6.1.8) to apply. In circumstances in which this condition is not met, as with appreciable swelling or shrinking, in transients or periodics, convection terms must be included in the balances, if such effects are relatively important. That is, there will then be a mass flow rate of region material across the bounding surfaces of the fixed element of volume, $dx\, dy\, dz$, chosen for analysis, as shown in Figs. 1.2.1 and 1.3.1. Typical convection terms in the x direction are

$u \, \partial C / \partial x$ and $u \, \partial t / dx$. A convection effect is included in the analysis of moving-source processes in Secs. 4.7, 8.3, and 8.4 and in Sec. 6.6.3.

The preceding formulations may also be applied to regions of simultaneous mass diffusion and heat conduction. The simplest circumstance arises when the two processes are completely independent. Then the two distributions, C and t, are determined, for example, from Eqs. (6.1.5) and (6.1.8). The concentration and temperature initial and boundary conditions are used separately.

However, such processes are often coupled, as when adsorption and desorption in the region are accompanied by a thermal effect. Then heat generation or absorption arises. As a result, $q''' \neq 0$ and Eq. (6.1.8) is coupled to the local substance source strength C'''. Then $q''' = f(C''')$, where $q''' = KC'''$ for a linear relation between these two effects. After C''' is determined from Eq. (6.1.5), the resulting q''' (C'''), in Eq. (6.1.8), determines the temperature distribution. The calculations are thus decoupled.

Complete coupling occurs, for example, when the local substance adsorption rate C''' also depends on the local temperature level. Then C''' and q''' interact directly. As a result, the equations must often be solved simultaneously, for the C and t distributions. Added complexity also arises when the diffusion coefficients D and k depend on local temperature or concentration. Examples of these kinds of interactions are given in Secs. 6.2 and 6.3.

6.1.1 Variable Diffusivity and Transformations

In many applications the diffusivity coefficient is not a single constant value over the region. It may vary with location, local concentration, time, and direction. There may also be local adsorption and desorption. The transformations for each of these effects were given in Sec. 1.3 and are summarized as follows, for convenience. When D is constant and uniform over an isotropic region, Eq. (6.1.5) becomes

$$D \nabla^2 C + C''' = \frac{\partial C}{\partial \tau} \tag{6.1.9}$$

In an orthotropic region, for D_x, D_y, and D_z constant and uniform, Eq. (6.1.4) is transformed by

$$X = x\sqrt{D_a/D_x} \qquad Y = y\sqrt{D_a/D_y} \qquad Z = z\sqrt{D_a/D_z} \tag{6.1.10a}$$

to give

$$\nabla^2 C + \frac{C'''}{D_a} = \frac{\partial C}{\partial (D_a \tau)} \tag{6.1.10b}$$

where ∇^2 is in terms of X, Y, and Z and D_a may be taken as $\sqrt[3]{D_x D_y D_z}$; see Chao (1963).

A uniform diffusivity across a region may change with time, as $D(\tau)$, due to chemical reactions and aging effects. Time τ is transformed to τ', as

$d\tau' = D(\tau)\,d\tau$, and

$$\tau' = \int_0^\tau D(\tau)\,d\tau \tag{6.1.11}$$

Then $C = C(x, y, z, \tau')$, and for $C''' = 0$ it is a solution of

$$\nabla^2 C = \frac{\partial C}{\partial \tau'} \tag{6.1.12}$$

Often the local level of adsorption rate c, of substance C, is proportional to the instantaneous local concentration C. This occurs in very fast adsorption processes. Then $c = R_r C(x)$, where R_r is the dimensionless proportionality constant and $C'''(x) = -\partial c/\partial \tau$. The equation is then

$$D' \nabla^2 C = \frac{D}{(1 + R_r)} \nabla^2 C = \frac{\partial C}{\partial \tau} \tag{6.1.13}$$

In this kind of process, $D' < D$, results in decreasing local mass flux, m'', in transients.

If the diffusivity is a function of local concentration, as $D(C)$, Eq. (6.1.5) is nonlinear in C. Then the Kirchhoff transformation is applied, as $dW = [D(C)/D(C_0)]\,dC$, where $W(x, y, z, \tau)$ is the new dependent variable. The resulting equation is

$$\nabla^2 W + \frac{C'''}{D_0} = \frac{1}{D}\frac{dW}{\partial \tau} \tag{6.1.14}$$

However, the right side is still nonlinear and this transformation applies in analyzing steady-state processes. Section 6.2.1 considers several specific kinds of dependence of D on C. Results are given for several characteristic examples. Section 6.2.2 concerns location-dependent diffusivity, as in inhomogeneous regions.

The preceding transformations are very important. In each one a more complicated differential equation is simplified, in the form of the diffusion, or conduction, term. In Eq. (6.1.10) the complexity of directional diffusivities is removed, to give $\nabla^2 C$. The time and local concentration effects on diffusivity are removed, with the same result. The effect of very rapid adsorption is removed, in terms of $[D/(1 + R_r)] \nabla^2 C$.

SUMMARY. Each of these transformations, in terms of concentration C, returns a more complicated diffusion process to a simpler general form. The following similar relation represents a very large range of heat conduction processes:

$$\nabla^2 t + \frac{q'''}{k} = \frac{1}{\alpha}\frac{\partial t}{\partial \tau} \tag{6.1.15}$$

Thereby, the principle of the similarity of mass diffusion to heat conduction is extended from one phenomenon to the other. As a result, a large portion of the analysis and results in Chaps. 2 through 5 may be directly applied to mass diffusion, merely by interchanging the symbols of the thermal and mass diffusion formulations.

Therefore, the results developed in this chapter mainly concern processes which involve numerous kinds of processes which are more particular to mass diffusion. This material also includes detailed analysis of some of the special dependencies of D on concentration and on location in the region of diffusion. Other transformations also arise. The next section considers a somewhat different formulation, as for gas diffusion in porous solids.

6.1.2 Permeable Regions

Several characteristic general kinds of general boundary conditions are given in Sec. 1.4.2. They are imposed surface concentration level, a surface convective process, and an imposed diffusing substance flux at a boundary of a region. These are seen in Fig. 6.1.1(a), (b), and (c). However, diffusion often occurs through a region because of an imposed difference of pressure, $p_1 - p_2$, or of partial pressure, in a gas phase with one or several diffusing components. This formulation is shown in Fig. 6.1.1(d). The mass flux in steady state is then written as

$$m'' = P(p_1 - p_2)/L = (p_1 - p_2)/R \qquad (6.1.16)$$

where P is the permeability, or permeability constant. This formulation assumes a linear relation between m'' and the pressure gradient $(p_2 - p_1)/L$. The resistance R of the region is L/P. The dimensions and units of P are those of $m''L/p$, in whatever system is used.

The permeability P may be related to the diffusion coefficient D, under sufficiently simple conditions. First, D is assumed uniform across the region. If there is a linear relationship between the external vapor pressure and the corresponding equilibrium concentration within the layer:

$$C = Sp \qquad (6.1.17)$$

where C is the local concentration and S is the solubility at pressure level p. That is, S is the fractional solubility of a gas at pressure p. Since

$$m'' = -D\frac{(C_2 - C_1)}{L} = -D\frac{S(p_2 - p_1)}{L} = P\frac{(p_1 - p_2)}{L} \qquad (6.1.18)$$

the permeability is proportional to the diffusivity

$$P = DS \qquad (6.1.19)$$

In this formulation, P and D are simply related under these conditions, in terms of solubility. Then the substance transport rate may be calculated in

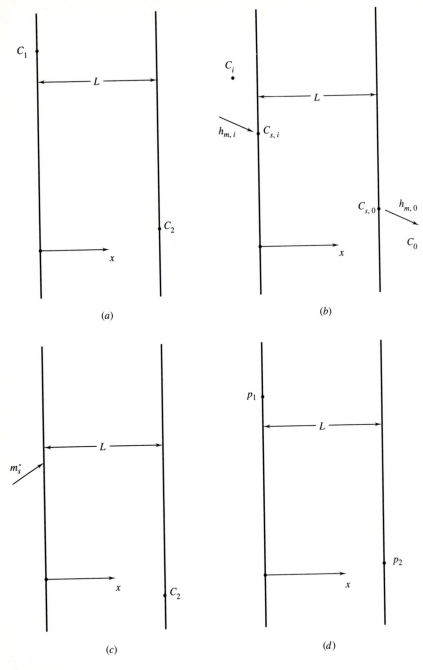

FIGURE 6.1.1
Characteristic boundary conditions: (a) concentration surface condition; (b) surface convective processes with adjacent regions at C_i and C_0; (c) an imposed rate of substance diffusion at a surface; (d) pressures p_1 and p_2 maintained at the two surfaces of a permeable region.

terms of either ΔC or Δp, as in Eq. (6.1.18). Thereby, permeability transport may often also be evaluated, in terms of the formulations in terms of concentration C, as in Sec. 6.1.1.

6.1.3 Heat Conduction and Mass Diffusion Analogies

The parallel development of heat conduction and mass diffusion in Chap. 1, and also in Chaps. 2 through 4, indicated that very many heat conduction mechanisms have an analog in mass transfer. As a consequence, the abundant techniques, methods, and results in heat conduction may be simply transformed into analogous mass diffusion results. Some examples are discussed in more detail as they appear.

The reverse is also useful, in terms of mass transfer results converted to heat conduction information. Much of the information in this chapter also has this characteristic. Section 6.2 concerns steady-state diffusion where D depends on concentration, $D(C)$, on location, $D(x)$, or on both as $D(C, x)$. This is analogous to $k(t)$, $k(x)$, and $k(t, x)$, as considered in Chap. 2, for one-dimensional steady-state processes. However, the typical forms of the dependencies, for k and D, are often quite different, as following from differences in characteristic physical effects which are most important in various applications.

Distributed sources, as q''' and C''', may also be analogous when each is completely specified in terms of imposed conditions. One example is energetic particle irradiation which produces internal energy dissipation or produces chemical species. Another would be internal microwave or Joulean energy dissipation. An analog might then be internal chemical species generation, or adsorption from the environment, due to aging effects. Similarity with a temperature effect, such as $q'''(t)$, would arise for $C'''(C)$. One mechanism, in transients, would be simple internal adsorption and desorption rates which depend on C locally.

The multidimensional and transient heat conduction results in Chaps. 3 through 5 are also useful in mass diffusion, when the equations, bounding conditions, and mechanisms result in analogous effects. For example, shape factors are applicable to mass diffusion when the differential equations and imposed surface conditions are of the same form. Thermal melting and freezing fronts have analogs in mass diffusion, as do moving sources. The integral method in Sec. 4.8 applies to comparable mass diffusion formulations, as do most aspects of the numerical methods developed in Secs. 3.9, 4.9, and 5.5.

However, there are many important mass diffusion mechanisms which do not have a simple and direct heat conduction analog. Distributed linear and nonlinear chemical rate reactions are examples. Irreversible reactions, constant reaction rates, and bimolecular reactions are often encountered. There are many kinds of processes in which several parallel mechanisms, expressed in individual forms, arise. Conditions of chemical catalysis often do not have a convenient heat conduction analog. These and other such processes are also considered in this chapter. The following section gives formulations for the Fickian mechanisms of diffusion in gases, liquids, and crystalline solids. The formulations of other mechanisms are given where used.

6.1.4   Diffusivity Mechanisms

The common fundamental thermal conductivity component mechanisms were discussed in Sec. 1.1.2 and listed in Table 1.1.1. However, mass diffusion was considered in Sec. 1.1.3 only in terms of a general formulation for the coefficient D, without regard to the many different kinds of fundamental processes whereby mass commonly diffuses across a region of material. Mass diffusion processes are often very different, since they amount to a migration of one or several distinct mass–species components. Mass diffusion is considered here, as through a stationary material barrier.

Often the process actually amounts to one species diffusing in one direction and another diffusing in the opposite direction. An example is two bodies of different gases put in contact at their interface. They then diffuse into each other. In mass diffusion through liquids and solids, the process appears more like a distinct component diffusing through a stationary region or matrix of the other material. This latter kind of process is often simpler to model and analyze. This picture leads directly to the simple expression of Fick's law given in Eqs. (6.1.1) and (6.1.2). However, this formulation also applies directly, or through transformations, to many counterdiffusion processes. The simpler point of view is followed in this chapter. In this section the basic mechanisms of diffusion which determine D are discussed for gases, liquids, and solids.

THE KINETIC THEORY RESULT FOR GASES. This is similar to the first model given in Table 1.1.1, for thermal conduction in gases. The diffusion region is shown in Fig. 6.1.2. Sample molecules of substances A and B are in random thermal motion. Typical collision diameters, d, are around 0.3 nm. A plane region of the gas of thickness λ, the local mean free path, is shown, where λ is about 10^{-5} cm or 100 nm at standard conditions. That is, λ is about $300d$. The molecules move at an average velocity of $\bar{v} \approx 300$ m/s. Therefore, the time between subsequent collisions is $\Delta\tau = \lambda/v$, in a dilute gas, and the collision frequency is the inverse, $f = \bar{v}/\lambda \approx 3 \times 10^9$ s^{-1}. See Chapman and Cowling (1939) for specific results, and also Cussler (1985).

The numbers of molecules, say A, associated with each region boundary, are taken as n_1 and n_2, per unit of boundary area. The resulting net molecule flux is then estimated as

$$m'' = f(n_1 - n_2)/3 = -\bar{v}(n_2 - n_1)/3\lambda \qquad (6.1.20)$$

where, in local equilibrium, an equal number of molecules are going in each of the x, y, z directions, at each location 1 and 2. Therefore, only 1/3 of any total sample is going in one direction, say x. Since n_1 and n_2 each represent a region

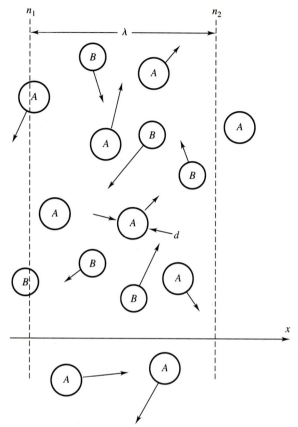

FIGURE 6.1.2
The interdiffusion of gas molecules
A and B.

λ thick, $n_1/\lambda = C_1$ and $n_2/\lambda = C_2$. Then Eq. (6.1.20) becomes

$$m'' = -\bar{v}\lambda\frac{(C_2 - C_1)}{3\lambda} = -\frac{\bar{v}\lambda}{3}\frac{(C_2 - C_1)}{\lambda} = -D\frac{\Delta C}{\Delta x} \approx -D\frac{\partial C}{\partial x} \quad (6.1.21)$$

and $D = \bar{v}\lambda/3$. This result is in good agreement with measured binary diffusion coefficients; see Kirkaldy and Young (1987). Values of D are given in App. E.1, E.2, and E.3.

Similar analysis also determines the average diffusion distance L of a particular molecule in time interval τ as approximately

$$L = 1.7\sqrt{D\tau} \quad (6.1.22)$$

Thus the diffusion distance does not depend linearly on the time of diffusion. The resulting effect is estimated by calculating a typical value of D for O_2 at standard conditions. For $\bar{v} \approx 300$ ms^{-1} and $\lambda \approx 90$ nm, $D \approx 0.1$ cm^2 s^{-1}. Therefore, an average molecule diffuses about 10 cm in the first second but only about 600 cm in the first hour. These results demonstrate that purely molecular

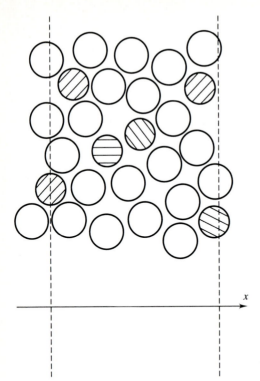

x

FIGURE 6.1.3
A model for a liquid mixture.

diffusion from a source, over long distances, may be a very slow process. Other mechanisms, such as fluid mixing, are usually very much faster.

DIFFUSION IN LIQUIDS. This phase is commonly much more compact, like a solid. It is modeled in Fig. 6.1.3. Thermal vibrations of the molecules allow, for example, the hatched molecules, to slip through the assembly in all directions. The result is their net preferential diffusion into regions of lower concentration. Again $\bar{v}$ and λ are estimated, but for different kinds of processes.

The kinetic energy per molecule, of mass m_0, is taken as $3kT/2$. The average velocity, see Bird et al. (1960), in terms of the Boltzmann constant k, is

$$\bar{v} = \sqrt{8kT/\pi m_0} \qquad (6.1.23)$$

The mean free path λ depends on the free volume in the structure and is estimated in Kirkaldy and Young (1987) as

$$\lambda \approx 3\alpha \, dT$$

where d is again diameter and α here is the linear thermal expansion coefficient of the material. Therefore,

$$D \approx \alpha \, d\sqrt{8kT^3/\pi m_0} \qquad (6.1.24)$$

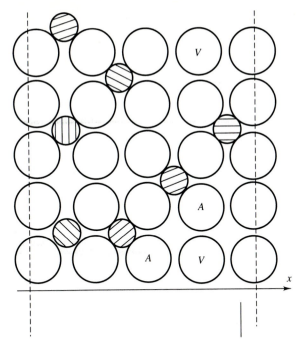

FIGURE 6.1.4
A crystalline structure with an interstitial impurity and vacant sites.

This temperature dependence is approximately applicable for many liquids. The value of D is commonly about 10^{-5} cm^2 s^{-1}. This is about four orders of magnitude less than in ideal gases. The resulting diffusion distance in 1 hr is about 3.5 mm, much smaller than in gases. The foregoing model applies most directly to approximately spherical molecules. Many of the values of D given in App. E.4, E.5, and E.6 are close to 10^{-5} cm^2 s^{-1}.

DIFFUSION IN SOLIDS. A model similar to those for gases and liquids also applies in crystalline solids, modeled as in Fig. 6.1.4. The hatched interstitial particles diffuse through interatomic channels. The diffusion coefficient for this mode approaches about 10^{-6} cm^2 s^{-1}, at high temperatures. Another diffusion route is through the locations of vacancies, V, in the crystal. These are more numerous at higher temperatures. Also, adjacent elements like A may diffuse, in interaction with a nearby vacancy. The coefficient for vacancy diffusion is about 10^{-7} cm^2 s^{-1} at high temperatures.

Statistical mechanics may be applied to solids consisting of particles in a random vibrational array. These considerations indicate that the particle kinetic energy distribution function is proportional to $\exp(-E'/kT)$. In order for particles to diffuse, for example, the interstitials shown in Fig. 6.1.4, they must have a sufficient activation energy level E to penetrate the barrier posed by the adjacent matrix particles. The probability P of this is

$$P \propto \exp(-E/RT)$$

Therefore, the frequency of crossing the barrier is proportional to P and the resulting diffusion coefficient is expressed as

$$D = D_0 \exp(-E/RT) \qquad (6.1.25)$$

where D_0 is the empirical constant of proportionality. As an example, for carbon in austenite, $D_0 = 0.175$ cm^2 s^{-1} and the barrier energy $E = 138.3$ kJ mol^{-1}. The gas constant is $R = 8.3143 \times 10^{-3}$ kJ (K mol)$^{-1}$. The preceding relation is then a very good fit of measured values of D over the range of $10^4/T$ from about 10 to 6. This range results in an increase in the ratio of diffusivities of 8000. Appendix E.7 lists values of D in crystalline solids. They span a very wide range.

SUMMARY. The preceding results characterize the diffusion coefficients, in gases, liquids, and solids, as the order of 10^{-1}, 10^{-5}, and 10^{-10} cm^2 s^{-1}. The resulting diffusion distances, L, in 10 s, are then about 1, 0.01, and 0.0001 cm. These dimensions are about the order of the size of precipitates which might evolve in 10 s in supersaturated air as snow crystals, as sugar crystals in water, and as carbide crystals in steel.

Diffusion coefficients in gases and liquids are not highly variable from one substance to another or with temperature. However, given the exponential form in Eq. (6.1.25), for a crystalline structure, very small values of D result at low temperatures. For example, the diffusion length of ferrite in steel at environmental temperatures, in about 3 years, is only about 10 nm. Nevertheless, even such effects may be important in strength considerations and in the reliability of increasing small electronic circuit elements and sensors.

The results in this section give perspective to the differing values of the diffusion coefficient, over a range of materials and conditions. However, the primary source of definite information remains physical measurements. Methods of measurement are discussed in detail by Crank (1975), Chap. 10.

The preceding formulations of diffusion mechanisms are very valuable. However, there are a very large number of important additional diffusion mechanisms. One of these is noncontinuum, or Knudsen, gas diffusion effects in porous regions. In many porous media, there are also several simultaneous parallel mass diffusion mechanisms. An example is H_2O diffusion in soil. One mechanism is vapor diffusion through the voids. Another is the capillary flow of water films over the soil particles. Yet another is molecular diffusion of thin surface-adsorped H_2O. The simplest Fickian formulation does not apply in these circumstances. Relevant formulations are given later in Sec. 6.6 as such mechanisms are considered.

6.2 VARIABLE DIFFUSION COEFFICIENTS IN STEADY STATE

Section 6.1.1 indicated transformations which remove the effects of several kinds of variation of D from the equations, under some conditions. The examples are: orthotropic regions, D_i, $i = 1, 2, 3$, $D(\tau)$, and $D(C)$. Another very important matter in many applications is often the spatial inhomogeneity of the region in which diffusion occurs as $D = D(x, y, z)$.

This section considers three aspects. The first is the concentration dependence of D, as $D(C)$. The transformation given in Eq. (6.1.14) then applies generally, in steady-state processes, to remove the nonlinearity in the term $\nabla \cdot D(C)\nabla C$. The general techniques, used for the analysis of time-dependent processes, wherein $\partial C/\partial \tau$ must be retained, are given in Chaps. 4 and 5, for the similar heat conduction effect.

Several simple and one-dimensional steady-state processes, for $k(t)$, are analyzed in Sec. 2.2.1 for one-dimensional conduction in plane, cylindrical, and spherical regions. Temperature distributions are determined and heat transfer rates are determined. These solutions also apply for mass diffusion, from the similarity of the following two equations, for example, for a plane region

$$\frac{d}{dx}k(t)\frac{dt}{dx} = 0 \quad \text{and} \quad \frac{d}{dx}D(C)\frac{dC}{dx} = 0 \qquad (6.2.1)$$

Section 6.2.1 concerns processes for which D is a function of local concentration, as $D(C)$. Results are given in Sec. 6.2.2 for D dependent on location in the diffusing region, as $D(x)$, for example. This arises from spatial inhomogeneity, as in regions of variable porosity or composition. The comparable equations for a plane region and for steady state are

$$\frac{d}{dx}k(x)\frac{dt}{dx} = 0 \quad \text{and} \quad \frac{d}{dx}D(x)\frac{dC}{dx} = 0 \qquad (6.2.2)$$

Results for heat conduction in plane, cylindrical, and spherical regions are also given in Table 2.3.1, in terms of integrals of $[k(x)]^{-1}$, $[rk(r)]^{-1}$, and $[r^2k(r)]^{-1}$, in plane, cylindrical, and spherical regions.

Comparable transients are discussed later in Sec. 6.3. Those examples are mostly for infinite and semiinfinite regions and are similar to the heat conduction analysis in Sec. 4.2.3 for k uniform. However, most results here are for D dependent on local concentration, for different forms of the variation of $D(C)$.

The third effect, considered in Sec. 6.2.3, is the diffusion coefficient dependent on both effects, as $D(C, x)$. Results are given in Table 2.3.1 for steady state in a plane layer for $k(t, x)$ equal to a product of linear functions in t and in x, as

$$k(t, x) = K(t)f(x) = K_0[1 + \beta(t - t_0)][1 + \gamma(x - x_0)] \qquad (2.2.40b)$$

where t_0 and x_0 are reference values. All of the results are again applicable to mass diffusion as expressed in Eq. (6.2.2). Section 6.2.3 considers processes in a plane layer, in which D depends on both concentration and location in the region, as $D = D(C, x)$.

6.2.1 Concentration-Dependent Diffusivity

This effect arises in diverse applications from many different mechanisms. A common cause of the variation is a change in the rate mechanism of diffusion in regions of gradients in concentration of the diffusing substance. The emphasis here concerns the kinds of variations, $D(C)$, which demonstrate common characteristics and are amenable to simple analysis. With definite solutions it is possible to determine the kinds of variations of transport which occur and to evaluate the net effects. Often the variations of $D(C)$ used are idealized in terms of the forms which will give simple solutions. Results given here are for steady-state processes. The effects in transients are considered in Sec. 6.3.2.

Consider first the plane region in Fig. 6.1.1(a), with the boundary conditions shown. The variation of D is taken as

$$D(C) = D_0[1 + f(C)] \tag{6.2.3}$$

where $f(C)$ is dimensionless and D_0 is, nominally, the value of $D(C)$ at a very low concentration level. For steady state and $C''' = 0$, Eq. (6.1.5) becomes

$$\frac{d}{dx} D(C) \frac{dC}{dx} = \frac{d}{dx} D_0[1 + f(C)] \frac{dC}{dx} = 0 \tag{6.2.4}$$

Integration yields

$$D_0 \int [1 + f(C)] \, dC = B_1 x + B_2 \tag{6.2.5}$$

The corresponding results for cylindrical and spherical regions are

$$D_0 \int [1 + f(C)] \, dC = B_1 \ln r + B_2 \tag{6.2.6}$$

and

$$D_0 \int [1 + f(C)] \, dC = -\frac{B_1}{r} + B_2 \tag{6.2.7}$$

where the common integral of $f(C)$ in the preceding equations is denoted by

$$F(C) = \int_0^C f(C) \, dC \tag{6.2.8}$$

These formulations are considered for the plane layer and the cylindrical and spherical shells in Fig. 6.2.1, with the boundary conditions shown.

The preceding relations are applied to determine the concentration distribution, C, for the geometries in Fig. 6.2.1, for the boundary conditions C_1 and C_2 shown. The similar results for the plane, cylindrical, and spherical barriers are, respectively,

$$\frac{C_1 + F(C_1) - C - F(C)}{C_1 + F(C_1) - C_2 - F(C_2)} = \frac{x}{L} \tag{6.2.9}$$

$$= \frac{\ln(r_1/r)}{\ln(r_1/r_2)} \tag{6.2.10}$$

$$= \frac{r_2(r_1 - r)}{r(r_1 - r_2)} \tag{6.2.11}$$

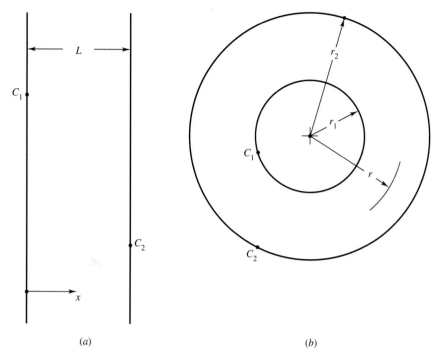

(a)   (b)

FIGURE 6.2.1
Plane, cylindrical, and spherical geometries: (a) a plane region; (b) cylindrica! and spherical shells.

These relations are analogous to those for heat conduction in Eqs. (2.1.11), (2.1.18b), and (2.1.24b), in terms of $\phi = (t - t_2)/(t_1 - t_2)$. They become the same, if $f(C)$ is taken as zero.

The effect of variable diffusivity in the form given in Eq. (6.2.3) is seen in Fig. 6.2.2. There $f(C)$ is a linear function of C, as $f(C) = bC$, where b has the dimensions of C^{-1}. The boundary conditions are normalized as $C_1 = 1$ and $C_2 = 0$. The condition at $x = L$ then assumes a perfect absorber at that location.

The diffusing mass flux across the region remains uniform in any steady-state process, for $C''' = 0$, that is, for any specific value of b. Therefore, D decreases with x, for a positive value of b, since the concentration decreases in Eq. (6.2.3). Therefore, the slope of the curves, the gradient in C, ∇C, increases in magnitude, with x, to maintain the flux m'' constant, at larger x. The reverse is true for b negative, as in curves 5 and 6. Curve 4 is for D uniform across the region.

OTHER VARIATIONS $D(C)$. Some other particular examples of variable diffusivity are given in Fig. 6.2.3, for $f(C)$ taken as dependent on a wide range of different kinds of different variations with C. Curves 1 through 8 are for $f(C)$

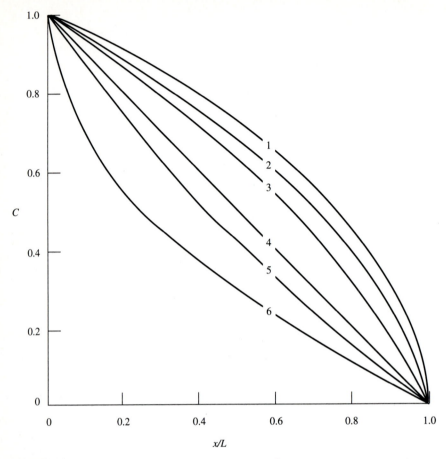

FIGURE 6.2.2
Steady-state concentration distributions across a region wherein $D = D_0[1 + bC]$. For curves 1–6, $b = 100, 10, 2, 0$ (D uniform), -0.5, and -1, respectively. [From Crank (1975).]

being

1. be^{gC}; $b = 1$, $g = 3$
2. $bC/(1 + gC)$; $b = 100$, $g = 1$
3. $-b\sqrt{C} + gC^2$; $b = 1$, $g = 2$
4. $bC/(1 + gC)$; $b = 1$, $g = 1$
5. D uniform
6. $-bC/(1 + gC)$; $b = 0.9$, $g = 1$
7. $-b\sqrt{C}$; $b = 1$
8. $-b\sqrt[4]{C}$; $b = 1$

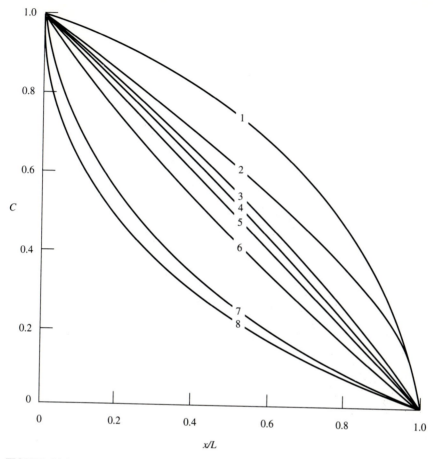

FIGURE 6.2.3

Steady-state concentration distributions across a region wherein $D = D_0[1 + f(C)]$, for the circumstance $C_1 = 1$ and $C_2 = 0$. [From Crank (1975).]

Again the slopes of the C distributions increase at larger x as D decreases. The decrease occurs for $f(C)$ decreasing at lower concentration levels.

RESULTS IN TERMS OF $\tilde{C}$. An alternative and often more useful formulation may be given when $D(C)$ may be formulated instead in terms of $\tilde{C}$, as a concentration excess ratio, as $\tilde{C}$, below

$$D(C) = D_0\left[1 + f(\tilde{C})\right] \quad \text{where } \tilde{C} = \frac{C - C_2}{C_1 - C_2} \qquad (6.2.12)$$

Here $C_1 - C_2$ is the imposed concentration difference across the region, and D_0 is a reference value. The relevant differential equation, (6.2.4), is then written as follows, where $\tilde{C}(0) = 1$ and $\tilde{C}(L) = 0$,

$$\frac{d}{dx}D\frac{d\tilde{C}}{dx} = \frac{d}{dx}D_0\left[1 + f(\tilde{C})\right]\frac{d\tilde{C}}{dx} = 0 \qquad (6.2.13)$$

This becomes

$$\int_0^x\left[1 + f(\tilde{C})\right]d\tilde{C} = B_1 x + B_2 \qquad \tilde{C}(0) = 1 \qquad \tilde{C}(L) = 0$$

The result, in terms of $\tilde{C}$ and x/L, is

$$\frac{(1 - \tilde{C}) - \int_1^{\tilde{C}}f(\tilde{C})\,d\tilde{C}}{\int_0^1 f(\tilde{C})\,d\tilde{C} + 1} = \frac{x}{L} \qquad (6.2.14)$$

where the limits on the integrals are in terms of $\tilde{C}$. Comparable relations apply for the cylindrical and spherical geometries in Fig. 6.2.1(b) (see page 319). All three relations in terms of $\tilde{C}$ are much simpler in form and require the evaluations of only two integrals. These may be called $F(\tilde{C})$ and $F(1)$ in the notation of Eq. (6.2.8).

MASS DIFFUSION RATE. The mass diffusion flux rate m'' may be calculated from each of the preceding formulations for a plane region, on the basis of

$$m'' = -D(C)\frac{dC}{dx} = constant \quad \text{from } x = 0 \text{ to } L$$

or $\qquad m''\,dx = -D(C)\,dC$

This is integrated from $x = 0$, $C = C_1$ to $x = L$, $C = C_2$ as

$$\int_0^L m''\,dx = m''L = -\int_{C_1}^{C_2}D(C)\,dC$$

or $\qquad m'' = -\frac{1}{L}\int_{C_1}^{C_2}D(C)\,dC = \frac{1}{L}\int_{C_2}^{C_1}D(C)\,dC \qquad (6.2.15)$

The results for cylindrical and spherical shells are

$$m' = \frac{2\pi}{\ln(r_2/r_1)}\int_{C_2}^{C_1}D(C)\,dC \qquad (6.2.16)$$

and $\qquad m = \frac{4\pi r_1 r_2}{r_2 - r_1}\int_{C_2}^{C_1}D(C)\,dC \qquad (6.2.17)$

The common integral may be calculated when $D(C)$ is specified in any particular application. The preceding results in this section, for local concentration-dependent diffusivity, are analogous to those in Sec. 2.2.1 for temperature-dependent thermal conductivity. See also Sec. 9.5.1 for the general numerical formulation of this effect.

6.2.2   Location-Dependent Diffusivity

In plane, cylindrical, and spherical regions, this effect is expressed as $D(x)$ and $D(r)$. This variation commonly arises as a result of a material inhomogeneity across the diffusion region. Examples are varying porosity or changing volume fractions of multiple crystalline components. The formulation and calculation for this effect are simpler, since the governing equation, for example, Eq. (6.2.19), remains linear.

First, for the plane region in Fig. 6.1.1(a), the variation is taken as

$$D(x) = D_0[1 + f(x)] \tag{6.2.18}$$

The differential equation

$$\frac{d}{dx} D(x) \frac{dC}{dx} = \frac{d}{dx} D_0[1 + f(x)] \frac{dC}{dx} = 0 \tag{6.2.19}$$

becomes

$$D_0[1 + f(x)] \frac{dC}{dx} = B_1 \quad \text{or} \quad [1 + f(x)] \frac{dC}{dx} = \frac{B_1}{D_0}$$

Integration yields

$$\int_{C_1}^{C} dC = C - C_1 = \frac{B_1}{D_0} \int_0^x \frac{dx}{1 + f(x)}$$

Then B_1/D_0 is determined from the condition $C = C_2$ at $x = L$ and the result is

$$C - C_1 = (C_2 - C_1) \frac{\displaystyle\int_0^x \frac{dx}{1 + f(x)}}{\displaystyle\int_0^L \frac{dx}{1 + f(x)}} \tag{6.2.20}$$

or

$$\tilde{C}(x) = \frac{C - C_2}{C_1 - C_2} = 1 - \frac{\displaystyle\int_0^x \frac{dx}{1 + f(x)}}{\displaystyle\int_0^L \frac{dx}{1 + f(x)}} \equiv 1 - \frac{I_p(x)}{I_p(L)}$$

The analogous relations for cylindrical and spherical shells are

$$\tilde{C}(r) = \frac{C - C_2}{C_1 - C_2}$$

$$= 1 - \frac{\displaystyle\int_{r_1}^{r} \frac{dr}{r[1 + f(r)]}}{\displaystyle\int_{r_1}^{r_2} \frac{dr}{r[1 + f(r)]}}$$

$$\equiv 1 - \frac{I_c(r)}{I_c(r_1, r_2)} \qquad (6.2.21)$$

$$\tilde{C}(r) = \frac{C - C_2}{C_1 - C_2}$$

$$= 1 - \frac{\displaystyle\int_{r_1}^{r} \frac{dr}{r^2[1 + f(r)]}}{\displaystyle\int_{r}^{r_2} \frac{dr}{r^2[1 + f(r)]}}$$

$$\equiv 1 - \frac{I_s(r)}{I_s(r_1, r_2)} \qquad (6.2.22)$$

In all three geometries, $\tilde{C}$ is evaluated from the integrals, for the particular form of $f(x)$ or $f(r)$ of interest. Figure 6.2.4 shows the concentration distribution across a plane region for the forms $f(x) = bx$ and $bx + cx^2$. As in Figs. 6.2.2 and 6.2.3, C_2 is taken as zero and C_1 as one. Then $\tilde{C}(x)$ in Eq. (6.2.20) becomes C/C_1.

The mass diffusion flux rate m'' is calculated, as for $D(C)$ as follows:

$$m'' = -D_0[1 + f(x)]\frac{dC}{dx}$$

$$\frac{m'' \, dx}{[1 + f(x)]} = -D_0 \, dC$$

$$\int_0^L \frac{m'' \, dx}{[1 + f(x)]} = m'' \int_0^L \frac{dx}{[1 + f(x)]}$$

$$= -\int_0^L D_0 \, dC = D_0(C_1 - C_2)$$

and $\quad m'' = D_0(C_1 - C_2) \Big/ \int_0^L \frac{dx}{[1 + f(x)]} \equiv D_0(C_1 - C_2)/I_p(L) \quad (6.2.23)$

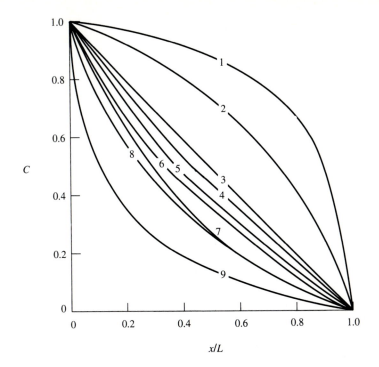

FIGURE 6.2.4
Steady-state concentration distributions across a region wherein $D = D_0[1 + f(x)]$. $C_1 = 1$, $C_2 = 0$. For $f(x) = bx$, $b = -0.99$, -0.90, 0, 1.0, 2.0, 9, and 99 for curves 1, 2, 3, 4, 5, 8, and 9, respectively. For curves 6 and 7, $f(x) = 2x + x^2$ and $f(x) = 3x + 2.25x^2$, respectively. [From Crank (1975).]

In this relation $I_p(L)/D_0$ is the diffusion resistance of the plane region R_C. This is seen to be the reciprocal of the local conductance, averaged across the region. This is the same result as in Eq. (2.2.27) and given in Table 2.3.1, for $k = k(x)$, in heat conduction. The analogous results for cylindrical and spherical shells are also given in that table.

6.2.3 Diffusivity Dependent on Both Concentration and Location

These effects arise, as $D(C, x)$, in a region where both of the mechanisms considered in Secs. 6.2.1 and 6.2.2 occur simultaneously over the region. The analysis for this effect in heat conduction, as $k(t, x)$, is in Sect. 2.2.3. Results are given there for the condition when the temperature and location effects in $k(t, x)$ arise from separate causes and remain independent of each other. An example is when the temperature effect is a property only of a particular heat-conducting material, for example, steel or rock, and the location effect is due to variable porosity across the region. When these effects cause only relatively small changes, $k(t, x)$ may be taken as $K(t)f(x)$, as in Eq. (2.2.40b). The comparable formulation here is

$$D(C, x) = \bar{D}(C)f(x) \tag{6.2.24}$$

This permits separation of the effects of C and x, as in Eqs. (2.2.43) and (2.2.44). As applied to mass diffusion, the results are

$$\int_{C_1}^{C} \bar{D}(C) \, dC = B_1 \int_0^x \frac{dx}{f(x)} \tag{6.2.25}$$

$$\int_{C_1}^{C_2} \bar{D}(C) \, dC = B_1 \int_0^L \frac{dx}{f(x)} = -m'' \int_0^L \frac{dx}{f(x)} \tag{6.2.26}$$

Defining average value of $\bar{D}(C)$, over the concentration range C_1 to C_2 as

$$\bar{D}_m(C) = \frac{1}{C_2 - C_1} \int_{C_1}^{C_2} \bar{D}(C) \, dC \tag{6.2.27}$$

the mass flux in Eq. (6.2.26) is

$$m'' = \bar{D}_m \frac{(C_1 - C_2)}{L} \frac{1}{L} \int_0^L \frac{dx}{f(x)} = \frac{(C_1 - C_2)}{R_m} \tag{6.2.28}$$

where R_m is the mass transfer resistance per unit area. This result is completely analogous to the result for heat conduction in Eq. (2.2.46), as listed in Table 2.3.1.

Neither of the averages of $\bar{D}(C)$ and $[f(x)]^{-1}$ in Eqs. (6.2.27) and (6.2.28) change if the boundary conditions $C_1(0)$ and $C_2(L)$ are applied in the opposite sense, that is, as $C_1(L)$ and $C_2(0)$. Therefore, m'' does not change. However, with other and more complicated variations of $D(C, x)$, the value of m'' may change as the direction of diffusion is changed by inverting C_1 and C_2. For further consideration of such effects, see Sternberg and Rogers (1968), Peterlin and Williams (1971), Rogers and Sternberg (1971), and Crank (1975). In proper form, the preceding considerations also apply in heat conduction, for $k(t, x)$.

6.3   TRANSIENT MASS DIFFUSION IN PLANE LAYERS

Transient processes in mass diffusion are often very similar to those in heat conduction. Often the relevant form of the Fickian equation is very similar to the simplest form of the Fourier equation. Internal species production C''' may often be modeled as was q''', in heat conduction. However, there are many important processes in each kind of mechanism which entail additional complexity. For some of these, the effect is physically similar in both mechanisms. For example, accounting for $k(t)$ and $D(C)$ involves similar considerations. However, in very many important applications in mass diffusion, the effect $D(C)$ must be taken into account. Also, a wide range of functional forms of $D(C)$

arise. Many idealized forms were considered for one-dimensional steady-state processes, in Sec. 6.2.1. Several variations of $D(x)$ and $D(C, x)$ are discussed in Secs. 6.2.2 and 6.2.3.

Section 6.3.1 discusses transient response for constant and uniform diffusivity. It gives several kinds of transient solutions in plane layers, which are in some ways more characteristic of concerns in mass diffusion processes. These include the determination of sorption and desorption in a layer, the effect of limited ambient media, and the inference of diffusivity from measurements. Section 6.3.2 then considers transients idealized as occurring in infinite and semiinfinite regions for $D = D(C)$. The effects of different forms are assessed. Section 6.3.3 concerns convective processes and indicates how diffusion properties may be inferred from other measurements.

6.3.1 Transients with Diffusivity Constant and Uniform

These processes are very much like those in Sec. 4.2 for plane regions and in Sec. 4.3 for cylindrical and spherical regions. Most of those results for heat conduction apply here. Therefore, the three other results given in the following discussion are more related to concerns in mass transfer.

A PLANE LAYER. A region of thickness $2s$ is shown in Fig. 6.3.1. It initially has a uniform internal solute concentration level $C_i = C(x, 0)$. The surface concentration level $C(0, \tau) = C_1(\tau)$ is raised on each side, to a final level C_f. Here $C_1(\tau)$ is taken to increase as follows:

$$C_1(\tau) - C_i = (C_f - C_i)[1 - e^{-\beta\tau}] \tag{6.3.1}$$

This increase is a better model of reality than the step change applied as shown in Fig. 4.2.1(a), for heat conduction. Here $\partial C_1/\partial \tau = \beta e^{-\beta\tau}$, which is bounded

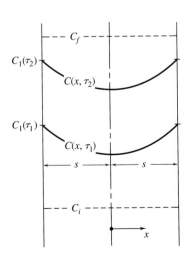

FIGURE 6.3.1
A region initially uniformly at C_i whose surface concentration level $C_1(\tau)$ increases with time.

initially. Then $C_1(\tau)$ approaches C_f asymptotically. Intermediate distributions, at τ_1 and τ_2, are sketched in Fig. 6.3.1.

The interest here is to determine the sorption–time curve. This is the total amount of diffusing substance C which has entered the region by time τ. The concentration variable $\tilde{C}$ is defined as follows and the transient solution for $\tilde{C}(x, \tau)$ is then given in terms of $\beta\tau$, $\beta s^2/D$, and x/s.

$$\tilde{C}(x, \tau) = [C(x, \tau) - C_i]/(C_f - C_i) \quad \text{where } \tilde{C}(x, 0) = 0 \tag{6.3.2}$$

$$\tilde{C}(x, \tau) = 1 - \frac{\cos\sqrt{\beta x^2/D}}{\cos\sqrt{\beta s^2/D}} e^{-\beta\tau}$$

$$-\frac{16}{\pi} \sum_{n=0}^{\infty} \frac{(-1)^n \exp\left[-(2n+1)^2\pi^2 \, \text{Fo}/4\right]}{(2n+1)\left[4 - (2n+1)^2\pi D/\beta s^2\right]} \cos\frac{(2n+1)\pi x}{2s} \tag{6.3.3}$$

This solution applies for β not equal to any values of $(2n+1)^2\pi^2 D/4s^2$. The parameters in this result are $\beta s^2/D$ for the surface condition, x/s and $\beta x^2/D$ for the location, and $\beta\tau$ and $\text{Fo} = D\tau/s^2$ for the time. That is, $\tilde{C} = \tilde{C}(\beta s^2/D, x/s, \beta x^2/D, \beta\tau, \text{Fo})$.

The sorption–time curve is in terms of $M(\tau)$, the increased amount of substance C, or the amount of solution, in the region of thickness $2s$ at time τ. From the preceding result

$$\frac{M(\tau)}{2s(C_f - C_i)} = 1 - \tan\sqrt{\beta s^2/D}\, e^{-(\beta\tau)\sqrt{D/\beta s^2}}$$

$$-\frac{8}{\pi^2} \sum_{n=0}^{\infty} \frac{e^{-(2n+1)^2\pi^2 \, \text{Fo}_s/4}}{(2n+1)^2\left[1 - (2n+1)^2\pi^2 D/4\beta s^2\right]} \tag{6.3.4}$$

In terms of Fo and $\beta s^2/D$, Eq. (6.3.4) may be written as

$$\frac{M(\tau)}{2s(C_f - C_i)} = \overline{M}(\beta s^2/D, \text{Fo}) \tag{6.3.5}$$

The uptake of the solute M is plotted in Fig. 6.3.2 versus Fo, for given values of $\beta s^2/D = 0.01$ to ∞. For $\beta \to \infty$, C_1 immediately becomes C_f. This is the heat conduction solution, Eq. (4.2.6), which has an infinite surface flux, $q''(0, \tau)$ in that formulation. The curves have inflexion points, except for large $\beta s^2/D$. They are often referred to as sigmoid sorption curves. These results are better models, for the gradual increase in $C_1(\tau)$ which occurs in actual processes.

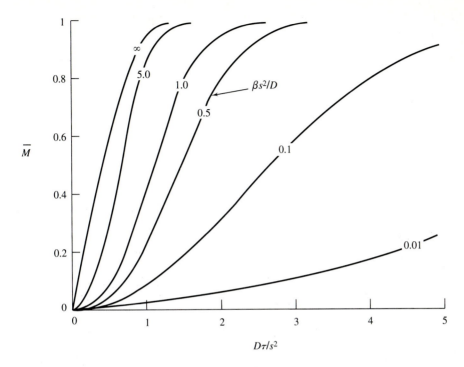

FIGURE 6.3.2
Sorption curves for the surface concentration increasing as $(1 - e^{-\beta\tau})$. [From Crank (1975).]

A LINEAR SURFACE CONCENTRATION INCREASE. Another useful model is the circumstance in which $C_1(\tau)$ increases linearly with time, as $C_1(\tau) - C_i = K\tau$. Figure 6.3.3 shows changed concentration distributions $C(x, \tau) - C_i$, over the region $x = 0$ to s, for various times, in terms of $D\tau/s^2 = \text{Fo}$. Figure 6.3.4 shows the sorption curve. The solute uptake, see Crank (1975), again in terms of $M(\tau)$, is

$$\frac{DM(\tau)}{Ks^3} = \overline{M}(\text{Fo}) = 2\text{Fo} - \frac{2}{3} + \frac{64}{\pi^2} \sum_{n=0}^{\infty} \frac{e^{-[(2n+1)^2\pi^2 \, \text{Fo}/4]}}{(2n+1)^4} \qquad (6.3.6)$$

A SURFACE IN A LIMITED AMBIENT MEDIUM. Another similar and important sorption effect also often arises in many applications. An absorbing layer of material, initially having a uniform solute concentration distribution C_i may be placed in an ambient environment at, say $C_a > C_i$, of limited extent. Then C_a decreases with time as the solute is absorbed in the layer. Eventually equilibrium is achieved. The ambient volume, per unit of surface area of the layer, is denoted by V. The concentration distribution in the layer, $C(x, \tau)$, satisfies the

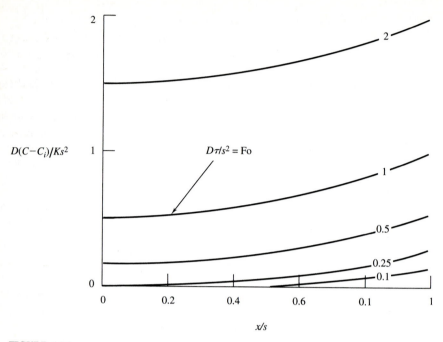

FIGURE 6.3.3
Concentration distributions $C(x,\tau) - C_i$ over the region $x = 0$ to s, at various times, in terms of $D\tau/s^2 = $ Fo. [From Crank (1975).]

following formulation:

$$D\frac{\partial^2 C}{\partial x^2} = \frac{\partial C}{\partial \tau} \qquad C(x,0) = C_i \qquad (6.3.7)$$

The ambient medium is added at $\tau = 0$ and well stirred thereafter. C_a is a function of time, as $C_a(\tau)$. For $C_a(\tau) > C_i$ the ambient concentration decreases with time. The instantaneous conditions at $x = -s$ and s, recall Fig. 6.3.1, are

$$V\frac{\partial C_a}{\partial \tau} = \mp D\frac{\partial C}{\partial x} \quad \text{or} \quad V\frac{\partial [C_a(\tau) - C_i]}{\partial \tau} = \mp D\left[\frac{\partial(C - C_i)}{\partial x}\right]_{x = \mp s} \qquad (6.3.8)$$

This formulation assumes that there is no added resistance to diffusion at the two surfaces of the layer. That is, $C_a(\tau) = C(-s, \tau) = C(s, \tau)$. A surface resistance may be incorporated in terms of a partition factor K, where

$$KC_a(\tau) = C(-s, \tau) = C(s, \tau) \qquad (6.3.9)$$

The partition may also result from different solubility characteristics of species C between the ambient and the material of the plane region; see Cussler (1985) and Crank (1975). The partition effect may be incorporated in Eq. (6.3.8) by replacing V by V/K.

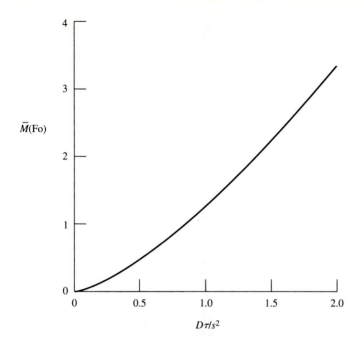

FIGURE 6.3.4
The sorption curve in terms of generalized variables $\overline{M}$ and Fo. [From Crank (1975).]

The solution of Eqs. (6.3.7) and (6.3.8) is given by Carslaw and Jaeger (1959), Sec. 3.13. The sorption at time τ, $M(\tau)$, is given by Crank (1975) in terms of the eventual total sorption $M(\infty)$ as

$$\frac{M(\tau)}{M(\infty)} = 1 - \sum_{n=1}^{\infty} \frac{2\alpha(1 + \alpha)}{1 + \alpha + \alpha^2 \beta_n} e^{-\beta_n^2 \, \text{Fo}} \qquad (6.3.10)$$

where $\text{Fo} = D\tau/s^2$, $\alpha = V/s$ or V/Ks, and the β_n are the roots of $\tan \beta_n = -\alpha\beta_n$. The final sorption, when equilibrium is achieved, is

$$M(\infty) = 2s[C_a(\infty) - C_i] = 2V[C_a(0) - C_a(\infty)] \qquad (6.3.11)$$

Then $M(\infty)$, and the fractional amount transferred into the layer, are

$$M(\infty) = \frac{2V[C_a(0) - C_i]}{1 + \alpha} \qquad (6.3.12)$$

and
$$\frac{M(\infty)}{2V[C_a(0) - C_i]} = \frac{1}{1 + \alpha} \qquad (6.3.13)$$

This quantity is 0.5 for $\alpha = 1$. A similar result follows if the ambient is of infinite extent, that is, $V \to \infty$ and $\alpha \to \infty$. Then $\beta_n = (2n + 1)\pi/2$,

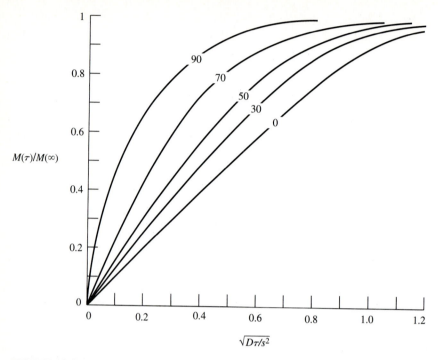

FIGURE 6.3.5
Increasing plane layer solute content, as $M(\tau)/M(\infty)$, as a function of time as $D\tau/s^2$. Curves are the fraction of M_∞ to the total initially available in the ambient region. [From Crank (1975).]

$C_s(-s, \infty) = C_s(s, \infty) = C_a(\infty)$, and

$$\frac{M(\tau)}{M(\infty)} = 1 - \frac{8}{\pi^2} \sum_{n=0}^{\infty} \frac{8}{(2n + 1)^2} e^{-[(2n+1)^2\pi^2 \text{Fo}/4]} \qquad (6.3.14)$$

The instantaneous uptake $M(\tau)/M(\infty)$ is shown in Fig. 6.3.5, as a function of $D\tau/s^2$. The numbers on the curves are the percentage of the available amount of the component initially in the solution, $V[C_a(0) - C_i]$, taken up in the layer of material, when diffusion ceases, as given in Eq. (6.3.13). Thus the behavior in any particular occurrence follows one of these curves, with increasing time, in terms of $D\tau/s^2$, as $M(\tau)/M(\infty) \to 1$. For example, if 90% is eventually taken up, as for V much less than s, the initial rate is large. The lowest curve shown, 0, is for $V \gg s$, as given in Eq. (6.3.14).

Solutions for other limited ambient regions are given in Carslaw and Jaeger (1959), Sec. 3.1.3. These are in terms of heat conduction for a plane layer of material in a stirred environment which is at a different initial temperature condition.

In the three examples given previously, the formulation results in solute diffusion into a layer of material. In each mechanism, the inverse process may arise, of solute diffusing out, as for $C_i > C_1(0, \tau)$ or $C_a(\tau)$. The preceding results apply, when properly interpreted. For example, $M(\tau)$ now becomes the amount of solute desorption from the plane region at time τ and $M(\infty)$ is still $M(\tau)$ at large time.

SUMMARY. The preceding examples are interesting results and are different from those characteristic of thermal transients in Sec. 4.2. In particular, the solution for the stirred ambient of limited extent may be used to determine D for diffusion in a plane layer. The rate of change of the concentration in the ambient, $C_a(\tau)$, may be measured. From this, $M(\tau)/M(\infty)$ is known from Eqs. (6.3.10) and (6.3.12). Then D may be inferred. See Crank (1975), Sec. 4.3.5, for a particular technique.

Multidimensional solutions for the kinds of effects analyzed previously may also be determined by the techniques given in Chap. 5. One-dimensional transients in cylindrical and spherical geometries may also be patterned from the results in Sec. 4.3. Surface convection conditions in mass diffusion may also be included, as in Chaps. 4 and 5. The Biot number there, $\mathrm{Bi} = hs/k$, becomes $h_m s/D$ called the Sherwood number Sh. Here h_m is the surface, or convection coefficient, for mass diffusion.

6.3.2 Concentration-Dependent Diffusivity

Section 6.3.1 concerns plane layers having uniform and constant diffusivity. Results are given for several common kinds of time-dependent boundary conditions. This section concerns the effects of $D(C)$ in both sorption and desorption processes. Results are given for quite different forms of the variation of $D(C)$, with local concentration C, to indicate the kinds of effects which arise.

THE GENERAL FORMULATION. Equation (1.3.3) applies for isotropic diffusion. For one dimension and the distributed source C''' absent, the form, sometimes called the Boltzmann equation, is

$$\frac{\partial}{\partial x} D(C) \frac{\partial C}{\partial x} = \frac{\partial C}{\partial \tau} \tag{6.3.15}$$

For D constant and uniform, this relation has a solution in the form of the error function, of the dependent variable η. This variable combines x and τ, as $x/\sqrt{\tau}$. The heat conduction solution is Eq. (4.2.15), in terms of $\eta = x/2\sqrt{\alpha\tau}$. For $D = D(C)$, Eq. (6.3.15) becomes, in general,

$$\frac{d}{d\eta'}\left(D\frac{dC}{d\eta'}\right) = -2\eta'\frac{dC}{d\eta'} \tag{6.3.16}$$

where $\eta' = x/2\sqrt{\tau}$ has dimensions.

Equation (6.3.16) results in a closed-form solution when the boundary and initial conditions may be expressed solely in terms of η. Here this arises for the following semiinfinite region conditions on $C(x, \tau)$:

$$C(0, \tau) = C_1 \quad \text{and} \quad C(x, 0) = C_i \qquad (6.3.17)$$

The solution, in terms of $\eta = x/2\sqrt{D\tau}$, is then, as in the thermal transient,

$$\tilde{C} = \frac{C - C_i}{C_1 - C_i} = 1 - \text{erf } \eta = \text{erfc } \eta \qquad (6.3.18)$$

As seen in many analyses in Chap. 4, the simple differential equation which applies for k uniform and constant, Eq. (4.2.16a), is very useful in analyzing transients in infinite and semiinfinite regions. It has also been widely used for similar circumstances in mass diffusion. The more general form, Eq. (6.3.16), for D variable, has been used in approximate analyses of conditions in which $D = D(C)$. This is analogous to $k = k(t)$ in heat conduction, where the differential equations become nonlinear, for most reasonable variations $k(t)$. Nonlinear heat conduction processes are considered in Sec. 9.5, in a numerical calculation.

However, comparable mass diffusion nonlinearity is discussed here, for a range of processes and of forms of $D(C)$. All examples apply to infinite or semiinfinite regions. Transient concentration distributions and sorption and desorption curves are given. The first results which follow are for $D = D_0(C/C_0)$. Then results for polynomial and exponential dependencies of D are given. Sorption and desorption results are also given for both of these variations of D.

D VARIES AS SIMPLE FUNCTIONS OF C / C_0. A semiinfinite region is taken and the subscript 0 denotes the reference condition. The first result given for

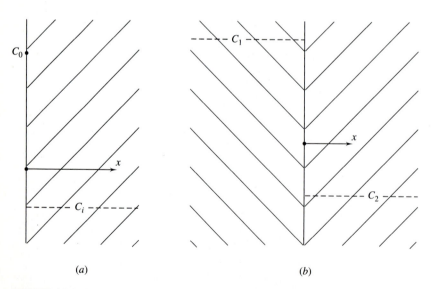

(a)   (b)

FIGURE 6.3.6
Diffusion regions and formulation of conditions, for $D = D(C)$.

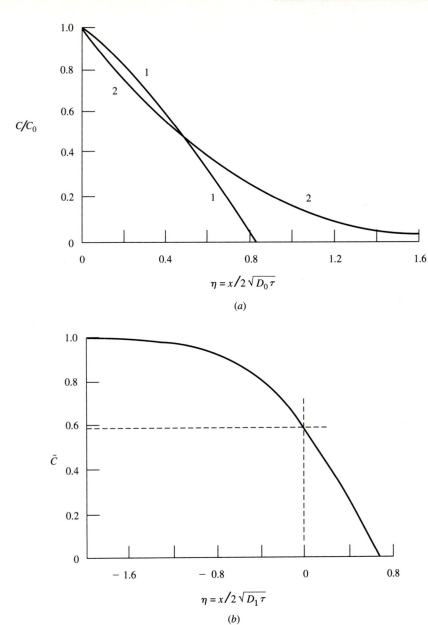

FIGURE 6.3.7

Concentration distributions C/C_0: (a) in a semiinfinite region for $D = D_0(C/C_0)$, where $C_0 = C(0, \tau)$; (b) in an infinite region where $C(x, 0) = C_1$ for $x < 0$ and $C(x, 0) = C_2$ for $x > 0$, $D = D_1(C/C_1)$. [From Crank (1975).]

$C(x, \tau)$ is for the initial concentration, C_i, in Fig. 6.3.6(a) zero.

$$C(x, 0) = 0 \quad \text{where} \quad C(0, \tau) = C_0$$

The resulting $C(x, \tau)$ distribution is shown in Fig. 6.3.7(a), where C/C_0 is plotted in terms of $\eta = x/2\sqrt{D_0\tau}$. The decay curve, 2, for comparison, is the error function solution given in Eq. (6.3.18). Curve 1 is the resulting distribution C/C_0, for $D = D_0(C/C_0)$. The form $\tilde{C}$ might be used instead. Curve 1 is everywhere concave downward because D decreases rapidly with $C(x, \tau)$. Whereas curve 2 goes asymptotically to zero, curve 1 goes directly to zero, at about $\eta = x/2\sqrt{D_0\tau} = 0.81$. This is where C and $D(C)$ become zero. Therefore, this location, or front, in terms of η, propagates to larger x with increasing time as $x_F = 1.62\sqrt{D_0\tau}$. The velocity of the front, v, decreases with time as $v = 0.81 D_0/\tau$.

The behavior shown in Fig. 6.3.7(b) in terms of $\tilde{C}$ is for the infinite region in Fig. 6.3.6(b) for $D = D_1(C/C_1)$. Note that the $\eta = 0$ location is the interface between the two regions. The process amounts to adsorption in the region $\eta > 0$, at the expense of the region $\eta < 0$. However, given the sharp rate of decrease of $D(C)$ with C, the desorption effect in the region $\eta < 0$ reaches only to about $\eta = -2$. The value of $\tilde{C}$ at $\eta = 0$ is about 0.59. This is the higher, due

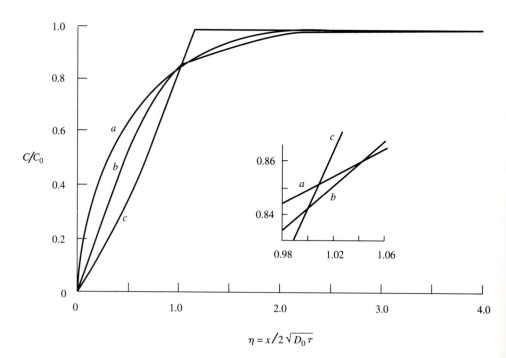

FIGURE 6.3.8

Desorption for $D(C)$ as: (a) $D_0(C/C_0)$; (b) D constant and uniform; (c) $D_0/[1 - (C/C_0)]$. [From Crank (1975).]

to $D(C) \propto C$, than for D uniform and constant. Recall the solution given for heat conduction in Eqs. (4.2.25a) and (4.2.25b), for two regions of equal properties suddenly placed in intimate contact. Then $\tilde{C}(0, \tau)$ would be 0.5. In Fig. 6.3.7(b) the diffusion effect is seen to reach only to about $\eta = 0.68$. This is because the value of D has decreased, reducing the mass flux generally.

The next example is for desorption, in the semiinfinite region in Fig. 6.3.6(a). Here $C(x, 0) = C_0$ and $C(0, \tau) = 0$. Results for three variations of $D(C)$ are compared in Fig. 6.3.8. Curve (b) is D constant and uniform and $C/C_0 = \text{erf } x/2\sqrt{D\tau}$. The other two curves lie above, for $D_0(C/C_0)$, and

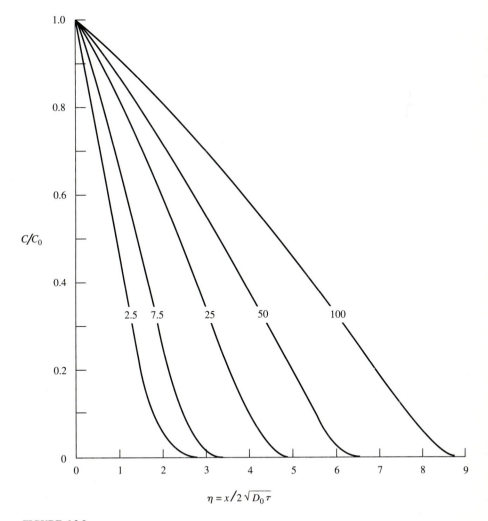

FIGURE 6.3.9
Concentration distributions for $D = D_0[1 + b(C/C_0)]$ for sorption, where the numbers on the curves are values of b. [From Crank (1975).]

below, for $D_0/[1 - (C/C_0)]$. For any given value D_0, the slope of the initially higher curve is greater than that of the lower. That is, D is less for $D(C) \propto C$, the direct proportionality, than for $D(C) \propto [1 - (C/C_0)]^{-1}$. Clearly, model (c) gives the highest desorption rate. The sharp knee in curve (c) is analogous to the behavior of curve 1 in Fig. 6.3.7(a).

The results of the model, for example, in Fig. 6.3.6(b), may be expressed more conveniently in terms of $\tilde{C}$. However, ratios of concentration levels were used here instead. Either form may be used in all examples, as long as C_0 and C_1 are properly interpreted in terms of boundary conditions. Several additional, numerically determined sorption and desorption curves are given later. Thereafter, some examples of sorption and desorption time curves are given, along with loss and uptake results.

OTHER SORPTION AND DESORPTION RESULTS FOR $D = D(C)$. Additional results are given in the following discussion for both linear and exponential forms of $D(C)$. For sorption and desorption the conditions are taken as $C(0, \tau) = C_0$ and $C(x, 0) = 0$ and $C(0, \tau) = 0$ and $C(x, 0) = C_0$, respectively. The instantaneous concentration level is written as $C(x, \tau)/C_0 = C/C_0$, for

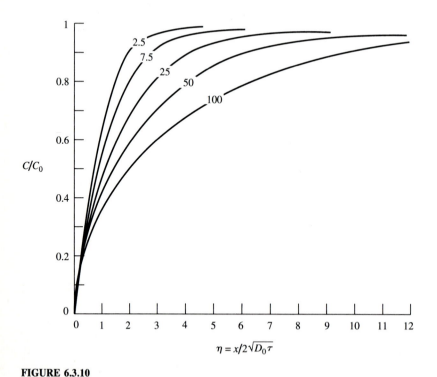

FIGURE 6.3.10
Concentration desorption distributions for $D = D_0[1 + b(C/C_0)]$ where the values of b are given on the curves. [From Crank (1975).]

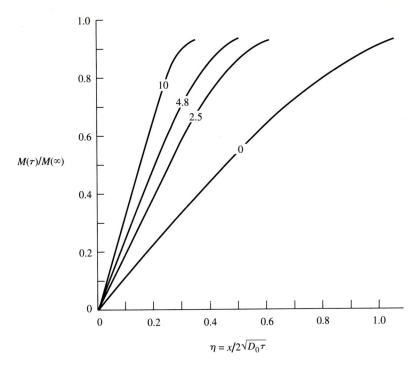

FIGURE 6.3.11
Sorption curves for $D(C) = D_0[1 + b(C/C_0)]$, for the values of b shown on the curves. [From Crank (1975).]

both sorption and desorption. Also given are sorption–time curves for both the linear and exponential variations of $D(C)$.

Figures 6.3.9 and 6.3.10 show sorption and desorption curves for $D(C) = D_0[1 + b(C/C_0)]$, with the formulation of conditions on $C(x, \tau)$ given before. The numbers on the curves are values of b. This form of $D(C)$ results in higher values of D for C nearer to C_0, increasingly so as the value of b increases. Therefore, all of the sorption distributions penetrate further out, in $x/2\sqrt{D_0\tau}$, than for D uniform and constant. This is seen in comparing erfc η, curve 2 in Fig. 6.3.7(a), with Fig. 6.3.9. The result in Fig. 6.3.10, for desorption, has the same property. However, smaller values of D now arise near the surface. This retards desorption, again thickening the active diffusion region.

The sorption and desorption curves for $D = D_0[1 + b(C/C_0)]$ are shown on Figs. 6.3.11 and 6.3.12, where the applicable values of $b = 0, 2.5, 4.8,$ and 10 appear on the curves. The fractional uptake parameter is $M(\tau)/M(\infty)$, where $M(\tau)$ is the amount of uptake by time τ and $M(\infty)$ is the value of $M(\tau)$ as $\tau \to \infty$. Desorption is the loss from $M(0)$ to $M(\tau)$. Increasing values of b enhance both sorption and desorption rates and levels at any given time. Note that high diffusivity, for any value of b, lies in the region of high local

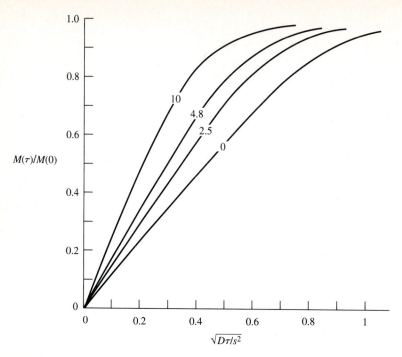

FIGURE 6.3.12
Desorption curves for $D(C) = D_0[1 + b(C/C_0)]$, for the values of b shown on the curves. [From Crank (1975).]

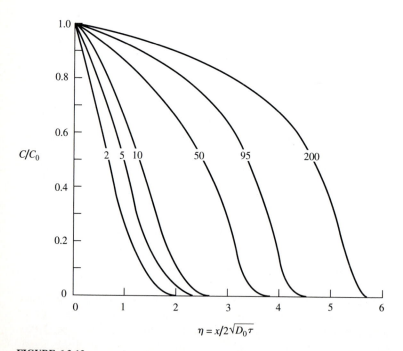

FIGURE 6.3.13
Sorption distributions for $D(C) = D_0 \exp K(C/C_0)$. The numbers shown are the value of e^k. [From Crank (1975).]

340

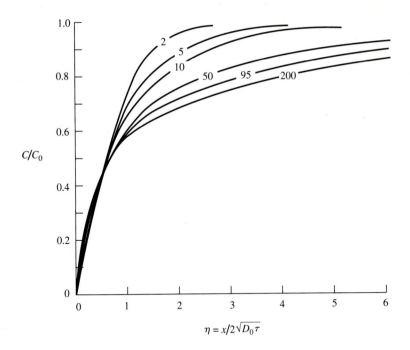

FIGURE 6.3.14
Desorption distributions for $D(C) = D_0 \exp k(C/C_0)$. The numbers shown are the values of e^k.
[From Crank (1975).]

concentration. This is near the region boundary for sorption and in the remote region for desorption.

The last example is for the exponential form, $D(C) = D_0 \exp k(C/C_0)$. With this form, the largest values of $D(C)$ occur where C is largest. For sorption, $C_0 = C(0, \tau)$ is the largest value of D. For desorption, $C_0 = C(\infty, \tau)$ is the largest value. Sorption and desorption are shown in Figs. 6.3.13 and 6.3.14, where the numbers on the curves are e^k. This is the ratio of D at C_0 to D at $C = 0$. For $e^k = 2$ and 200, k is about 0.69 and 5.3, respectively. The effect of k is very large in both processes. For sorption, large $D(C)$ occurs at small $\eta = x/2\sqrt{D_0\tau}$. For desorption, $D(C)$ is large at large η. The slopes of the curves in these regions show these large effects.

The sorption and desorption curves are shown in Fig. 6.3.15, in terms of $M(\tau)/M(\infty)$. The different behavior again shows the same effect of regions of large $D(C)$, as seen in the concentration distributions.

SUMMARY. The preceding results concern transients in several different geometries and for differing kinds of boundary conditions. Concentration-dependent mass diffusion coefficients $D(C)$ were considered. Large effects were shown to arise. Behavior was also demonstrated in terms of sorption– and desorption–time curves. Also see in Crank (1975) the additional kinds of $D(C)$

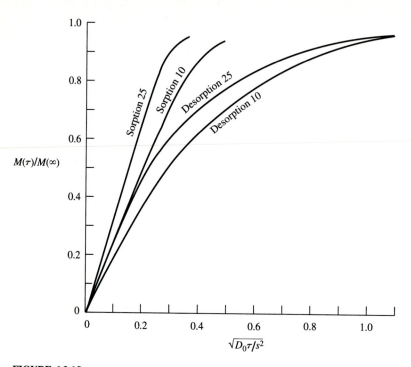

FIGURE 6.3.15
Sorption and desorption time curves for $D(C) = D_0 \exp k(C/C_0)$. The numbers on the curves are e^k. [From Crank (1975).]

variations considered by Shampine (1973). The next section considers evaporation and methods of measuring diffusivity.

6.3.3 Some Additional Mechanisms

Transients are commonplace in mass diffusion mechanisms of importance. The foregoing sections consider relatively simple processes, for D constant and uniform as well as dependent on concentration, as $D(C)$. Here, several other aspects are considered. The first is the surface convection or evaporation process, whereby drying or curing occurs. The second matter is the experimental determination of mass diffusivity for diffusing species in solid materials. Several basic methods are given.

SURFACE VAPOR LOSS OR GAIN. This is a very common occurrence in drying, humidifying, or curing a material. An example is circulating dry air over a damp material. Such a process is a transient. It may be in a thin or thick layer. There commonly is a convective resistance at any interface. This is the boundary condition C.2 in Sec. 1.4.2. The relation at the surface is then

$$h_m[C_e - C_s] = -D\left(\frac{\partial C}{\partial n}\right)_s = m''_s(\tau) \tag{6.3.19}$$

where h_m is the vapor mass transfer coefficient, C_e and C_s are the ambient and surface concentration levels, and n is the inward-normal space coordinate.

For this formulation, the solutions given in Secs. 4.2.2, 4.2.3, 4.3.3, and 4.4.1 apply, in plane, cylindrical, and spherical regions, for both transients and periodics. The result given in the following discussion is more directly related to typical concerns in mass diffusion.

Consider the process at the surface of a very thick solid region, initially at C_i throughout, suddenly exposed to an atmosphere of concentration C_e, where $C_s(\tau)$ at the interface and C_e are concentrations in the vapor phase. Energy effects are ignored. Also, thermal equilibrium is assumed at the interfaces. The solution to this formulation for heat conduction is Eq. (4.2.17), in terms of $\theta/\theta_e = (t - t_i)/(t_e - t_i)$. The rate of gain or loss of vapor from the region, at the surface, is

$$m''_s(\tau) = -D\left(\frac{\partial C}{\partial x}\right)_{x=0} = h_m[C(0, \tau) - C_e] \tag{6.3.20}$$

Equation (4.2.17), converted to $\tilde{C}$ and D, is used to calculate the gradient in Eq. (6.3.20). The mass loss per unit surface area, for $C_e < C_s$, is then calculated as the integral of $m''_s(\tau)$ from 0 to τ. The result is

$$\frac{h_m M(\tau)}{D(C_e - C_i)} = \left[e^{h_m^2\tau/D} \operatorname{erfc}\sqrt{\frac{h_m^2\tau}{D}} - 1 + \frac{1}{\sqrt{\pi}}\sqrt{\frac{h_m^2\tau}{D}}\right] \tag{6.3.21}$$

where the time variable is $h_m^2\tau/D$. Thus, for $C_e < C_i$, the diffusing material is lost from the medium to the ambient. For $C_e > C_i$, it is gained.

MEASUREMENT OF DIFFUSIVITY. Variable diffusivity effects were considered in Secs. 6.2 and 6.3.2 in steady state and in transient processes. Many diffusion processes are very complicated; often consisting of simultaneous parallel processes or resulting from diffusion traversing regions of different local property variations. Postulating the net diffusion characteristic is then more complicated and is often only the best approximation that may be made. Some of these effects are discussed in Sec. 6.6.

However, there are many processes which may be approximated in terms of a single value of D locally. Then D may be determined experimentally. Both steady and transient experiments are used. The solutions in Secs. 6.1 and 6.2 suggest many opportunities in steady state. The flux m'' is measured and D is

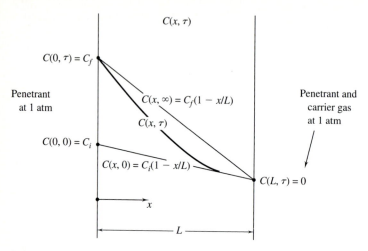

FIGURE 6.3.16
A thin layer or membrane, initially with steady-state diffusion, eventually at a new steady state.

inferred from an analytical solution. However, many transient methods have also been used. Three methods are given in the following discussion. The first two use plane region transients and the third uses a plane region periodic.

A transient, $C(x, \tau)$, is shown in Fig. 6.3.16. The diffusing substance is applied to the left face. A rapidly flowing carrier gas, at equal pressure on the other side, removes the penetrant emerging from the region, and holds the concentration $C(L, \tau)$ at zero. The transient mass flux $m''(\tau)$ emerging on that side is continually determined from the carrier gas concentration. The experiment is begun in steady state with the penetrant at $C(0, 0) = C_i$ on the left face. This steady-state distribution, $C(x, 0)$, is shown. At $\tau = 0$ the concentration at the left face is raised to C_f and held at that level, as $C(0, \tau) = C_f$. The eventual new steady state $C(x, \infty)$ is also shown.

The intervening transient, $C(x, \tau)$, using the Fickian formulation, may be determined by the method shown in Sec. 4.2.1. From such a solution, the instantaneous mass flux at the surface, at $x = L$, that is, $m''(L, \tau)$, may be determined from

$$m''(L, \tau) = -D(\partial C/\partial x)_{x=L} = M(\tau) \tag{6.3.22}$$

In terms of $D\tau/L^2$, this becomes

$$M(\tau) = \frac{m''(L, \tau)}{DC_i/L} = 1 + \frac{2}{\sqrt{\pi\,\mathrm{Fo}}}\left[\frac{C_f}{C_i} - 1\right]\sum_{n=0}^{\infty} e^{-[(2n+1)^2/4\mathrm{Fo}]} \tag{6.3.23}$$

The change in $M(\tau)$ between the two steady states is known in terms of D, as

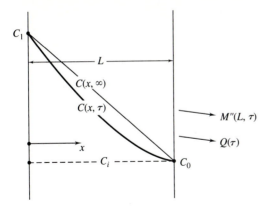

FIGURE 6.3.17
The transient "time-lag" method applied in a plane layer.

$M(\infty) - M(0) = D(C_f - C_i)L$. Also taking only the first term in Eq. (6.3.23):

$$M(\tau) - M(0) = [M(\infty) - M(0)]\frac{2}{\sqrt{\pi\, \text{Fo}}}e^{-(4\text{Fo})^{-1}}$$

$$(6.3.24)$$

or $$\frac{M(\tau) - M(0)}{M(\infty) - M(0)} = \overline{M}(\tau) = \frac{2}{\sqrt{\pi}}\frac{1}{\sqrt{\text{Fo}}}e^{-(4\text{Fo})^{-1}}$$

Experimental values of $\overline{M}(\tau)$ are then used to determine $\text{Fo} = D\tau/L^2$ and, therefore, D. See Pasternak et al. (1970).

Diffusion coefficients, permeability, solubility, and even variations of $D(C)$ may also be determined from a transient "time-lag" method. The example given here is for a layer L thick. The experiment is idealized in Fig. 6.3.17. The initial condition is C_i, which remains constant as $C_i = C_0$ at $x = L$. The method amounts to determining D from the measurement of the amount of the diffusing substance C which has diffused through the layer at time τ, after $C(0, \tau)$ was changed from C_i to C_1. This quantity is called $Q(\tau)$.

$$Q(\tau) = \int_0^\tau m''(L, \tau)\, d\tau = -D\int_0^\tau \left(\frac{\partial C}{\partial x}\right)_{x=L} d\tau \qquad (6.3.25)$$

This diffusion configuration also has a solution, $C(x, \tau)$, which may be as determined by the method in Sec. 4.2.1. The appropriate solution evaluates Eq. (6.3.25) as

$$\frac{Q(\tau)}{LC_1} = \text{Fo} - \frac{1}{6} - \frac{2}{\pi^2}\sum_{n=1}^{\infty}\frac{(-1)^n}{n^2}e^{-[n^2\pi^2\,\text{Fo}]} \qquad (6.3.26)$$

This is the curve in Fig. 6.3.18. As Fo becomes large the preceding relation

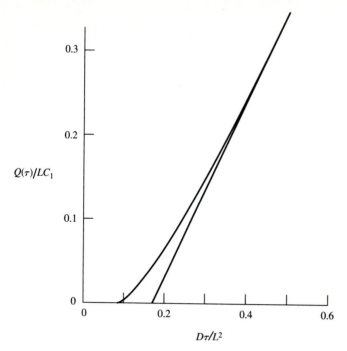

FIGURE 6.3.18
Transient "time-lag" response and the eventual response trend.

becomes the line on Fig. 6.3.18,

$$\frac{Q(\tau)}{LC_1} = \text{Fo} - \frac{1}{6} \tag{6.3.27}$$

This long-time form of the curve is extrapolated back to $Q(\tau) = 0$. The intercept, at time T, is

$$\text{Fo} = \frac{DT}{L^2} = \frac{1}{6} \quad \text{or} \quad D = \frac{L^2}{6T} \tag{6.3.28}$$

Thus T is the time delay, or lag, of this intercept from the time $\tau = 0$, when the process was begun.

This method is applied by measuring $m''(L, \tau)$ or $Q(\tau)$ directly, during an experiment. $Q(\tau)$ is then plotted against time τ. The asymptotic tendency is then extrapolated to $Q(\tau) = 0$, giving the value of Fo. Then D is calculated from Eq. (6.3.28). This kind of method has also been used, based on additional considerations, to determine permeabilities and solubilities, as well as $D(C)$. See Crank (1975).

Another method uses a steady-periodic process to determine gas permeability. A sample of the material is placed in a volume of the permeating gas initially at p_m. The initial steady-state concentration in the sample is then Kp_m, where K is the partition factor. The gas pressure p is then varied sinusoidally at frequency $f = 2\pi\omega$, around p_m. After a sufficiently long time the process is a steady periodic. The surface concentration on each side varies as

$$C(\pm s, \tau) = Kp(\tau) = K[\, p_m + \Delta p_{m,a} \sin \omega\tau + \epsilon\,] \qquad (6.3.29)$$

where $\Delta p_{m,a}$ is the amplitude and ϵ is the phase lag. Recall the conduction analysis in Sec. 4.4.2. Using the periodic solution, as in Eq. (4.4.11), the amplitude of the sample weight change and the phase lag with respect to the periodic input p determine both the diffusion coefficient and the solubility.

6.4 INTERFACES AND MOVING BOUNDARIES

These kinds of mechanisms occur very commonly in mass diffusion, in many different kinds of regions. They are caused by many different physical effects. The circumstances considered are those in which an appreciable process, affecting the local concentration, occurs over a region of relatively limited physical extent. Examples in heat conduction are freezing and melting fronts and also moving sources, as in Secs. 4.6 and 4.7. Mass transfer also occurs when a freezing front advances into water containing dissolved salt. Since salt is very insoluble in ice, most of it is rejected back into the water as the ice interface advances. When idealized, this phase interface of discontinuous density change is also a location of a discontinuity of salt concentration. There is also a discontinuous change in diffusivity there, since $D \neq 0$ in the water and $D \approx 0$ in the ice.

In mass diffusion, many additional similar processes arise in which concentration or diffusivity distributions are nominally discontinuous. Many common causes are concentrated chemical and other combining reactions, which locally consume or immobilize the diffusing species. An example is the local formation of a stable chemical compound. Another is a local precipitation reaction. If the local immobilization reaction rate is very fast, compared to the concentration diffusion rate, the idealization of the processes amounts to a moving front. Behind the front, the diffusion of unused species C continues. Ahead of the front, $C = 0$, and there is no diffusion.

The idealized model is shown in Fig. 6.4.1(a), where $C(x, \tau)$ applies for $x \leq X(\tau)$ and $C = 0$ for $x > X(\tau)$. The process amounts to a plane sink of strength m'', moving at a decreasing velocity $u(\tau)$. In effect, $D \neq 0$ applies for $x \leq X(\tau)$ and $D = 0$ for $x > X(\tau)$. This process is completely analogous to the ice-melting-front mechanism shown in Fig. 4.6.2(a), where t and k replace C and D. However, the concept of $k = 0$ for $x > X(\tau)$ is commonly not a reasonable approximation in heat conduction analysis.

Many other processes are found in which such a sudden change in D also occurs in the diffusion region. Common examples are liquid diffusion in porous media. Macroscopic local diffusion processes then frequently arise which, al-

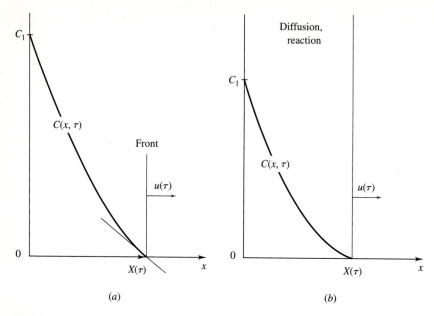

FIGURE 6.4.1
Regions with spatially concentrated mass diffusion effects: (*a*) a localized immobilization process front; (*b*) a distributed process.

though still treated by Fick's first law, require some threshold concentration level for diffusion. Both these, and the immobilization reactions, are often analyzed with the postulation of discontinuities in D, over the region.

However, the diffusing concentration front is often not so sharply defined. This may arise when the immobilization rate is much less than or comparable to the diffusion rate. Then the front broadens into a region.

An example of this is the oxidation of a metal in air. The thickness of the surface oxide layer increases with time as oxygen diffuses inward to react, over the surface region, with the metal locally. This is often a slow process. It is idealized in Fig. 6.4.1(*b*),.where the instantaneous oxide layer thickness is $X(\tau)$. The reaction occurs throughout the region $x = 0$ to $X(\tau)$ and often is a quasi-static process.

Combustion waves, either in diffusion flames or in rapidly moving detonation waves, are similar. They may be idealized as fronts, to determine several overall properties. However, analysis of their internal structure must be in terms of a finite layer, like that in Fig. 6.4.1(*b*).

This section considers both kinds of analysis, each for several specific mechanisms in which they are the most suitable formulation. Section 6.4.1 considers discontinuous diffusion coefficients. These analyses then amount to a moving sharp-boundary transient. Section 6.4.2 analyzes a group of particular applications which involve other aspects of adsorption and moving-boundary effects.

6.4.1 Concentrated Effects as Moving Boundaries

A discontinuity of the diffusion coefficient arises as a diffusing substance passes from one region to another of different properties. This occurs in composite barriers, as shown in Fig. 2.1.2 for steady-state heat conduction. Comparable transients were considered in Sec. 4.5. Those analyses were for constant and uniform thermal properties k_j and α_j, in each of the layers of the composite barrier. These results also apply directly to mass diffusion through composite barriers when the D_j are also constant and uniform.

The transient analyses given here apply to a different kind of physical effect. It arises when the diffusion of a particular substance in a given material may have different values of D, in different ranges of the local concentration level C. For example, at high concentration C, in region A, $D = D_A$. At concentrations below C_x, in region B, $D = D_B$. That is, two different diffusion mechanisms, D_A and D_B, may operate in the two regions where $C \geq C_X$ and $C_X < C$, respectively. This circumstance is sketched in Fig. 6.4.2, assuming both a particular location $X(\tau)$, at which C_X occurs across the field of diffusion, at time τ, and $D_A > D_B$. Also shown is a transient distribution of $C(x,\tau)$. The levels of D_A and D_B are also shown in the regions A and B in which they each apply, at the time level for the distribution of $C(x,\tau)$ shown.

Since $\partial^2 C/\partial x^2$ is everywhere positive in this particular example, the concentration $C(x,\tau)$ is increasing at all locations across the region. Therefore, C_X occurs further out, at larger x, at later times. That is, $X(\tau)$ increases with

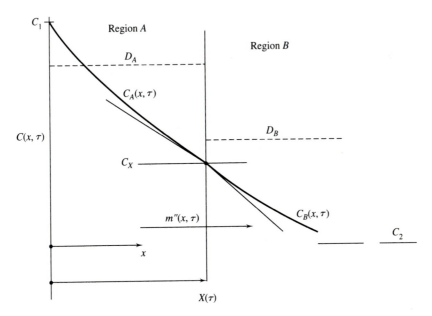

FIGURE 6.4.2
A discontinuity of D, at the concentration level C_x, in a transient in a semiinfinite region.

time, from $X(0) = 0$. Thus the region in which D_A applies increases in extent and this is a process with a moving boundary determined by C_X.

This circumstance is considered in the following discussion. Then the results for the simpler circumstance of $D_A \neq 0$ and $D_B = 0$ are given. Thereafter, the effects of the three different levels, D_A, D_B, and D_C, are briefly discussed. The last matter considered is the inclusion of an immobilizing reaction at the location where C_X occurs.

A DISCONTINUOUS CHANGE OF D FROM D_A TO D_B. Referring again to Fig. 6.4.2, the boundary conditions, in terms of $C_A = C_A(x, \tau)$ and $C_B = C_B(x, \tau)$, are

$$C_A(0, \tau) = C_1 \qquad C_B(x, 0) = C_2 \tag{6.4.1}$$

$$C_A(X, \tau) = C_B(X, \tau) \tag{6.4.2}$$

$$m''(x, \tau) = -D_A \frac{\partial C_A}{\partial x} = -D_B \frac{\partial C_B}{\partial x} \quad \text{at } x = X(\tau) \tag{6.4.3}$$

This last condition matches the mass flux at $x = X(\tau)$, in the two regions. The preceding formulation is similar to the ice melting and freezing processes diagrammed in Fig. 4.6.1. However, it is different in that Eq. (6.4.2) does not have the latent heat term, $\rho u H$, in Eq. (4.6.1b). The presence of this additional effect resulted in the transcendental relation Eq. (4.6.23), for the parameter λ as a function of $N_1 = C_S(t_{PE} - t_0)/\sqrt{\pi} H$. The equations for determining C_A and C_B in the two regions are

$$D_A \frac{\partial^2 C_A}{\partial x^2} = \frac{\partial C_A}{\partial \tau} \quad \text{and} \quad D_B \frac{\partial^2 C_B}{\partial x^2} = \frac{\partial C_B}{\partial \tau} \tag{6.4.4}$$

The solutions for C_A and C_B, in regions A and B, respectively, are written as

$$C_A(x, \tau) = C_1 + B_A \operatorname{erf}\left(x/2\sqrt{D_A \tau}\right) = C_1 + B_A \operatorname{erf} \eta_A \tag{6.4.5}$$

and

$$C_B(x, \tau) = C_2 + B_B \operatorname{erfc}\left(x/2\sqrt{D_B \tau}\right) = C_2 + B_B \operatorname{erfc} \eta_B \tag{6.4.6}$$

where $\eta_A = x/2\sqrt{D_A \tau}$ and $\eta_B = x/2\sqrt{D_B \tau}$. The interface condition, Eq. (6.4.2), gives

$$B_A \operatorname{erf} \frac{X}{2\sqrt{D_A \tau}} = C_X - C_1 = B_A \operatorname{erf} \frac{K'}{2\sqrt{D_A}} \tag{6.4.7}$$

$$B_B \operatorname{erfc} \frac{X}{2\sqrt{D_B \tau}} = C_X - C_2 = B_B \operatorname{erfc} \frac{K'}{2\sqrt{D_B}} \tag{6.4.8}$$

when $X(\tau)$ is taken as $K'\sqrt{\tau}$. The value K' remains to be determined.

The preceding equations, in conjunction with the differential equations, yield the following relationship for the unknown parameter K':

$$\frac{C_X - C_1}{F_A(K_A)} + \frac{C_X - C_2}{F_B(K_B)} = 0 \qquad (6.4.9)$$

where F_A and F_B are the following functions of $K_A = K'/2\sqrt{D_A}$ and $K_B = K'/2\sqrt{D_B}$, respectively:

$$F_A(K_A) = \sqrt{\pi}\, K_A e^{K_A^2}\, \text{erf}\, K_A \qquad (6.4.10)$$

$$F_B(K_B) = \sqrt{\pi}\, K_B e^{K_B^2}\, \text{erfc}\, K_B \qquad (6.4.11)$$

Equations (6.4.9)–(6.4.11) are solved numerically for K_A and K_B. Then K' is known and B_A and B_B may be evaluated from Eqs. (6.4.7) and (6.4.8) and the distributions of $C_A(x, \tau)$ and $C_B(x, \tau)$ in Eqs. (6.4.5) and (6.4.6) are known. Crank (1975), Sec. 13.2, gives figures of $F_A(K_A)$ and $F_B(K_B)$ which facilitate the determination of K_A, K_B, and, therefore, K'.

This procedure is very similar to that of determining λ for the moving phase change fronts which result in latent heat absorption or release. Here, the preceding solution also applies for a layer of thickness L, initially at C_2, for short times when $C_B(L, \tau)$ is very small compared to $C_1 - C_2$.

A ZERO DIFFUSION CONSTANT AT LOW CONCENTRATION. This is the same formulation given previously, except that $D_B = 0$ beyond the location $X(\tau)$ where $C_A(x, \tau) = C_X$. Then C_2 remains zero and the concentration distribution is given in terms of $C_A(x, \tau)/C_1$ over the range $x = 0$ to $x = X(\tau)$, that is, in region A. The physical result is that the diffusion region A extends to the location where $C_A(X, \tau)/C_1 = C_X/C_1$. The initial gradient at $x = 0$ is infinite. As the concentration there increases to C_X, the growth of the layer, for which $C(x, \tau) > C_X$, begins. Then the diffusion, $m''(X, \tau)$, arriving at $X(\tau)$, continually extends region A. Applying the condition $K_B \to \infty$, as $D_2 \to 0$, in Eq. (6.4.11), Eq. (6.4.9) in terms of $F_A(K_A)$ in Eq. (6.4.11) becomes

$$\frac{C_X - C_1}{F_A(K_A)} + C_X = 0 \quad \text{or} \quad F_A(K_A) = \left(\frac{C_1}{C_X} - 1\right) \qquad (6.4.12)$$

Therefore, $F_A(K_A)$ and $K' = 2K_A\sqrt{D_A}$ are known directly from the imposed condition C_X/C_1. The solution for $C_A(x, \tau)$ is then determined as before.

Distributions are given in Fig. 6.4.3 for particular values $C_X/C_1 = 5/6$ to $1/3$. Recall that $\eta_A = x/2\sqrt{D_A \tau}$ applies at all times. The curve for $5/6$ is essentially linear, for the relatively thin region which results, in terms of η_A. All curves go abruptly to zero at the intercept, called η_i.

These results again apply in a layer of thickness L, from $\tau = 0$ until the time $\tau = L^2/4\eta_i^2 D_A$. Thereafter, the later distribution $C(x, \tau)$ may be obtained using the relevant result given previously as an initial distribution for the

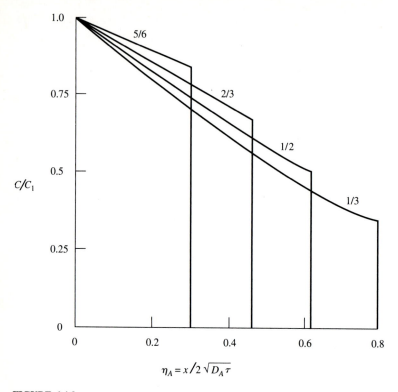

FIGURE 6.4.3
Concentration distributions C/C_1, in terms of η_A, for four values of the diffusion cutoff condition C_2/C_1. [From Crank (1975).]

continuing transient, when $D = D_A$, over the whole region. The sorption curves for the time $\tau = L^2/4\eta_i^2 D_A$, and thereafter, is given by Crank (1975).

THREE LEVELS OF DIFFUSIVITY D_A, D_B, AND D_C. This arises when three different mechanisms of diffusion arise over the concentration range C_1 to C_2 imposed across a region. There are then two dividing concentration levels, $C_{X, AB}$ and $C_{X, BC}$, and three distinct concentration distributions, $C_A(x, \tau)$, $C_B(x, \tau)$, and $C_C(x, \tau)$, in general. The two boundaries, should they both arise, are $X_{AB}(\tau)$ and $X_{BC}(\tau)$. Solutions are again constructed of erf η and erfc η as before; in each region, two parameters K'_{AB} and K'_{BC} arise. The relations which relate the resulting K_{AB} and K_{BC} to $C_{X, AB}$ and $C_{X, BC}$ are given by Crank (1975). The procedure is similar thereafter.

AN IMMOBILIZING REACTION. Crank (1975) shows that the analysis and the resulting local concentration distributions in Fig. 6.4.3, for $D_1 \neq 0$ and $D_2 = 0$ for $C < C_X$, may be reinterpreted in terms of such a local reaction. For this, C

is taken as the sum of the local diffusing and immobilized concentrations. The value of $m''(X, \tau)$ is related to increasing $X(\tau)$ as

$$m''(X, \tau) = -D\frac{\partial C}{\partial x} = C_X \frac{dX}{d\tau} \qquad (6.4.13)$$

If this relation is used in place of the first equation in Eq. (6.4.4), for $D_2 = 0$, the result is again Eq. (6.4.12). Therefore, immobilization may be reinterpreted as a pure diffusion process in which $D = 0$ for C over the range $0 \le C < C_X$ and D changes abruptly to greater than zero at $C = C_X$.

SUMMARY. The preceding results indicate the effects of concentrated changes of D across diffusion regions. These occur when different physical diffusion mechanisms arise at different concentration levels, for a given substance over the diffusion region. For example, a mechanism at high concentration D_1 may be different from that elsewhere in the region, at low concentration levels. In steady-state processes, this effect may be treated as a composite of the two regions, at D_1 and D_2. However, in transients, the interface between the regions where D_1 and D_2 apply, moves. If it is assumed that the change of D is abrupt, a moving-front mechanism results, as analyzed previously. Many actual similar processes are more complicated. The following section considers a group of these, some of which are not simply modeled as a concentrated moving front.

6.4.2   Other Heterogeneous Effects

Several commonly occurring surface effect mechanisms are considered here. The first is the determination of the thickness growth rate of an oxidized surface layer on metal. A simple quasi-static model is used. Then a transient analysis is given for a surface skin or coating on an extensive region, which has a much higher diffusivity than the internal region. Transient sorption results are given. Zone refining is then analyzed. Such processes arise when the substance C has higher solubility in the liquid phase. As a result, substance C is expelled from an advancing solidification front, ahead into the liquid.

SURFACE LAYER OXIDATION. Oxide layers grow into metals and other materials. The growing surface layer formed consists of a chemical combination, MO, of oxygen and a metal, for example. The effect seen in Fig. 6.4.4 arises from the diffusion of cations outward, and oxygen anions inward. If the solubility of oxygen in the pure metal is taken as zero, the concentration in the region $X(\tau) > 0$ remains zero and diffusion occurs only in the growing surface oxide layer.

   If the surface layer process is assumed to be very slow, a linear distribution of MO may be assumed, across the oxide layer of thickness $X(\tau)$. Therefore,

$$\frac{dX}{d\tau} = V_{MO}D\frac{\Delta C}{X} \quad \text{and} \quad X(0) = 0 \qquad (6.4.14)$$

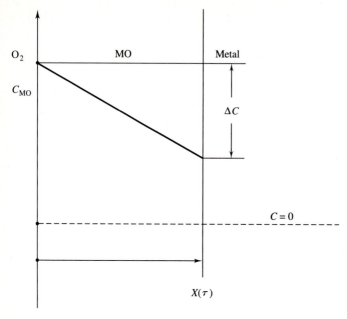

FIGURE 6.4.4
A growing oxidized surface layer on an extensive region.

where V_{MO} is the volume of oxide formed per unit of oxygen and metal combined, per unit of surface area. The result is

$$X^2(\tau) = 2V_{MO} D\tau \, \Delta C = k_p \tau \qquad (6.4.15)$$

where k_p is termed the parabolic rate constant. This equation may be also written in terms of η as

$$\frac{X^2(\tau)}{4D\tau} = \eta^2 = \frac{V_{MO} \, \Delta C}{2} \qquad (6.4.16)$$

Many oxidation processes involve much more complicated mechanisms than assumed in the simple model given previously. An analysis, in terms of the detailed components of migration, is reviewed by Kirkaldy and Young (1987).

SURFACE SKINS. These are surface layers on regions which have different diffusion properties than the bulk of the region. This layer often has very different properties than the bulk of the material due to the way the material was formed and in its physical response to the inward diffusion of particular materials.

This mechanism is similar to that of heat conduction into a region with a surface layer. An example is seen in Fig. 4.5.2, with the solution given in Eq. (4.5.5). Another example is in Fig. 4.5.3, with the solution in Eq. (4.5.7).

An example is shown in Fig. 6.4.5 where the region, initially at $C_i = 0$ throughout, suddenly has the concentration at $x = 0$ raised to C_1. A uniform

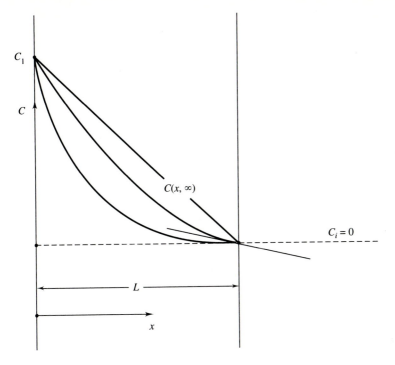

FIGURE 6.4.5
A resistive skin on the surface of an extensive region.

and constant value D is used in the surface layer, $x = 0$ to L. In the region beyond, D_∞ is taken as much larger.

The beginning of this transient is again the semiinfinite region solution $\tilde{C} = \text{erfc } \eta$. Later, the concentration effect penetrates to $x = L$. However, the concentration there remains at zero since $D_\infty \gg D$. Thereafter, the distribution in $x = 0$ to L approaches the steady state shown as $C(x, \infty)$. The total sorption at time τ, in terms of $D\tau/L^2 = \text{Fo}$, is

$$\frac{M(\tau)}{2LC_1} = \sqrt{\text{Fo}} \left[\frac{1}{\sqrt{\pi}} + 2 \sum_{n=1}^{\infty} i \, \text{erfc}(n/\sqrt{\text{Fo}}) \right] \tag{6.4.17}$$

This is linear in terms of $\sqrt{\text{Fo}}$ at short times. However, it does not reach a limit as $\tau \to \infty$, because of the infinite capacity of the region $x > L$. This circumstance is a special example of a discontinuous adsorption coefficient, considered in more detail in Sec. 6.4.1.

SOLIDIFICATION OF A LIQUID. The rate of such a process is governed by the transient rate of the heat removal process which causes the solidification. These effects are considered in Sec. 4.6. A principal result there is the propagation rate of the phase interface, at $x = X(\tau)$ at time τ. The interface velocity is

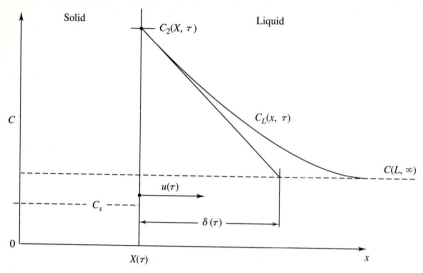

FIGURE 6.4.6
Solidification of a liquid into a crystalline solid, with resulting segregation.

$u(\tau) = \partial[X(\tau)]/\partial\tau$. Those results showed that $X(\tau) \propto \sqrt{\tau}$ and that $u(\tau) \propto 1/\sqrt{\tau}$, commonly.

The solidification rate sets the time scale in which the resulting mass diffusion effects occur. These effects result from the differing solubility of some component in the two phases. Solubility is commonly less in crystalline solids than in the liquid phase, as discussed in Sec. 4.6. In a phase front, advancing at a rate $u(\tau)$, this results in a partition ratio at the interface. That is, the concentration in the solid being formed is less than in the liquid, at the interface. The ratio is defined, for equilibrium partition, as

$$\frac{C_S}{C_L(X,\tau)} = K < 1 \tag{6.4.18}$$

An example is seen in Fig. 6.4.6, in a similar formulation to that given in Fig. 6.4.7, for a periodic zone-refining process. The value of K for the conditions in Fig. 6.4.6 is 0.2.

The following approximate analysis is given in Kirkaldy and Young (1987), including a possible varying convection effect at the phase interface. The diffusing flux in the liquid, at the interface, is determined by the tangent shown in Fig. 6.4.6, as $D_L[C_L(X,\tau) - C_{L,\infty}]/\delta(\tau)$. This is equal to the loss of C in the volume rate of solid production, u, per unit area. This rate is $u(\tau)[C_L(x,\tau) - C_S]$. The result is

$$u(\tau)[C_L(X,\tau) - C_S] = D_L[C_L(X,\tau) - C_{L,\infty}]/\delta \tag{6.4.19}$$

or

$$\delta(\tau) = \frac{D}{u(\tau)} \frac{1 - K(C_{L,\infty}/C_S)}{1 - K} \tag{6.4.20}$$

This kind of formulation is used for a transient beginning, for example, when a cold mold is quickly filled with a melt. For a tall mold, the solidification process may often be approximated as planar. Convection is here assumed absent. The initial conditions are $C_L(x, 0) = C_{L,\infty}$, $C_S = KC_L(X, \tau)$, and the position of the phase interface is estimated as $X(\tau) = \sqrt{\alpha\tau}$, where α is the thermal diffusivity. The interface condition is

$$u(\tau)[C_L(X, \tau) - C_S] = u(\tau)C_L(X, \tau)(1 - K) = -D\left(\frac{\partial C_L}{\partial x}\right)_X \quad (6.4.21)$$

The resulting instantaneous distribution in the liquid for $x \geq X$ is

$$\frac{C_L(x, \tau)}{C_{L,\infty}} = 1 + \frac{(1 - K)\text{erfc }\eta_D}{2\sqrt{D\pi/\alpha}\, e^{-\alpha/4D} - (1 - K)\text{erfc }\eta_D} \quad (6.4.22)$$

where $\eta_D = x/2\sqrt{D\tau}$. The level of segregation in the liquid phase is

$$\frac{1}{K}\frac{C_L(X, \tau)}{C_{L,\infty}} = \left[1 - \sqrt{\frac{\pi\alpha}{4D}}\frac{\text{erfc}\sqrt{\alpha/4D}}{e^{-\alpha/4D}}(1 - K)\right]^{-1} \quad (6.4.23)$$

or, for $\alpha \gg D$,

$$\frac{C_L(X, \tau)}{C_{L,\infty}} \approx \frac{1}{1 + 2(1 - K)D/\alpha K} \quad (6.4.24)$$

This last result indicates more clearly the magnitude of the effects of partition, and of α and D, on the level of segregation ahead of the advancing solidification front.

ZONE REFINING. An important use of this method arises in preparing very pure semiconductor material for compact integrated circuitry. The method amounts to placing a bar of impure material, for example silicon, in a "boat," which holds it when molten. The refining is accomplished by a propagating induction heating melting front, which moves back and forth through the material. Since the solubility of the impurities in the solid phase is less, they are being excluded by the following freezing interface, advancing into the melt.

The process is diagrammed in Fig. 6.4.7(a). The compact induction heater moves back and forth, over the range L, leaving liquid behind. This is solidified at time τ at a particular following moving location $\xi = 0$, in moving coordinates. In effect, the impurities are pushed ahead in the liquid, toward the right end, in Fig. 6.4.7(b). When the motion is reversed, the process is repeated, toward the left. Thereby, the impurities are concentrated in the material at the two ends.

This process is an example of the moving-source formulation in Sec. 4.7. In one dimension, Eq. (4.7.4) applies. However, there are lateral heat losses along the whole length, to produce solidification. This is suggested by the moving temperature distribution $t(\xi)$ in Fig. 6.4.7(b).

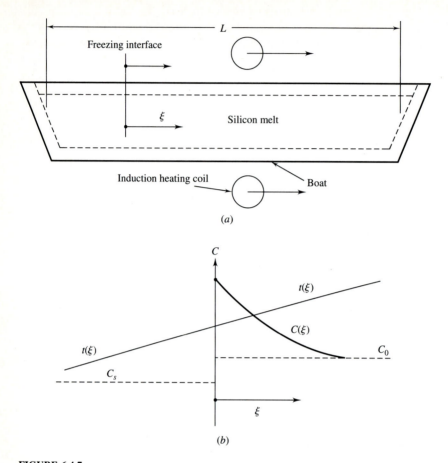

FIGURE 6.4.7
Zone refining of a solid material by induction heating: (*a*) the heating arrangement; (*b*) exclusion of the impurity back into the liquid.

An approximate steady periodic eventually arises. The impurity concentration distribution in the liquid ahead of the advancing front is

$$(C - C_0)/C_0 = [(1 - K)/K]e^{-(v/D)\xi} \qquad (6.4.25)$$

where K is a constant partition factor of the impurity between the liquid and the solid phases. It indicates the efficiency of the separation process across the advancing interface. For $K = 1$, $(C - C_0)/C_0 = 0$, and for $K = 0$, $(C - C_0) \to 0$, at $\xi = 0$. Further considerations are found in Chalmers (1964).

SUMMARY. These examples concern several different kinds of moving concentrated mass diffusion regions. The processes amount to a region over which the concentrated effect occurs. They are therefore different than effects idealized as occurring instantaneously at a plane front. Many of the mechanisms which arise in practice may be idealized by one of these methods.

6.5 SOME EFFECTS OF CHEMICAL REACTIONS

The preceding considerations in this chapter have concerned only diffusion and sorption mechanisms. Steady and transient concentration distributions have been determined, for various forms of the diffusion coefficient D and with moving boundaries. With few exceptions, the diffusing substance was assumed to be mobile, conserved, and entirely controlled by local concentration gradients. These were simple conservative processes. However, there are many processes in which the local adsorption in a region may both diffuse and also react chemically. Also, many common reactions are irreversible. That is, the adsorption amounts to irreversible binding of substance C into the local material.

This section considers several additional different kinds of processes which occur as distributed local adsorption at sites or on internal surfaces distributed over the region of diffusion. That is, some of the diffusing material is adsorbed by the material of the region. Thereby, diffusing material disappears into the structure of the region. The local level of immobilization often depends upon the instantaneous local concentration, $C(x, \tau)$, of the diffusing material. See Sec. 1.3.3, which relates such effects to the distributed species source, defined as C'''.

6.5.1 Reaction Mechanisms

The actual behavior depends upon the chemical rate of the adsorption or capture mechanism. A very simple idealization is that the adsorption reaction is very fast, compared to the local diffusing mass flux rate $m''(x, \tau)$, at the instantaneous local concentration level $C(x, \tau)$. Then the local level of adsorption $c(x, \tau)$, per unit volume, depends only on the total concentration $C(x, \tau)$. If the dependence of c on C is linear, then

$$c(x, \tau) = R_r C(x, \tau) \tag{6.5.1}$$

where R_r is the proportionality constant. However, this relation provides for both adsorption and release, as $C(x, \tau)$ increases or decreases. The relation between $c(x, \tau)$ and $C'''(x, \tau)$, the local diffusing substance source rate, per unit volume and time, is

$$C''' = -\frac{\partial c}{\partial \tau} = -R_r \frac{\partial C}{\partial \tau} \tag{6.5.2}$$

or

$$D \nabla^2 C - R_r \frac{\partial C}{\partial \tau} = \frac{\partial C}{\partial \tau} \tag{6.5.3}$$

This is rearranged as

$$\left(\frac{D}{1 + R_r}\right) \nabla^2 C = D' \nabla^2 C = \frac{\partial C}{\partial \tau} \tag{6.5.4}$$

Thus fast capture, with a linear dependence on local concentration, is also subject to Eq. (1.3.6), with D replaced by $D' = D/(1 + R_r)$.

However, many adsorption processes are not this simple. If the proper relation is nonlinear as

$$c = R_r C^n \quad \text{or} \quad \frac{\partial c}{\partial \tau} = nR_r C^{n-1} \frac{\partial C}{\partial \tau} \tag{6.5.5}$$

The resulting relation is then nonlinear in C as

$$D \nabla^2 C - nR_r C^{n-1} \frac{\partial C}{\partial \tau} = \frac{\partial C}{\partial \tau} \tag{6.5.6}$$

However, if R_r is large, the transient term on the right may be neglected and Eq. (6.5.3) is then written in one dimension as

$$D \frac{\partial^2 C}{\partial x^2} = R_r \frac{\partial C}{\partial \tau} = \frac{\partial c}{\partial \tau} \tag{6.5.7}$$

This may be transformed to

$$\frac{\partial}{\partial x} \left[\frac{D}{n} \left(\frac{1}{R_r} \right)^{1/n} c^{(1-n)/n} \frac{\partial c}{\partial x} \right] = \frac{\partial c}{\partial \tau} \tag{6.5.8a}$$

Thus $c(x, \tau)$ is the solution of a diffusion process having the following equivalent variable diffusion coefficient D_e:

$$D_e = \frac{D}{n} \left(\frac{1}{R_r} \right)^{1/n} c^{(1-n)/n} \tag{6.5.8b}$$

This is the same kind of $D(C)$ dependence as analyzed in Sec. 6.3.2. The change here is that D depends instead on $c(x, \tau)$.

The nonlinear adsorption effect is seen in Fig. 6.5.1, for a long solid cylinder of radius R. The curves, for $n = 0$, $1/4$, $1/2$, and 1, are of total adsorption $M(\tau)$ as a function of time versus $D\tau/R^2$, divided by $M(\infty)$. For $n = 0$, the adsorption level is $c = R_r$, a constant. Since C''' does not increase with $C(x, \tau)$ the response is slow. For $n = 1$, Eq. (6.5.1) applies, with a much faster response. The other two curves are a weaker dependence of c on C, with intermediate results. All curves approach $M(\infty)$ slowly. This occurs because the surface flux, $m''(0, \tau)$, is taken by adsorption, leaving less for onward diffusion.

6.5.2 Irreversible Reactions

The preceding formulation, $c = R_r C^n$, relates the instantaneous local level of adsorption $c(x, \tau)$ directly to the instantaneous local concentration $C(x, \tau)$. Thereby, there is a take-up and release, as $C(x, \tau)$ increases and decreases. In a completely irreversible reaction, the adsorbed substance is locally consumed or locked into the local structure by other processes. Examples are the hydration of concrete as it cures or the carburization of steel at high temperatures.

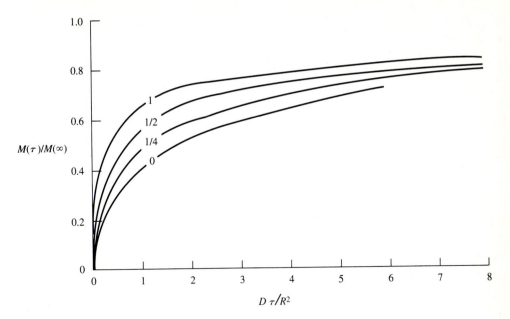

FIGURE 6.5.1
Adsorption increase with time, in a long cylinder, for $n = 0, 1/4, 1/2$, and 1.

LINEAR ADSORPTION. If the adsorption rate depends linearly on local concentration, the rate relation is $C''' = -KC(x, \tau)$. The minus sign indicates a local loss of the diffusing species C. The equation is then

$$D \nabla^2 C - KC = \frac{\partial C}{\partial \tau} \tag{6.5.9}$$

where K here is a constant of proportionality. The diffusion gain minus the adsorption rate, locally, is equal to the rate of change of local concentration.

Equation (6.5.9) is similar to the differential equations used in Chaps. 3 and 4 for heat conduction with internal generation and surface convection. For example, the term $-KC$ in Eq. (6.5.9) is analogous to the distributed source term effect q''' analyzed in Sec. 3.6 in steady-state heat conduction. The surface convection formulation Eq. (3.6.11) is the same as Eq. (6.5.9), applied to steady-state diffusion. A solution is given in Sec. 3.6.2 for a two-dimensional region. One-dimensional transients, including the q''' effect, are given in Sec. 4.2.4. However, q''' is taken as uniform or merely a function of location. However, Carslaw and Jaeger (1959), Sec. 4.2, give a transient solution for $t(z, \tau)$ in a long and thin rod, with convective loss from the surface, as $h[t(z, \tau) - t_e]$.

A general method of transforming Eq. (6.5.9) for transients, to simpler forms for solutions, was given by Danckwerts (1951). The following two conditions at the surface are considered, along with the initial condition $C(x,0) = 0$:

$$C(0,\tau) = C_0 \tag{6.5.10}$$

and
$$-D\frac{\partial C}{\partial x} = h[C(0,\tau) - C_e] \tag{6.5.11}$$

where C_e, for the convective condition, is the concentration level in the environment. The solution is constructed in terms of $C_1(x,\tau)$, as the solution of Eq. (6.5.9) without $-KC$, that is, the Fourier equation.

Given this solution, $C_1(x,\tau)$, the solution of Eq. (6.5.9), in terms of $-K\tau$, may be shown to be

$$C(x,\tau) = K\int_0^\tau C_1 e^{-K\tau}\,d\tau + C_1 e^{-K\tau} \tag{6.5.12}$$

Also, the initial condition, $C_1(x,0)$, becomes the initial condition on $C(x,0)$, from Eq. (6.5.12). The two different surface conditions in Eqs. (6.5.10) and (6.5.11), applied to $C_1(0,\tau)$, may also be seen in Eq. (6.5.12) to amount to the same condition on $C(0,\tau)$. Therefore, given the much simpler solution for $C_1(x,\tau)$, the solution for $C(x,\tau)$ is determined from Eq. (6.5.12). Although this transformation was discussed in terms of a one-dimensional process in a plane region, it also applies to an equation of the form of Eq. (6.5.9).

This technique has been applied to many processes in plane, cylindrical, and spherical geometries; see Crank (1975). This method may also be applied to heat conduction in regions where the distributed source strength varies as $q''' = K(t - t_r)$, where t_r is the reference temperature in the formulation.

ABSORPTION RATE WITH AN IRREVERSIBLE REACTION. The result given here is the one-dimensional transient in a semiinfinite solid initially at $C(x,0) = 0$ with $C(0,\tau) = C_0$. This solution may also be interpreted as a long thin rod with convective mass transfer along its surface. The solution, using Eq. (6.5.9), is

$$\frac{C(x,\tau)}{C_0} = \frac{e^{-\sqrt{x^2 K/D}}}{2}\,\text{erfc}\left[\eta_D - \sqrt{K\tau}\right] - \frac{e^{\sqrt{x^2 K/D}}}{2}\,\text{erfc}\left[\eta_D - \sqrt{K\tau}\right] \tag{6.5.13}$$

where $\eta_D = x/2\sqrt{D\tau}$ and K has the dimensions of inverse time.

The absorption up to time τ, as $M(\tau)$, is seen in Fig. 6.5.2, in terms of $(M(\tau)/C_0)\sqrt{K/D} = \overline{M}(\tau)$ versus $K\tau$, as given by

$$\overline{M}(\tau) = \left[\left(K\tau + \tfrac{1}{2}\right)\text{erf}\sqrt{K\tau} + \sqrt{K\tau/\pi}\,e^{-K\tau}\right] \tag{6.5.14}$$

At long times $\text{erf}\sqrt{K\tau} \to 1$ and the concentration distribution and $\overline{M}(\tau)$ become

$$\frac{C(x,\tau)}{C_0} = e^{-\sqrt{x^2 K/D}} \tag{6.5.15}$$

$$\overline{M}(\tau) = K\tau + \tfrac{1}{2} \tag{6.5.16}$$

Thus the concentration at any x becomes constant and the total sorption $\overline{M}(\tau)$

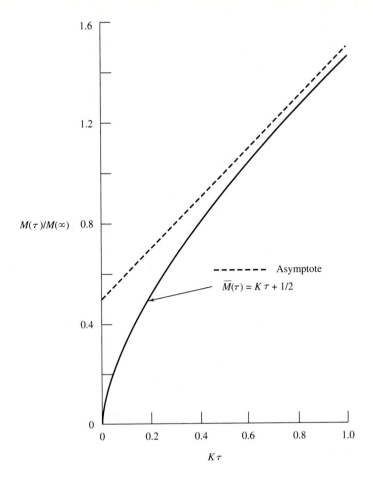

FIGURE 6.5.2
Increasing adsorption with time, with irreversible adsorption.

eventually increases linearly with time as seen in Fig. 6.5.2. The short-time result, in terms of $\mathrm{erf}\sqrt{K\tau}$, reduces to sorption without appreciable irreversible adsorption.

6.5.3   Reversible Reactions

The kind of adsorption reactions envisioned are those in which the immobilization reaction at higher concentration levels is reversed at lower concentrations. Equilibrium arises locally when the local adsorption rate is balanced by an equal release rate. The forward and release rates are equal in steady-state processes. In transients, there is net adsorption at sufficiently high local concentration levels. There is net release at sufficiently low concentration levels.

If diffusion is extremely rapid, compared to the adsorption rate, the concentration and level of adsorption may both be uniform over the region. This is the other extreme condition from that of very rapid adsorption, formulated as $c(x, \tau) = R_r C(x, \tau)$ in Eq. (6.5.1).

The circumstance considered here is the range of diffusion and adsorption rates between these two extremes. The governing relation is still Eq. (6.5.3), written as follows in terms of $C(x, \tau)$ and $c(x, \tau)$:

$$D\frac{\partial^2 C}{\partial x^2} - \frac{\partial c}{\partial \tau} = \frac{\partial C}{\partial \tau} \tag{6.5.17}$$

where $c(x, \tau)$ is the net local adsorption level. The rate of change of $c(x, \tau)$ is governed by the balance between the instantaneous rate of adsorption $R_F C$ and the inverse release rate $R_R c$, as

$$\frac{\partial c}{\partial \tau} = R_F C(x, \tau) - R_R c(x, \tau) \tag{6.5.18}$$

where R_F and R_R are the rate constants of the forward and release rates and $R = R_F/R_R$ is the partition factor between adsorbed and free solute. That is, there are two competing effects and $\partial c/\partial \tau$ is the net adsorption rate. Equation (6.5.18) assumes that both effects are linear in the driving concentrations, $C(x, \tau)$ and $c(x, \tau)$, respectively. The extreme circumstance formulated in Eq. (6.5.1) amounts to $R_F \gg R_R$.

The effects of reversible reactions are discussed later for the plane geometry in Fig. 6.5.3. The thickness is $2s$ and a well-stirred ambient region on each side, initially at $C_a(0)$, is of volume V per unit of surface area. The levels of initial local concentration, $C(x, 0)$, and of adsorption, $c(x, 0)$, in the region, are zero. At long times the ambient is depleted to $C_a(\infty)$, to provide uniform concentration $C(x, \infty)$ and adsorption level $c(x, \infty)$, interior to the region of thickness $2s$.

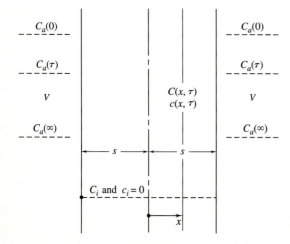

FIGURE 6.5.3
A plane layer transient with a reversible chemical reaction, immersed in an ambient solvent on each side.

The solution may be simply obtained using Laplace transforms of $C(x,\tau)$ and $c(x,\tau)$ as

$$\overline{C} = \int_0^\infty C(x,\tau)e^{-p\tau}\,d\tau \quad \text{and} \quad \overline{c} = \int_0^\infty c(x,\tau)e^{-p\tau}\,d\tau \quad (6.5.19)$$

Equations (6.5.17) and (6.5.18) become

$$D\frac{\partial^2 \overline{C}}{\partial x^2} - p\overline{c} = p\overline{C} \quad (6.5.20)$$

and

$$p\overline{c} = R_F\overline{C} - R_R\overline{c} \quad (6.5.21)$$

The conditions at the two exposed surfaces, for the partition factor $K = 1$, are

$$\mp D\frac{\partial C}{\partial x} = V\frac{\partial C}{\partial \tau} \quad \text{at } x = -s, s \quad (6.5.22)$$

or

$$\mp D\frac{\partial \overline{C}}{\partial x} = -VC_a(0) + pV\overline{C} \quad (6.5.23)$$

The solution for $C(x,\tau)$ is given in Crank (1975) as

$$\frac{C}{C_a(0)} = \cfrac{1}{1 + \cfrac{s}{V}(R + 1)}$$

$$+ \sum_{n=1}^\infty \cfrac{C_a(0)e^{p_n\tau}}{1 + \left[1 + \cfrac{R_F R_R}{(p_n + R_R)^2}\right]\left[\cfrac{s}{2V} + \cfrac{p_n}{2Dk_n^2} + \cfrac{p_n^2 sV}{2D^2 k_n^2}\right]} \cdot \frac{\cos k_n x}{\cos k_n s}$$

$$(6.5.24)$$

where the p_n are the nonzero roots of

$$\frac{Vp_n}{D} = k_n \tan k_n V \quad \text{and} \quad k_n^2 = -\frac{p_n}{D}\frac{p_n + R_F + R_R}{p_n + R_R} \quad (6.5.25)$$

This solution, for $C/C_a(0)$, is also that for the adsorbed solute $c(x,\tau)$, if the terms in the summation are multiplied by $R_F(p_n + R_R)$. The ratio of the sum of the internal amount of solute free to diffuse and immobilized, $M(\tau)$, to the amount in the eventual equilibrium, is

$$\frac{M(\tau)}{M(\infty)} = 1 - \sum_{n=1}^\infty \cfrac{(1 + \alpha)e^{p_n\tau}}{1 + \cfrac{R_F R_R}{(p_n + R_R)^2}\left[\cfrac{s}{2V} + \cfrac{p_n}{2Dk_n^2} + \cfrac{p_n^2 sV}{2D^2 k_n^2}\right]} \quad (6.5.26)$$

where

$$\alpha = V/(R + 1)s \quad \text{and} \quad M(\infty) = VC_a(0)/(1 + \alpha) \quad (6.5.27)$$

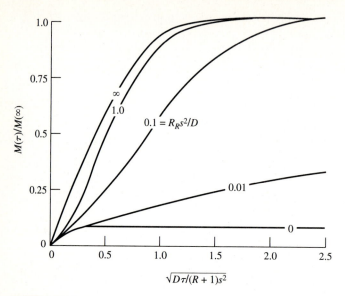

FIGURE 6.5.4

Uptake $M(\tau)/M(\infty)$ for a reversible internal reaction, for $R = R_F/R_R = 10$. The contours are each a constant value of $R_R s^2/D$.

Crank (1975) gives tabulated results and graphical contours for total absorption $M(\tau)/M(\infty)$ versus time as $\sqrt{D\tau/(R+1)s^2}$, for $R_R s^2/D$ from 0 to ∞. Information is given for the plane, cylindrical, and spherical regions, but for an infinite extent of ambient solute, that is, for $V/s \to \infty$.

A plot of $M(\tau)/M(\infty)$ is seen in Fig. 6.5.4, for $R = R_F/R_R = 10$. That is, $R_F = 10R_R$. Therefore, the adsorption is very fast compared with the release rate. The contours are total adsorption response, in terms of $R_R s^2/D$. Consider any given value of D/s^2 in the abscissa and in $R_R s^2/D$. The total absorption rate, the slope of $M(\tau)/M(\infty)$ at $\tau = 0$, increases as does $R_R s^2/D$. This is because the adsorption rate increases at 10 times the release rate. Therefore, adsorption is very much faster. This consumes C more rapidly, maintaining a very high concentration gradient near the surface, for inward diffusion. The value of $M(\tau)/M(\infty)$ increases as the front advances to 1.0. The 0.1 and 0.01 values provide a much lower rate of adsorption.

These same results are given in Fig. 6.5.5, in terms of $\sqrt{R_R \tau}$, with contours again in terms of $R_R s^2/D$. This plot shows the total adsorption effect variation, as $M(\tau)/M(\infty)$, as a function of $R_R s^2/D$. These curves may be interpreted as the effect of increasing D, for any particular choice of R_R. The range of D is from a small value, $R_R s^2/D = 1$, to $D = \infty$ for $R_R s^2/D = 0$. Increasing diffusivity comes to completely dominate the process, in terms of very quick penetration of solute C, compared to the immobilization process.

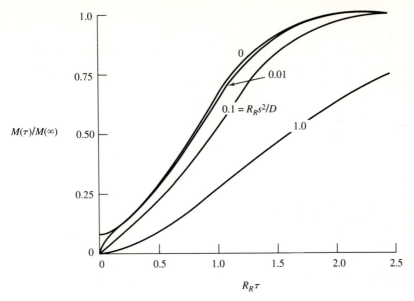

FIGURE 6.5.5
Uptake $M(\tau)/M(\infty)$, again for $R = R_F/R_R = 10$ with $R_R s^2/D$ contours, vs. $R_R s^2/D$.

Solutions for the cylindrical and spherical geometries are also given by Crank (1975). Additional explanations and other physical aspects of the results are also discussed.

DESORPTION PROCESSES IN A REVERSIBLE REACTION. The inverse of the preceding process in Fig. 6.5.3 is the plane region initially uniformly at a concentration level $C(x,0) = C_i$ with the solute in equilibrium at $c(x,0) = c_i$. The ambient is at $C_a(\tau) = C_a(0) = 0$. Then the internal reaction proceeds with desorption at each face of the plane layer, into a volume V of ambient solvent. At equilibrium, $R_F C_i = R_R c_i$. Desorption response curves are given by Crank (1975) for different values of $R_R s^2/D$, for $R = 1$ and 10.

6.6   INTERACTING DIFFUSION AND CONDUCTION, AND OTHER EFFECTS

All of the foregoing considerations of heat and species diffusion have treated each mechanism as independent of any effect arising from the other. However, in many processes the two effects occur together and each may affect the other. Hot air or steam heating of wood increases interior water vapor pressure levels, enhancing outward diffusion. The surface region diffusion carburization of steel, by heating it to a high temperature when immersed in a carbonaceous material, involves other effects. Higher temperature levels enhance D, and, thereby, the inward carbon diffusion rate. The local reaction rate with the steel is also increased.

These are examples of the application of heat to enhance diffusion. There are many other processes in which diffusion and the resulting internal reactions create a temperature field. The adsorption of gaseous ammonia, NH_3, in silica gel is an exothermic reaction. The humidification of many fibrous materials such as wool is also strongly exothermic and under some conditions this initiates combustion. In that circumstance $C''' < 0$ results in $q''' > 0$. Curing of materials often simultaneously results in both the generation of diffusing material and heat.

Several characteristic kinds of such effects are formulated. Section 6.6.1 concerns a porous material, with the diffusion of species C through the porosity. There is local adsorption, over the region, into the solid matrix. This adsorption rate may be accompanied by a release of energy. These two effects are written as $C'''(x, \tau)$ and $q'''(x, \tau)$, for a plane region. Also given is an example of the effect of condensation of a vapor on both surfaces of a plane region. Internal mass and heat transfer, along with surface heat loss to the environment, are considered.

Section 6.6.2 considers an example of flame propagation in a mixture of oxidant and fuel, such as methane gas and air. A common example is a standing conical flame front, on a Bunsen burner, as diagrammed in Fig. 6.6.3(*a*) (see page 373). This is a diffusion flame in that heat diffuses ahead of the actual flame front into the unburned mixture to preheat it, largely in the region δ_p, in Fig. 6.6.3(*b*). Also the ensuing chemical reaction results in the diffusion of any fuel intermediates produced during the process, and of the combustion products, internal to the structure of the front. The flame front propagates into the moving air–fuel mixture at a relative velocity of $U = u \sin \theta$, where U is called the flame speed, S. A typical value of S, for hydrocarbon diffusion-controlled combustion in air, is around 40 cm/s. A simplified analysis is given for this kind of process.

In many circumstances, a premixed stream of material, as pictured in Fig. 6.6.3, may instead undergo an explosive process. This could occur as a pressure shock wave propagates through the combustible mixture. If the resulting rise in temperature is sufficiently great, very rapid combustion occurs and is completed in a very thin detonation layer. Such waves may propagate at very high velocity. These thin layers are customarily treated as gas-dynamic discontinuities.

However, in the same general configuration, deflagration waves may arise. These moving waves are a propagation flame front. They arise at low flow velocities and very small pressure differences are generated. See Glassman (1987) for a consideration of these two kinds of processes.

Section 6.6.3 concerns transpiration cooling. This technique is often used to protect a surface or a region from a very hostile environment. An example is a gaseous medium at very high temperatures. These conditions arise in fires and in the very high velocity entry of space devices into planetary atmospheres. The resulting very high heat flux, imposed on a bounding surface, often results in destructively high surface temperatures.

Intensive cooling on the back side of such a surface is often not sufficient. An alternative is to use a porous material. A cool liquid or gas may then be supplied on the back side. It may either diffuse or be forced through the porosity by back-side pressure. Other examples of such fluid diffusion processes arise in fiber wicks in heat pipes and in water vapor motion in soils. A typical example of transpiration cooling is given in Sec. 6.6.3.

Section 6.6.4 concerns processes in which the diffusion mechanisms may not be completely formulated in terms of the Fickian model, for example, as $\overline{m}'' = -D\,\nabla C$. That form applies for single-phase or single-mechanism diffusion. It is also restricted to continuum processes. That is, the characteristic diffusion length, the mean free path for a gas, must be small compared to the geometric lengths in the region which interact with the process.

Analysis and results are given for two common processes. The first is the drying of green wood. It contains fiber-bound water, free water, and gas bubbles containing water vapor. Increasing internal temperature levels drive water vapor out, by a Fickian process. However, liquid water also moves along the internal surfaces, by capillary effects. This migration is not modeled by a Fickian formulation.

The second circumstance concerns the diffusion of a gas into a porous layer, subject to ordinary diffusion as well as to local adsorption on the surfaces. An application is moisture removal in a bed of porous silica gel pellets. Diffusion over the solid surfaces is an important mechanism. Also, when the pore size is comparable to the gas molecular mean free path, noncontinuum effects arise. Both of these effects must be modeled by formulations which are not of a Fickian form.

6.6.1 Coupled Heat and Mass Diffusion

The basic equations and initial and boundary conditions remain those developed in Chap. 1. However, these must be modified here to apply to a porous region. The porosity, defined as σ, is the fraction of the total region volume which is pores. The solid fraction is $1 - \sigma$. Diffusion is assumed here to occur only through the gaseous contents of the pores. Thermal conduction is accounted for only in the solid matrix. The mass diffusion relation, in terms of D, the diffusivity of C in the gas-filled pores, is

$$\sigma(T'D)\frac{\partial^2 C}{\partial x^2} = \sigma\frac{\partial C}{\partial \tau} + (1 - \sigma)\rho_s\frac{\partial c}{\partial \tau} \tag{6.6.1}$$

where T' is a tortuosity factor. This accounts for the indirect internal path for mass diffusion provided by the interconnecting elements of porosity. Also ρ_s is the solid matrix density and $c(x, \tau)$ is, as before, related to C''' as

$C''' = \partial[c(x, \tau)]/\partial \tau$. The energy equation is

$$\frac{\partial^2 t}{\partial x^2} + \frac{H}{k}\frac{\partial c}{\partial \tau} = \frac{1}{\alpha}\frac{\partial t}{\partial \tau} \tag{6.6.2}$$

where $\alpha = k/\rho c$ is based on the average properties, including the pores, and H is the heat evolved per unit of adsorption into the matrix. In this relation $H(\partial c/\partial \tau) = q'''(x, \tau)$.

The coupling between the previous two relations is the term $\partial c/\partial \tau$. In general, the two effects must be considered together, along with all of the initial and boundary conditions which apply. Several additional assumptions are inherent in Eqs. (6.6.1) and (6.6.2). All of the coefficients are assumed constant. The diffusion through the structure is through gaseous mechanisms. That is, there is no liquid-phase capillary migration. The local adsorption and desorption by the matrix material is assumed to follow $C(x, \tau)$ reversibly. That is, there is no hysteresis.

The remaining matter which must be specified is the relation of $(\partial c/\partial \tau)$ to the local instantaneous concentration and temperature levels $C(x, \tau)$ and $t(x, \tau)$. The simplest assumption is that they are linear effects, as in Eq. (6.5.1) for $C(x, \tau)$ in relation to $c(x, \tau)$. The rate of adsorption into the matrix is then

$$\frac{\partial c(x, \tau)}{\partial \tau} = R_r \frac{\partial C(x, \tau)}{\partial \tau} + R_{th}\frac{\partial \theta(x, y)}{\partial \tau} \tag{6.6.3}$$

where R_{th} is the thermal rate constant and $\theta(x, y) = t(x, y) - t_r$, where t_r is a reference temperature. With this assumed linear relationship, the mass diffusion and heat conduction relations, Eqs. (6.6.1) and (6.6.2), are written in terms of $w = \rho_s(1 - \sigma)/\sigma$ as

$$(T'D)\frac{\partial^2 C}{\partial x^2} = \frac{\partial}{\partial \tau}[C + wR_rC + wR_{th}\theta] \tag{6.6.4a}$$

or

$$\frac{T'D}{1 + wR_r}\frac{\partial^2 C}{\partial x^2} = \frac{\partial}{\partial \tau}\left[C + \frac{wR_{th}}{1 + wR_r}\theta\right] \tag{6.6.4b}$$

or, in terms of β_1 and β_2,

$$\beta_1\frac{\partial^2 C}{\partial x^2} = \frac{\partial}{\partial \tau}[C + \beta_2\theta] \tag{6.6.4c}$$

Also,

$$\frac{\partial^2 \theta}{\partial x^2} = \frac{\partial}{\partial \tau}\left[\frac{\theta}{\alpha} - \frac{H}{k}R_{th}\theta - \frac{H}{k}R_{th}C\right] \tag{6.6.5a}$$

or

$$\frac{\alpha}{\left[1 - \dfrac{\alpha H}{k}R_{th}\right]}\frac{\partial^2 \theta}{\partial x^2} = \frac{\partial}{\partial \tau}\left[\theta - \frac{\alpha HR_r/k}{1 - \dfrac{\alpha H}{k}R_{th}}C\right] \tag{6.6.5b}$$

or, in terms of β_3 and β_4 as

$$\beta_3 \frac{\partial^2 \theta}{\partial x^2} = \frac{\partial}{\partial \tau}[\theta - \beta_4 C] \qquad (6.6.5c)$$

Equations (6.6.4c) and (6.6.5c) remain coupled, for the dependent variables $C(x, y)$ and $\theta(x, y)$. Solutions may be examined in terms of a linear combination of C and θ, as $\beta_5 C + \beta_6 \theta$. This form must satisfy the simple Fourier equation as follows:

$$\frac{\partial^2}{\partial x^2}(\beta_5 C + \beta_6 \theta) = \frac{\partial}{\partial \tau}(\beta_5 C + \beta_6 \theta) \qquad (6.6.6)$$

Resulting diffusion rates are calculated, as by Henry (1948) and reproduced by Crank (1975). See Sec. 15.6. Expressions for changes in C and θ are also discussed and results are given for initial conditions and changed boundary conditions. These results permit the assessment of the complicated effects which arise in these interacting processes.

A TEMPERATURE EFFECT DUE TO SURFACE CONDENSATION. This simpler circumstance is shown in Fig. 6.6.1. A uniform plane region of t_e is suddenly placed in contact with a saturated vapor also at t_e. Heat is assumed to be continuously evolved by condensation at each surface. The heat release, per unit mass, is H. The heat release rate at the surface of the region is assumed balanced by the processes of inward heat conduction and also by convection from the surface back to the vapor region at t_e. See Fig. 6.6.1.

The formulation is as follows in terms of $\theta(x, \tau) = t(x, \tau) - t(x, 0) = t(x, \tau) - t_e$. The internal conduction equation is

$$\frac{\partial^2 \theta}{\partial x^2} = \frac{1}{\alpha}\frac{\partial \theta}{\partial \tau} \qquad (6.6.7)$$

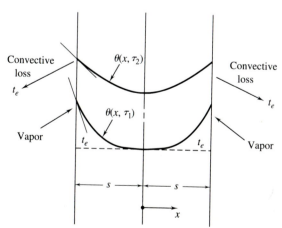

FIGURE 6.6.1
A plane region with vapor condensation on both surfaces.

The condition at $x = 0$ is $\partial\theta/\partial x = 0$, because of symmetry, as in the very similar analysis in Sec. 4.2.2. The surface condition equates inward conduction, at $x = s$, and the surface heat loss to the environment $h\theta(s, \tau)$, to the rate of surface heat generation due to condensation, again, as

$$-k\frac{\partial\theta}{\partial x} + h\theta(s, \tau) = H\frac{dM(\tau)}{d\tau} \tag{6.6.8}$$

where $M(\tau)$ is the total mass of vapor taken up by time τ, per unit surface area of the plane region, as condensate. This is the sum over the region $x = 0$ to s. The value of $M(\tau)$ is taken to be of the form

$$M(\tau) = sC_0\sqrt{Fo_s/\pi} \tag{6.6.9}$$

where $Fo_s = D\tau/s^2$ and C_0 is the eventual uniform equilibrium concentration across the layer and D is the appropriate layer diffusivity. Note that $\sqrt{Fo_s}$ in Eq. (6.6.9) is equivalent to the form found in Sec. 6.3 to apply to transients.

The results of this formulation are given in graphical form, by Crank (1975), as shown in Fig. 6.6.2. The surface temperature response $\theta(s, \tau)$ is plotted as θ/θ_{ch}, where θ_{ch} is a characteristic temperature difference for the process. This quantity may be derived from the coordinates given by Crank

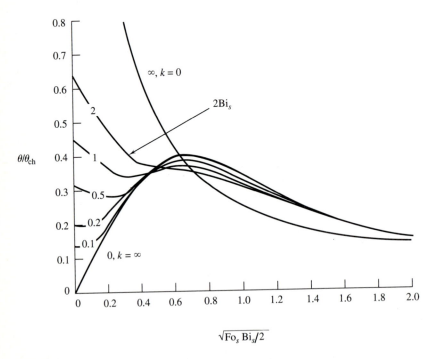

FIGURE 6.6.2
The variation of surface temperature as $\phi = \theta(s, \tau)/\theta_{ch}$ with time in terms of $Fo_s = \alpha\tau/s^2$ and $Bi_s = hs/k$. The contours are for values of $2Bi_s$. [From Crank (1975).]

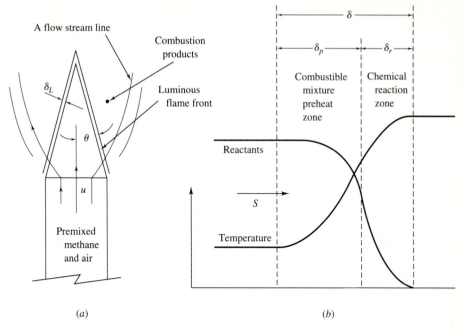

FIGURE 6.6.3
A diffusion flame: (*a*) the example of a luminous Bunsen burner flame of thickness δ_L; (*b*) the structure of the diffusion processes in a combustion front of overall thickness δ.

(1975). At any given time, as $\sqrt{Fo_s\,Bi_s/2}$, the resulting surface temperature response θ/θ_{ch} is a function of $2Bi_s = 2hs/k$. Contours are given over the range $2Bi_s = 0$ to ∞. The two limits are for $hs/k = 0$ and ∞, as for k very large and very small, compared with h. Small internal conductivity results in a lower rate of heat penetration into the layer. The heat liberated, $H\,dM(\tau)/d\tau$, is then mainly lost to the environment at t_e. The inverse applies for high conductivity.

6.6.2 Diffusion-Controlled Reactions

The mechanism of laminar flame propagation is diagrammed in Fig. 6.6.3. The mechanisms are discussed briefly in Sec. 6.6.1. They are conduction preheat in region δ_p followed by chemical processes, largely in region δ_r, as shown in Fig. 6.6.3(*b*). The reaction results in an internal and location-dependent distributed heat source, q'''. The reaction also amounts to distributed sinks of fuel and oxidant, as C'''_{CH_4} and C'''_{O_2}, and sources of the products, as C'''_{CO_2} and C'''_{H_2O}. There are also additional intermediate chemical effects during the process.

In a steady flame front, as in the example of the Bunsen burner flame, the processes may be analyzed as a quasi-static process in a coordinate system moving with respect to the gas. The general transformation is given in Sec. 4.7.

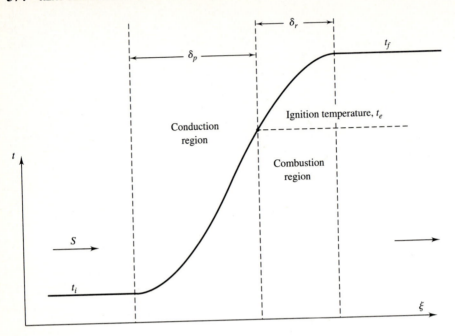

FIGURE 6.6.4
A model of a diffusion-controlled reaction.

This procedure, applied to a plane front, results in Fig. 6.6.3(b). It is also seen in Fig. 6.6.4, as a model for analysis. The combustible mixture approaches the reaction zone at t_i and leaves the zone after the completion of combustion processes at t_f.

The processes which occur result from the conduction toward the left. There is also diffusion of whatever chemical radicals are formed during combustion. These diffuse in both directions in the zone, to form the final combustion products, eventually at t_f.

A simple conduction model was given by Mallard and LeChatier (1983), in terms of the ignition temperature t_e. The heat flux required to preheat the combustible mixture from t_i to the ignition temperature, t_e, is $m''c_p(t_e - t_i)$, where $m'' = \rho_i u$ is the mass flux into the burning region. This energy is supplied by conduction from the burning region. The average slope across this region is $(t_f - t_e)/\delta_r$. The conduction flux is then $k(t_f - t_e)/\delta_r$. Then

$$m''c_p(t_e - t_i) = \rho_i u c_p(t_e - t_i) \approx k(t_f - t_e)/\delta_r \qquad (6.6.10)$$

Since the combustion front remains at a constant location, in terms of the coordinate ξ, u is equal to the speed of the stationary flame, S.

$$u = S = \frac{k(t_f - t_e)}{\rho_i c_p \delta_r (t_e - t_i)} = \frac{\alpha(t_f - t_e)}{\delta_r(t_e - t_i)} \qquad (6.6.11)$$

However, S may not be calculated directly from this result, since δ_r is not known. This analysis is carried further by relating δ_r to S in terms of the chemical reaction rate as

$$\delta_r = ST = \frac{S}{(d\epsilon/d\tau)} = \frac{S}{R_R} \tag{6.6.12}$$

where T is a characteristic reaction time. Also $d\epsilon/d\tau = R_R$ is the reaction rate, in terms of the fraction ϵ of chemical conversion. Eliminating δ_r between Eqs. (6.6.11) and (6.6.12) gives

$$S = \sqrt{\alpha\left(\frac{d\epsilon}{d\tau}\right)\frac{(t_f - t_e)}{(t_e - t_i)}} \tag{6.6.13}$$

This is further simplified by taking the ignition temperature as $(t_f + t_i)/2$, to result in the very approximate result

$$S = \sqrt{\alpha R_R} \tag{6.6.14}$$

The value of δ_r is estimated by eliminating R_R between Eqs. (6.6.14) and (6.6.12) as

$$\delta_R = \alpha/S \tag{6.6.15}$$

For a flame speed of 40 cm/s in a combustible mixture at 1300 K, $\delta_r \approx 1$ mm. The characteristic time α/S^2 is several milliseconds.

More comprehensive models have been developed. These account for additional important component mechanisms in the overall process. Among the most important aspects are the accurate modeling of the mass diffusion and heat conduction aspects, including distributed sources and sinks. A discussion of such a model, and of the results, is given in Glassman (1987), Chap. 4.

6.6.3 Transpiration Cooling

This mechanism is discussed in the introduction of this section. It is a means of protecting intensely heated barrier surfaces by making the barrier porous and forcing a cooling fluid through from the other side. The fluid cools the material directly and also supplies a relatively cool layer over the exposed surface. Such parallel heat conduction and mass diffusion processes are also extremely common in other circumstances.

An example of transpiration cooling is seen in Fig. 6.6.5. The objective is taken to be to limit the temperature on the exposed right face, at $x = L$, to some safe level t_2, given the expected flux loading there, of q_2''. The transpiration cooling effect may also be augmented at the inside face, as q_1'', for example, by refrigerant coils or by radiant emission there.

Interior to the plane region, there is inward conduction through the porous matrix and outward moving fluid. The combined conductivity of the barrier material is taken as k_e. Given that $t_2 > T_\infty$, the fluid is heated as it flows

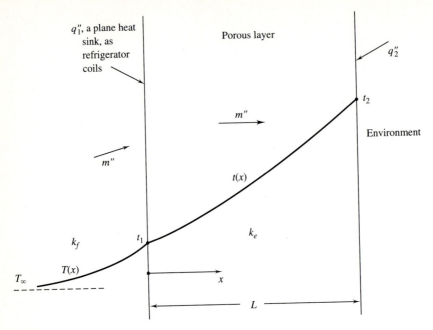

FIGURE 6.6.5
A transpiration-cooled wall, subject to an imposed surface heat flux q_2'' at $x = L$ and a plane sink q_1'' at $x = 0$.

from the condition t_1, through the porous barrier. This amounts to a distributed energy sink q''' in the barrier. The coolant, remotely at T_∞, is heated to t_1 as it approaches the left face of the barrier, if the plane sink strength q_1'' is sufficiently small. For very large values of q_1'', t_1 may be less than T_∞. The general differential equation in steady state, in the barrier, is

$$\frac{d^2t}{dx^2} + \frac{q'''}{k_e} = 0 \qquad (6.6.16)$$

In the fluid preheating region, at the left, k is the fluid conductivity. In the porous barrier, k_e is the effective thermal conductivity of the fluid–solid material. The strength of the sink is equal to the product of the mass flow rate m'', per unit area, the fluid specific heat, c_p, and the local rate of fluid temperature rise, per unit distance x. If the fluid is locally everywhere in equilibrium with the solid matrix, this local rate is dt/dx. Then

$$q''' = -m''c_p\frac{dt}{dx} \qquad (6.6.17)$$

Therefore, Eq. (6.6.16) becomes

$$\frac{d^2t}{dx^2} - \frac{m''c_p}{k_e}\frac{dt}{dx} = 0 \quad \text{for } 0 < x < L \qquad (6.6.18)$$

The fluid approaches the inside surface of the barrier, at $x = 0$, at a remote temperature T_∞. It is heated by the conduction flux coming from the left face of the barrier. The relevant differential equation is again (6.6.16) with the sink term in Eq. (6.6.17). The temperature distribution in this region is denoted as $T(x)$. The equation is

$$\frac{d^2T}{dx^2} - \frac{m''c_p}{k}\frac{dT}{dx} = 0 \quad \text{for } x < 0 \tag{6.6.19}$$

where k is the fluid conductivity.

The general solutions for $t(x)$ and $T(x)$, which apply in the porous layer and on the left side, are

$$t(x) = A + Be^{(m''c_p/k_e)x} = A + Be^{Mx} \tag{6.6.20}$$

$$T(x) = C + De^{(m''c_p/k)x} = C + De^{Nx} \tag{6.6.21}$$

where $M = m''c_p/k_e$ and $N = m''c_p/k$. The constants are determined from the conditions at $x = L$ and 0 and at $x = -\infty$. These are

$$T(x) = T_\infty \quad \text{as } x \to -\infty \tag{6.6.22a}$$

$$T(x) = t(x) = t_1 \quad \text{at } x = 0 \tag{6.6.22b}$$

$$-k\frac{dT}{dx} = -k_e\frac{dt}{dx} + q_1'' \quad \text{at } x = 0 \tag{6.6.22c}$$

$$t(L) = t_2 \tag{6.6.22d}$$

The specific temperature distributions are then found to be

$$t(x) = T_\infty - \frac{q_1''}{m''c_p} + \frac{q_2''}{m''c_p}e^{(m''c_p/k_e)(x-L)} \tag{6.6.23}$$

$$T(x) = T_\infty + \frac{q_2''}{m''c_p}e^{(m''c_p/k_e)(xk_e/k-L)} - \frac{q_1''}{m''c_p}e^{(m''c_p/k)x} \tag{6.6.24}$$

A characteristic temperature difference is defined as $\Delta t_{ch} = q_2''/m''c_p$. This would be the fluid temperature difference across the whole region for q_1, the energy sink at $x = 0$, being absent. However, retaining $q_1'' \neq 0$, the equations for $t(x) - T_\infty$, for the barrier, and $T(x) - T_\infty$, for the approaching fluid, are

$$\phi_B = \frac{t(x) - T_\infty}{\Delta t_{ch}} = e^{-M(L-x)} - \frac{q_1''}{q_2''} \quad \text{for } x = 0 \text{ to } L \tag{6.6.25}$$

$$\phi_A = \frac{T(x) - T_\infty}{\Delta t_{ch}} = e^{-M(L-xk_e/k)} - \frac{q_1''}{q_2''}e^{Nx} \quad \text{for } x = 0 \text{ to } -\infty \tag{6.6.26}$$

The resulting surface temperatures t_2, resulting from the imposed surface flux

q_2, and $t_1 = T(0)$, are given by

$$\frac{t_2 - T_\infty}{\Delta t_{ch}} = 1 - \frac{q_1''}{q_2''} \tag{6.6.27}$$

$$\frac{T_1 - T_\infty}{\Delta t_{ch}} = e^{-ML} - \frac{q_1''}{q_2''} \tag{6.6.28}$$

The relation for the temperature of the exposed surface, as $t_2 - T_\infty$, is given in terms of q_1, q_2, and mc_p as

$$t_2 - T_\infty = \frac{q_2'' - q_1''}{mc_p} \tag{6.6.29}$$

That is, the net purely transpiration effect $q_2'' - q_1''$ is equal to the transpiration energy absorption rate $mc_p(t_2 - T_\infty)$.

SUMMARY. The general results, in Eqs. (6.6.27) and (6.6.28), reduce to simpler forms for $q_1'' = 0$. Then $t_2 - T_\infty = \Delta t_{ch}$ and $t_1 - T_\infty = \Delta t_{ch} e^{-ML}$. Thus the ratio R of the heat absorption in the porous layer $(t_2 - T_1)$ to that conducted into the approaching fluid, $(T_1 - T_\infty)$, is

$$R = \frac{t_2 - T_1}{T_1 - T_\infty} = \frac{(t_2 - t_\infty) - (T_1 - T_\infty)}{T_1 - T_\infty} = \frac{t_2 - T_\infty}{T_1 - T_\infty} - 1 = e^{ML} \tag{6.6.30}$$

Since $ML = m'' c_p L / k_e$, this effect is large for $m'' c_p$ large compared to k_e/L. Thus the porous region transpiration effect would be large for water and small for air, for the same mass flow rate m''. See Schneider (1955) and Edwards et al. (1979) for further consideration of transpiration effects. There are many circumstances of surface flux loadings which are too large for transpiration cooling. The cooling mechanism may then become ablation of the exposed surface material. See Rohsenow et al. (1985), pages 4-22, 4-28, and 4-135. Plots are given of the surface flux which arises during several very vigorous processes of this kind. See also page 8-88 for analyses of transpiration gas cooling.

6.6.4   Non-Fickian Diffusion in Porous Media

Many processes occur in which the Fickian diffusion formulations, developed in Sec. 1.3 and used previously in this chapter, are not applicable. That formulation assumes that the diffusion of species C through a region is driven by a concentration gradient of species C, present in only a single phase, for example, as a vapor or a liquid. In very many drying processes, the diffusing substance moves as a vapor through the pores of the region. In unsaturated liquid regions there is also a parallel process of liquid layer diffusion along the surfaces which bound the internal volumes of porosity. This liquid diffusion is often driven by surface tension effects and resulting capillary motion.

Examples are in drying wood, clay objects, soap, paper pulp, soil, and in the curing of layers applied initially with a solvent carrier. The vapor phase concentration C may be contained in an inert gas, such as air. The vapor is diffused through the air due to a vapor concentration gradient, as $\overline{m}''_v = -D_v \nabla C_v$, where the molecular mean free path λ is small, compared to the pore size δ. This is a Fickian diffusion mode.

However, there may also be an additional motion of the liquid phase along the pore surfaces. This is due to capillary effects in regions in which there are also gradients in the liquid density ρ_l, as $\nabla \rho_l$. Here ρ_l is the local liquid density per unit volume of the region. The resulting diffusion rate $\overline{m}''_l$ is not directly related to $\nabla \rho_l$ by the vapor diffusivity D_v. This physical process may not be affected by the vapor diffusion processes. An independent formulation is necessary. This is to be based on the particular characteristics of the prevailing diffusion mechanism. The first specific result given in the following discussion applies to a region of regular lamellar porosity, a softwood. Then the liquid diffusivity, D_l, may be formulated in terms of the porous structure.

The preceding combined mechanism is relatively simple. Many more complicated effects also commonly arise. Consider first the previous example, with only a diffusing vapor present. Due to chemical affinity, the vapor may be adsorbed on the internal surfaces of the porous structure. With a gas-phase concentration gradient, there may also be a simultaneous adsorbed layer concentration gradient. Then the adsorbed molecules will also surface-diffuse to regions of lower surface concentration. Models different from those for liquid layer capillarity effects then arise.

Another different kind of effect also frequently arises in gaseous diffusion in porous media and in packed beds. If the mean free path of the molecules, λ, is comparable to the pore size, δ, then noncontinuum or Knudsen effects arise in the gas diffusion process. The result is decreased mass transfer, compared to that following from a continuum process, for which D_v would apply for $\lambda \ll \delta$. Then the adsorbed surface layer diffusion process may become relatively much more important. This is the second specific kind of result given in the following discussion.

A recent development, using synthetic membranes, permits increasingly effective separation of mixed gas species from each other, for example, O_2 from N_2 in air. The active regions of these membranes may consist of very small aligned and hollow polymer tubes. For large Knudsen number λ/δ, the diffusion is by a free-molecule mechanism. A pressure gradient, that is, a gas density gradient, results in faster diffusion of the lighter, or lower mass, molecular species. The resulting species separation may then be approximated as proportional to the inverse ratio of the species' molecular weights. A similar selective diffusion and separation process may also be achieved with solid synthetic polymer membranes. These may be formed with free volumes in the material. Such regions, both separately, and in conjunction with tubular structures, may result in very effective gaseous pressure-driven species separation, at high efficiency. See, for example, Michaels (1989) and Spillman (1989).

General reviews of modeling for the simpler mechanisms discussed previously are given by Whitaker (1977) and DeVries (1987). The first of the following two specific examples is of that more conventional form. The second includes noncontinuum effects which arise in the take-up and desorption in a porous bed of dessicant.

THERMALLY DRIVEN WATER DIFFUSION IN WOOD DRYING. The model is of a fresh long softwood log dried in hot air or in steam. The process is idealized in terms of a single radial coordinate r in Fig. 6.6.6(a). However, for simplicity in analysis the region will instead be taken as planar, as in Fig. 6.6.6(b), with the coordinate denoted as x. The log is initially at t_i. At $\tau = 0$, the environment temperature is increased to t_e. The log is assumed not to change in volume.

The internal structure of the wood consists of long and narrow longitudinal hollow cells, or passages. There are frequent small orifices between adjacent cells. These permit the outward passage of fluid during drying. The cell structure consists of three physical regions. The cell wall material may contain bound water up to the fiber saturation point, FSP. The cells may contain free water. There are also gas bubbles in the cells. These are a mixture of air and water vapor.

In principle, a set of conservation equations may be written for each such region. However, during this drying transient, both the free water and bubble

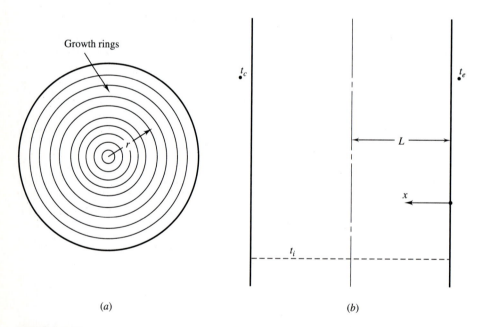

(a)   (b)

FIGURE 6.6.6
Model of a long log undergoing drying: (a) the log cross section; (b) the planar equivalent used in analysis, in terms of coordinate x, with all fluxes taken as zero at $x = L$.

regions change. Such effects are also very common in other similar kinds of processes. A common alternative to using separate regional formulations is to average the phase equations over a representative region. This average then includes the overall effect of each phase. The procedure of Whitaker (1977) was used for a wood-drying process by Plumb et al. (1985). Within the local averaging volume, the thermal conductivity and specific heat are assumed uniform.

Moisture migration arises through liquid and vapor phase motion, due to internal heating. The equations for these two effects are added to obtain a continuity equation in terms of the total local free moisture content. It is in terms of liquid and vapor component velocities u_l and u_g, as

$$\Phi \frac{\partial S}{\partial \tau} + \frac{\partial}{\partial x}\left(u_l + \frac{\rho_v}{\rho_l}u_g\right) = \frac{\partial}{\partial x}\left(\frac{\rho_v}{\rho_l}D\frac{\partial}{\partial x}\frac{\rho_v}{\rho_g}\right) \qquad (6.6.31)$$

where Φ is the void fraction, and S is the void space saturation, that is, the fraction of the void space filled with water. The subscripts l, v, and g refer to the liquid, the water vapor, and the combined air and vapor gas phase. In terms of volume fractions in the void space V_l and V_v:

$$S = (V_l\rho_l + V_g\rho_v)/\rho_l\Phi \qquad (6.6.32)$$

The internal pressure field which results from heating is relatively small. Then $\rho_l \gg \rho_v$. For V_l comparable to V_g, $S \approx V_l/\Phi$. Equation (6.6.31) is seen to consist of three effects. The first two are changing local water storage and material velocity, carrying liquid and vapor outward. These effects are equal to the water vapor diffusion, in terms of an effective diffusion coefficient D, on the right side.

The thermal energy equation is a volume average over the phases in the same way. The result is

$$\rho c_p \frac{\partial t}{\partial \tau} + (\rho_l c_{p,l}u_l + \rho_g c_{p,g}u_g) + m''H = \frac{\partial}{\partial x}\left(k\frac{\partial t}{\partial x}\right) \qquad (6.6.33)$$

where H is the latent heat, c_p is the specific heat, and m'' is the local mass transfer rate. Three of the terms are analogous to those in Eq. (6.6.31) and $m''H$ amounts to an energy sink arising from local heat absorption during vaporization. Also, ρ and ρc_p are determined from instantaneous void fractions, including that of the solid, V_s, as

$$\rho = V_s\rho_s + V_l\rho_l + V_g(\rho_v + \rho_a) \qquad (6.6.34)$$

and
$$\rho c_p = V_s\rho_s c_{p,s} + V_l\rho_l c_{p,l} + V_g(\rho_v c_{p,v} + \rho_a c_{p,a}) \qquad (6.6.35)$$

This energy formulation also assumes thermal equilibrium between the phases, at all locations over the region, at all times.

The convection velocity terms internal to the region, in Eq. (6.6.31), are in terms of u_l and u_g, the local velocities of the mobile liquid and gas phases. The region is a porous medium. Darcy's law is used to relate the local velocities to the pressure gradients. These gradients arise in the region because of temperature gradients. That is, the local two-phase equilibrium temperature changes across the region. Darcy's law for the two mobile phases is

$$u_g = -\frac{K_g}{\mu_g}\frac{\partial p_g}{\partial x} \quad \text{and} \quad u_l = -\frac{K_l}{\mu_l}\frac{\partial p_l}{\partial x} \tag{6.6.36}$$

These apply for horizontal flows, where μ and K are the viscosities and relative permeabilities of the two media, respectively.

The Darcy equations then provide the basis for eliminating u_l and u_g, thereby coupling the instantaneous local temperature field $t(x, \tau)$, in Eq. (6.6.33), to the mass continuity relation, Eq. (6.6.31). The result is two coupled equations, in $S(x, \tau)$ and $t(x, \tau)$. These equations are then written in terms of a temperature variable ϕ and M, the sum of the local vapor and free water, where M_{max} is the maximum possible value of M in the given wood structure:

$$\phi(x, \tau) = \frac{t - t_i}{t_e - t_i} \tag{6.6.37}$$

$$M(x, \tau) = S(M_{max} - \text{FSP}) + \text{FSP} = S\,\Delta M + \text{FSP} \tag{6.6.38}$$

The resulting equations, for the formulation in Fig. 6.6.6, are

$$\frac{\Phi}{\Delta M}\frac{\partial M}{\partial \tau} + \frac{\partial}{\partial x}\left(\frac{K_l C_s}{\mu_l \Delta M}\frac{\partial M}{\partial x}\right) = D\frac{\partial}{\partial x}\left(\frac{\partial M}{\partial x}\right) \tag{6.6.39}$$

$$\rho c_p\frac{\partial \phi}{\partial \tau} + \rho_l c_p\frac{K_l C_s}{\mu_l \Delta M}\frac{\partial M}{\partial x}\frac{\partial \phi}{\partial x} = \frac{\partial}{\partial x}\left(k\frac{\partial \phi}{\partial x}\right) \tag{6.6.40}$$

where $C_s = \partial p_c/\partial S$, is called the capillary transport coefficient, in terms of capillary pressure p_c. This accounts for the changing capillary pressure level effect as the radius of liquid–gas interface curvature changes during the process. The bounding conditions are

$$\phi(x, 0) = 0 \qquad m''(L, \tau) = 0 \qquad (\partial \phi/\partial x)_{x=L} = 0 \tag{6.6.41}$$

The mass and energy balances between the instantaneous conditions across the layer and at the surface, at $x = 0$, are written in integral form, for convenience, as

$$\frac{d}{d\tau}\int_0^L M(x)\,dx = -m''(0, \tau) \tag{6.6.42}$$

and

$$\frac{d}{d\tau}\int_0^L \rho c_p\phi\,dx = h[t_e - t(0, \tau)] - Hm''(0, \tau) \tag{6.6.43}$$

where $m''(0, \tau)$ is the rate of water loss to the hot environment, per unit of surface area. It is written in terms of a mass transfer coefficient h_m and the water vapor pressure difference between the surface, $p(0, \tau)$, and the ambient,

p_e, as

$$m''(0, \tau) = h[p(0, \tau) - p_e] \qquad (6.6.44)$$

Numerical calculations of these formulations were made by Plumb et al. (1985). The results were also compared with experimental measurements of the air drying of plane southern pine-boards. The principal values needed for the calculation were

$$D = \exp[3.746 - 5.12\gamma - 4317/T(K)] \quad \text{in cm}^2/\text{s} \qquad (6.6.45)$$

$$C_s = 0 \quad \text{for } S > 0.802 \qquad (6.6.46)$$

$$C_s = \frac{\sigma}{2.7 \times 10^{-3}} \left(\frac{0.802}{S}\right)^{3/2} \quad \text{for } S < 0.802 \text{ in g/s}^2 \text{ cm} \qquad (6.6.47)$$

$$\sigma = 75.64 - 0.144T \quad \text{in gm/s}^2 \qquad (6.6.48)$$

$$K = 1 \times 10^{-12} \text{ and } 5 \times 10^{-12} \quad \text{in cal/cm s K} \qquad (6.6.49)$$

where γ is the specific gravity of dry wood compared to water, K is the permeability of the wood structure, and C_s is the capillary transport coefficient.

The calculations and measurements for one experiment are compared in Fig. 6.6.7. A growth ring curvature effect is apparent. Since radial permeability is much greater than tangential permeability, this curvature results in added resistance for this kind of wood sample. Using a lower permeability would produce better data agreement in the upper sets of curves, which show temperature response with time. However, the calculated moisture content response, for a permeability of 1×10^{12}, is in very good agreement with the data. An interesting result in Fig. 6.6.7 is that, at a drying time of about 2000 min, the value of M_R is about 0.1 or 10%. Therefore, 93% of the water was removed, including about $8\#/\text{ft}^3$ of bound water.

The preceding initial formulation and subsequent additional assumptions involve what would appear to include far-reaching approximations. Nevertheless, the realism of the results, compared to experiments, indicates that the residual effects may be small in the resulting overall transport response. Although the importance of the several measures of approximation is not apparent, such models may represent the physical reality well. A similar analysis is given by Perre (1987). Reference is also made to DeVries (1987), for a recent review of the kind of analysis and approximations used previously. Wang and Yu (1988) discuss the evaluation of the effective diffusivity in heat and mass transfer in moist porous media. Also see Arnaud and Fohr (1988) for a formulation appropriate to slow drying. Models are discussed.

NONCONTINUUM AND SURFACE DIFFUSION. The introduction to this section briefly discusses two kinds of mass diffusion processes which may not be realistically modeled in a purely Fickian form, that is, as $m'' = -D\,\nabla C$. One process arises in gas-phase diffusion processes, when the characteristic size of internal voids δ are comparable to or smaller than the mean free path of the gas

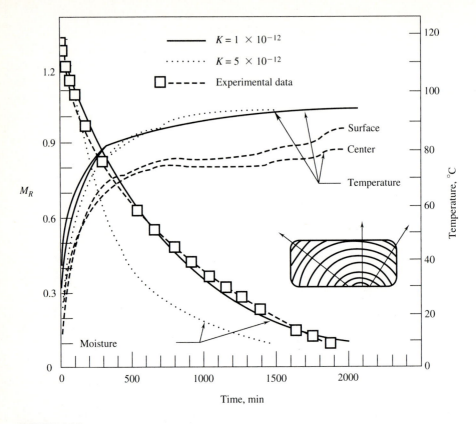

FIGURE 6.6.7

The comparison of numerical calculations, for both $K = 1 \times 10^{-12}$ and 5×10^{-12}, with experimental measurements, where M_R is the ratio of the mass of water content to the mass of completely dry wood, where the density of the dry wood is $\rho_{dw} = 35$ to $40\#/ft^3$. [From Plumb et al. (1985).]

molecules λ. Then the molecules more frequently collide with the surfaces than with each other. Thus the body of gas is not a region of local equilibrium, as achieved through intramolecular collisions. Such a condition is called a noncontinuum process.

Another common mechanism which is not modeled in Fickian form arises from the surface adsorption of diffusing molecules. A surface layer is formed. Imposed gradients of gas-phase concentration also result in gradients of concentration across the gas phase in the individual pores. Therefore, pore surface layer concentration gradients also arise along the surfaces. Surface adsorption was idealized in Sec. 6.5, as arising from distributed internal chemical reaction on the surfaces.

The gradients of surface layer concentration result in the diffusion of molecules in the direction of decreasing layer concentration. The surface diffusion rate, therefore, also depends upon the particular physicochemical processes which arise in the adsorbed molecular layer.

These two kinds of effects are discussed in terms of their characteristics. The formulations of the diffusion equations are given. A simple result follows for noncontinuum effects. Local chemical effects may also be accommodated in analysis. First, an analysis is given of the performance of a catalyst bed which oxidizes the carbon monoxide in automotive emissions. Then the mechanisms of vapor removal from air, in silica gel beds, are considered. The diffusion of surface-adsorbed material is a principal effect in this application.

NONCONTINUUM EFFECTS. The contribution of the noncontinuum or Knudsen effect is approximately modeled in terms of an effective diffusivity D_e, or resistance $1/D_e$, for the diffusion of a species C through the porosity of a gas-filled region. The value of D_e^{-1} is taken as the sum of the continuum and noncontinuum resistances, $1/D_C$ and $1/D_{NC}$. Using the form of Fick's law as

$$m'' = -D_e \nabla C \qquad (6.6.50)$$

D_e is expressed as

$$\frac{1}{D_e} = \frac{1}{D_C} + \frac{1}{D_{NC}} \qquad (6.6.51)$$

The component diffusivities are written in terms of the void fraction or porosity, V_p, and the tortuosity factor, T', as

$$D_C = V_p D/T' \quad \text{and} \quad D_{NC} = V_p D_{NC,I}/T' \qquad (6.6.52)$$

where D is the ordinary diffusion coefficient. The tortuosity factor accounts for the increased diffusion length resulting from the indirect diffusion paths in actual porous media. The fundamental noncontinuum component $D_{NC,I}$ is determined from molecular theory. Edwards et al. (1979), page 83, estimate its value in terms of effective pore size r_e and average molecular speed $v = \sqrt{8RT/\pi}$, where R is the gas constant per unit mass, as

$$D_{NC,I} = 2r_e v/3 = \sqrt{32RTr_e^2/9\pi} \qquad (6.6.53)$$

Therefore, D_{NC} is weakly dependent on local temperature, as $\sqrt{T}$. If D_C and V_p/T' are also uniform over the region, Eq. (6.6.50) applies in a very simple form. Then the simplest analyses, for D uniform, given before in this chapter and previous ones, apply directly. If the variation of T across the region is large, due to simultaneous heat conduction or chemical reaction, the thermal and mass diffusion processes are directly coupled in a much more complicated formulation.

An idealization of the catalytic reaction removal of CO in engine emissions, in a fixed bed of spherical catalyzing pellets, is seen in Fig. 6.6.8. The limiting process is taken to be CO mass diffusion into the porous pellet. That is, the CO concentration around each pellet of radius R is taken as C_s. The catalyst material is assumed to remain equally effective through time. The diffusion into the pellet reacts at the particle surfaces. This amounts to a distributed sink of CO, $C'''_{CO}(r)$, in the pellet volume. The flow resistance of the pellet material is taken as sufficiently high that there is no forced through-flow.

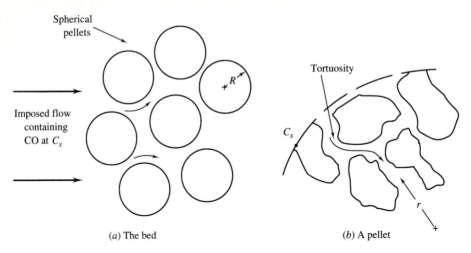

FIGURE 6.6.8

Oxidation of CO in a fixed bed of pellets carrying a catalyst through which exhaust gases flow: (*a*) the bed; (*b*) the surface region of a pellet.

The relevant equation is in terms of local concentration $C(r)$ and the reaction rate $C'''_{CO}(r)$. The reaction rate is equal to the product of the reaction rate per unit area, R'', the catalyst surface area per unit volume of catalyst, A_c, and the local concentration, $C(r)$. Therefore, $R''A_c C(r)$ is the CO chemical reduction rate, per unit of fixed-bed volume. This form is similar to that in Sec. 6.5. The equation for $C(r)$ and the boundary conditions in steady state are

$$\frac{d^2C}{dr^2} + \frac{2}{r}\frac{dC}{dr} - \frac{R''A_c}{D_e}C = 0 \tag{6.6.54}$$

$$C = C_s \text{ at } r = R, C \text{ is bounded at } r = 0 \tag{6.6.55}$$

This formulation assumes that the catalyst continually catalyzes the reduction reaction.

The preceding equation is of the form given in Sec. 2.4.1, for heat conduction with a uniformly distributed energy source q'''. Here, C'''_{CO} depends on location r. The solution, in terms of $b^2 = R''A_c/D_e$, is

$$\frac{C(r)}{C_s} = \frac{R\sinh(br)}{r\sinh(bR)} \tag{6.6.56}$$

The total rate of CO catalysis by the pellet, M_{CO}, is the sum of the surface flux $m''(R) = -D_e(\partial C/\partial r)$, evaluated at $r = R$ from Eq. (6.6.56):

$$M_{CO} = 4\pi R^2 m''(R) = 4\pi RD_e C_s[1 - bR/\tanh bR] \tag{6.6.57}$$

In the absence of diffusive resistance in the pellet, the catalysis, M_{CO}, is simply calculated as

$$M_{CO,i} = 4\pi R^3 R'' A_c C_s / 3 \qquad (6.6.58)$$

That is, C_s penetrates to all areas of the pellet surface, as for $D_e \to \infty$. The actual pellet efficiency is then defined as the ratio of M_{CO} to $M_{CO,i}$ as

$$\eta = \frac{3}{bR} \left[\frac{1}{\tanh(bR)} - \frac{1}{bR} \right] \qquad (6.6.59)$$

This efficiency is similar to that for heat transfer enhancement, achieved by putting fins on heat transfer surfaces. That analysis is given in Sec. 8.2. The analogy there to the preceding result is infinite thermal conductivity of the fin material. However, the foregoing analysis also includes the noncontinuum effect formulated in Eq. (6.6.51). This decreases the continuum diffusion, characterized by D_C^{-1}, by the additional resistance component D_{NC}^{-1}, to obtain the effective value D_e^{-1}.

THE FORMULATION OF SURFACE DIFFUSION. As discussed previously there are many processes in which surface-adsorbed layers of molecules diffuse. The adsorbed layer may have a concentration gradient due to the concentration gradient of the diffusing material in the pores. This pore diffusion may be either continuum or free molecule, as modeled previously. In the particular circumstance considered here, the pores are to be taken as very small, that is, $\lambda > \delta$. Therefore, the mechanism involves only surface and noncontinuum diffusion.

The process analyzed is that of water vapor adsorption in a packed bed of porous silica gel dessicant particles. The model for a particular particle is seen in Fig. 6.6.9. Surface-adsorped water molecules move inward along the surfaces of the particle porosity. At the same time, water vapor also diffuses through the

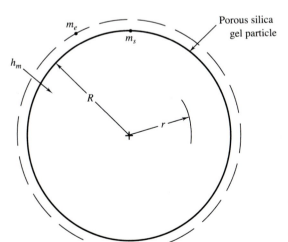

FIGURE 6.6.9
A porous silica gel particle, during the dehumidification of air, where m is the vapor mass fraction.

pores by the Knudsen mechanism. The results given next follow the analysis of Pesaran and Mills (1987).

A spherical gel particle of radius R is analyzed, assuming that the water vapor concentration is uniform around and out from the surface of the particle, as m_e, as shown in Fig. 6.6.9. The surface mass transfer coefficient is h_m. This is a batch process, in that the bed must be regenerated by heating, to drive off the adsorbed vapor, to begin a new cycle. Therefore, this analysis is of the transient adsorption phase, starting with a uniform initial gel particle temperature of t_i and water content in general, as $W_i = f(r)$, where $W(r, \tau)$ is the water/dry dessicant mass ratio distribution over the volume. The particle is suddenly exposed to humid air with a water vapor/humid air mass fraction, in general, as $m_e = F(\tau)$. The surface mass flux condition, $m''(0, \tau)$, is related to the mass fraction transfer coefficient h_m and m_e as follows, where $m''(R, \tau) < 0$:

$$m''(R, \tau) = h_m(m_s - m_e) \qquad (6.6.60)$$

Water molecules move through the porous material by both Knudsen and surface diffusion, with adsorption occurring along the pore walls. The adsorption and desorption processes are taken to be very fast compared to diffusion, for the small pores common in this application. Therefore, the local transient vapor concentration ρm and gel water content W are assumed to be in equilibrium. The water conservation equation, in terms of local water vapor mass flux $m''(r, \tau)$ and local mass fraction $m(r, \tau)$, is

$$\frac{1}{r^2}\frac{\partial}{\partial r}(r^2 m'') + \epsilon_p \frac{\partial(\rho m)}{\partial \tau} + \rho_p \frac{\partial W}{\partial \tau} = 0 \qquad (6.6.61)$$

where ϵ_p and ρ_p are the particle porosity and density and ρ is the density of the humid air. The mass flux $m''(r, \tau)$ results from the combination of Knudsen and surface diffusion and from any convective effects.

The Knudsen and surface diffusion components are each modeled, in terms of flux, as

$$m''_K(r, \tau) = -\rho D_K \frac{\partial m}{\partial r} \qquad (6.6.62a)$$

$$m''_S(r, \tau) = -\rho D_S \frac{\partial W}{\partial r} \qquad (6.6.62b)$$

Recall that $m(r, \tau)$ and $W(r, \tau)$ are defined differently.

These two diffusion effects are in parallel and are additive, to obtain $m''(r, \tau)$. Substitution in Eq. (6.6.61) gives

$$\frac{1}{r^2}\frac{\partial}{\partial r}r^2\left[D_K\rho\frac{\partial m}{\partial r} + D_S\rho_p\frac{\partial W}{\partial r}\right] + \epsilon_p\frac{\partial(\rho m)}{\partial \tau} + \rho_p\frac{\partial W}{\partial \tau} = 0 \quad (6.6.63)$$

where the first term is the sum of the two mass flux terms. The second and third terms are the changing local storage of vapor and water. The initial and boundary conditions are

$$W(r,0) = W_i(r) \quad \text{and} \quad \rho m(r,0) = \rho m_i(r) \qquad (6.6.64a)$$

$$m''(0,\tau) = 0 \quad \text{and Eq. (6.6.60)} \qquad (6.6.64b)$$

Local equilibrium between the vapor and the surface phase is assumed

$$\rho m(r,\tau) = g[W(r,\tau),t] \qquad (6.6.64c)$$

where g is the equilibrium isotherm $\rho m = g(W,t)$. The continuity of m at the interface amounts to the following condition:

$$m(R,\tau) = m_s(\tau) \qquad (6.6.64d)$$

Clearly, the preceding formulation also applies for either the Knudsen or the surface mechanism dominant in a particular process.

The preceding general formulation may be simplified in many ways, in the application to silica gel dehumidification. For example, the surface effect h_m is usually small compared to the internal conductance, D/s, of the gel particles of characteristic size s. This is the local diffusion Biot number approximation. This assumes a uniform internal particle concentration at each time level during the process. Also, the overall dehumidification process is sufficiently slow that several components of the transient terms in Eq. (6.6.63) may be deleted.

Several other approximations, along with values of the thermodynamic, diffusion, and materials properties, complete the formulation for the calcula-

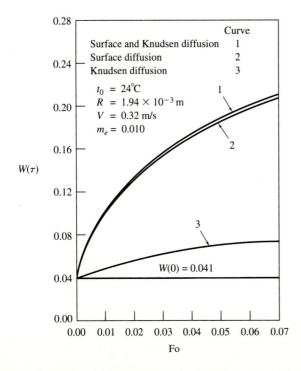

FIGURE 6.6.10

Increase of the average of dessicant-adsorbed water content, as $W(\tau)$, as a function of time as Fo = $D\tau/R^2$, for regular density particles; for an 1800-s run at $m_e = 0.01$.

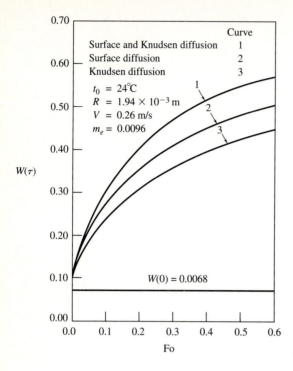

FIGURE 6.6.11

Increase of $W(\tau)$ for intermediate density particles; for a 1200-s run at $m_e = 0.0096$.

tions of the water take-up. Results are given in Figs. 6.6.10 and 6.6.11, in terms of the average dessicant-adsorbed water content of the bed, $W(\tau)$. This is the mass ratio of water content and dry dessicant. These figures are for regular- and intermediate-density silica gel particles. These have pore radii of 11 and 68 Å, respectively. The H–O bond length for the water molecule is 0.958 Å.

As a result of the approximations made to reach the final simpler formulation, these results apply to isothermal particles. The initial value of $W = W(0)$ is 0.041 and 0.0068, for the results shown. The Fourier number is $Fo = D\tau/R^2$, where D is a relevant initial equivalent value of the diffusion coefficient. Also, V is the superficial air velocity and m_e is the suddenly imposed mass of water vapor, per unit mass of humid air, that is, the vapor mass fraction. This is held constant after the step increase. The three response curves 1, 2, and 3 in Figs. 6.6.10 and 6.6.11 are: for curve 1, combined surface and noncontinuum; and for curves 2 and 3, for each surface and noncontinuum alone.

Comparing the curves in Fig. 6.6.10 indicates that the effect of noncontinuum diffusion is very small, compared to surface diffusion, for the 11-Å pore radius material. Figure 6.6.11, for 68-Å pores, shows that these two effects are comparable. This results from the longer mean free path, that is, the larger diffusion length in the larger pores. Curves 2 and 3 do not sum to curve 1 since these two effects are not simply additive, since these are nonlinear and intercon-

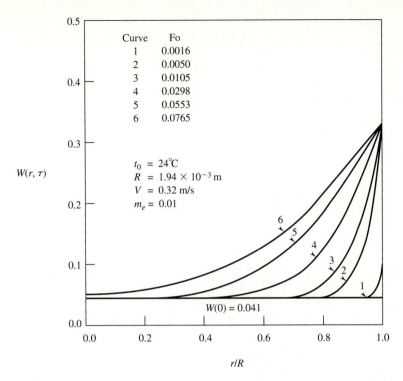

FIGURE 6.6.12

Distributions of $W(r, \tau)$ for regular density particles, at various times, in terms of Fo $= D\tau/R^2$, for an 1800-s run.

nected processes. Figure 6.6.12 shows the distribution of $W(r, \tau)$ at six time levels, for the conditions in Fig. 6.6.10.

The analysis by Pesaran and Mills (1987) also includes results for the transient response of a fixed bed of dessicant. The results of measurements are also given. These are in reasonable agreement with the calculated results. However, an uncertainty arises in the proper value of the effective diffusion coefficient. Also, the measurements suggest that hysteresis arises in cyclic-bed operation.

PROBLEMS

6.1.1. For transient diffusion in an orthotropic rectangular region, write out the relevant differential equation and bounding conditions in terms of concentration levels. Develop this formulation in terms of an equivalent diffusivity.

6.1.2. A plane region of thickness $2s$ has an initial concentration level of C_i. At $\tau = 0$ the surface concentrations are raised to C_0. The diffusivity is taken as time dependent as $D(\tau) = A\tau$.

(a) Write the simplest formulation of this transient response, taking $x = 0$ at the midplane.

(b) Determine the local concentration in terms of $\tilde{C} = (C - C_i)/(C_1 - C_i)$.

6.1.3. For the geometry and conditions in the preceding problem, D is constant and uniform. However, there is local adsorption at a rate of $c(x, \tau) = R_r C(x, \tau)$.

(a) Determine the midplane response $C(0, \tau)$ as $\tilde{C}(0, \tau) = (C(0, \tau) - C_i)/(C_1 - C_i)$.

(b) Plot this response as it depends on $D/(1 + R_r)$.

6.1.4. A layer of thickness L has surface layer concentrations of C_1 and C_2. The material has a location-dependent diffusivity of $D(x) = D_0(L/x)$.

(a) Write the governing equation and other conditions.

(b) Determine the concentration distribution across the layer.

(c) Calculate the mass diffusion rate across the layer.

6.2.1. Consider the plane layer in Fig. 6.2.1(a), for the diffusivity variable as $D(C) = D_0[1 + f(C)] = D_0[1 + bC]$.

(a) Calculate $C(x)$ as a function of b, C_1, C_2, and x/L.

(b) Plot the result as $\tilde{C} = (C(x) - C_2)/(C_1 - C_2)$ vs. x/L, for $b = 1$.

(c) Determine the diffusion rate m'' across the barrier.

6.2.2. For the circumstance in the preceding problem, determine the diffusion rate, as a function of b.

6.2.3. Consider the tubular region in Fig. 6.2.1(b), for the diffusivity variable as $D = D_0[1 + f(C)] = D_0[1 + bC]$.

(a) Calculate $C(r)$ as a function of b, C_1, C_2, and r.

(b) Determine the diffusion rate per unit of inside area.

(c) Determine the diffusion rate per unit of tube length.

6.2.4. The plane layer in Fig. 6.2.1(a) has the left and right surfaces maintained at C_1 and at $C_2 = 0$, respectively. Plot the concentration level at $x/L = 0.5$, as $\tilde{C}$, for the values of b shown in Fig. 6.2.2.

6.2.5. Consider $D(C) = D_0[1 + f(C)] = D_0[1 + be^{gC}]$, across a plane layer of thickness L. The surface conditions are $C(0) = C_0$ and $C(L) = 0$.

(a) Calculate the concentration distribution in terms of $\tilde{C}$.

(b) Determine the diffusion mass flux rate through the region.

(c) Compare the result of part (a) with that in Fig. 6.2.3.

6.2.6. Consider a plane layer of thickness $2s$. There is a plane source of species C at the midplane. This source keeps the concentration there constant at C_0. The surfaces are maintained at C_1. For the left and right halves of the region $D_L = D_0[1 + b_1 C]$ and $D_R = D_0[1 + b_2 C]$.

(a) Determine the concentration distribution across the layer.

(b) Determine the diffusion rate in both regions.

6.2.7. A plane layer, as shown in Fig. 6.2.1(a), has a location-dependent diffusivity, as $D(x) = D_0[1 + f(x)]$, where $f(x) = bx$.

(a) Determine the concentration distribution across the layer in terms of $\tilde{C}$.

(b) Calculate the diffusion rate.

(c) For the result in part (b), determine the value of the "effective" diffusivity of the region.

(d) Determine the diffusion resistance, per unit area.

6.2.8. For the spherical shell shown in Fig. 6.2.1(b), $D(r) = D_1(1 + r/r_1)$.
(*a*) Determine the concentration distribution.
(*b*) Calculate the diffusion rate through the shell.

6.2.9. For the plane region in Fig. 6.2.1(a), the diffusivity is both concentration- and location-dependent as $D(C, x)$. This dependence is to be approximated as $\bar{D}(C)f(x) = D_0(1 + aC)(1 + bx)$.
(*a*) Determine $C(x)$.
(*b*) Calculate the mass diffusion rate and the mass transfer resistance of the region.

6.3.1. A 1-cm-thick layer of water at 25°C is exposed to hydrogen, H_2, on each side. This causes a very rapid rise in surface concentrations, to C_{H_2}, at each surface.
(*a*) Determine the time at which this effect first reaches the midplane.
(*b*) Determine the total concentration of H_2 in the layer at that time, in dimensionless form.
(*c*) Repeat part (*b*) if the surface concentration is increased exponentially as in Eq. (6.3.1).

6.3.2. A thin plane region of thickness $2s$ has an initial concentration level of C_i. An external source of species C, on each surface, increases the surface concentration rate as given in Eq. (6.3.1).
(*a*) Evaluate the concentration level response at the midplane, in terms of $\tilde{C} = (C - C_i)/(C_f - C_i)$, at a time when $\beta\tau = \beta s^2/D = \mathrm{Fo}_s = 1$. Evaluate only the first few terms.
(*b*) Determine the total sorption $\bar{M}$ at this time.

6.3.3. The plane region in the preceding problem is a 0.2-mm-thick foil of nickel, subject to the inward diffusion of hydrogen. Calculate the time at which $\mathrm{Fo}_s = 1$ and also the value of β.

6.3.4. Repeat the calculation in the preceding problem, neglecting any convection effects, for:
(*a*) A layer of water at 25°C.
(*b*) A layer of air at 1 atm, for $C_f - C_i$ being, relatively, a very small concentration difference.

6.3.5. A plane layer of iron of 1-mm thickness at a temperature of 1100°C initially contains no carbon. At $\tau = 0$ a carburization effect is imposed at each surface. This amounts to a surface concentration level of $C_1(\tau)$ which increases linearly with time as $C_1(\tau) = K\tau$, where $K = 1$ for time τ in units of hours.
(*a*) Calculate the time at which this effect would reach the center of the layer.
(*b*) Determine the total sorption at this time.

6.3.6. A steel rod of radius R is to be case-hardened in the surface layer by the diffusion of carbon into the surface at high temperature. The rod is packed in a high-carbon solid material. The initial carbon concentration in the steel is 0.3% by weight. This assembly is quickly heated to a high temperature, at which the carbon diffusion coefficient in steel becomes 6×10^{-6} cm^2/s. Surface equilibrium considerations indicate that the concentration in the immediate surface region of the rod will be 1.8%.
(*a*) Calculate the required time at high temperature for the carbon concentration at a distance of 0.06 cm below the rod surface to reach 1%.
(*b*) Determine the carbon flux at the rod surface at this time.

6.3.7. A plane layer is immersed in a well-stirred limited ambient medium at $C_a, C_i = 0$.
  (a) Plot the sorption in dimensionless terms as a function of time, for the large and small volume fractions 10 and 0.1.
  (b) Compare the results in part (a) with that for an infinite ambient.

6.3.8. Consider the circumstance in the preceding problem, in part (b). Plot the ratio of sorption with a partition factor $K = 0.5$ to that for $K = 1$, for an infinite ambient on each side.

6.3.9. Consider the infinite solid transient in Fig. 6.3.6(b), formed of regions uniformly at C_1 and C_2, respectively, placed in contact, with $D = D_1(C/C_1)$.
  (a) Determine the penetration distance of the diffusion effect as a function of time.
  (b) Calculate the velocity of the front and compare this result with that in Fig. 6.3.7(a), in terms of differences in the gradients generated in the two processes.

6.3.10. From Fig. 6.3.8 estimate the location and velocity of the desorption effect front for each of the three formulations of diffusivity. Relate these results to the variations of $D(C)$.

6.3.11. Consider sorption and desorption in a semiinfinite region when $D(C)$ varies as $D_0(1 + bC/C_0)$. For each effect, draw a plot of:
  (a) The penetration distance effect as a function of b, that is, two curves.
  (b) The propagation velocity of the front, two curves.
  (c) Explain the differences between the two curves in parts (a) and (b).

6.3.12. Consider the water vapor loss in air drying at the exposed surface of a very thick region of porous wood, at 30°C, containing water vapor. The initial vapor internal concentration is in thermodynamic equilibrium at 30°C. The outside air is also at 30°C but at 50% relative humidity. For the surface convective process, $h_m = 1$ cm/s.
  (a) Calculate the total mass of the vapor losses at 1 and at 5 hr after the beginning of the process.
  (b) Determine the surface concentration levels at these times.

6.3.13. Consider determining the diffusivity of CO_2 through a bed of thickness L, of fine glass beads, using the transient method shown in Fig. 6.3.16. Experimental test times are to be limited to 3 hr. Determine the relation between the bed thickness and the expected diffusivity, D, which would yield results within this time limit.

6.4.1. Consider a semiinfinite region which has a zero local level of diffusivity D at concentrations less than C_X. The initial concentration is zero. The surface is suddenly exposed to a concentration level of $C_1 > C_X$.
  (a) Determine the rate of front propagation for $C_1/C_X = 5/6$, 2/3, 1/2, and 1/3.
  (b) For $C_1/C_X = 5/6$ estimate the amount of sorption which has occurred, $M(\tau)$, at time τ into the transient.

6.4.2. A newly formed plane layer of iron of 0.2-mm thickness is exposed to oxygen which diffuses inward and oxidizes the iron through a local chemical reaction. This occurs on both sides. Assume that $V_{MO} = 0.5$.
  (a) Using the approximate analysis, calculate the time at which the oxidation effect has penetrated over the whole region, for $D = 1.2 \times 10^{-8}$ cm^2/s.
  (b) Compare this result for the diffusion time of helium into quartz (as SiO_2), to the midplane, without chemical reaction.

6.4.3. Consider a plane porous layer of thickness $2L$ consisting of layers A and B, each of thickness L, in intimate contact. Gas d is diffusing to the right and gas e to the left. The diffusivity of gas d is $D_{d,A} > D_{d,B}$ in the region $x = 0$ to L and $x = L$ to $2L$, respectively. For gas e, $D_{e,A} < D_{e,B}$. The pressures at the two surfaces at $x = 0$ and $2L$ are maintained at $p_{d,0} > p_{d,2L}$ and $p_{e,0} < p_{e,2L}$.

(a) Determine the internal pressure distribution of each gas, d and e, and the sum of the two.

(b) Calculate the transport rate of each gas through the layer.

(c) For $D_{d,A}/D_{d,B} = D_{e,B}/D_{e,A} = 2$, plot the distributions of $(p_d - p_{d,2L})/(p_{d,0} - p_{d,2L}) = P_d$, P_e, and $P_d + P_e$.

6.4.4. Consider a resistive skin, of thickness L on an extensive very high diffusivity region. The initial concentration is C_i and $C(0, \tau) = C_1$.

(a) Determine the mass of diffusing solute contained in the skin, compared to that over the whole region, as a function of time.

(b) Evaluate this relation for short times and indicate the significance of the result.

6.4.5. A molten metal, having a contaminant concentration C_c, is quickly poured into a large cold mold. The advancing solid phase has no appreciable solubility for the contaminant.

(a) Determine the propagation velocity of the front, using the properties of molten steel.

(b) Determine the phase interface concentration, in terms of C_c, for $D = 10^{-9}$ cm^2/s in the melt.

(c) Does the approximation in Eq. (6.4.24) apply for the values used previously?

6.4.6. Consider the zone-refining method shown in Fig. 6.4.7. Assume a partition factor of 0.5 and a diffusion coefficient of 10^{-8} cm^2/s.

(a) For silicon, with a heating coil velocity of 0.5 cm/hr, calculate the ratio of the impurity difference at the phase interface to that in the melt ahead.

(b) Determine the number of cycles necessary to lower the initial concentration C_i to 10% of its initial value.

6.5.1. For the general reaction rate formulation, $c = R_r C''$, determine the surface flux, per unit area, in terms of the parameters of the formulation.

6.5.2. As a comparison of linear, constant, and nonlinear local adsorption dependence:

(a) Compare the levels of adsorption at Fo $= 1$ and 2.

(b) Interpret these results in terms of the effects of the several values of n.

6.5.3. For linear sorption show that Eq. (6.5.12) satisfies Eq. (6.5.9).

6.5.4. Consider linear sorption as formulated in Eq. (6.5.9). The temporal response, in terms of $K\tau$, is shown in Fig. 6.5.2, as well as the asymptotic trend.

(a) Plot the ratio of the full calculation to the approximate one, over the $K\tau$ range shown.

(b) In terms of the actual process, indicate the physical effects which make the full calculation response slower at smaller $K\tau$.

6.5.5. Consider a plane layer in which immobilization, as R_F, is countered by a reverse reaction which frees species C. Take the ambient region on each side as of infinite extent.

(a) For a partition of $R = R_F/R_R = 10$, determine the progress to equilibrium achieved, at a given time, as $D\tau/(R + 1)s^2 = 1$, for different rates of the reverse reaction.

(b) Make the same determination at a value of 0.5 of the time variable and explain the physical effects, which cause the differences which arise.

(c) Indicate the direction in which these responses would differ in a limited extent of ambient material.

6.6.1. Consider a dry plane layer at a temperature of t_i and thickness $2s$, placed in a vapor at t_e. Condensation at the surfaces heats it. The condensate diffuses inward in a Fickian mode, with a coefficient D. Assume that the region is not saturated at the time when complete equilibrium is attained, at t_e. Neglect the energy effect of condensate diffusion inward, as appropriate for a vapor with a large latent heat release at the surface.

(a) Write the differential equations which govern the thermal and condensate diffusion processes.

(b) Formulate the initial and boundary conditions which apply, for an initial condensate concentration of zero.

(c) Indicate the kinds of coupling which arise, in terms of this formulation, between the local temperature and condensate distributions, in terms of $t(x, \tau)$ and $C(x, \tau)$.

6.6.2. Show that Eqs. (6.6.20) and (6.6.21) are solutions of the differential equations given for the two regions of the transpiration process.

6.6.3. For transpiration cooling, determine the temperature distributions in both the approach and barrier regions for both $q_1 = 0$ and $q_1 \neq 0$.

6.6.4. For transpiration cooling, plot the generalized temperature profiles across the whole region for both $q_1 = q_2$ and for $q_1 = 0$ for each ML and $NL = 0.2$ and 2.0 and $k_e/k = 1$.

6.6.5. A porous concrete structural wall 20 cm thick is subject to a hot gas, at 100°C on one side. A convection coefficient of 100 W/m² °C applies on that side. Air at 25°C is to be supplied on the other side at the rate necessary to keep the hot side of the wall at 75°C for this design condition. Assume a porosity of 20% for the concrete and calculate the barrier conductivity on the basis of a volume average of air at 25°C and ordinary concrete.

(a) Find the necessary flow rate of air in steady state, per unit of wall area.

(b) Determine the temperature level at the cool side of the barrier.

6.6.6. Consider the transient response of a transpiration-cooled plane barrier initially at t_i, placed in an environment at t_0 with a relatively very high convection coefficient h_0. The transpiring fluid remains at t_i, with also a very high convection coefficient.

(a) Write the equations and relevant bounding conditions.

(b) Determine the time dependence of the outside surface of the barrier, as a function of time.

6.6.7. Consider the porous barrier heat protection model in Fig. 6.6.5, for a long cylindrical tube with the transpiring air in the inside. Assume that the inside convection coefficient is very large, compared to the value h_0 to the air on the outside at t_0. Also neglect the axial temperature increase of the inside air flow.

(a) Develop the appropriate differential equation and boundary conditions for the temperature distribution in the tube wall, based on r_i and r_0.

(b) Determine the temperature distribution through the wall in terms of t_i, t_0, and other parameters of the circumstance.

6.6.8. Consider the air drying of wood. Show that the general relations, Eqs. (6.6.31) and (6.6.33), with two additional formulations, result in Eqs. (6.6.39) and (6.6.40).

6.6.9. For the results in Fig. 6.6.7, the density of dry wood may be taken as $\rho_{dw} = 40\#/ft^3$. The initial value of M_R is seen to have been about 1.4. Determine:
   (*a*) The initial wood density and the fractional water density.
   (*b*) The percentage of the water removed and the amount of bound water removed, per unit volume of the wood.

6.6.10. For CO gas diffusion in a catalyst pellet, determine, for a temperature level of 150°C:
   (*a*) The average molecular speed.
   (*b*) The diffusion coefficient for an average pore size of 5 μm.

6.6.11. Consider a CO catalyst in a bed of pellets, the parameter of the process may be taken as $R''A_c/D_eR$.
   (*a*) Plot the local concentration in a pellet, as $C(r/R)_{CO}/C_s$ vs. r/R, for the values of the dimensionless parameter $R''A_cR^2/D_e$ of 1 and 10.
   (*b*) Determine the pellet efficiency for each of these values.

6.6.12. Compare the adsorption concentration time response, for regular and intermediate density particles, at Fo = 0.07, in terms of $W(\tau)/W(0)$. Discuss the effects of pellet density and air velocity in accounting for this difference.

6.6.13. Compare the form of the transient responses shown in Fig. 6.6.12, with the comparable responses, in terms of ϕ, for inward transient heat conduction in a spherical solid subject to a step increase in environment temperature t_e. Scale $t(r, \tau) - t_e$ with a suitable Δt_{ch}. Choose a Biot number and α, in Fo = $\alpha\tau/R^2$, which gives good agreement at R at τ = 1800 s. Comment on reasons for the differences.

REFERENCES

Arnaud, G., and J.-P. Fohr (1988) *Int. J. Heat Mass Transfer*, **31**, 2517.

Bird, R. B., W. E. Stewart, and E. N. Lightfoot (1960) *Transport Phenomena*, John Wiley and Sons, New York.

Carslaw, H. S., and J. C. Jaeger (1959) *Conduction of Heat in Solids*, Clarendon Press, Oxford.

Chalmers, B. (1964) *Principles of Solidification*, John Wiley and Sons, New York.

Chao, B. T. (1963) *Appl. Sci. Res.*, **A12**, 134.

Chapman, S., and T. G. Cowling (1939) *The Mathematical Theory of Non-Uniform Gases*, Cambridge Univ. Press, Cambridge.

Crank, J. (1975) *The Mathematics of Diffusion*, 2nd ed., Oxford Science Publications, Clarendon Press, Oxford.

Cussler, E. L. (1985) *Diffusion Mass Transfer in Fluid Systems*, Cambridge Univ. Press, Cambridge.

Danckwerts, P. V. (1951) *Trans. Faraday Soc.*, **47**, 1014.

DeVries, D. A. (1987) *Int. J. Heat Mass Transfer*, **30**, 1343.

Edwards, D. K., V. E. Denny, and A. F. Mills (1979) *Transfer Processes*, 2nd ed., Hemisphere, New York.

Glassman, I. (1987) *Combustion*, 2nd ed., Academic Press, New York.

Henry, P. S. H. (1948) *Discuss. Faraday Soc.*, No. 3, 243.

Kirkaldy, J. S., and D. J. Young (1987) *Diffusion in the Condensed State*, The Institute of Metals, London.

Mallard, E., and H. L. LeChatier (1983) *Ann. Mines*, **4**, 379.

Michaels, A. S. (1989) *Chemtech*, **19**, 162.

Pasternak, R. A., J. F. Schimscheimer, and J. Heller (1970) *J. Polymer Sci.*, **A2**, 8, 167.

Perre, P. (1987) *Int. Comm. Heat Mass Transfer*, **14**, 519.

Pesaran, A., and A. F. Mills (1987) *Int. J. Heat Mass Transfer*, **30**, 1037 and 1051.

Peterlin, A., and J. L. Williams (1971) *J. Appl. Polymer Sci.*, **15**, 1493.

Plumb, O. A., G. A. Spolek, and B. A. Olmstead (1985) *Int. J. Heat Mass Transfer*, **28**, 1669.

Rogers, C. E., and S. Sternberg (1971) *J. Macromol. Sci. B*, **5**(1) 189.

Rohsenow, W. M., J. P. Hartnett, and E. Ganić (1985) *Handbook of Heat Transfer Fundamentals*, McGraw-Hill, New York.

Schneider, P. J. (1955) *Conduction Heat Transfer*, Addison-Wesley, Reading, MA.

Shampine, L. F. (1973) *Q. Appl. Math.*, **30**, 441.

Spillman, R. W. (1989) *Chem. Eng. Prog.*, **85**, 41.

Sternberg, S., and C. E. Rogers (1968) *J. Appl. Polymer Sci.*, **12**, 1017.

Wang, B. X., and W. P. Yu (1988) *Int. J. Heat Mass Transfer*, **31**, 1005.

Whitaker, S. (1977) *Adv. in Heat Transfer*, **13**, 119.

CHAPTER
7

COMPOSITE TRANSPORT REGIONS

7.1 MECHANISMS

Chapter 1 formulates heat conduction and mass diffusion in terms of general and special differential equations. Both the thermal conductivity and the diffusion coefficient are formulated in their tensor forms, Eqs. (1.2.3a) and (1.3.2a), that is each may have nine coefficients. These equations are then reduced to the simpler forms which apply to the large majority of mechanisms of most interest. Chapters 2 through 5 then apply these latter equations to one-dimensional and multidimensional steady and unsteady regions, including variable properties, internal generation, and mass diffusion. More general and complicated mass diffusion processes are then analyzed in Chap. 6.

This chapter considers several very important mechanisms which lie beyond those simpler formulations. The first is contact resistance between adjacent conduction regions, as briefly formulated and applied in Sec. 2.1.4. Recall Figs. 2.1.4 and 2.1.5. This kind of resistance is described there and the effects are assessed in a very simple model. Section 7.2 discusses the constituent mechanics in some detail and shows the basis for calculating its effects in an assembly of conduction regions. Section 7.3 then considers the component mechanisms of contact resistance, in terms of their calculation and estimation.

The second application in this chapter, in Sec. 7.4, concerns what are often called composite materials. These include a wide range of commonly used arrangements, wherein a matrix material contains inclusions of another material. A solid with pores is an example. However, the generalization is a matrix

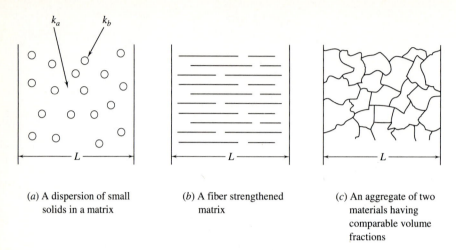

(a) A dispersion of small
solids in a matrix

(b) A fiber strengthened
matrix

(c) An aggregate of two
materials having
comparable volume
fractions

FIGURE 7.1.1
Characteristic composite regions, of effective conductivity k_e.

material containing inclusions of another material. There is a convention to think of both materials as a solid. However, they need not be for conduction analysis, if internal convective effects do not arise. Common examples of composites are a dispersion of small solids or voids in the matrix, fiber-strengthened structural members, and aggregates like concrete and granular solids. See Fig. 7.1.1. The composite in Fig. 7.1.2 is formed of irregular continuous regions of comparable extent, or fractions. The analysis of such conduction is then complicated by the requirement of matching heat flow conditions across all interfaces between regions of different conductivity.

FIGURE 7.1.2
A region formed of materials having directional conductivity effects.

Composites are first considered which are dilute in the embedded material. That is, the volume fraction of this component, defined as Φ, is small compared to that of the matrix, of volume fraction $1 - \Phi$. Results are then given for composites having more comparable volume fractions for the two components, as in many aggregates and granular materials.

Section 7.5 concerns composite insulation. There are innumerable circumstances in which insulation is interposed between hot and cold regions, to reduce the heat transfer rate. The object is either to reduce the heat loss from a hot region, or to reduce the heat gain from a cold region. Examples are a heated space in winter and a cryostatic vessel containing a liquified gas, such as LN_2.

The objective of the insulation, in both circumstances, is to reduce the heat flux across the barrier. Many simple and complicated examples arise such as fur, building insulation, and evacuated layers. In most applications and configurations, several barrier region heat transfer modes operate simultaneously. One is conduction through the material of the barrier, which may be a fibrous, granular, or porous material. Then conduction is through the material, and across the direct contact areas between material elements. Contact conduction is discussed in Secs. 7.2 and 7.3, in terms of the mechanism of constriction resistance.

Another mode is heat transfer across any gas which occupies the voids in the material. This is similar to gap conduction considered in Sec. 7.3. It may be convective transfer, for high-void fraction materials, in a buoyancy force field. For very small voids, or for an evacuated barrier, this gap component may be pure conduction, perhaps even noncontinuum conduction, as discussed in Sec. 7.3.4 for gaps in contact regions.

An often larger heat leak arises from thermal radiant emission from the surfaces of the filler material in the barrier. Net radiant transport arises in all regions, due to the temperature gradient in the material. This effect may be decreased by making the filler opaque and/or by supplying thin highly reflective metalized layers, across the barrier. A simple example of this measure is the simple evacuated silvered-glass Dewar flask, or vacuum bottle.

Section 7.5.1 first summarizes the performance of typical kinds of insulating barriers, principally in terms of their effective conductivity k_e. Then the individual heat transfer modes are discussed in Sec. 7.5.2, as though they operate independently. Section 7.5.3 evaluates radiation, gas, and filler conduction components. Section 7.5.4 then analyzes overall heat flux for an evacuated multilayer barrier, with optically thin spacers. This is one of the most superior insulations for cryogenic liquid storage. For example, the complete boil-off time for a 34-liter container of LN_2 at 77 K may be as long as 200 days.

7.2   CONTACT RESISTANCE PROCESSES

Assemblies of conduction elements into a composite barrier to heat conduction, or to mass diffusion, commonly result in appreciable intraelement resistances,

R_c, across the contact regions between the layers. These contact regions usually arise as irregular gaps, between adjacent elements, as modeled in Fig. 2.1.4. The gap region may contain a material of relatively low conductivity, compared to those of the adjacent regions, k_a and k_b in Fig. 2.1.4, as with an air space between solids. Sometimes the gap contains a material of higher conductivity. In critical applications, such as in the conduction cooling path for compact electronics, a deforming soft material of higher conductivity is often provided in the space between the adjacent pressure-mated conduction regions.

Therefore, the importance of the contact region resistance R_c depends on the relative magnitude of the effective gap conductivity, k_g, compared to k_a and k_b. If all the values of k are about the same, the contact resistance effect is relatively small. If k_g is much greater than k_a and k_b, conduction through the composite region is enhanced. However, for k_g considerably less than k_a and k_b, the value of R_c becomes relatively much more important in the total conduction path. In composite insulating walls, a low k_g in the contact region is an advantage.

The mechanisms of contact resistance are discussed initially in Sec. 2.1.4, in conjunction with composite barriers. Contact conductances, $C_c = R_c^{-1}$ are given for a group of typical materials; see Table 2.1.1 and Fig. 2.1.6. The effect of mating pressure is shown. The present and following sections consider the components of contact resistance in detail.

7.2.1   The Components of Contact Resistance

The evaluation of any particular contact resistance depends on many additional aspects of the actual contact region, between adjacent conduction regions such as a and b in Fig. 2.1.4. In any subregions of actual physical contact, the conduction is usually direct and very effective between the two adjacent conduction regions a and b. The relative magnitude of this contact region effect, overall, depends on both the direct contact area sizes and the number of these direct contact areas, that is, on their area and their density, per unit area of the contact region. The density is often high when each surface is relatively flat. Also, the extent of each direct contact area is larger when a higher pressure is applied to the assembly. Then the regions in contact deform into larger areas. This deformation may be elastic. However, in many applications, plastic deformation is a much more important effect.

CONSTRICTION RESISTANCES. Another completely related and often very large effect arises in the adjacent conduction regions, in their interaction with the individual contact areas. It is called constriction resistance. This is interpreted in terms of the contact region in Fig. 2.1.4, enlarged in Fig. 7.2.1. Three direct contact areas are shown. For a relatively small gap space conductivity, k_g, much of the total heat transfer is across these direct contact areas, if they are numerous. The heat flow across each of these three contact areas, for $t_a > t_b$,

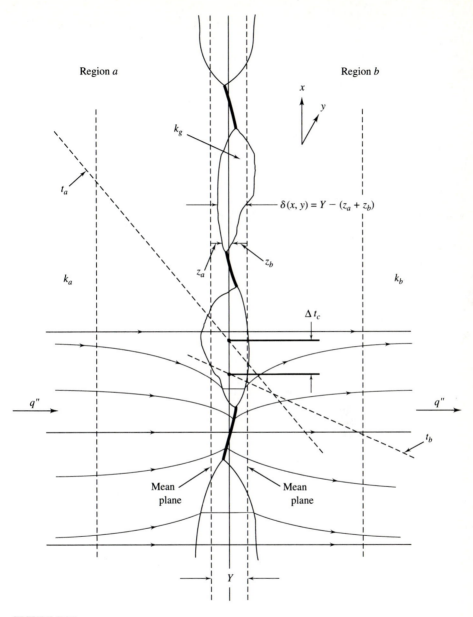

FIGURE 7.2.1
Direct contact areas between regions a and b, showing the contact region temperature difference, Δt_c.

contracts from region a, crosses the area, and expands outward again into region b. This amounts to a constriction resistance.

Some approximate paths of heat flow are sketched to demonstrate this process. Some of these paths supply the heat flow across the remaining gap regions. However, if k_a and k_b are considerably greater than k_g, most paths converge as shown, to cross through the direct contact areas. The local heat flux increases in these regions, with increasing convergence, as does the temperature gradient. On the other side, in region b, a similar effect arises.

These resulting increases in temperature gradient are the effect which causes, in part, the net temperature drop Δt_c in the region across the contact resistance. This effect arises because of the constriction of the heat flow as it approaches and is conducted away from the contact areas. This might be modeled, as shown in Fig. 7.2.2. Each contact area $A_{c,i}$ is associated with a

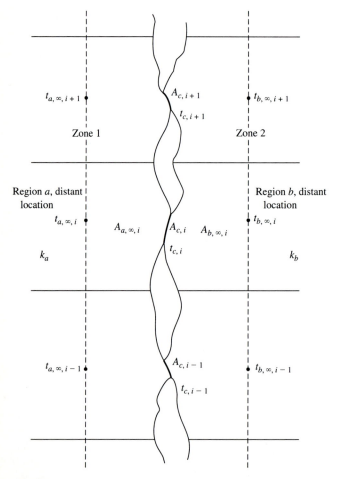

FIGURE 7.2.2
The constriction model of contact resistance, for direct contact areas $A_{c,i}$.

more-or-less distant extensive region of area $A_{\infty,i}$ on each side, depending on k_a and k_b. The temperature in region a just before constriction is $t_{a,\infty,i}$, and in region b, after constriction, the temperature is $t_{b,\infty,i}$. Here, ∞ denotes the relatively short distance away, normal to the contact region, where the temperature fields in regions a and b are little affected by the constrictions and are essentially one dimensional.

The constriction resistances then become the conduction resistances R_{ac} and R_{bc} between the direct contact area $A_{c,i}$ and the two locations designated as $t_{a,\infty,i}$ and $t_{b,\infty,i}$ in regions a and b. These are evaluated in Sec. 7.3. At the interface of these two regions, that is, across the actual direct contact area $A_{c,i}$, there may also be a resistance R_{dc}, due to imperfect contact. This is then added to R_{ac} and R_{bc}, to give the total resistance associated with the contact area $A_{c,i}$, as shown in the resistance diagram in Fig. 7.3.1(b) (see page 407).

GAP RESISTANCE. Heat is also transferred across the remaining gap, in Figs. 7.2.1 and 7.2.2. This effect is more important when the spatial density of direct contact areas, and their sizes, are not very large or when k_g is very much smaller than k_a and k_b. The flux across this gap may depend on some combination of conduction, radiation, and convection, depending on the particular circumstance.

In general, the temperature distributions along each of the exposed surfaces in the gap are not uniform, as suggested by the heat flow paths sketched in Fig. 7.2.1. Also, the thickness of the gap is usually not uniform between contact areas. Therefore, the local heat flux across a gap is variable. It may depend on a number of relatively independent effects which are frequently difficult to evaluate.

7.2.2 Characteristics of Contact Regions

The foregoing discussion of the series and parallel components of contact resistance indicates several of the principal mechanisms, and their interactions, which result in the overall contact region resistance, R_c. However, additional specifications of materials, geometry, and other aspects are necessary to begin a determination of actual values. These characteristics must be clarified for each particular kind of application.

The additional information which is required is conveniently divided into categories in terms of: surface geometries, at both large and small scale; deformability; applied pressure level; and the addition of treatments in the contact region. The considerations and measures in these categories include:

1. Geometry:
   (a) Conforming surfaces, such as two surfaces locally parallel over their extent.
   (b) Nonconforming or bent surfaces, when local nonparallelisms result in sizable local regions of no direct contact areas.

(c) Surface roughness and texture, in terms of the distributions of surface area peaks, their heights, and solidity.

2. Deformability in direct contact regions, $A_{c,i}$:
   (a) As related to the hardness and ductility of the solid materials used.
   (b) In terms of the underlying material support against deformation of the peaks, in case (1c), which will form contact areas.

3. Pressure:
   (a) Changing mating pressure effects on direct contact area density and surface area $A_{c,i}$.
   (b) How general deformation may improve local conformity.
   (c) Effect of deformation on local gap thickness, $\delta(x, y)$.

4. Treatments in the contact region, such as:
   (a) An added sheet of soft material.
   (b) A conductive grease layer.
   (c) A deposited soft layer on one or both sides, to increase contact areas $A_{c,i}$.

The preceding factors interact with other quantitative physical considerations to determine the net conductive behavior. A principal consideration is the constriction resistance. Another is the deformation at locations of direct physical contact. These factors are a part of the model of the parallel conduction paths across the gaps and the direct contact areas. The effect of thermal transients and cycling is also often very important. Such questions are considered in the next section, in the determination of the net contact resistance R_c and how it depends on several of the most important effects.

7.3   ANALYSIS OF CONTACT RESISTANCE

Section 7.2 describes many aspects of the mechanisms and considerations which are basic to the rational evaluation of contact resistance and to the assessments of the changing effects in different applications. The principal features which control the process are:

1. The surface density and the geometry of the direct contact areas, $A_{c,i}$, which result from roughness.

2. The elastic and plastic deformation which occurs at these contacts.

3. The constricted conduction which occurs in the two conduction regions adjacent to each local direct contact area.

4. The conductance across the gap between adjacent direct contact areas.

5. Any treatments added to the contact region, during assembly, to reduce overall contact resistance R_c.

7.3.1 The Contact Region Resistance Model

The surface region, $A_{T,i}$, associated with each local direct contact area $A_{c,i}$ is seen in Fig. 7.3.1. The total parallel–series resistance R_i is approximated as

$$\frac{1}{R_i} = \frac{1}{R_{ac,i} + R_{dc,i} + R_{bc,i}} + \frac{1}{R_{g,i}} \qquad (7.3.1)$$

where $R_{ac,i}$ and $R_{bc,i}$ are the two constriction resistances adjacent to $A_{c,i}$, $R_{dc,i}$ is any resistance remaining between regions a and b across their direct contact area $A_{c,i}$, and $R_{g,i}$ is the resistance of the associated gap of area $A_{T,i} - A_{c,i}$.

This calculation of R_i is an approximation. The temperatures along the two gap surfaces, $t_{ag}(x, y)$ and $t_{bg}(x, y)$, between adjacent contacts, are each taken uniform as t_a and t_b, to write R_i as in Eq. (7.3.1). This is often an

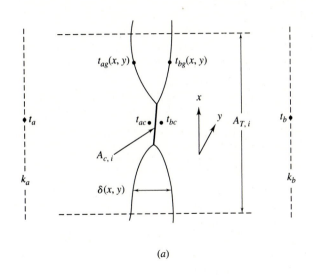

(a)

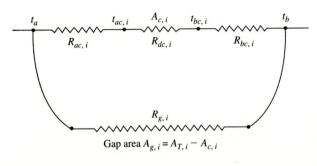

(b)

FIGURE 7.3.1
A simplified resistance model of the region A_T associated with a direct contact area $A_{c,i}$: (a) the characteristic contact area $A_{T,i}$; (b) the overall resistance of area $A_{T,i}$.

accurate approximation when R_g is much greater than both R_{ac} and R_{bc} or R_{dc}.

The heat flow, Q_i, across the area $A_{T,i}$, associated with $A_{c,i}$, is

$$Q_i = (t_a - t_b)_i / R_i \qquad (7.3.2)$$

The heat flux q'' per unit area of contact region is then the sum over the regions i lying in a unit area of the contact region:

$$q'' = \sum_i Q_i = (t_a - t_b) \sum_i \frac{1}{R_i} = \frac{(t_a - t_b)}{R_c} \qquad (7.3.3)$$

where $(t_a - t_b)$ in Eq. (7.3.2), defined in Fig. 7.3.1(a), is taken as constant along the total contact region of unit area. This would apply for one-dimensional conduction processes in each region a and b, at the locations shown. In Eq. (7.3.3), $\sum_i (1/R_i)$ is the total conductance, C_c, per unit area of the contact region and $1/C_c = R_c$ is the total resistance.

The following sections consider components of the preceding listing of the five principal features which affect R_c. First, the characterization of surface roughness and the deformation of the direct contact regions $A_{c,i}$ are discussed. The modeling of constriction resistance is then set forth. Evaluation of gap resistance is then in terms of the heat transfer processes and local thickness, $\delta(x, y, p)$, at applied pressure p. The last section concerns additional aspects of gas conduction and the noncontinuum heat transfer effects which commonly arise in gas-filled gaps. A review is also given of common methods of reducing gap resistance.

7.3.2   Roughness and Deformation

The pressure contact of a rough surface, with an idealized flat surface, is seen in Fig. 7.3.2. Interference at several of the peaks has caused local deformations and also plastic deformations of some asperities. In many applications, most of the actual contacts undergo plastic deformation. As a result of their greater

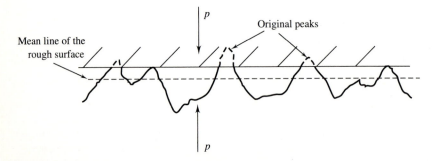

FIGURE 7.3.2
An idealized pressure-deformed contact region.

number and larger direct contact area, they usually contribute a large fraction of the conductance component due to direct contact areas.

ELASTIC DEFORMATION. This effect may be approximately calculated, as between the two contacting rough surfaces, shown in Fig. 7.2.1. This is commonly done for two corresponding asperities: in terms of their elasticities, E_a, E_b; the Poisson ratios μ_a, μ_b; the two radii of curvature which describe each asperity, ρ_a, ρ'_a and ρ_b, ρ'_b; for different angles of alignment of the two aspertities. See Seely and Smith (1963).

An elliptical direct contact area results. This area, $A_{i,e}$, depends, at least approximately, on the load carried elastically, $p_{i,e}$, as $p_{i,e}^{2/3}$. The ratio of the major and minor axes of the elliptical area depends on the initial radii of curvature and the other solid properties listed previously. The contact areas which result, the $A_{i,e}$, and their distribution over the contact region, are sometimes a part of the information necessary to determine constriction resistance. Elastic deformation is considered in Sec. 7.5.3, in connection with constriction resistance arising from direct contact areas internal to thermal insulation.

PLASTIC DEFORMATION. In many applications, the need for high contact area region conductance is achieved by applying a considerable mating pressure p to the contact region. This commonly results in plastic deformation at a large fraction of the resulting direct contact areas. Approximate modeling of this mechanism, as discussed by Fenech and Rohsenow (1963), Mikic and Rohsenow (1966), Greenwood and Williamson (1966), Cooper et al. (1969), and Mikic (1971), indicates that the actual resulting direct contact area A_c, with metallic materials, may be approximately related to the overall geometric contact region area A_a, as

$$\frac{A_c}{A_a} = \frac{p}{H} \tag{7.3.4}$$

where p is the applied pressure, H is the contact microhardness of the material, and A_c/A_a is the fractional direct contact area.

In calculating the plastic deformation, it is assumed that the $A_{c,i}$ are circular and also that the distribution of surface roughness heights is of Gaussian form. These considerations result in the following relation for A_c, when the two surfaces are first brought together under load:

$$\frac{A_c}{A_a} = \frac{1}{2}\operatorname{erfc}\left(\frac{\eta}{\sqrt{2}}\right) \qquad \eta = \frac{Y}{\sigma} \tag{7.3.5}$$

Here, Y is the resulting distance between the mean planes of the two surfaces, as sketched in Fig. 7.2.1, idealized in a two-dimensional plane. The standard deviation σ is that of the local roughness of the combined profile ($z_a + z_b$), also defined in Fig. 7.2.1.

Analysis by Cooper et al. (1969) indicates that the resulting direct contact region density n, that is, the number of contacts per unit area, is approximately

$$n = \frac{m^2}{16\sigma^2} \frac{e^{-\eta^2}}{\text{erfc}(\eta/\sqrt{2})} \qquad (7.3.6)$$

where m is the mean of the absolute value of the asperity slope m, of the combined profile $(z_a + z_b)$, on first approach, and σ is the standard deviation. See Cooper et al. (1969) and Mikic (1971). For example, the combined profile, $(z_a + z_b)$ in Fig. 7.3.2, is that of the bottom surface z_b. The absolute average of $\tan \theta = m$ is determined from the varying surface inclination seen. Another example is the contact region idealization in Fig. 7.3.7. In practice, there is seldom sufficient information to make a definite evaluation of this average. Then estimates must be made based on the surface properties Y and σ. Otherwise, conventional values must be issued.

The contact resistance parameter is $\Sigma_i (a_i'/A_a)$, where the a_i are the contact radii. The sum of the radii of all direct contact areas over a total contact A_a is

$$\sum_i \frac{a_i'}{A_a} = \frac{m}{4\sigma\sqrt{2\pi}} e^{-\eta^2/2} \qquad (7.3.7)$$

The average radius $\bar{a}'$ is

$$\bar{a}' = \frac{\sigma}{m} \sqrt{\frac{8}{\pi}} e^{-\eta^2/2} \, \text{erf}\left(\frac{\eta}{\sqrt{2}}\right) \qquad (7.3.8)$$

Eliminating A_c/A_a from Eqs. (7.3.4) and (7.3.5) yields a relation between the geometric factor $\eta = Y/\sigma$ and the loading property p/H,

$$\frac{2p}{H} = \frac{2A_c}{A_a} = \text{erfc}\frac{Y}{\sigma\sqrt{2}} \quad \text{and} \quad \frac{Y}{\sigma} = \sqrt{2}\,\text{erfc}^{-1}\left(\frac{2p}{H}\right) \qquad (7.3.9)$$

where $2p/H$ expresses the ratio of the applied pressure to the contact micro-hardness. It relates to the direct contact spot density and radius, and the mean separation Y.

Consider the above relation first for $Y/\sigma\sqrt{2}$ large, as for generally smooth contacting surfaces, σ small, having infrequent relatively high peaks, that is, Y large. Then the direct contact area A_c/A_a is small. Clearly, A_c may be increased by increasing the mating pressure p. This decreases the mean spacing Y, by plastic deformation, while σ, the standard deviation, remains largely unchanged. Without the large peaks, that is, for $Y/\sigma\sqrt{2}$ smaller, there are many more contacts and A_c/A_a is very much larger. The variations of the parameters

$$A_c/A_a \qquad \bar{n} = n\sigma^2/m^2 \qquad \bar{p} = (\rho m^2/\sigma)$$

with $\eta = Y/\sigma$ are shown in Fig. 7.3.3, where ρ is the asperity or prominence radius of curvature, before loading. This is function of $\eta = Y/\sigma$. Very large effects are seen over the range $1 \le (Y/\sigma) \le 4$. For example, A_c/A_a changes by about four orders of magnitude.

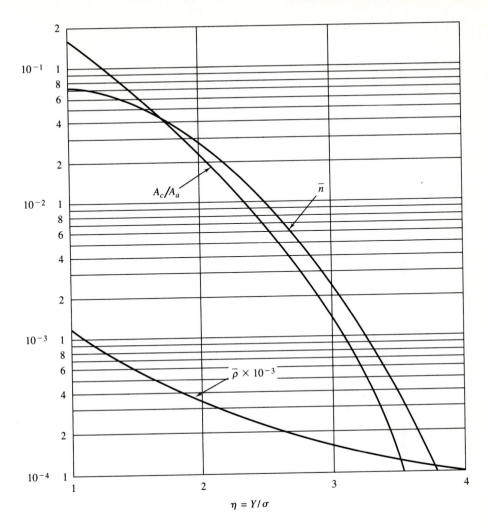

FIGURE 7.3.3
Contact region surface interaction parameters.

CONTACT REGION LOADING CYCLES. The preceding treatment of plastic deformation determines A_c/A_a as in Eq. (7.3.4), from the metrological characteristics of the two surfaces in contact. The principal initial geometric parameters are the average spacing Y and the standard deviation σ of the local roughness. Other interacting effects are the mating pressure p and the contact microhardness H.

Therefore, the direct contact area plastic deformations are for the initial assembly and application of pressure to the contact region. Any subsequent reassembly or pressure release may result in dislocation of direct contacts. There may also be further plastic deformation and a change of A_c/A_a, when pressure is reapplied. Nominally, changed estimates of Y and σ are then required. However, this is seldom practical and recourse may be to the consideration of bounding values of the range of A_c/A_a, in particular applications.

SUMMARY. The foregoing analysis of plastic deformation is independent of any prior and parallel processes of elastic deformation, which also occur at direct contact areas. Further information on such effects is discussed by Mikic (1974). Also, the effects of deformation are not related to the remaining gaps between the two contacting areas, after deformation, $\delta(x, y, p)$. This is discussed briefly after constriction resistances, $R_{ac, i}$ and $R_{bc, i}$ in Fig. 7.3.1, as idealized in Fig. 7.2.2, are determined.

7.3.3 Constriction Resistance

This effect is described in Fig. 7.2.1, in terms of the convergence of heat transfer paths, from the more remote locations in regions a and b, to $A_{c, i}$. The remote area, $A_{T, i}$, associated with each direct contact area, $A_{c, i}$, is shown in Fig. 7.3.4. The effects of the resulting two constriction resistances, $R_{ac, i}$ and $R_{bc, i}$, on the total resistance, R_i, of the region associated with $A_{c, i}$, is as shown in Fig. 7.3.1(b). The calculated value of $R_{T, i}$ is given in Eq. (7.3.1). This result follows from the assumption that the gap surfaces of regions a and b, which are associated with $A_{c, i}$, are at uniform temperatures t_a and t_b, respectively. These surfaces are diagrammed in Fig. 7.3.1(a). The model in Fig. 7.3.1(b) also assumes, in postulating $R_{dc, i}$, that any different temperatures on the two sides of the direct contact area $A_{c, i}$ are uniform over $A_{c, i}$.

THE CONSTRICTION EFFECT. The physical effect resulting in the two added constriction resistances, $R_{ac, i}$ and $R_{bc, i}$, is the contraction of the heat flow rate, q_i, across $A_{c, i}$. For heat flow toward the right in Fig. 7.3.4, it is a contraction on the left, over a distance of length $L_{a, i}$, in region a. On the right, it is an

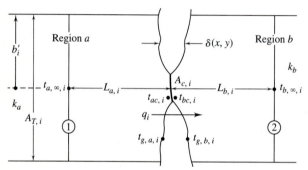

FIGURE 7.3.4
A typical contact element, modeled as a circular contact area $A_{c, i} = \pi a_i'^2$ between two cylindrical areas of radius b'.

expansion, in region b, of length $L_{b,i}$. These two effects cause an added conductive resistance between locations 1 and 2 in Fig. 7.3.4.

These constriction effects may be approximated in terms of the total resistances $R_{a,i}$ and $R_{b,i}$ from $A_{T,i}$ to $A_{c,i}$, in terms of q_i, as follows:

$$q_i = \frac{t_{a,\infty,i} - t_{ac,i}}{R_{a,i}} = \frac{t_{bc,i} - t_{b,\infty,i}}{R_{b,i}} \qquad (7.3.10)$$

where the three temperatures are first taken as uniform over the respective areas, for simplicity. The total resistances, $R_{a,i}$ and $R_{b,i}$, are each composed of two components, assumed to be in series. One component is the two constriction resistances very near the contact areas, $R_{ac,i}$ and $R_{bc,i}$, as shown in Fig. 7.3.1(b). The other component is the approximately one-dimensional conductive resistances from $L_{a,i}$ to $A_{c,i}$ and $A_{c,i}$ to $L_{b,i}$, as $L_a/k_a A_{T,i}$ and $L_b/k_b A_{T,i}$. Therefore,

$$R_{ac,i} = R_{a,i} - \frac{L_{a,i}}{k_a A_{T,i}} \qquad R_{bc,i} = R_{b,i} - \frac{L_{b,i}}{k_b A_{T,i}} \qquad (7.3.11)$$

If these one-dimensional conductive resistance effects are small, relative to $R_{a,i}$ and $R_{b,i}$, the constriction resistances are approximated directly, in terms of the temperatures $t_{a,\infty,i}$, $t_{b,\infty,i}$, $t_{ac,i}$, and $t_{bc,i}$ in Fig. 7.3.4, as follows:

$$q_i = \frac{t_{a,\infty,i} - t_{ac,i}}{R_{ac,i}} = \frac{t_{bc,i} - t_{b,\infty,i}}{R_{bc,i}} \qquad (7.3.12)$$

This condition would apply more accurately for smaller values of $A_{c,i}/A_{T,i}$, that is, for a smaller density of direct contact areas in the contact region.

The preceding formulation is in terms of uniform bounding temperatures, $t_{a,\infty,i}$, $t_{b,\infty,i}$, $t_{ac,i}$, and $t_{bc,i}$, as assumed in the model in Fig. 7.3.1(b). Recall that if $R_{dc,i} = 0$, then $t_{ac,i} = t_{bc,i} = t_{c,i}$ in Fig. 7.3.1(b). However, the converging flux lines shown in Fig. 7.2.1 suggest strong temperature gradients over the area $A_{c,i}$. Therefore, as a further approximation, a better estimate of $t_{c,i}$ in Eq. (7.3.12) is

$$\bar{t}_{c,i} = \frac{1}{A_{c,i}} \int_{A_{c,i}} t_{c,i} \, dA_{c,i} \qquad (7.3.13)$$

Consider now the condition $A_{c,i} \ll A_{T,i}$, in Fig. 7.3.4, as for the $A_{c,i}$ being widely spaced compared to their dimension. Then the distances L_a and L_b, out to the location of approximately one-dimensional heat conduction, would be relatively small. There are two consequences. One is that $t_{a,\infty,i}$ and $t_{b,\infty,i}$ may each be taken as uniform, as assumed previously. The other is that the gap surface temperatures $t_{g,a,i}$ and $t_{g,b,i}$ in Fig. 7.3.4 may be approximated as uniform and equal to $t_{a,\infty,i}$ and $t_{b,\infty,i}$, respectively. Then the constriction resistances are approximated as follows from Eq. (7.3.12), as $q_i R_{ac,i} = t_{a,\infty,i} - \bar{t}_{c,i}$:

$$q_i R_{ac,i} = \frac{1}{A_{c,i}} \int_{A_{c,i}} (t_{a,\infty,i} - t_{c,i}) \, dA_{c,i} = t_{a,\infty,i} - \bar{t}_{c,i}$$

$$= (\Delta t_{c,i})_a \qquad (7.3.14a)$$

where $(\Delta t_{c,i})_a$ is an average value over $A_{c,i}$. Also,

$$q_i R_{bc,i} = \frac{1}{A_{c,i}} \int_{A_{c,i}} (t_{c,i} - t_{b,\infty,i}) \, dA_{c,i} = \bar{t}_{c,i} - t_{b,\infty,i} = (\Delta t_{c,i})_b \quad (7.3.14b)$$

The values $(\Delta t_{c,i})_a$ and $(\Delta t_{c,i})_b$ are the constriction region temperature differences on the two sides. These depend upon q_i and the conductivities of regions a and b. The constriction resistances are

$$R_{ac,i} = \frac{(\Delta t_{c,i})_a}{q_i} \qquad R_{bc,i} = \frac{(\Delta t_{c,i})_b}{q_i} \qquad (7.3.14c)$$

where the equal heat flows in the two adjacent regions, q_i, are proportional to the Δt_c values which arise in each region for that heat flow rate.

THE EVALUATION OF CONSTRICTION RESISTANCE. This effect arises from the increased temperature gradients which arise as the heat flow constricts to smaller cross-sectional area, as sketched in Fig. 7.2.1. Figure 7.2.2 shows the spatial model of these constrictions in the contact region. Figure 7.3.4 is in terms of a model of a specific circular direct contact area $A_{c,i} = \pi a_i'^2$, at $t_{c,i}$, with the associated cylindrical regions on each side, of area $\pi b_i'^2$. The constriction resistance depends on a_i'/b_i'. For $a_i' = b_i'$ there is no constriction. For $a_i' \ll b_i'$ the temperature field is very constricted, unless the gap resistance, $R_{g,i}$, over area $A_{T,i} - A_{c,i}$, is small.

A very approximate estimate of constriction resistance in the limit $a_i' \ll b_i'$ is simply obtained by the idealization seen in Fig. 7.3.5. A contact region A_c is idealized as a small sphere of radius a', of infinite conductivity, at temperature t_c. It is embedded between infinite regions at $t_{a,\infty}$ and $t_{b,\infty}$ on the two sides. The

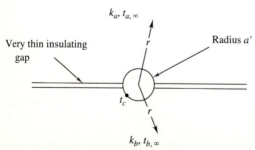

FIGURE 7.3.5
A simple model of constriction resistance.

conduction heat transfer q is calculated from Eq. (2.1.41) as

$$\text{Nu} = \frac{ha'}{k} = 1 \quad \text{or} \quad h = \frac{k}{a'} \qquad (2.1.41)$$

then

$$q = hA(t_\infty - t_c) = \frac{k}{a'}A(t_\infty - t_c)$$

and the constriction resistance is

$$R_{cr} = \frac{(t_\infty - t_c)}{q} = \frac{a'}{kA_c}$$

where A_c is the contact area to each side. It is evaluated as $\pi a'^2$, the equatorial plane of the sphere, not as $2\pi a'^2$, the area of the hemisphere on each side. Then the value of the constriction resistance on each side is

$$R_{ac} = \frac{t_{\infty,a} - t_c}{q} = \frac{1}{\pi k_a a'} \quad \text{and} \quad R_{bc} = \frac{t_c - t_{\infty,b}}{q} = \frac{1}{\pi k_b a'}$$

Since R_{ac} and R_{bc} are in series, the contact region resistance from region a to region b in Fig. 7.3.4, for $k_a = k_b = k$, is

$$2R_{cr} = \frac{2}{\pi k a'}$$

or, for the two regions having different conductivities, k_a and k_b,

$$R_{ac} + R_{bc} = \frac{1}{\pi a'}\left(\frac{1}{k_a} + \frac{1}{k_b}\right)$$

This approximate result indicates that $ka'R_c$ is dimensionless.

A MORE ACCURATE EVALUATION. The actual direct contact condition is much more accurately represented by a flat surface $A_c = \pi a'^2$. The calculation of constriction resistance is then much more complicated than for the spherical inclusion assumed previously. The temperature is then generally not uniform over such a direct contact area A_c. See the discussion in Cooper et al. (1969). Numerical and other calculations, by Cetinkale and Fishenden (1951), Clausing (1965), Mikic and Rohesnow (1966), Cooper et al. (1969), and Veziroglu and Chandra (1969), evaluate R_{cr} for $a'_i \ll b'_i$.

The result for a single contact area between two infinitely extensive regions a and b, given by Cooper et al. (1969), is written as

$$R_{ac} + R_{bc} = \frac{\Delta t_{ab}}{q} = \frac{1}{2k_s a'} \qquad (7.3.15a)$$

and

$$k_s = 2k_a k_b/(k_a + k_b) \qquad (7.3.15b)$$

where k_a and k_b are defined in Fig. 7.3.1 and $\Delta t_{ab} = t_a - t_b$. The quantity k_s is sometimes called the harmonic mean of k_a and k_b. Recall Eq. (3.3.12).

For the realistic circumstance of multiple direct contact regions $A_{c,i}$, the amount of constriction is less than given previously. It decreases with increasing a_i'/b_i'. Then Eq. (7.3.15), for $R_{cr,i}$, is modified with a function $\psi(a_i'/b_i')$ as

$$R_{cr,i} = R_{ac,i} + R_{bc,i}$$

$$= \frac{\Delta t_{ab,i}}{q_i} = \frac{1}{2k_s a_1'} \psi\left(\frac{a_i'}{b_i'}\right) = \frac{\psi_i}{2k_s a_i'} \qquad (7.3.16a)$$

or

$$q_i = \frac{2k_s a' \Delta t_{ab,i}}{\psi_i} = \frac{\Delta t_{ab,i}}{R_{ac,i} + R_{bc,i}} \qquad (7.3.16b)$$

where ψ_i is called a contact resistance or constriction parameter. This parameter is unity for a single contact area between two infinite surfaces. It decreases, as does the constriction and $1/R_{cr,i}$, as the density of direct contact areas increases. Cooper et al. (1969) indicate that ψ decreases from unity, in the

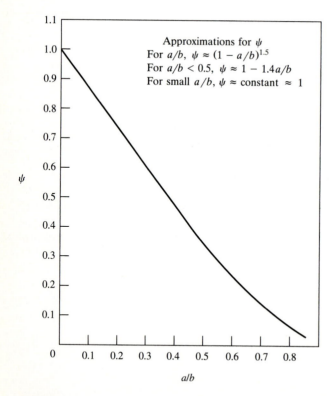

FIGURE 7.3.6
Constriction resistance as a function of direct contact area and remote cylinder equivalent radii, a_i' and b_i'. [From Cooper et al. (1969)].

range $0 \le a'/b' \le 0.4$, as

$$\psi_i = \left(1 - \frac{a'_i}{b'_i}\right)^{1.5} \approx \left[1 - \left(\frac{a}{b}\right)_{i,\,av}\right]^{1.5} = (1 - \sqrt{A})^{1.5} \qquad (7.3.17)$$

where $(a/b)_{i,\,av}$ is an average value and A here is the fractional area of direct contact, $A_c/A_a = A$. The upper limit of this approximation is $\sqrt{A} = 0.4$. This amounts to 16% direct contact. The preceding variation of ψ, and results from Clausing (1965), are plotted in Fig. 7.3.6, adapted from Cooper et al. (1969). At $a'_i/b'_i = 0.8$, the direct contact is 64% of the total area.

CONSTRICTION RESISTANCE. The preceding results for q_i, in Eq. (7.3.16), may be written in terms of the contact region heat flow $Q = \Sigma_i\, q_i$ over a total contact area interface of A_a, as

$$q''_c = \frac{Q}{A_a} = \sum_i \frac{q_i}{A_a} = 2k_s \sum_i \frac{a'_i\, \Delta t_{ab,i}}{A_a \psi_i} = \frac{\Delta t_{ab}}{A_a} \sum_i \frac{1}{R_{ac,i} + R_{bc,i}} \qquad (7.3.18)$$

where q''_c is the heat flow across the direct contact areas which are present in a unit area of contact region. This sum is to include a typical surface area A_a. Also, $\overline{\Delta t}_{ab,i}$ is an average of $\Delta t_{ab,i}$ in Eq. (7.3.16b). This formulation assumes that each direct contact area, $A_{c,i}$, is characterized by Eq. (7.3.16), as having distant regions like those shown in Fig. 7.2.2. In Eq. (7.3.18), $\overline{\Delta t}_{ab,i}$ is nominally the direct contact area weighted average of $\Delta t_{c,i}$.

The total conductance h_d and resistance R_d of the direct contact region, per unit contact area, is written from Eq. (7.3.18) in terms of the parallel constrictive conductances h_i across the direct contact surfaces $A_{c,i}$ as

$$q''_c = \Delta t_{ab} \sum_i \frac{1}{R_{ac,i} + R_{bc,i}} = \overline{\Delta t}_{ab} \sum_i h_i = \overline{\Delta t}_{ab} h_d = \frac{\overline{\Delta t}_{ab}}{R_d} \qquad (7.3.19)$$

where R_d is to be augmented by the total gap resistance, R_g, to determine the contact resistance R_c.

However, the $\Delta t_{ab,i}$ given previously are all approximately the same, since this quantity is the difference in the remote temperature levels, from region 1 to 2 in Fig. 7.2.2 or 7.3.4. Then Eq. (7.3.19) becomes

$$R_d^{-1} = h_d = 2k_s \sum_i \frac{a'_i}{A_a \psi_i} \qquad (7.3.20)$$

If ψ_i is assumed constant over the interface area, as for very regular roughness, Eq. (7.3.20) becomes

$$R_d^{-1} = h_d = \frac{2k_s}{\psi} \sum_i \frac{a'_i}{A_a} \qquad (7.3.21)$$

Eliminating the summation between Eqs. (7.3.21) and (7.3.27), and evaluating ψ

as in Eq. (7.3.17), gives

$$R_d^{-1} = h_d = \frac{k_s m}{2\sigma\sqrt{2\pi}\,(1 - \sqrt{A}\,)^{1.5}}e^{-\eta^2/2} \qquad \eta = Y/\sigma \qquad (7.3.22)$$

This relation evaluates R_d, or h_d, entirely from mean spacing Y, the standard deviation of the local roughness σ and the area A_c resulting from plastic deformation. The deformation is evaluated from Eqs. (7.3.4) and (7.3.5) as

$$\frac{A_c}{A_a} = A = \frac{p}{H} = \frac{1}{2}\mathrm{erfc}\left(\frac{\eta}{\sqrt{2}}\right) \qquad (7.3.23)$$

An approximate analysis by Mikic (1974) eliminates η and A from Eqs. (7.3.22) and (7.3.23) to yield the following relation:

$$R_{dc}^{-1} = h_{dc} = 1.13\frac{k_s m}{\sigma}\left(\frac{p}{H}\right)^{0.94} \qquad (7.3.24)$$

Allowing for elastic deformation below the region of plastic effects changes the last term in Eq. (7.3.24) from p/H to $p/(H + p)$. Accounting for this added deformation is seen to increase R_{dc}, as p becomes larger. That is, the additional elastic deformation decreases the area of direct contact A_c by decreasing the a_i at contact locations.

DIRECT CONTACT AREA A_c AND HEAT TRANSFER. The foregoing formulation involves the contact region plastic deformation in terms of the effects of mean spacing Y, standard deviation σ, the asperity property m, and the contact microhardness H. A simpler correlation, in terms of the direct contact area heat flux q_c'', was given by Yovanovich (1982) as

$$\frac{q_c''}{\Delta t} = \frac{1.25 k_s m}{\sigma}\left(\frac{p}{H}\right) = \frac{1}{R_d} \qquad (7.3.25)$$

where H is the contact microhardness of the softer material and k_s is the harmonic mean conductivity of the two regions a and b. Recall Eq. (7.3.15b). The range of application of this result was given as $10^{-6} \le (p/H) \le 2.3 \times 10^{-2}$. This amounts to four orders of magnitude variation of p, for a given value of H.

The formulations in Eqs. (7.3.4)–(7.3.9) also require the value of H, the material contact microhardness. This is also related to the surface roughness, the asperity property m, the kind of material, and the applied pressure p. These may be related to the indicated value of H given previously through Vickers microhardness H_v, and tests and correlations in terms of the bulk hardness H_b. Using these formulations and the relations for p/H and a' in Eqs. (7.3.5) and (7.3.8), the following relation between H and H_b is given by Song and Yovanovich (1988):

$$\frac{p}{H} = \left[\frac{p}{H_b[1.62\sigma/(md)]^e}\right]^{(1+0.071e)^{-1}} \qquad (7.3.26)$$

TABLE 7.3.1
Microhardness correlation parameters for several materials
1 Pa = 1 N / m², where 1 psi = 6894. 7 N / m² and 1 bar = 10^5 Pa
[from Song and Yovanovich (1988)]

Material	e	$d \times 10^6$, m	H_b, MPa
Zr-4	−0.26	53	1913
Zr-2.5 wt % Nb	−0.26	102	1727
Ni 200	−0.26	157	1668
SS 304	−0.26	3887	1427

For the range $10^{-6} \le p/H \le 2 \times 10^{-2}$, the two correlation parameters d and e are given in Table 7.3.1, for several materials.

With the values of H_b, e, d, and the contact pressure p, in a particular application, p/H may be determined from Eq. (7.3.26). This may then be used in Eq. (7.3.25) to determine the direct contact area contribution, q_c, to the total contact region heat flux.

SUMMARY. This section concerns the constriction resistance effect, contact microhardness, and the effect of pressure. These are especially important in direct contacts between good conductors, when the remaining gap is not highly conductive. The constriction effects are evaluated and related to the direct contact area deformation. Thereby, the component of contact resistance R_d, associated with direct contact areas, is determined. The other component of R_c, the gap resistance R_g, is considered in Sec. 7.3.4.

Omitted from the evaluations of R_d given previously was the surface resistance component $R_{dc,i}$ in the model shown in Fig. 7.3.1(b). This is a residual resistance which remains after asperities are deformed. This resistance may arise from contact surface roughness, films, and other effects. Its effect is also area dependent, as $\Sigma_i \, a'_i / A_a$. Since it is also in series with R_d, it may be used directly to increase R_d.

The principles of heat conduction constriction resistance developed here also have other important applications. Examples are fins protruding from flat heat transfer surfaces. The resistance constricts the supply of heat to the fins. Additional bond contact resistance effects are discussed in Sec. 8.2.5.

7.3.4 Gap Region Heat Transfer and Other Effects

The analysis in Sec. 7.3.3 concerns the characteristics of the direct contact component, R_d, of the contact resistance, R_c. The parallel component of R_c, R_g contributed by the gap, is considered here. A characteristic subregion, associated with a direct contact area, has an area of $A_{T,i}$, as seen in Fig. 7.3.1(a). The associated gap area is $A_{T,i} - A_{c,i}$. The local gap thickness is designated there as $\delta(x, y)$, where the coordinates x and y are in the local plane of the gap. The actual gap area is that without load $\delta_0(x, y)$, minus the decrease $\Delta(x, y)$ which follows the application of any mating pressure p. That is,

$$\delta(x, y) = \delta_0(x, y) - \Delta(x, y) \tag{7.3.27}$$

GAP HEAT TRANSFER. Among the heat transfer processes which commonly arise in such gaps are radiation and conduction, through the material in the gap. Convective motions are not common in small-separation assemblies intended to have small overall contact resistance, R_c. In gaps containing air or other weakly absorbing gases, the radiation is often simply accounted for as between flat parallel surfaces, having the temperature distributions $t_{ag}(x, y)$ and $t_{bg}(x, y)$ shown in Fig. 7.3.1.(a). The simplest circumstance is when t_{ag} and t_{bg} may each be approximated as uniform. This follows when the constriction resistances and k_a and k_b are large compared to the effective conductivity of the gap material, k_g. Then the radiation contribution may be calculated as in Eq. (2.1.37).

Accounting for a variation of t_{ag} and t_{bg} raises much more complicated considerations. The resulting distributions of $t_{ag}(x, y)$ and $t_{bg}(x, y)$ are then additional interacting effects in the heat transfer characterization of the total contact region. That is, they interact directly with both direct contact area conductance and gap conduction. When the radiation effect is small, compared to these two effects, as with small constriction resistance and gap width, it may be more simply approximated.

Conduction through the gap is commonly a major effect in determining R_c. Also, across the range of important applications, gap conduction may involve both continuum and noncontinuum conduction mechanisms. This mechanism is considered in a circumstance in which $\delta(x, y)$ is known, as in a very regular geometry. An example would be a corrugated surface of region a, in contact with a region b. See Fig. 7.3.7. Surface b is taken as flat, to the scale of the corrugation H. The gap conduction across area dx, per unit length normal to the plane shown in Fig. 7.3.7, is

$$dq_g = \frac{k(x)(t_a - t_b)\,dx}{\delta(x)} \tag{7.3.28}$$

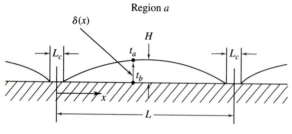

FIGURE 7.3.7
A regular periodic gap region between a corrugated and flat surface in contact, with t_a and t_b both assumed uniform.

If $k(x)$ and $(t_a - t_b)$ do not vary, q_g, the gap conduction for area $L - L_c$, is

$$q_g = k(t_a - t_b) \int_{L_c/2}^{L - L_c/2} \frac{dx}{\delta(x)} = \frac{t_a - t_b}{R_g} \qquad (7.3.29)$$

Two important considerations arise here. First, Eq. (7.3.29) may not be integrated for many completely ordinary forms of the function $\delta(x)$. This arises because of the singularities, of $1/\delta(x)$, like that at $x = L_c/2$ in this example. This is an unrealism of the model, which assumes that $t_a - t_\infty$ in Eq. (7.3.29) remains constant over the whole gap. Very effective conduction, in the thin-gap region around $x = L_c/2$, would drastically decrease the temperature difference there. Then Eq. (7.3.28) is unreasonable at such locations.

NONCONTINUUM GAS BEHAVIOR. Another very important consideration arises in many applications. The local spacing δ and mean spacing Y, between rough surfaces, see Fig. 7.2.1, is made very small in many assemblies, when a low contact resistance is necessary. A representative range in many applications, of the mean spacing, is $Y = 1$ to 50 μm. Therefore, the resulting range of $\delta(x, y)$ would be from almost 0 near direct contact areas $A_{c,i}$, to about 1 to 50 μm elsewhere. The molecular mean free path in atmospheric air is about $\lambda = 6 \times 10^{-6}$ cm $= 0.06$ μm. Therefore, in the conduction model postulated in Eq. (7.3.28), $k(x)$ may not in general be taken as the conventional continuum thermal conductivity k, near the direct contact areas.

The regimes of noncontinuum behavior gas are characterized in terms of the Knudsen number, Kn. These effects are discussed in Sec. 6.6.4 for mass diffusion and in Sec. 7.5 as related to processes in thermal insulation. The relevant definition here is

$$\text{Kn} = \lambda / \delta(x, y) \qquad (7.3.30)$$

In terms of δ from 0 at direct contact areas, to spacings of 50 μm in air, the resulting range is $\infty \geq \text{Kn} > 0.0012$. For example, for Kn > 2 there are very few collisions internal to the gas. Most collisions occur at the gap boundaries. This regime is called free-molecule transport. Between this regime and the continuum regime, for Kn < 0.01, is a transition regime. This regime lies between the two very different asymptotic transport processes for Kn > 2 and Kn < 0.01. Transport in the transition regime is often modeled in terms of temperature discontinuities at the two gas–solid interfaces. See Wesley and Yovanovich (1986) for a formulation. In both noncontinuum regimes, that is, for Kn > 0.01, the effective conductivity k_N is less than the continuum value k.

These considerations indicate that noncontinuum effects are very common at some locations in gas-filled gaps. Then it is necessary to use a location-dependent thermal conductivity $k'(x, y, t, \delta)$ in a formulation of gap heat transfer, to replace Eq. (7.3.28). The local relation, at point x, y, and the general relation for $q_{g,i}$, the gap conduction associated with a direct contact area $A_{c,i}$, are as follows:

$$dq_{g,i} = \frac{k'(x, y, t, \delta)[t_a(x, y) - t_b(x, y)]\, dx\, dy}{\delta(x, y)} \tag{7.3.31}$$

$$q_{g,i} = \int_{(A_T - A_c)_i} \frac{k'(x, y, t, \delta)[t_a(x, y) - t_b(x, y)]\, dx\, dy}{\delta(x, y)} \tag{7.3.32}$$

The gap conductance $C_{g,i}$ and resistance $R_{g,i}$ are defined as

$$q_{g,i} = C_{g,i}\overline{\Delta t}_{g,i} = \overline{\Delta t}_{g,i}/R_{g,i} \tag{7.3.33}$$

where $\overline{\Delta t}_{g,i}$ is some average of $t_{ag} - t_{bg}$, defined in Fig. 7.3.1(a).

THE NONCONTINUUM EFFECT IN GAP CONDUCTION. The conductivity k' in Eq. (7.3.32) is modeled in terms of the continuum conductivity k and the Knudsen number in Eq. (7.3.30) as follows:

$$k_N = \frac{k}{(1 + \alpha\beta\, \mathrm{Kn})} = k'(x, y, t, \delta) = \frac{k\,\delta(x, y)}{\delta(x, y) + \alpha\beta\lambda} \tag{7.3.34a}$$

This conductivity, k_N, is seen to decrease rapidly for $\delta(x, y)$ small, with increasing mean free path λ, that is, with decreasing $\delta(x, y)$; see Eq. (7.3.30). In the preceding equations, $\alpha \leq 1$ and β and Kn are the thermal accommodation coefficient, a gas property and Kn, defined as

$$\alpha = \frac{2 - \alpha_{ag}}{\alpha_{ag}} + \frac{2 - \alpha_{bg}}{\alpha_{bg}} \tag{7.3.34b}$$

$$\beta = \frac{1}{\mathrm{Pr}}\left(\frac{2\gamma}{\gamma + 1}\right) \tag{7.3.34c}$$

$$\mathrm{Kn} = \frac{\lambda}{\delta(x, y)} \tag{7.3.34d}$$

The ideal-gas mean free path is

$$\lambda = \frac{1}{\sqrt{2}\,\pi n \sigma^2} = \frac{16}{5}\frac{\mu}{\rho\sqrt{2\pi RT}} \tag{7.3.34e}$$

where α_{ag} and α_{bg} are the two surface accommodation coefficients and σ is the collision diameter. See Patterson (1958) for other measures of molecular motion.

The preceding formulation of k_N applies for a one-dimensional process, across the gap locally. To illustrate the effect of the formulation of k_N in

Eq. (7.3.34a), it is written as follows:

$$k_N = \left(\frac{k}{1 + \alpha\beta \, \mathrm{Kn}}\right) = \frac{k}{1 + \dfrac{N}{\delta(x, y)}} \qquad N = \alpha\beta\lambda \qquad (7.3.35)$$

where N is a constant which depends only on gas properties and the surface accommodation coefficients. This formulation of a noncontinuum k_N has the characteristic that the actual gap conductivity approaches zero with $\delta(x, y)$. Equation (7.3.35), substituted into the general relation for $q_{g,i}$ in Eq. (7.3.32), yields

$$
\begin{aligned}
q_{g,i} &= \int_{(A_T - A_c)_i} q_{g,i}''(x, y) \, dx \, dy \\[2mm]
&= \int_{(A_T - A_c)_i} \frac{k[t_a(x, y) - t_b(x, y)] \, dx \, dy}{\delta(x, y)\left[1 + \dfrac{N}{\delta(x, y)}\right]} \\[2mm]
&= \int_{(A_T - T_c)_i} \frac{k[t_a(x, y) - t_b(x, y)] \, dx \, dy}{\delta(x, y) + N} = \frac{\overline{\Delta t_{g,i}}}{R_{g,i}} \qquad (7.3.36)
\end{aligned}
$$

where $q_{g,i}''$ is the local heat flux. Comparing this form with that in Eq. (7.3.29), for a continuum conduction process, indicates that any singularities arising in the integrand are removed, by the postulate of noncontinuum conductivity in Eq. (7.3.34).

LOCAL GAP NONCONTINUUM EFFECTS. Equation (7.3.36) evaluates the gas layer resistance $R_{g,i}$, where the noncontinuum effect is N in the denominator. The local gap heat flux variation, $q_g''(x, y)$, depends on $\delta(x, y)$. Therefore, it also depends on the variation of the noncontinuum effect along the gap region between adjacent direct contact areas. Equation (7.3.35) indicates that this effect varies with $N \propto \lambda$.

A calculation which shows this effect is given by Ogniewicz and Yovanovich (1978). It applies for elastic deformation of the direct contact region between two spherical regions of radius R. See also Yovanovich and Kitscha (1974) for a sphere in contact with a flat surface. Calculations were carried out for a range of values of $M = \alpha\beta\lambda/R$, where λ/R is a Knudsen number based on the sphere radius. The model of gap conductance included an allowance for the variation of the local gap temperature difference $(t_{g,a} - t_{g,b})$, between the two surfaces, locally caused by solid conduction toward regions of large $q_{g,i}''(x)$.

Results were in terms of a nondimensional gap heat flow I_n. This is expressed in terms of the heat flow per unit of the radial coordinate r. This coordinate is measured outward from the center of the contact region. The variation of I_n is shown in Fig. 7.3.8, plotted against $x = r/a$, where a is the

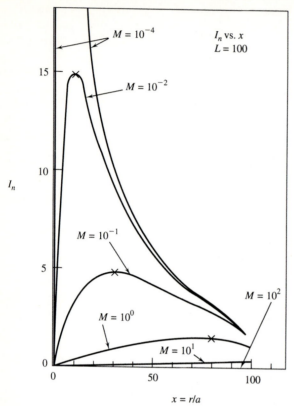

FIGURE 7.3.8
Local variation of heat flow, per unit length of x. [From Ogniewicz and Yovanovich (1978)].

radius of the direct contact area. These results are for a plastic deformation level of $L = R/a = 100$.

These results show several effects, related to the Knudsen number parameter M, which are determined by the gas density in the gap. For M small, that is, for continuum transport, the heat flux is very large near the direct contact area, that is, for $x/a \geq 1$ small. However, the value of I_n is still bounded, since $\delta(x, y)$ becomes much less than λ, as direct contact is approached. Noncontinuum effects, at $M = 10^{-1}$, have reduced the gap conductance substantially in this region. For $M = 10$, a very large local noncontinuum effect, the gas conductivity is very small at all x. The consequences are both a realistic gap conductivity $C_{g,i}$ and a large Knudsen number effect on the magnitude of $C_{g,i}$.

SOME OTHER CONTACT REGION EFFECTS. There are many important applications and resulting particular considerations which cannot be idealized, for example, as steady state, as regularly spaced direct contact areas, or as gas-filled gaps. Mention is made here of several additional studies which treat such other effects.

Contact resistance R_c may be reduced by interposing other materials between contacting regions during assembly. The use of a thermal grease and a soft metal insert is very common, with effects like those discussed by Fry (1965) and Fried and Kelley (1965). The effect of a lubricant was determined in experiments by Kitscha and Yovanovich (1975). Experimental results for gold foils are given by Mølgaard and Smeltzer (1970). For increasing pressure, to about 900 kg/cm^2, a large decrease in R_c was found. It is often more convenient to apply soft-metal surface coatings than to install foils. O'Callaghan et al. (1983) studied contact resistance between stainless-steel surfaces. One surface was coated with tin.

An interesting study of competing effects, by Peterson and Fletcher (1990), concerned anodizing one of the two surfaces, in a contact joint of aluminum 6061-T6. The coatings tested ranged in thickness from 60.9 to 163.8 μm. Measurements to 14 MPa showed that joint conductance decreased, with coating thickness, but increased with load. Thus the anodizing layer of lower conductivity tended to increase resistance, while the thicker layer improved direct contact conductance.

The decrease in R_c with both foils and soft-metal layers is a microhardness effect h on the resulting direct contact area A_c, as formulated in Eq. (7.3.4). Antonetti and Yovanovich (1985) developed a model for a metallic layer coating which agreed with data for silver layers on a nickel substrate. Measurements of the contact resistance of bounded low-roughness aluminum surfaces were reported by Lewis and Sauer (1965). Values in the range $C_c = 80$ to 2000 Btu/hr ft^2 °F were found, over a range of adhesives, at moderate pressure. See also Kang et al. (1990) for additional studies of surface coating effects.

Another method of reducing R_c is explosive bonding of the two contacting regions. Yovanovich et al. (1980) summarized the bonding mechanisms. Electrical measurements of constriction resistance in such bonds indicated that the bond resistance between aluminum and iron or copper may be uniform over the area of the bond. The resistance may be interpreted in terms of a concentrated electrical potential drop across the region of the bond.

Electronic component cooling, using large and regular contact areas, is increasingly important. Heat spreader and large area effects are considered by Peterson and Fletcher (1988) and by Negus et al. (1989). Other relevant references are given.

Most measurements of contact resistance R_c are done in steady-state heat flow fields. The contact region temperature difference Δt_c is often then inferred by extrapolation from temperature gradients in the remote connected regions. The inverse method would be to infer values of R_c, directly in terms of the measured temperatures away from the contact region. A formulation of such a procedure is given by Beck (1988) and applied to compare such results with those determined by other means.

For comprehensive reviews of recent research concerning contact resistance, see Fletcher (1988, Madhusudana and Fletcher (1988), and Snaith et al.

(1986). A broad range of classical concerns and modern applications are considered. Other more detailed reviews and applications are also cited.

7.4   COMPOSITE MATERIALS

Such mechanisms are briefly discussed in Sec. 7.1. Several common kinds of composites are shown in Figs. 7.1.1 and 7.1.2. These often arise in contaminated, strengthened, and fibrous materials. They also arise in layered or laminated assemblies of isotropic or anisotropic regions. However, the same kinds of transport mechanisms arise commonly, in both thermal and mass diffusion, in porous solids, like insulation, and with precipitates in crystalline materials.

7.4.1   General Considerations

The common feature in such composites is that the overall region of material is composed of two or more different materials, in divided form. Usually, the thermal conductivity and mass diffusivity are different in the two materials a and b, as k_a and k_b or D_a and D_b. Thereby, the temperature field and heat flow paths are altered. For example, in the plane layers in Fig. 7.1.1, with an imposed overall temperature difference $t_1 - t_2$, the internal temperature field would not be one dimensional. Although the Laplace conduction equation applies in both regions of the material, the temperature field $t(x, y, z)$ must, in principle, satisfy a heat flux conservation condition at all interface locations, between materials a and b, throughout the whole region. That is, the temperature distributions in materials a and b are to be simultaneously matched throughout the region. The equations in regions a and b are

$$\nabla^2 t_a = 0 \quad \text{and} \quad \nabla^2 t_b = 0 \tag{7.4.1}$$

The solutions are also matched over the boundaries between the phases, in terms of both the local temperature and of heat flux q'' equality, as

$$t_a = t_b \qquad k_a \frac{\partial t_a}{\partial n} = k_b \frac{\partial t_b}{\partial n} = -q'' \tag{7.4.2}$$

where n is the local normal direction. If there is a contact resistance R_c at the boundaries, then $t_a \neq t_b$ there and $t_a - t_b$ must be determined as in Sec. 3.7, from

$$t_a - t_b = q'' R_c \tag{7.4.3}$$

The preceding requirements directly imply the complexity which arises in accurately determining, for example, the effective conductivity k_e of a composite material.

As discussed in Chap. 1, the similar forms of the Fourier law of conduction and the Fick law of mass diffusion, with comparable boundary and internal conditions, result in the same solutions. Both mechanisms follow the same flux

TABLE 7.4.1
Diffusive-like processes in random dispersed structures

Composite region mechanism	Flux, $\bar{F}$, or its equivalent	Gradient, $\bar{G}$, or its equivalent	Transport coefficient
1. Heat conduction	$\bar{q}''$	∇t	Thermal conductivity
2. Mass diffusion	$\bar{m}''$	∇C	Mass diffusivity
3. Electrical conduction	Electric current	Electric field	Electrical conductivity
4. Electrical insulation	Electric displacement	Electric field	Dielectric constant
5. Diamagnetic or paramagnetic material	Magnetic induction	Magnetic field	Magnetic permeability

See Batchelor (1974) for additional examples of similar mechanisms.

and conservation rule, for processes near equilibrium in Eq. (7.4.4).

$$\bar{F} = \bar{K} \cdot \bar{G} \quad \text{and} \quad \nabla \cdot \bar{F} = 0 \qquad (7.4.4)$$

where $\bar{F}$ is the flux, $\bar{q}''$ or $\bar{m}''$, K is k or D, and the gradient $\bar{G}$ is ∇t or ∇C. These, and some other physical mechanisms which also follow Eq. (7.4.4), are compared in Table 7.4.1.

The correspondences in Table 7.4.1 are of more than simple conceptual value. They indicate the generality and broad importance of these mechanisms. The general formulations have led to analysis, by variational methods, to formulate upper and lower bounds of the kinds of transport coefficients defined in Table 7.4.1. For example, k_e may often be given bounds, in terms of composite region shape, detailed internal geometry of the two phases, and their respective values k_a and k_b. Another feature is that any specific result in any of the five mechanisms in Table 7.4.1 may commonly be used for each of the other four.

Many of the composites which arise are at least approximately random. Others are very well defined, as in processing equipment, structures, and reactors. Many composites are very dilute in one component. This is like a matrix of conductivity k_a with widely dispersed small inclusions of conductivity k_b. Then the volume fraction Φ_b is very small compared to Φ_a, where $\Phi_a + \Phi_b = 1$. Composites of two comparable phases, as in aggregates, commonly have $\Phi_a \approx \Phi_b$.

The specification of the composite region includes an adequate geometric description, the transport coefficient in each phase, any interfacial resistances at all internal phase boundaries, and the overall region boundary conditions. In these terms, the regimes of transport, the nature of the analysis, the approximations used, and the accuracy of the results relate to the following kinds of considerations:

1. Random or regular spacing
2. Geometry and orientation of inclusions or of mixed phases
3. The value of the conductivity ratio $k_b/k_a = K$
4. Touching, close or distant spacing, in dispersions
5. Inclusion characteristic size L_i, compared to region size L, as L_i/L
6. Contact resistance between the phases R_c, as $L_i/k_b R_c$, a Biot number, Bi

The results are expressed in terms of the effective conductivity k_e, for any composite structure, as

$$\frac{k_e}{k_a} = f\left(\frac{k_b}{k_a}, \Phi_b, \frac{L_i}{L}, \text{Bi}\right) \tag{7.4.5}$$

Regimes arise in terms of these parameters. Consider k_a/k_b near 1.0 and Φ_b, L_i/L, and Bi very small. Then the effects of the inclusions are relatively small. Then they may often be estimated with high accuracy.

The following material in this section considers some of the available specific results for several of the most important regimes which commonly arise, in terms of the preceding considerations. Section 7.4.2 first gives results in terms of effective conductivity for low-volume fraction composites, that is, for Φ_b small compared to Φ_a, the volume fraction of the matrix. Then results for the more common circumstance of larger Φ_b are given. These results are for spheres of several geometrical patterns of packing. Consideration of the various regimes, in terms of Φ and k_b/k_a, is discussed, along with bounds of k_e, determined from variational analysis. Section 7.4.3 considers other geometries and approximate analysis techniques.

7.4.2   Small Inclusions

Consider a continuous matrix, k_a, with a dispersion of small inclusions, k_b, as shown in Fig. 7.1.1(a). The inclusions may be of regular shape, as spheres, spheroids, or as cylinders normal to the heat flow direction. They may be in a regular array, such as the cubic, face-centered, or body-centered relationship. They might also be randomly spaced or form local agglomerations. They may also be overlapping, as with gas pores formed on initially randomly spaced active nuclei in the region.

LOW VOLUME FRACTION. This is the simplest regime. At sufficiently low Φ_b, the disturbances which two adjacent inclusions cause in the matrix temperature field will not interact with each other. Maxwell (1873) analyzed this circumstance and found the following result:

$$\frac{k_e}{k_a} = \frac{1 + (d - 1)\beta \Phi_b}{1 - \beta \Phi_b} \qquad (7.4.6)$$

where β is a thermal polarizability and $d = 3$, for spheres, and $d = 2$, for parallel thin cylinders normal to the flux direction. The latter are sometimes called disks.

$$\beta = \frac{k_b - k_a}{k_b + (d - 1)k_a} = \frac{\dfrac{k_b}{k_a} - 1}{\dfrac{k_b}{k_a} + (d - 1)} = \frac{K - 1}{K + (d - 1)} \qquad (7.4.7)$$

where $K = k_b/k_a$. A comparable result for mass diffusion for spherical inclusions is given by Crank (1975), page 271.

Equation (7.4.6) was subsequently found to be completely correct only in the first-order term in an expansion of Eq. (7.4.6) in terms of $\beta \Phi_b$. That is,

$$\frac{k_e}{k_a} = 1 + d\beta \Phi_b \qquad (7.4.8)$$

HIGHER VOLUME FRACTION. These considerations suggest expressing k_e/k_a, in general, as follows where $C_1 = 3\beta$ for spheres:

$$\frac{k_e}{k_a} - 1 = C_1 \Phi_b + C_2 \Phi_b^2 + \cdots = \sum_{n=1} C_n \Phi_b^n \qquad (7.4.9)$$

The role of $k_a/k_b = K$ is seen very clearly in the simple relation, Eq. (7.4.7). One limiting value of $k_b/k_a = K$ is zero, that is, $k_b = 0$. This arises when k_b is negligible compared to k_a, as in heat conduction in a highly conductive matrix containing small gas-filled pores. Then $\beta = -1/(d - 1)$ and for spheres and cylinders the results from Eq. (7.4.8) are

$$\frac{k_e}{k_a} = 1 - \frac{d}{(d - 1)}\Phi_b = 1 - \frac{3}{2}\Phi_b \text{ and } 1 - 2\Phi_b \qquad (7.4.10a)$$

On the other hand, for $k_b = k_a$, $\beta = 0$, and $k_e = k_a$ as in a homogeneous material. For $k_b \gg k_a$, $\beta = 1$ and the two results are

$$\frac{k_e}{k_a} = 1 + d\Phi_b = 1 + 3\Phi_b \text{ and } 1 + 2\Phi_b \qquad (7.4.10b)$$

A further calculation by Jeffrey (1973) determined the next term in the expansion in Eq. (7.4.9), that is, C_2, the coefficient of Φ_b^2. The result was a very slowly converging series involving $(K - 1)/(K + 2)$. Values of C_2 accurate to three significant figures are listed in Table 7.4.2, along with $3\beta = C_1$ from

TABLE 7.4.2
Values of the coefficients in the expansion of k_e/k_a in Eq. (7.4.9)

$k_b/k_a = K$	$C_1 = 3\beta$	C_2
0	-1.500	0.588
0.02	-1.455	0.558
0.1	-1.287	0.450
0.5	-0.600	0.110
1.0	0	0
2.0	0.75	0.208
5.0	1.713	1.23
50.0	2.826	3.90
∞	3	4.51

Eq. (7.4.7). The variation of $C_2(K)$, over the whole range between the two asymptotic behaviors, discussed previously, is also plotted in Fig. 7.4.1. These results were compared with the bound obtained by Hashin and Shtrikman (1962) by variational methods. Evaluation of further coefficients in the formulation in Eq. (7.4.9) will apparently require numerical analysis.

Landauer (1978) considered an equivalent mechanism, the electrical conductivity in inhomogeneous media. The analysis expresses the averages of local effects in the region, in terms of the applied fields. The result, in terms of k_e and Φ_b, called effective medium theory, is as follows:

$$(1 - \Phi_b)\frac{k_e - k_a}{k_a + (d-1)k_e} + \Phi_b\frac{k_e - k_b}{k_b + (d-1)k_e} = 0 \qquad (7.4.11)$$

This result is apparently most accurate when Φ_b is small and k_a and k_b are comparable. Torquato (1985) used a modified approach to include the effect of

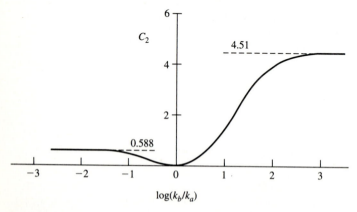

FIGURE 7.4.1
The variation of C_2 in Eq. (7.4.9), as a function of $k_b/k_a = K$.

the inclusion structure and this is an improvement at higher values of Φ_b. See also Chiew and Glandt (1982).

PERIODIC ARRAYS OF SPHERES. Rayleigh (1892) determined the effective conductivity of a composite containing a simple cubical array of spheres. The result is

$$\frac{k_e}{k_a} = 1 - 3\Phi_b \left[\frac{2+K}{1+K} + \Phi_b - \frac{(1-K)}{(4+3K)} m\Phi_b^{10/3} + O\left(\Phi_b^{14/3}\right) \right]^{-1} \quad (7.4.12)$$

where $K = k_b/k_a$ and the last term in the bracket is the order of the next term of higher order in Φ_b which arose. The value of m in Eq. (7.4.12) was corrected to be 1.57 by Runge (1925). A similar method was used, for example, by McKenzie et al. (1978) to calculate the effective electrical conductivity for both body- and face-centered cubic lattices of spheres.

Sangani and Acrivos (1983) investigated higher-order corrections to the effective conductivity for regular arrays. The resulting relation is similar to Rayleigh's result, as follows:

$$\frac{k_e}{k_a} = 1 - 3\Phi_b \left[-L_1^{-1} + \Phi_b + C_1'L_2\Phi_b^{10/3} \frac{1 + C_2'L_3\Phi_b^{11/3}}{1 - C_3'L_2\Phi_b^{7/3}} \right.$$

$$\left. + C_4'L_3\Phi_b^{14/3} + C_5'L_4\Phi_b^6 + C_6'L_5\Phi_b^{22/3} + O\left(\Phi_b^{25/3}\right) \right]^{-1} \quad (7.4.13)$$

The constants L_n are as follows and the values of $C_1', \dots, C_6'$ for the three arrays are listed in Table 7.4.3.

$$L_n = (K-1)/[K + 2n/(2n-1)] \quad (7.4.14)$$

TABLE 7.4.3
The constants in Eq. (7.4.13)

	Simple cubic	Body-centered cubic	Face-centered cubic
C_1'	1.3047	1.29×10^{-1}	7.529×10^{-2}
C_2'	2.305×10^{-1}	-4.1286×10^{-1}	6.9567×10^{-1}
C_3'	4.054×10^{-1}	7.6421×10^{-1}	-7.4100×10^{-1}
C_4'	7.231×10^{-2}	2.569×10^{-1}	0.4195×10^{-1}
C_5'	1.526×10^{-1}	1.13×10^{-2}	2.31×10^{-2}
C_6'	1.05×10^{-2}	5.62×10^{-3}	9.14×10^{-4}

TABLE 7.4.4
**The effective conductivity, k_e / k_a, for closely
packed arrays of spherical inclusions**

k_b / k_a	Simple cubic	Body-centered cubic	Face-centered cubic	Random arrays $(\Phi_{b,m} = 0.62)$
0	0.344	0.217	0.160	0.27
1	1.0	1.0	1.0	1.0
2	1.46	1.60	1.69	1.5
5	2.42	3.04	3.36	2.8
10	3.47	4.69	5.47	4.1
20	4.81	6.89	8.49	5.8
30	5.7	8.6	10.7	7.0
40	6.4	9.6	12.4	7.8
50	6.9	10.5	13.8	8.4

DENSE ARRAYS OF SPHERES. Keller (1963) gave results for a dense array of infinitely conducting spheres, that is, $k_b/k_a \gg 1$, as

$$\frac{k_e}{k_a} = -\frac{\pi}{2}\ln[\pi - \Phi_b] + \cdots \quad \text{for} \quad \frac{\pi}{6} - \Phi_b \ll 1 \quad (7.4.15)$$

where the upper limit results from a singularity where the spheres touch. Tabular results are also given by Sangani and Acrivos (1983) for closely packed regular spherical arrays and for random arrays. The values of k_e/k_a are given in Table 7.4.4 for $k_b/k_a = 0$ to 50, where $\Phi_{b,m} = 0.62$ denotes the maximum packing. Sangani and Acrivos note that these values for regular packing, and for nonconducting spheres, $K = k_b/k_a = 0$, are 0.344, 0.217, and 0.160. These values are in close agreement with the upper bounds, 0.378, 0.239, and 0.189, given by Hashin and Shtrikman (1962). For a general discussion of such bounds, see Strieder and Aris (1973).

7.4.3 Other Geometries and Considerations

The preceding section summarizes the results of the analyses of the simplest inclusion geometries. Limited other information on related geometries is given in the following discussion. Another usually very approximate method, which may be applied quite generally, is to model subregions containing inclusions as series–parallel circuits. However, recent study has increasingly used direct numerical analysis. These matters are considered briefly in the following discussion.

OTHER GEOMETRIES. A result directly related to spherical inclusions is that of Fricke (1924) for spheroids. The form of Maxwell's equation was modified as follows:

$$\frac{k_e - k_b}{k_e + xk_b} = \frac{k_a - k_b}{k_a + xk_b} \tag{7.4.16}$$

where x is a given function of k_b/k_a and of the ratio of the spheroidal axes. See Crank (1975). Measurements have indicated that Eq. (7.4.16) correlates data for many other kinds of inclusion geometries in rubber matrices, for a wide range of K. Then x in Eq. (7.4.16) is taken as

$$x = 1 - 3/S \quad \text{where } S = A_s/A_p \tag{7.4.17}$$

where S is the sphericity of a particle. It is taken as the ratio of the area A_s of a sphere, having the same volume as the particle, to the actual surface area of the particle, A_p.

Keller (1963) also calculated k_e for square arrays of cylinders, normal to the temperature gradient in the composite. The result, for $k_b \gg k_a$, is

$$\frac{k_e}{k_a} = \frac{\pi^{3/2}}{2[(\pi/4) - \Phi_b]^{1/2}} + \cdots \quad \text{for } \frac{\pi}{4} - \Phi_b \ll 1 \tag{7.4.18}$$

These considerations also indicated that $(k_e/k_a)_L$, for very low cylinder conductivity, that is, for $k_b \ll k_a$, is very simply related to that in Eq. (7.4.18), for $k_b \gg k_a$, as

$$\left(\frac{k_e}{k_a}\right)_L = \frac{k_a}{k_e} \tag{7.4.19}$$

where the right side is the inverse of Eq. (7.4.18). See also Perrins et al. (1979) for close-packing results for cylinders.

A SERIES–PARALLEL APPROXIMATION. There are very many practical circumstances in which the inclusions may not be approximated in terms of spheres, or by cylinders, normal to the heat or mass transfer average flux direction. Common examples are fibers or other long geometries parallel to the average flux direction. Several examples are shown in Fig. 7.4.2. The equivalent resistance circuits are also shown for the total sub-region volumes, $2LW^2$ and $(L_1 + L_2)W^2$, respectively, between locations 1 and 2. It is assumed that the spacing in the direction normal to the figure is also W.

The resistances in Fig. 7.4.2(a) are as follows. First, R_A is the resistance of the region lying entirely in material k_a, $(W^2 - \pi D^2/4)$ in area, and $2L$ in length. The parallel branch is two half-fiber lengths, amounting to L, of area $\pi D^2/4$, for $2R_C$, in material a. Then R_B is in the fiber material resistance, of area and length $\pi D^2/4$ and L.

This series–parallel model must assume, as implied in the resistance diagrams in Fig. 7.4.2, that the temperatures across the two planes, at locations 1 and 2, t_1 and t_2, are each uniform. It also assumes that there is no lateral

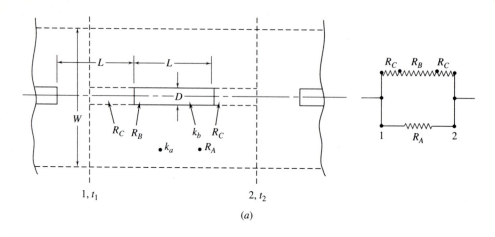

(a)

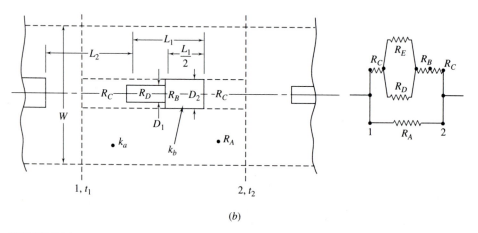

(b)

FIGURE 7.4.2
Composites having inclusions of several characteristically different geometries, also with the approximate conductive resistance equivalents: (a) round fibers of dimensions D and L, spaced at W; (b) round plugs, characterized as D_1, D_2, and L_1, spaced at W.

thermal contact between the two resistance paths, as assumed in the network diagram at the right.

Each of the component resistances are related to the heat flow by $R = \Delta t/q = L/kA$, Eq. (1.1.1). Therefore, the resistances in Fig. 7.4.2(a) are

$$R_A = \frac{2L}{k_a(W^2 - \pi D^2/4)} \qquad R_B = \frac{L}{k_b \pi D^2/4} \qquad 2R_C = \frac{L}{k_a \pi D^2/4} \qquad (7.4.20)$$

The total resistance, R_T, for the area W^2 and length $2L$, between locations at t_1 and t_2, is

$$\frac{1}{R_T} = \frac{1}{R_A} + \frac{1}{R_B + 2R_C}$$

The heat flow, $q = (t_1 - t_2)/R_T$, over area W^2, amounts to an average heat flux of

$$q'' = \frac{q}{W^2} = \frac{(t_1 - t_2)}{W^2}\left[\frac{1}{R_A} + \frac{1}{R_B + 2R_c}\right] \equiv \frac{k_e(t_1 - t_2)}{2L} \qquad (7.4.21)$$

Therefore, k_e/k_a is calculated and the values of R_A, R_B, and R_C in Eq. (7.4.20) are used.

$$\frac{k_e}{k_a} = \frac{2L}{k_a W^2}\left[\frac{1}{R_A} + \frac{1}{R_B + 2R_C}\right]$$

$$= 1 - \frac{\pi D^2}{4W^2} + \frac{\pi D^2}{2W^2}\left(\frac{k_a}{k_b} + 1\right)^{-1}$$

The volume fraction Φ_b is $\pi D^2 8W^2$ and the result is

$$\frac{k_e}{k_a} = 1 - 2\Phi_b\left[1 - 2\left(\frac{k_a}{k_b} + 1\right)^{-1}\right] \qquad (7.4.22)$$

For the geometry in Fig. 7.4.2(b), the analysis is more complicated. Resistance R_A is of the volume $(W^2 - \pi D_2^2/4)(L_1 + L_2)$. The remaining volume $(\pi D_2^2/4)(L_1 + L_2)$ is made up of five different resistances; R_A to R_E, related as shown in Fig. 7.4.2(b). R_C, at each end, is $L_2/(2k_a\pi D_2^2/4)$. R_B is $L_1/(2k_a\pi D_2^2/4)$. Then R_D and R_E, for the annulus, are similarly calculated. Again the flux q'' is calculated as in Eq. (7.4.21), in terms of the resistances.

A similar kind of an analysis, for a cubical arrangement of spherical inclusions, is given in Crank (1975). This is somewhat more difficult since the resistance of a hemisphere must be calculated. That result has value as a basis for comparing the series–parallel approximation method with the results from more comprehensive analyses. Also given in this reference is a further extension of this method to composites containing randomly distributed particles of irregular size and shape. See also Rohsenow et al. (1985), page 4-163, for additional series–parallel approximation results.

SUMMARY. The preceding considerations in this section, of calculations and results, apply to both heat conduction and simple mass diffusion. It also applies equally to the other processes in Table 7.4.1, as do the limitations in the kinds of methods and results discussed in this section. Therefore, any further improvements of the accuracy and generality of the information concerning composites will apply in many applications.

Further advances for more varied and practical composite configurations will come increasingly from transformations, in conjunction with numerical formulations and field calculations. Among references especially relevant to these matters are Garland and Tanner (1978) and Crank (1975).

A survey of the analysis of composite materials, including conductivity, is given by Hashin (1983). Durand and Ungar (1988) report a method of analysis and results for a dense composite of cylinders. Related references are given. A general and comprehensive review of the mechanisms and advances concerning transport in random heterogeneous media is given by Torquato (1991).

7.5   INSULATION

There are many applications and devices in which the incidental loss or gain of heat in a region is a penalty. There are a tremendous number of insulation schemes in use in technology, in the environment, and also by living organisms. The aim is to thermally isolate one region from others, by a barrier of poor heat transfer characteristics.

Common examples are fibrous insulation layers like fur and an evacuated space with silver-surfaced walls, as in the simple Dewar. The Dewar flask is evacuated, whereas fur is not. However, fibrous layers suppress natural convective and other force-field-driven currents which might arise to degrade an insulating barrier. These effects are often very important, even though gas conductivities are low. Since the region is subject to a temperature gradient, there will also be heat transfer by thermal radiation. Depending on the opacity of the insulating material and its radiation characteristics, there will be transmission, local absorption, and reradiation across the layer. For an evacuated insulating space, the gas is at a low density. Thereby, the convective effect is attenuated, since the buoyancy effect is proportional to the square root of the gas density. Silver-surfaced bounding walls have both low emissivity and high reflectivity. This reduces the net radiant transport component.

These two forms of insulation, a fibrous or porous material and an evacuated space, are a large fraction of all applications of insulation. However, rapidly increasing needs to transport and to store cryogens, such as liquid O_2, N_2, H_2, and He, required the development of much better cryogenic insulation. See an early consideration of improved insulation development by Glaser et al. (1967). The best resulting evacuated "superinsulation" may have an effective conductivity, k_e, of over three orders of magnitude below even that of a gas at rest, in a continuum conduction process. See App. B.7 for characteristic values. The consequence is a very low boil-off rate in ordinary cryostats, arising from the heat leak through the insulation layer. These superinsulations, or cryogenic insulators, have also made possible the long-term storage of liquified gases in space devices and in over-the-road tank trucks. For example, the boil-off may be as low as a few percent, over a period of many days.

Many types of such insulation layers have been developed, in terms of their internal structures and uses. All are evacuated, usually to extremely low gas densities. This density is often maintained, during use, by a local internal region or surface of sufficiently low temperature. This will condense or freeze out in-leaking gases or gases generated internally through time during use. In tankage used in space devices, the insulation may be vented to space, to allow gases to escape. Vacuum pumping is also used.

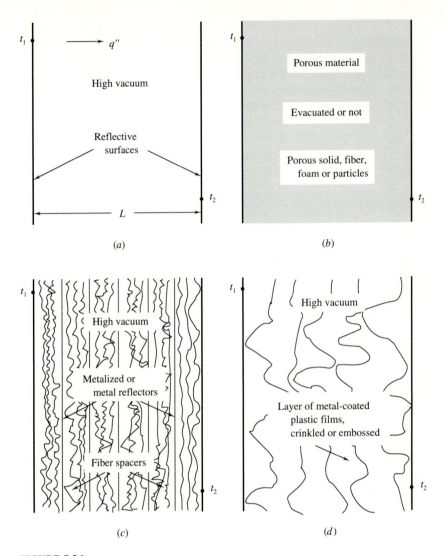

FIGURE 7.5.1
Several common forms of evacuated cryogenic-insulation barriers: (*a*) a simple Dewar flask; (*b*) a porous material; (*c*) metal or metalized foils, separated by fibrous spacers; (*d*) as in configuration (*c*) but with the metalized foils separated by being embossed or crinkled, to separate the foil layers.

Common types of evacuated barriers are shown in Fig. 7.5.1. A porous layer is shown in Fig. 7.5.1(*b*). The porous material may be a solid, a fibrous matrix, a powder, or a foam. It is sometimes a filler of particulate matter such as a granular material or powder. It may also be of regular shape, as small solid or hollow metalized spheres. One component of heat transfer is then conduction through the solid material and across any direct contact areas, for a divided filler material. The other principal heat transfer component is radiation across the gaps between the elements of solid material. This effect may be reduced by providing reflective properties on the surfaces of the filler material.

The multilayer insulations in Fig. 7.5.1(c) and (d) are commonly much more effective barriers. Given n individual layers, the net radiant heat flux component q_r'' across the barrier decreases as $1/n$. Radiation transport is further reduced by providing reflective properties on the surfaces of the filler material. The heat flux also decreases as $\epsilon/2(1 - \epsilon)$, where ϵ is the emissivity of the surfaces. These radiation shields may be very closely spaced, to perhaps 50 per centimeter of barrier thickness. Calculation of these effects is given in Sec. 7.5.4.

However, all direct contacts between the adjacent metal or metalized fibers or foils result in a direct conduction effect through the elements of the barrier. The separation of the active radiation shields is commonly achieved by the two methods shown in Fig. 7.5.1(c) and (d), at some expense of insulating quality. In the scheme in Fig. 7.5.1(c), thin fiber spacer sheets are placed between the radiation shields. These are often paper, a woven fabric sheet, or a tufted bonded layer, having an average thickness of about 0.1 mm.

All contacts between the spacer and shield materials result in conduction paths. A similar and often simpler measure is to emboss or crinkle the radiation shields. Then the paths for conduction are along the radiation shields, to the local contacts between adjacent shields. Of course, the advantage of low effective conductivity k_e is realized in heat conduction normal in the direction across the layer. These barriers are much poorer insulators in each of the other two coordinate directions.

The following section indicates the performance characteristics of different kinds of insulation layers. The measures are: the effective conductivity, k_e; the thermal diffusivity of the insulation, α_i, and the insulation mass, $\rho_i L$, per unit area of insulated wall. These are the kinds of measures which are important in design, and in optimizing the insulation effectiveness, in terms of a specific application.

7.5.1   Insulation Performance

The commonly important properties are the effective barrier conductivity k_e, the thermal diffusivity α_i, and the mass density ρ_i, per unit volume of the insulation. For an insulation layer of thickness L, the heat transfer resistance is $R_e = L/k_e$, per unit area. The weight or mass $\rho_i L = \rho_i k_e(\Delta t/q'')$ has separate importance, for example, in determining the launch requirements of space craft. For given values of Δt, the working overall temperature difference, and q'' the design heat leak, $\rho_i k_e$, are to be as small as practical. The volumetric specific heat $\rho_i c_i$ is also important in determining the length of required response time, to imposed transient changes in temperature bounding conditions.

Performance measures k_e, $k_e c_i$, and $k_e/c_i \rho_i$ are shown in Fig. 7.5.2 for common insulating schemes, where c_i is the specific heat, per unit mass of

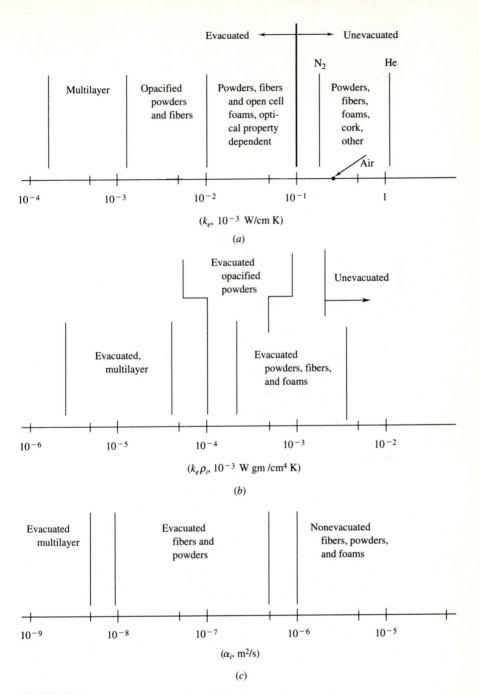

FIGURE 7.5.2
The performance of unevacuated and evacuated insulation: (*a*) effective conductivity, k_e, as 10^{-3} W/cm K; (*b*) k_e times insulation mass density ρ_i, as 10^{-3} W gm/cm^4 K; (*c*) thermal diffusivity, $k_e/c_i\rho_i = \alpha_i$, as m^2/s. [Adapted from Tien and Cunnington (1973).]

insulation. In each of these performance measures, evacuated insulations lie to the left. The classes of insulation include evacuated multilayer, the best in each property category, evacuated continuous materials such as powders and foams, and unevacuated continuous materials. The continuous and evacuated materials are less effective than a multilayer because of higher conduction, due to many direct contacts and continuous conduction paths. The unevacuated insulations are also subject to conduction through the gas in the voids. Also shown in Fig. 7.5.2(a) is the conductivity of air. Its value of k is greater than that shown for N_2, in a porous material, because of radiation shielding provided by the porous material.

Figure 7.5.2(a) indicates that multilayer values of k_e may be about three orders of magnitude less than that for stationary air. This is the combined effect of evacuation and radiation shielding. However, the insulating properties are very directional. On the other hand, porous insulations may have an almost isotropic conductivity and are commonly used in irregular regions of an insulating wall.

7.5.2 Heat Transfer Mechanisms in Insulation

The foregoing performance measures indicate the role of internal material configuration and other factors in optimizing a particular design. There are also other considerations, such as cost, reliability, and useful life. However, the general limit of performance in any kind of insulation is in terms of the sum of all of the internal conduction, radiation, and any other processes which contribute to heat flow across an insulating barrier, subject to a temperature gradient. The common such mechanisms of transport, and the measures which affect them, are listed as follows. These components may operate in parallel or may actively interact locally within the insulation layer.

1. Thermal radiation; the emissivity and absorptivity of the internal material and the number of local absorption–reflection processes across the layer.
2. Gas conduction; small internal cavity sizes and low density to suppress convection, also low pressure to result in noncontinuum gas conduction. This arises for the Knudsen number $Kn = \lambda/\delta > 0.01$ and as large as practical. Here, λ is the gas molecule mean free path and δ is the local pore size, or the internal spacing of dividers, as with multilayers. Recall the considerations in Sec. 7.3.4 relevant to contact resistance, R_c.
3. Conduction across local contacts; layered material between adjacent radiation shields as in Fig. 7.5.1(c) and (d). This is often like a constriction resistance. Such conduction is discussed in Secs. 3.3.3 and 7.3.3.
4. Conduction across a more continuous material; as through a porous solid, foam, packed particles, or fibrous materials.

In many kinds of insulation layers, the transport length of a process, say δ, is much less than the overall barrier thickness L. Examples are the three configurations in Fig. 7.5.1(b), (c), and (d), if the porosity and divider spacings are small. With small heat exchange length δ, and local interaction between the parallel modes of radiation and conduction, the temperature level may be approximately the same over each plane normal to the heat flow direction. Then one temperature distribution, across the layer, represents the whole region. This is equivalent to assuming that all of the local modes of heat transfer are independent and in parallel at all locations across the barrier. This is analogous to the analysis of composite barriers in Secs. 2.1.3 and 4.5.2. This assumption is used later in relatively simple representations of heat transfer, through insulation. The radiation, gas, and conduction components are summed as

$$q'' = q_r'' + q_g'' + q_s'' \qquad (7.5.1)$$

7.5.3 Heat Transfer Components Across a Multilayer Barrier

The radiation, gas, and solid conduction modes are evaluated in the following subsections. These brief summaries indicate some of the principal aspects of the mechanisms which determine insulation performance. In Sec. 7.5.4 the total heat flux q'' is analyzed for a layered insulation and the performance of a barrier of layers is formulated.

THERMAL RADIATION FLUX q_r''. This flux is an important component even in cryogenic insulation, since the total effective conductivity k_e is very small. Recall Fig. 7.5.2(a). The radiation transport is in terms of local emission, absorption, reflection, and scattering. These processes may depend very heavily on the nature of the radiant emission and its interaction with the other heat transfer modes present.

At $T = 300$ K, Wien's displacement law predicts a peak monochromatic emission rate at 10 μm wavelength. At $T = 10$ K the value is 290 μm. These values are sometimes comparable to the characteristic size δ of the spaces or voids. Then many more complicated considerations arise in the evaluation of radiant exchanges. The macroscopic properties of radiant interactions may not be reliable formulations of such transport. See the general discussion by Tien and Cunnington (1973), related to other appropriate radiation models.

Over the very wide range of insulations of importance, many different kinds of radiation transport models apply. The simplest is for the empty vacuum layer in Fig. 7.5.1(a) (see page 437), for gray diffuse surfaces 1 and 2, and for the radiation wavelengths at T_1 and T_2 being much less than the layer thickness L. Then the net radiant exchange between the bounding surfaces is

$$q_r'' = \frac{\sigma\left(T_1^4 - T_2^4\right)}{\dfrac{1}{\epsilon_1} + \dfrac{1}{\epsilon_2} - 1} \qquad (7.5.2)$$

This relation may also be applied for L divided into n layers by radiation shields. Then q_r'' in Eq. (7.5.2) is reduced by the factor n.

Another limiting condition of radiation behavior applies for gray and diffuse radiation when the space is filled with a material which has a very short absorption length, denoted here as μ_a^{-1}, compared to the layer thickness L. Refer to the simple analysis of extinction, condition T.6 in Sec. 1.4.1. A large value of μ_a means that local emission is immediately reabsorbed. This process is denoted as conduction model 4 in Table 1.1.1. For isotropic scattering, the resulting conductivity locally varies as $k \propto T^3$. This is the result written as follows, in terms of the index of refraction n and the extinction coefficient β. The effective radiative conductivity is approximated as

$$k = 16n^2\sigma T^3/3\beta \tag{7.5.3}$$

Across an optically thinner region, of isotropic absorbing and scattering material, as for β small, of thickness L and of optical thickness $\tau = \beta L$, the net radiant transfer is

$$q_r'' = \frac{n^2\sigma\left(T_1^4 - T_2^4\right)}{\dfrac{3\tau}{4} + \dfrac{1}{\epsilon_1} + \dfrac{1}{\epsilon_2} - 1} \tag{7.5.4}$$

For an optically thick or opaque region, that is, $\tau \to \infty$, the radiation transport across the layer may also be expressed as follows. See Sparrow and Cess (1978), Viskanta (1965), and Wang and Tien (1967).

$$q_r'' = 4n^2\sigma\left(T_1^4 - T_2^4\right)/3\tau \tag{7.5.5}$$

These results apply in special circumstances in which radiant transport may be evaluated by very simple models. In many applications, the wavelengths are comparable to local spacing δ. Some layered insulations are opaque. Some have some level of transparency. In many conditions, the radiation effects are not gray, diffuse, or isotropic. A thorough summary is given by Tien and Cunnington (1973). Also, with highly rereflective radiation shields, the diffuse radiation effects may result in appreciable lateral heat transfer. See, for example, Tien et al. (1969). The following subsections consider gas and also filler conduction, accounting for increased direct contact conduction arising from the stress level applied to the layer.

GAS CONDUCTION FLUX q_g''. Natural convection heat transfer across the voids internal to close-packed insulation is negligible, under essentially all conditions. This results because the Rayleigh number, $\text{Ra} = \text{Gr Pr} \propto \delta^3$ is always small for δ small. Therefore, gas conduction is the model. The magnitude of this effect depends on the interaction between the gas pressure level p and the characteristic void size δ, in terms of continuum or noncontinuum behavior.

These effects are discussed in Sec. 7.3, in relation to the conduction across gas spaces, between adjacent solids, as related to contact resistance, R_c. A continuum condition arises for $Kn = \lambda/\delta \leq 0.01$, where μ_v is the viscosity and λ is the gas molecular mean free path,

$$\lambda = \frac{16}{5} \frac{\mu_v}{p\sqrt{2\pi RT}} \qquad (7.5.6)$$

The other extreme regime is called free molecule, where λ is larger than δ, say as $Kn = \lambda/\delta > 10$. Then the molecular density is low and there are very few collisions between gas molecules. They travel back and forth only as a result of collisions with the two bounding surfaces. Between these two limits, in the range $10 > Kn > 0.01$, are two regimes in which both kinds of collisions occur. These are called the transition and the slip or jump regimes. The effective conductivity is a maximum for continuum conditions and decreases across the whole range of increasing Kn, at a given value of δ.

Another physical effect which arises is the extent of thermal accommodation of the molecules to the surface temperature level, as they interact with it during each encounter. For a plane gas layer and free-molecule transport, the overall accommodation coefficient between two parallel surfaces A_1 and A_2 is

$$\alpha = \frac{\alpha_1 \alpha_2}{\alpha_2 + \alpha_1(1 - \alpha_2)} \qquad (7.5.7)$$

where α_1 and α_2 are the perhaps different levels of accommodation at two adjacent surfaces. The resulting net free-molecule heat flux, for diffuse reflection, is then

$$q''_{FM} = \alpha \left[\frac{\gamma + 1}{\gamma - 1}\right] p(t_2 - t_1)\sqrt{R^0/8\pi MT} \qquad (7.5.8)$$

and where p is in N/cm^2, R^0 is in $gm\ cm^2/s^2$, M is in gm/mol and $\sqrt{R^0/MT}$ is in $cm/s\ K$. Then q''_{FM} is in W/cm^2, γ is the specific heat ratio of the gas, and $T(K) = (T_1 + T_2)/2$. See also Springer (1971).

This heat flux is independent of the spacing between the surfaces δ, assuming free-molecule transport. The required vacuum level for this condition, for a large spacing of $\delta = 1$ cm, is about 5×10^{-4} Torr $= 5 \times 10^{-4}$ mm Hg $\approx 0.7 \times 10^{-7}$ atms. The use of ultrafine particle insulation, for $\delta \approx 100\text{Å}$, see Tien and Wang (1988) and Yarborough et al. (1985), reduces the need for ultra-low pressures to result in noncontinuum gas conduction.

As for radiation transport, see the introduction in this section, the value of q''_{FM} is reduced by dividing the $\delta = 1$ cm layer into n layers, by intermediate surfaces. Then the overall conductive flux in Eq. (7.5.8) is reduced by a factor of n. However, a higher level of vacuum is required to result in $Kn \geq 10$, for the resulting thinner layers, of thickness δ/r.

For the transition and slip regimes, which lie in the range 10 > Kn > 0.01, the conductive heat flux across a layer is given approximately by Springer (1971) and Tien and Cunnington (1973) as

$$
\frac{q''_g}{q''_{FM}} = \left[1 + \frac{q''_{FM}}{q''_c} \right]^{-1} = \left[1 - \frac{K\delta}{\lambda} \right]^{-1}
\tag{7.5.9}
$$

where q''_{FM} is from Eq. (7.5.8), K results from combining Eqs. (7.5.7)–(7.5.9), and q''_c is the conduction under continuum conditions, where k is approximated from kinetic theory, as follows:

$$
q''_c = \frac{k(t_1 - t_2)}{\delta} = \frac{(t_1 - t_2)[(9\gamma - 5)\mu c_v/4]}{\delta}
\tag{7.5.10}
$$

where δ is the layer thickness.

These relations then determine q''_g from the values of q''_c and q''_{FM}, Eqs. (7.5.8) and (7.5.10). These apply for the asymptotic processes of free-molecule and continuum flow, respectively, that is, for Kn > 10 and Kn < 0.01. Then Eq. (7.5.9) evaluates the actual flux q''_g.

A sometimes useful alternative to the preceding procedure is in terms of a noncontinuum gas conductivity k'. An approximate relation for k' was given by Verschoor and Greebler (1952), as $k' = \alpha k[\delta/(\delta + \lambda)]$. Here, α is the accommodation coefficient, k and λ are continuum values, and δ is the characteristic length of the noncontinuum conduction path.

CONDUCTION THROUGH FILLERS, q''_s. This is commonly a large component of the total heat transfer across such an insulating layer. Powder and fiber fillers are used to reduce both gas heat transfer, as discussed previously, and radiation transfer, by shielding. However, fillers also offer conductive paths through their material and also across direct contact areas between adjacent material. Recall from Fig. 7.5.2(a) that powder and fiber insulators have an effective conductivity of one or two orders greater than multilayer insulations.

Given any combination of insulated layer internal structure, conduction is reduced by lengthening the many conduction paths across the region. Other effective measures are the use of porous solid filler, and hollow spheres, and keeping the direct contact areas between adjacent filler material small. The crinkled shield in Fig. 7.5.1(d) is a configuration which seeks these ends. Of course, packing any such shields tightly increases both the number of the direct contacts and their area. Tortuous paths are also often provided by particles and foams and by foil separators of fibers, screens, and woven layers.

The conduction is commonly through the material, to the direct contact areas, and then away on the other side. Contact resistance, $R_c = 1/2ka'$, and the effects of contact region constriction resistance, are discussed in Sec. 7.3.3. However, this process is not a simple series conduction path. As in contact region conduction, there is parallel conduction and radiation across the gaps. However, the gas conduction effect is very small at high vacuum. The best materials are those with low conductivity. However, these materials have relatively high absorptivity. They are often opacified by either coating them with highly reflective thin metallic films or by including interspersed reflective or absorbing particles, to provide effective radiative attenuation.

The analysis of solid conduction, and the constrictive resistances in series, is very difficult, except as idealized in a regular geometry, such as packed spheres. If a highly evacuated condition is considered, as for $Kn = \lambda/\delta > 10$, the gas conduction effect is small. Then radiation is augmented only by conduction through the filler material. There is constriction resistance at the direct contact areas.

A simple example of an ordered filler is a cubic packing of spherical particles of equal diameter D_s. See Fig. 7.5.3. Refer to Chan and Tien (1973). The solid fraction in such a layer is 0.52. It is assumed that the temperature gradient isotherms are normal to the planes of the spheres. Then the solid conduction, along with the constriction resistances, amount to equal parallel resistances across each layer of spheres. That is, there is no lateral heat transfer. The number of successive layers, per unit distance normal to the insulation layer, is $N_s = 1/D_s$. Only constriction resistance R_c is considered in the following discussion, as for spheres of relatively high conductance, of order k_s/D_s, compared to R_c^{-1}. The series resistance R_s, interior to such a layer, that is, over N_s contacts of the same individual resistance R_c, is $R_s = N_s R_c$. The number of parallel paths per unit area, normal to the heat flow, is $N_p = N_s^2$.

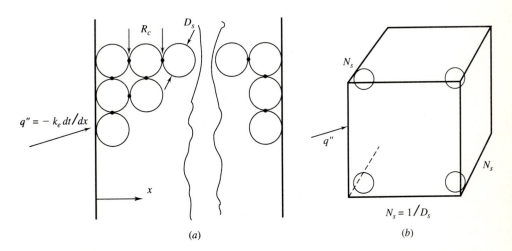

FIGURE 7.5.3
An insulation layer composed of cubically packed solid spheres: (*a*) contact conductance between adjacent spheres; (*b*) a unit cube of the region, to determine k_e.

The resistance of a unit cube of this material, R_e, is then

$$R_e = N_s R_c / N_s^2 = R_c / N_s \tag{7.5.11}$$

Since $N_p = N_s^2$, $R_e = R_c / N_s$. The equivalent conductivity k_e is

$$k_e = \frac{1}{R_e} = \frac{N_s}{R_c} \tag{7.5.12}$$

For a layer of area A having a thickness L, and subject to a temperature difference of $t_1 - t_2$, the heat flow is

$$q = \frac{k_e A(t_1 - t_2)}{L} = \frac{N_s A(t_1 - t_2)}{R_c L} \tag{7.5.13}$$

The resistance R_c, of each single contact area, depends on the pressure applied to the bed of spheres. This determines the resulting contact area radius $a' = r_c$, between adjacent spheres, per unit of lateral area. Referring to the result in Sec. 7.3.3, for constriction resistance, the relation between R_c and r_c is

$$R_c = \frac{1}{2kr_c} \tag{7.5.14}$$

where k applies for the material of the spheres. Next, R_c and k_e must be found from the relation between applied force F on each sphere and the resulting contact area deformation, of radius r_c. For spheres, and only elastic deformation, the Hertz equation is

$$r_c = \left[\frac{3(1 - \mu^2)r_s F}{4E} \right]^{1/3} \tag{7.5.15}$$

where μ is Poisson's ratio, $r_s = D_s / 2$, E is Young's modulus, and $F = p/N_s^2$, where N_s^2 is the number of contacts, per unit of lateral area, and p is the applied pressure. Combining Eqs. (7.5.14) and (7.5.15) results in the following relation for the effective conductivity:

$$k_e = 0.909k \left[\frac{1 - \mu^2}{r_s^2 E} F \right]^{1/3} = 0.909k \left[\frac{(1 - \mu^2)p}{r_s^2 E N_s^2} \right]^{1/3} \tag{7.5.16}$$

Calculations similar to these may also be used for hexagonal packing, a packing fraction of 0.74, and for the random packing of spheres or other idealized shapes. See Yang et al. (1983) for an analysis of radiant heat transfer through a randomly packed bed of spheres. Duncan et al. (1989) report measurements of the effects of thermal conductivity, the loading, and the interstitial gas.

The preceding analysis and results apply for cubic packing. Gravity effects were not included. The direct contact area was calculated from elastic deformation considerations. Most applications in insulation avoid high pressure. Should

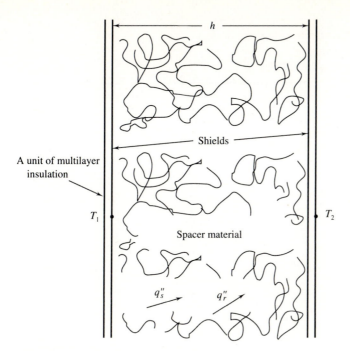

FIGURE 7.5.4
A unit of multilayer insulation in a layered barrier.

plastic deformation arise, the determination of the direct contact area is as given in Sec. 7.3.2.

7.5.4 Multilayer Overall Heat Transfer

Recalling Eq. (7.5.1), the individual effects of thermal radiation, gas conduction, and direct contact, or of continuous material in a porous region, may often be summed to give the total thermal energy flux q''. The foregoing results evaluate individual components for several kinds of conditions. The particular example given here concerns the multilayer insulation in Fig. 7.5.1(c). A typical layer is shown in Fig. 7.5.4.

Gas conduction is assumed negligible, as it would be for pressures less than the order of 10^{-6} Torr. The remaining heat transfer modes are then conduction through the spacer layer and the net radiation heat transfer from one bounding shield to the other, through the spacer material itself. It is assumed that the optical thickness of the spacer material for radiation is very small. That is, $\tau = (\kappa + \gamma)h \ll 1$, where κ and γ are the absorption and scattering coefficients and h is the thickness of the spacer layer. See Table 7.5.1 for characteristic properties of several spacer materials.

TABLE 7.5.1
**Total absorption and scattering coefficients for
several insulating materials
[From Tien and Cunnington (1973)]**

Material	Density (kg/m^3)	Diameter (m)	Source Temperature (K)	Absorption coefficient (m^{-1})	Scattering coefficient (m^{-1})
Borosilicate glass fibers	200	1×10^{-6}	500	1,300	26,000
			650	1,100	27,000
			800	1,100	28,000
			1,000	700	3,100
			1,700	600	2,500
Silica fibers	50	1.5×10^{-6}	500	200*	3,300
			650	200*	5,000
			800	100*	710
			1,000	100*	740
Silica fibers	50	1×10^{-5}	500	200*	3,800
			650	200*	5,700
			800	100*	730
			1,000	100*	760
Carbon fibers	65	1×10^{-5}	775	400	38,500
			923	200	26,000
			1,123	200	18,500
			1,273	400	20,000
Polyurethane foam	35	Random pore size	500	200	2,850

*These listed values are upper limits.

Under the conditions of the optical thickness $\tau = \beta h$ very small, the net radiation transport is assumed to be unaffected by the presence of the spacer. See Wang and Tien (1966). Then the heat flux between adjacent shields is the sum of the conduction and radiation effects, as

$$q'' = \frac{k(T_1 - T_2)}{h} + \frac{\sigma(T_1^4 - T_2^4)}{\dfrac{3\tau}{4} + \dfrac{1}{\epsilon_1} + \dfrac{1}{\epsilon_2} - 1} = q_s'' + q_r'' \qquad (7.5.17)$$

where k is, nominally, the effective conductivity of the spacer layer. If the shield surfaces are taken as metalized, the total hemispherical emittances ϵ_1 and ϵ_2 are small compared to 1.0. Then Eq. (7.5.17) is, see Tien and Cunnington (1973),

$$q'' = \frac{k(T_1 - T_2)}{h} + \frac{\epsilon_1 \epsilon_2 \sigma(T_1^4 - T_2^4)}{\epsilon_1 + \epsilon_2} \qquad (7.5.18)$$

For spacers of refractive index $n > 1$, the black-body surface emissive power is $n^2 \sigma T^4$. See Domoto and Tien (1970). Taking $\epsilon = \bar{\epsilon}$, the total hemispherical emittance to a vacuum, Eq. (7.5.18), becomes

$$q'' = \frac{k(T_1 - T_2)}{h} + \frac{\bar{\epsilon}_1 \bar{\epsilon}_2 n^3 \sigma (T_1^4 - T_2^4)}{\bar{\epsilon}_1 + \bar{\epsilon}_2} \qquad (7.5.19)$$

The metalized foils are commonly very thin, on the order of 5 μm. They are made of a material which has very high conductivity, compared to the spacer layer, which is on the order of 250 μm thickness. Therefore, the two metalized foils are assumed to be isothermal, at T_1 and T_2, respectively. The layer in Fig. 7.5.4 represents one layer of the multilayer barrier.

The spacer conductivity k given previously is the equivalent conductivity. This may consist only of contact interfaces present between the two bounding interfaces. Recall the similar model developed in the preceding subsection, for a filler of solid spheres. The conduction effect may be modeled as follows, see Tien and Cunnington (1973):

$$k = Chp^d / N_C \qquad (7.5.20)$$

where $N_C = 1$ or 2 is the number of spacer contacts with the adjacent foils, p is the imposed pressure, and d is an exponent which expresses the rate of change of k due to the imposed pressure. $N_C = 1$ applies for one side of the spacer being bonded to a foil, d may be of order 0.5 and C is a constant of proportionality. Equation (7.5.19) then becomes

$$q'' = \left[\frac{Chp^d}{N_C} \right] \frac{(T_1 - T_2)}{h} + \frac{\bar{\epsilon}_1 \bar{\epsilon}_2 n^3 \sigma (T_1^4 - T_2^4)}{\bar{\epsilon}_1 + \bar{\epsilon}_2} \qquad (7.5.21)$$

or
$$q'' = A(T_1 - T_2) + B(T_1^4 - T_2^4) \qquad (7.5.22)$$

This equation applies across any particular layer i, as seen in Fig. 7.5.5, where T_1 and T_2 are the absolute bounding temperatures for that particular layer. To determine the heat flux across an assembly of M such layers in series, it is necessary to meet the requirement that the flux across each layer is the same.

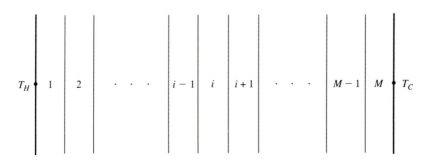

FIGURE 7.5.5
A barrier of multilayer insulation, with bounding temperatures T_H and T_C.

This requirement is the same as the one imposed to determine the overall conductance of a plane composite barrier, in Sec. 2.1.3.

The analysis is more complicated here. One aspect is that Eq. (7.5.21) or (7.5.22) involve both ΔT and ΔT^4. Another is that A and B may be different at different locations across the barrier. Consider a barrier consisting of many layers, M, the barrier having bounding temperatures T_H and T_C, as seen in Fig. 7.5.5. With M large, ΔT across the ith layer, ΔT_i, is small compared to $T_H - T_C$. This suggests expressing $\Delta(T^4)$ as a differential, as $d(T^4) = 4T^3\, dT$, or $\Delta(T^4)_i \approx 4T_i^3\, \Delta T_i$. Then T_i^3 is an average value over the ith layer, between the shields which bound layer i, and ΔT_i is the difference in temperature. Then Eq. (7.5.22) for the ith layer is

$$q'' = \left(A_i + 4T_i^3 B_i \right) \Delta T_i = \left(A_i + B_i' \right) \Delta T_i \qquad (7.5.23)$$

where $\bar{\epsilon}_i$ here is taken equal to $\bar{\epsilon}_2$, for simplicity. The coefficients are

$$A_i = \left(Cp^d/N_C \right)_i \quad \text{and} \quad B_i' = \left(2n^3\sigma\bar{\epsilon}T^3 \right)_i \qquad (7.5.24)$$

The other condition, to be satisfied by the ΔT_i, is

$$\sum_i^M \Delta T_i = T_H - T_C \qquad (7.5.25)$$

Equations (7.5.23)–(7.5.25) are the formulation for the calculation of the heat flux q'' through the barrier. The formulation is complicated, in accounting for the initially unknown variations of the A_i, B_i or B', and ΔT_i across the barrier. Given any variations of C, d, N_C, n, and $\bar{\epsilon}$, with location or temperature level, an iterative procedure is indicated. This might be begun using assumed values of the T_i, over the range from T_H to T_C. Convergence is then sought in terms of the values of the q_i''. All of the q_i'' values must approach a uniform value q'' over the whole region.

SUMMARY. Sections 7.5.1 through 7.5.3 summarize insulation performance, indicate aspects of radiative and conductive modes of heat transfer, and analyze several aspects of radiation and gas and solid conduction. Several common aspects of the net insulation heat leak are then considered in Sec. 7.5.4, for multilayer insulation. The results indicate the relative effects of radiation and spacer conduction, and of contact conductances, acting in parallel. A comprehensive model is given, including these effects. This may be used for approximate estimates.

Many other insulation techniques are in use which offer additional possibilities in general or in specific applications. Reference is given to some additional information. Recent measurements were made of extinction coefficients, by Chu et al. (1988), in ultra-fine Aerosil powder, of size 70 Å. For insulation temperatures in the range of 320–400 K, the radiation component amounted to only about 10% of q'', where gas conduction was the rest. These results show excellent performance of radiation shielding, in ordinary temperature level applications. Tien and Vafai (1989) briefly review radiative transport results in porous beds. Extensive references are given and many relate solely to radiative heat transfer. See also Wolf et al. (1990) for a review of radiation conductivity in a material containing dispersed voids. A variational principle determines upper bounds.

PROBLEMS

7.3.1. Two extensive regions are in idealized direct contact at raised flat circular areas of 20-μm radius spaced at 1 mm in both directions. Both conduction regions are 1% C steel. The intervening gaps, filled with air at 1 atm, are of average thickness 100 μm.

(a) Calculate the total constriction resistance for a single direct contact of area $A_{c,i}$.

(b) Determine the resistance of the air gap associated with each direct contact area, assuming only air conduction and continuum transport.

(c) Find the total resistance, per contact area subregion, consisting of one direct contact area $A_{c,i}$ and the associated air gap area, $A_{g,i}$.

(d) Determine the contact resistance R_c, per unit area.

(e) If the heat flux across the contact region is 1 W/cm^2, determine the contact region temperature difference Δt_c.

(f) Calculate the radiant energy flux across the gaps, per unit of total contact region area, for clean surfaces.

(g) Calculate the relevant value of Kn in this configuration.

7.3.2. In the regular contact region geometry shown in Fig. 7.3.7, regions a and b are copper and aluminum. The gaps are filled with air at 1 atm and $L = 20L_c$.

(a) Determine, either by calculation or by estimation, the constriction resistance per unit area of the contact region, assuming that the direct contact areas are isothermal.

(b) For $L_c = 20$ μm, determine the heat flux across the direct contact areas, per unit of contact area, per unit of temperature difference.

(c) For $L_c = 20$ μm, approximate the air gap as triangular, with $H = 40$ μm. Calculate the gap resistance per unit area, accounting approximately for noncontinuum effects.

7.3.3. Consider the regular contact region geometry given in Prob. 7.3.1.

(a) Determine the constriction factor.

(b) Find the air gap thickness which would result in the first noncontinuum conduction effects.

(c) What thickness would result in free-molecule transport?

7.3.4. Consider a contact region between two extensive flat stainless-steel plates. The pressure applied at the interface is $p = 10$ atm. The contact microhardness of the steel is 1000 MPa. Assume Y and σ to be 20 and 5 μin., respectively.

(a) Calculate the direct contact area, per unit of overall contact area.

(b) Determine η and the number of contacts, per unit area, in terms of an estimate of the value of m as 0.3.

(c) Calculate the value of R_{dc}.

(d) Determine the approximate constriction factor.

(e) Calculate the average gap thickness and the gap conductance, for air at 1 atm.

(f) Determine the total contact resistance.

(g) Compare this result with the values in Table 2.1.1 and Fig. 2.1.6.

7.3.5. A contact region between two metal layers is characterized by $Y = 20$ and $\sigma = 5\ \mu$in., respectively. The imposed pressure is 50 atm. A comparison is to be made between the contact resistance achieved using layers, of 1000 MPa and 300 MPa contact hardnesses, respectively.

(a) For each material, calculate the resulting fraction of direct contact area.

(b) Assuming a value of $m^2 = 0.15$, determine the number of contact areas.

(c) Determine the constriction factors.

(d) Calculate each constriction resistance and the direct contact resistance.

7.3.6. A contact region between an aluminum region and a hardened-steel region is characterized by an average spacing and roughness of 65 and 10 μin., respectively. Take the relevant contact microhardness to be 300 MPa. Determine the direct contact resistance, per unit of total contact area, over an estimated reasonable range of the mean absolute asperity slope, m.

7.3.7. A contact region between two regions of SS 304 has a roughness standard deviation of 300 μm. Consider a range of applied contact pressure p and take an absolute mean asperity slope of 0.10.

(a) Calculate the extent of this range, in MPa, in terms of the limit on Eq. (7.3.26).

(b) Plot the value of H over this range of pressure.

(c) Also plot the direct contact area and the direct contact resistance per unit area.

7.4.1. A solid metallic material consists of two distinct crystalline phases, phase 1 having three times the volume of phase 2. The two thermal conductivities are 0.6 and 0.3 W/m°C. Estimate the effective thermal conductivity of the solid, taking the crystals as approximately spherical.

7.4.2. A plastic material k_a, contains parallel cylindrical rods, k_b, of diameter D and length L. The conduction direction is parallel to the rods. The rod midlength points are on a cubical array, spaced $2L$ by $2L$, in all three directions.

(a) Sketch a typical subregion of this arrangement.

(b) Calculate the volume fractions for $L = 10D$.

(c) Sketch a minimum typical subregion which would be chosen to calculate the effective conductivity of this region, using the parallel–series method.

(d) Draw the equivalent series–parallel circuit and evaluate each resistance.

7.4.3. If the rods in Prob. 7.4.2 are spheres, instead, each having the same volume and center spacing:

(a) What is the volume fraction?

(b) Calculate the effective conductivity from the simplest relationship.

7.4.4. A plastic matrix, $k_1 = 0.4$ Btu/hr ft °F, has spheres of 0.1-in. diameter embedded in it, with a cubical center spacing s.

(a) For copper spheres, $k = 224$ Btu/hr ft °F and $s = 0.4$ in., determine the conduction regime and the effective conductivity.

(b) Repeat part (a) for Pyrex spheres, $k = 0.63$ Btu/hr ft °F and $s = 0.2$ in.

(c) Compare the results in part (b) with the "exact" result of Maxwell and with the EMT theory for two packed phases.

7.5.1. A radiative insulation barrier, like the Dewar flask in Fig. 7.5.1.(a), is to be improved by subdividing the thickness L into n layers, by partitions. The radiative behavior is to be assumed gray with diffuse reflection, neglecting conductive resistance across the partitions.

(a) Calculate the radiant transport as a ratio to that without subdivision, if the layer is totally evacuated.

(b) Plot the radiant flux as a function of the surface emissivity, for values of ϵ across the range 0 to 1.0, for the boundary surfaces maintained at $T_H = 100°C$ and $T_C = 0°C$.

7.5.2. For the configuration and kinds of surface conditions in Prob. 7.5.1:

(a) Calculate the heat flux, with no subdivision, if the space is filled with air at 1 atm, neglecting any convection, for $\epsilon = 0.1$.

(b) Repeat this calculation if the space is divided into two layers.

7.5.3. Consider the circumstance in Prob. 7.5.2, for the space divided into many layers n, for a very small difference $T_H - T_C$, and linearizing $\Delta(T^4)$.

(a) Develop the formulation for the heat flux across the barrier in general form, under evacuated conditions.

(b) Repeat part (a), including the effect of atmospheric air, making any appropriate simplifications in the model.

7.5.4. Consider the evacuated layer in Fig. 7.5.1(a), with a thickness of 1 cm. Use the radiative behavior given in Prob. 7.5.1 with air, at pressure p and $T_H = 20°C$ and $T_C = 0°C$. Neglect convection.

(a) Calculate the heat flux in the absence of air, for $\epsilon = 0.05$.

(b) Calculate the heat flux $q''(p)$ for selected values of air pressure, down to the lower limit of free-molecule transport, Kn $= 10$. Express the results as a plot of q''_{FM}/q'' (1 atm) vs. p (in atm).

7.5.5. Estimate the trend of the flux ratio q''_{FM}/q'' (1 atm), at yet higher vacuum levels, for the conditions in Prob. 7.5.4.

7.5.6. An evacuated insulation layer of thickness L, with bounding surfaces at T_H and T_C, is filled with an optically thin material, $\tau = \beta L$ less than 1.0. Neglecting any conduction, calculate and plot the ratio of the radiant flux to that across the same layer, without any filler, for $\tau = 0$ to 1. Assume $\epsilon_1 = \epsilon_2 = 0.04$ and $n = 1.6$.

7.5.7. Consider an evacuated multilayer insulation with radiation shields spaced at 50 μm. The absorption coefficient of the filler is 0.2 mm^{-1}. Neglect any conduction effects. Take $\bar{\epsilon}_1 = \bar{\epsilon}_2 = 0.02$ and $n = 1.6$.

(a) Calculate βL in the layers.

(b) For the appropriate regime of radiation, calculate the radiation attenuation across a layer, from a surface at 300 K.

(c) Calculate the net radiant transport across the layer, for a temperature difference of 1°C.

7.5.8. Consider a 1-cm-thick layer like that in Fig. 7.5.1(a), evacuated of air to a pressure level p (Torr). The bounding surface temperatures are $T_H = 20°C$ and $T_C = 0°C$ and the internal surface values are $\epsilon_H = \epsilon_C = 0.1$, $\alpha_H = \alpha_C = 0.7$. Neglect gravity.

(a) Calculate the pressure ranges, in Torr, for which the continuum and the free-molecule regime would govern air conduction flux, q''_g.

(b) Determine the mean free path for each condition.

(c) Calculate the conductive heat flux for each bounding condition.

(d) Determine how the convective flux varies in the range Kn > 10.

(e) For the Kn limit of free-molecule transport, calculate the ratio of air conduction flux to q_r''.

7.5.9. For the layer of air and surface conditions in Prob. 7.5.8:

(a) Determine the pressure range, in Torr, which would encompass both the free molecule to slip and the slip to continuum transport regimes.

(b) Plot the gas conduction flux at conditions of Kn = 0.01, 0.1, 1, and 10, as q_g'' vs. Kn.

(c) Also plot the radiation flux q_r''.

(d) Calculate the effective conductivity of this layer, over this range of conditions.

7.5.10. Consider evacuated multilayer insulation consisting of aluminized Mylar layers. Each pair of layers is separated by n layers of cubically stacked glass spheres, of diameter D_s. The imposed packing pressure on the insulating barrier assembly is p. $E = 10.4 \times 10^6$ psi, $D_s = 40$ μm, $\mu = 0.16$.

(a) Determine the number and mass of spheres, per cm^3.

(b) For $p = 0.1$ and 1 atm, calculate the radius and size of the contact areas and the constriction resistances which arise.

(c) Calculate the effective conduction conductivities at each pressure level.

(d) Compare the preceding results with the conduction regimes and values in Fig. 7.5.2.

7.5.11. Consider a thick multilayer insulation barrier, evacuated to 10^{-7} Torr. It consists of very thin aluminized films, $\bar{\epsilon} = 0.02$ and $n = 1.5$, separated by spacer layers of $h = 200$ μm thickness, having very small optical thickness. The barrier consists of 400 layers. The bounding surface temperatures are $T_H = 300$ K and $T_C = 77$ K. The value of Chp^d/N_c may be taken as 50×10^{-6} W/m K.

(a) Determine the linearized formulation which applies for calculating the barrier heat flux.

(b) Assuming a linear variation of T_i across the insulation, calculate the flux q''.

(c) Is the result in part (b) an over- or understatement of q''?

REFERENCES

Antonetti, V. W., and M. M. Yovanovich (1985) *J. Heat Transfer*, **107**, 513.

Batchelor, G. K. (1974) *Ann. Rev. Fluid Mech.*, **4**, 227.

Beck, J. V. (1988) *J. Heat Transfer*, **110**, 1046.

Cetinkale, T. N., and M. Fishenden (1951) *Int. Conf. Heat Transfer*, The Institution of Mechanical Engineers, London, p. 271.

Chan, C. K., and C. L. Tien (1973) *Trans. ASME, J. Heat Transfer*, **95**, 302.

Chiew, Y. C., and E. D. Glandt (1982) *J. Coll. Int. Sci.*, **94**, 90.

Chu, H. S., A. J. Stretton, and C. L. Tien (1988) *Int. J. Heat Mass Transfer* **31**, 1627.

Clausing, A. M. (1965) NASA Report ME-TN-242-2, Univ. of Illinois.

Cooper, M. G., B. B. Mikic, and M. M. Yovanovich (1969) *Int. J. Heat and Mass Transfer*, **12**, 279.

Crank, J. (1975) *The Mathematics of Diffusion*, Clarendon Press, Oxford.

Domoto, G. A., and C. L. Tien (1970) *ASME J. Heat Transfer*, **92**, 299.

Duncan, A. B., S. P. Peterson, and L. S. Fletcher (1989) *ASME J. Heat Transfer*, **111**, 830.

Durand, P. P., and L. H. Ungar (1988) *Int. J. Numerical Methods in Engineering*, **26**, 2487.

Fenech, H., and W. M. Rohsenow (1963) *J. Heat Transfer*, **85**, 15.

Fletcher, L. S. (1988) *J. Heat Transfer*, **110**, 1059.

Fricke, H. (1924) *Phys. Rev.*, **24**, 575.

Fried, E., and M. J. Kelley (1965) "Thermal Conductance of Metallic Contacts in a Vacuum," AIAA-65-661, AIAA Thermophysics Specialists Conference, Monterey, CA.

Fry, E. M. (1965) "Measurements of Contact Coefficients of Thermal Conductance," AIAA 65-662, AIAA Thermophysics Specialists Conference, Monterey, CA.

Garland, J. C., and D. B. Tanner (1978) "Electrical Transport and Optical Properties of Inhomogeneous Media," Amer. Inst. Phys. AIP Conf. Proc. No. 40.

Glaser, P. E., I. A. Black, R. S. Lindstrom, F. E. Ruccia, and A. E. Wechsler (1967) "Thermal Insulation Systems," NASA SP-5027.

Greenwood, J. A., and J. B. P. Williamson (1966) *Proc. Roy. Soc. London*, **A295**, 300.

Hashin, Z. (1983) *J. Appl. Mech.*, **105**, 481.

Hashin, Z., and S. Shtrikman (1962) *J. Appl. Phys.* **33**, 3125.

Jeffrey, D. J. (1973) *Proc. Roy. Soc. London*, **A335**, 355.

Kang, T. K., G. P. Peterson, and L. S. Fletcher (1990) *ASME J. Heat Transfer*, **112**, 894.

Keller, J. (1963) *J. Appl. Phys.*, **34**, 991.

Kitscha, W. W., and M. M. Yovanovich (1975) *Prog. Astro. Aero.*, **39**, 93.

Landauer, R. (1978) *Electrical Conductivity in Inhomogeneous Media*, J. C. Garland and D. B. Tanner, Eds., *Amer. Inst. Phys. Proc. No. 40*.

Lewis, D. M., and H. J. Sauer, Jr. (1965) *J. Heat Transfer*, **87**, 310.

Madhusudana, C. V., and L. S. Fletcher (1986) *AIAA J.*, **24**, 510.

Maxwell, J. C. (1873) *Electricity and Magnetism*, 1st ed., Clarendon Press, New York.

McKenzie, D. R., R. C. McPhedran, and G. H. Derrick (1978) *Proc. Roy. Soc. London*, **A362**, 211.

Mikic, B. B. (1971) *ASME J. Lubrication Technol.*, **93**, 451.

Mikic, B. B. (1974) *Int. J. Heat Mass Transfer*, **17**, 204.

Mikic, B. B., and W. M. Rohsenow (1966) "Thermal Contact Resistance," MIT Report 4542-41.

Mølgaard, J., and W. W. Smeltzer (1970) *Int. J. Heat Mass Transfer*, **13**, 1153.

Negus, K. J., M. M. Yovanovich, and J. V. Beck (1989) *ASME J. Heat Transfer*, **111**, 804.

O'Callaghan, P., W. B. Snaith, S. D. Propert, and F. R. Al-Astrabadi (1983) *AIAA J.*, **21**, 1325.

Ogniewicz, Y., and M. M. Yovanovich (1978) *Heat Transfer and Thermal Control Systems*, L. S. Fletcher, Ed., *Prog. Astro. Aero.*, **60**, 209.

Patterson, G. N. (1956) *Molecular Flow of Gases*, John Wiley & Sons, New York.

Perrins, W. T., D. R. McKenzie, and R. C. McPhedran (1979) *Proc. Roy. Soc. London*, **A369**, 207.

Peterson, G. P., and L. S. Fletcher (1988) *ASME J. Heat Transfer*, **110**, 996.

Peterson, G. P., and L. S. Fletcher (1990) *ASME J. Heat Transfer*, **112**, 579.

Rayleigh, R. S. (1892) *Philos. Mag.*, **34**, 481.

Rohsenow, W. M., J. P. Hartnett, and E. Ganić, Eds. (1985) *Handbook of Heat Transfer Fundamentals*, McGraw-Hill, New York.

Runge, I. (1925) *Z. Tech. Phys.*, **6**, 61.

Sangani, A. S., and A. Acrivos (1983) *Proc. Roy. Soc. London*, **A386**, 263.

Seely, F. B., and J. O. Smith (1967) *Advanced Mechanics of Materials*, 2nd ed., John Wiley and Sons, New York.

Snaith, B., S. D. Probert, and P. W. O'Callaghan (1986) *Appl. Energy*, **22**, 31.

Song, S., and M. M. Yovanovich (1988) *J. Thermophys.*, **2**, 40.

Sparrow, E. M., and R. D. Cess (1978) *Radiation Heat Transfer*, Hemisphere, New York.

Springer, G. S. (1971) *Adv. in Heat Transfer*, **7**, 163.

Strieder, W., and R. Aris (1973) *Variational Methods Applied to Problems of Diffusion and Reaction*, Springer-Verlag, New York.

Tien, C. L., and G. R. Cunnington (1973) *Adv. in Heat Transfer*, **9**, 350.

Tien, C. L., P. S. Jagannathan, and B. F. Armaly (1969) *AIAA J.*, **7**, 1806.

Tien, C. L., and K. Vafai (1989) *Adv. in Appl. Mech.*, **27**, 260.

Tien, C. L., and K. Y. Wang (1988) Thermal Insulation Heat Transfer, U.S.–China Heat Transfer Workshop, *J. Eng. Thermophys.*, 1–11.

Torquato, S. (1985) *J. Appl. Phys.*, **58**, 3790.

Torquato, S. (1991) *ASME Appl. Mech. Rev.*, **44**.

Verschoor, J. D., and P. Greebler (1952) *Trans. ASME*, **74**, 961.

Veziroglu, T. N., and S. Chandra (1969) *Prog. Astro. Aero.*, **21**, 591.

Viskanta, R. J. (1965) *Trans. ASME, J. Heat Transfer*, **87C**, 143.

Wang, L. S., and C. L. Tien (1966) *Proc. Int. Heat Transfer Conf.*, **5**, 190.

Wang, L. S., and C. L. Tien (1967) *Int. J. Heat Mass Transfer*, **10**, 1327.

Wesley, D. A., and M. M. Yovanovich (1986) *Nucl. Tech.*, **72**, 70.

Wolf, J. R., J. W. C. Tseng, and W. Strieder (1990) *Int. J. Heat Mass Transfer*, **33**, 725.

Yang, Y. S., J. R. Howell, and D. E. Klein (1983) *Trans. ASME, J. Heat Transfer*, **105**, 325.

Yarbrough, D. W., T. W. Wong, and D. L. McElroy (1985) *High Temp. Sci.*, **19**, 213.

Yovanovich, M. M. (1982) *Prog. Astro. Aero.*, T. E. Horton, Ed., AIAA, New York.

Yovanovich, M. M., and W. W. Kitscha (1974) *Prog. Astro. Aero.*, **35**, 293.

Yovanovich, M. M., G. E. Schneider, and V. S. Cecco (1980) AIAA 18th Aerospace Sciences Meeting, January, Pasadena, CA.

CHAPTER
8

OTHER
APPLICATIONS

8.1 INTRODUCTION

Chapters 1 and 2 concern the mechanisms of heat conduction and mass diffusion transport. The fundamental differential equations are developed and some useful transformations are given. Then general kinds of boundary conditions are formulated.

Chapter 2 analyzes idealized one-dimensional heat and mass transfer, including variable properties, composite barriers, and internal energy generation. Chapter 3 concerns multidimensional steady-state conduction in plane, cylindrical, and spherical geometries, including the effects of distributed sources and variable and directional conductivity. Conduction shape factors and contact resistance are also included. Numerical modeling is discussed, in its simplest forms.

Chapter 4 treats a large number of different kinds of one-dimensional transient processes, including some effects in each plane, cylindrical, and spherical geometry. Several periodic processes are also analyzed. Transients in interacting conduction regions are analyzed, including moving fronts of phase change, as in freezing and melting. Moving sources are considered, along with the transient integral method. This is a convenient alternative to full analysis. The particular characteristics of numerical methods, stability analysis, and calculation procedures are first given, for relatively simple one-dimensional processes.

Chapter 5 concerns multidimensional transients. Solution methods are discussed and the general and valuable technique of product solutions is reviewed. Results are given. Much attention is then given to the establishment of the most direct kinds of numerical formulation for multidimensional transients, including stability considerations.

Chapter 6 concerns mass diffusion. Emphasis there concerns processes which are not simple analogs of heat conduction. Variable mass diffusivity is very commonly encountered and analyses of characteristic examples of behavior are given. Concentrated sources and moving boundaries are treated. Chemical reaction mechanisms and their consequences are considered. Several typical kinds of interacting mass diffusion and heat conduction processes are examined.

Following the earlier brief consideration of contact resistance, between conducting regions, Chap. 7 first examines the components of overall contact resistance. An analysis then quantifies the components of direct contact and gap resistance. Then mechanisms in composite conductive materials are discussed. The last matter concerns heat transfer in a very important kind of conduction–radiation barrier, superinsulation.

The preceding summary of the material in Chaps. 1 through 7 introduces the additional topics in this chapter. These are a group of important applications, related to that material. However, they are not conveniently included there. Each of these applications consist of one or more features which involve the same physical mechanisms. However, each concerns additional aspects of behavior. Some also relate to important processes which are formulated differently than in preceding chapters.

The first matter, in Sec. 8.2, concerns extended surfaces, or fins. They amount to solid conduction, with lateral surface convection and radiation. However, special assumptions are made for analysis. These make it convenient to optimize specific kinds of fin designs for different uses.

Section 8.3 concerns moving concentrated energy sources in conduction regions. The quasi-static approximation is made. Then the results for line and point sources are given. These results relate particularly to various kinds of welding processes.

Thermal stresses are very important in many designs. Section 8.4 discusses the mechanisms. Then quantitative results are given for two different kinds of processes. One is an idealized geometry in which the tendency to strain, due to local relative thermal expansion, is completely suppressed. The other examples are radial geometries in which the dimensions of the region change, due to thermal stresses.

Section 8.5 concerns the effects of random disturbances applied to a conduction region. These may displace a heating surface from one location of a conductive region, to another. A new transient conduction process then begins there. The resulting net heat transfer rate is an average over such a sequence of stochastically distributed conduction episodes. The formulation for heat transfer is developed, in terms of the length of subsequent individual episodes, τ_c. The length of these episodes are taken to be randomly distributed through time. The analysis given simulates some of the kinds of effects which arise under conditions of low and zero gravity.

Section 8.6 indicates how conduction modeling has been widely applied to model transport processes of fluid flow, and even to flows with velocity gradients. There are many important applications in which such conduction analysis yields relatively simple and very accurate results. The advantage is that these formulations are usually very direct. Such analysis has been used over a broad range of applications.

The material in this chapter supplements that of Chaps. 1 through 7, in terms of the conception of the preceding transport processes. It also indicates some of the typical additional kinds of analyses which are very effective in some applications.

8.2 FINS AND EXTENDED SURFACES

Heat dissipation from bounding solid base surfaces is very commonly enhanced by adding solid extensions, or fins. These are added on the side of the base surface on which the convective and/or radiative processes are least effective. Effective fin surfaces increase the dissipation area, even though the fin material may appear to add resistance between the base surface and the environment. Examples of extended surface geometries are shown in Fig. 8.2.1, for plane and cylindrical base surfaces. They are of both plate and pin form. Figure 8.2.2 is a pin or rod fin of length L, with a surface convective process, h, which is assumed uniform over the whole surface.

8.2.1 Fin Performance Measures

Performance is evaluated in terms of several distinct measures. Fin efficiency η expresses the conduction resistance in the fin itself. For example, consider the geometry in Fig. 8.2.2(a). If there is no contact or bond resistance at the base, where the fin is attached, the fin base will be approximately uniformly at t_1. If the conductivity of the fin material, k, were infinite, the whole fin, and its surface, would also be at t_1. This would be an efficiency of 100%. For a finite fin conductivity, compared to the convection and radiation surface effects, there is instead a temperature distribution along the fin, shown in Fig. 8.2.2(b). It is assumed to be a one-dimensional internal conduction process inside the fin, as formulated in Fig. 8.2.2(a). This fin has an efficiency of less than 100%. If the average fin surface temperature is t_f where $t_1 > t_f > t_e$, the efficiency is

$$\eta = \frac{t_f - t_e}{t_1 - t_e} \quad \text{or} \quad \eta = \frac{t_f - t_2}{t_0 - t_2} \tag{8.2.1}$$

in the plate fin array shown in Fig. 8.2.3.

The other commonly used evaluation of fin performance is called the effectiveness e. It is the ratio of the actual heat transfer from the fin surface, to the heat transfer, q_0, from a bare surface element of the same surface area as

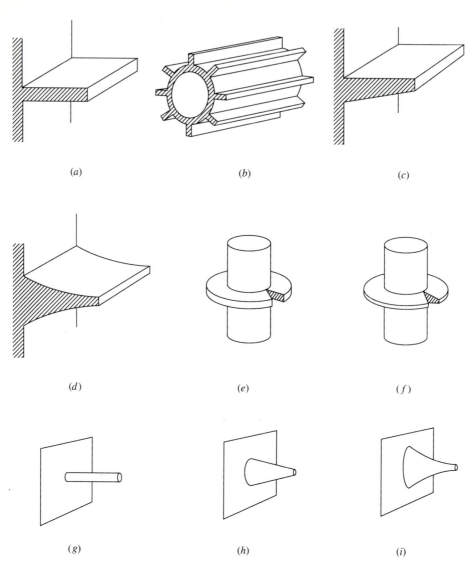

FIGURE 8.2.1
Some typical examples of extended surfaces: (a) longitudinal fin of rectangular profile; (b) cylindrical tube equipped with fins of rectangular profile; (c) longitudinal fin of trapezoidal profile; (d) longitudinal fin of parabolic profile; (e) cylindrical tube with a radial fin of rectangular profile; (f) cylindrical tube with a radial fin of truncated conical profile; (g) cylindrical spine; (h) truncated conical spine; (i) parabolic spine. [From Kern and Kraus (1972).]

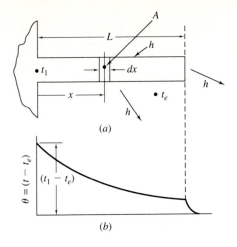

FIGURE 8.2.2
A pin fin attached to a solid base surface at t_1, with both lateral and end losses: (a) the geometry; (b) the fin temperature distribution.

that covered by the base of the fin. This area is of width δ and of unit depth in Fig. 8.2.3. This heat flow rate is evaluated, in terms of Fig. 8.2.3, as $q_0 = \delta h_0(t_0 - t_2)$. Here h_0 is taken as the coefficient which would apply for the bare surface around point 0. The fin effectiveness is then

$$e = \frac{hA(t_f - t_e)}{q_0} = \frac{h(2L + \delta)(t_f - t_e)}{h_0\delta(t_0 - t_e)} \tag{8.2.2}$$

where $(2L + \delta)$ is the exposed surface area, per unit of fin length normal to the figure.

The preceding evaluation does not account for the convective loss from the two edge faces, each of area $L\delta$, of a fin of finite length in the direction normal to the plane shown in Fig. 8.2.3.

The effectiveness must be greater than 1.0 for any benefit to arise from the fin. For an infinite fin conductivity, and $h_0 = h$, e in Eq. (8.2.1) would be $(2L + \delta)/\delta$. This is merely the area ratio. Other measures such as the weighted fin efficiency and fin resistance are also sometimes used.

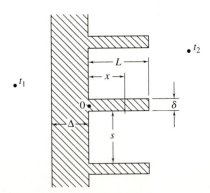

FIGURE 8.2.3
A flat plate with rectangular fins.

The remainder of this section considers several of the most important kinds of considerations which arise with extended surfaces. The first matter concerns a simple pin fin of uniform cross section, as shown in Fig. 8.2.2. The heat transfer analysis shows the interaction of convection, fin geometry, and fin material choice on fin performance. Fin effectiveness and efficiency are also calculated for plate fins.

The geometry in Fig. 8.2.3 is not the most efficient use of material, since there is decreasing heat flow at increasing $x < L$. The tapering of fins, as shown in some examples in Fig. 8.2.1, is often an economical and weight-saving alternative. Results are given for several tapered plane fin geometries, for both convective and radiative lateral loss mechanisms.

The preceding considerations all assume that the fin surface heat loss may be modeled in terms of a coefficient h times a temperature difference. However, if the external fluid is a gas and the surface temperatures are high, radiation effects may be an important part of the process. For large temperature differences, radiation transfer may not always be taken as linear in Δt. The more general formulation, including radiation, is given in Sec. 1.4. This additional effect is analyzed for a plate fin.

Fins are sometimes formed on thick walls by externally rolling or knurling the outside surface. The resulting region of area increase may then retain a good metal bond with the barrier. However, most fins in use are additional material shapes applied to the base surface. Processes such as pressed fits, soldering, and welding are common. A contact or bond resistance then arises. Also, when different materials are used, changing temperature levels may change these resistances due to differential thermal expansion. Results given in Sec. 8.2.5 are similar to the formulation in Secs. 7.2 and 7.3 for contact resistances.

8.2.2   A Fin of Uniform Cross Section

The geometry in Fig. 8.2.2 is assumed to have the same periphery at each x, of length P, independent of x. Examples are: a square rod $L \times L$, $P = 4L$; a round rod $P = 2\pi R$, or any uniform cross-sectional area A and periphery P. It is assumed here that the conductivity of the fin material is large compared to the lateral convection effects; for example, kR/h is much greater than 1, where h is the effective convection or surface coefficient. Then the radial temperature gradients are much smaller than the axial ones and the temperature may be assumed to be uniform across the rod material at each x. That is, $t(x, y) \to t(x)$. This set of assumptions concerning heat transfer was early given by Murray (1938) and Gardner (1945). For an analysis which also includes conduction across a square fin, see Eqs. (3.2.19) and (3.2.20). For pin fin analyses, see Sec. 3.3.

The differential relation between t and x may be derived by writing an energy balance for the element dx having a face area, normal to the rod axis, of A, and a lateral surface for convection, on its perimeter, of area $P\,dx$. The rate of heat conduction at x is $-k(dt/dx)A$, and the difference across the two faces of element dx is

$$-\frac{d}{dx}\left(-k\frac{dt}{dx}A\right)dx = k\frac{d^2t}{dx^2}A\,dx \tag{8.2.3a}$$

where heat flow into the element of volume $A\,dx$ is taken as positive. This energy gain is set equal to the rate of loss on the lateral surface, of extent $P\,dx$, as

$$k\frac{d^2t}{dx^2}A\,dx = hP\,dx(t-t_e) \tag{8.2.3b}$$

This reduces to the following equation when the temperature excess, $\theta = (t-t_e)$, is introduced. Note that $dt = d\theta$.

$$\frac{d^2\theta}{dx^2} - \frac{hP}{kA}\theta = 0 \tag{8.2.3c}$$

One boundary condition applies at each end:

$$\text{At } x = 0: \qquad \theta = \theta_1 = t_1 - t_e$$

At $x = L$, the exposed end face, the heat lost by convection, $kA(t-t_e) = hA\theta$, must be equal to the heat arriving by conduction from the inside of the rod, $-kA(d\theta/dx)$. Therefore,

$$\text{At } x = L: \qquad -kA\frac{d\theta}{dx} = hA\theta \quad \text{or} \quad \frac{d\theta}{dx} = -\frac{h}{k}\theta \tag{8.2.3d}$$

Equation (8.2.3), with the two boundary conditions, may be solved for an expression for temperature excess in terms of x. However, the solution may be made more useful by first changing the variables θ and x to the new ones, $\phi = \theta/\theta_1$ and $X = x/L$, where θ_1 is the temperature excess at $x = 0$, that is, $(t_1 - t_e)$. These variables vary only in the restricted range 0 to 1.0. The differential equation, written in terms of ϕ and X, becomes

$$\frac{d^2\phi}{dX^2} - \frac{hPL^2}{kA}\phi = 0 \tag{8.2.4}$$

For simplicity, the square root of the constant coefficient of ϕ in Eq. (8.2.4) will be set equal to a new constant n. That is,

$$n = \sqrt{\frac{hPL^2}{kA}} \tag{8.2.5}$$

With these changes, the differential equation and boundary conditions become

$$\frac{d^2\phi}{dX^2} - n^2\phi = 0 \tag{8.2.6}$$

$$\text{At } X = 0: \qquad \phi = 1.0$$

$$\text{At } X = 1.0: \qquad \frac{d\phi}{dX} = -\frac{hL}{k}\phi$$

Equation (8.2.6) amounts to a conduction term and an equivalent distributed sink term $n^2\phi$. Thus lateral convection losses amount to a distributed sink in this model. This kind of effect was analyzed in Sec. 2.4, with $n^2\phi$ as a constant. The general solution of Eq. (8.2.6) is

$$\phi = C_1 e^{nX} + C_2 e^{-nX} \tag{8.2.7}$$

The boundary conditions are applied to determine the constants:

At $X = 0$: $\phi = 1 = C_1 + C_2$

At $X = 1.0$: $\dfrac{d\phi}{dX} = C_1 n e^{nX} - C_2 n e^{-nX} = -\dfrac{hL}{k}\phi$

and, therefore,

$$C_1 n e^n - C_2 n e^{-n} = -\frac{hL}{k}(C_1 e^n + C_2 e^{-n})$$

The two equations for C_1 and C_2 may be combined to give

$$C_2 = \cfrac{1}{1 + \cfrac{1 - hL/nk}{1 + hL/nk}e^{-2n}} \qquad \text{and} \quad C_1 = 1 - C_2$$

For simplicity, the constant hL/nk is denoted as m:

$$\frac{hL}{nk} = \sqrt{\frac{hA}{kP}} = m \tag{8.2.8}$$

With this change, Eq. (8.2.7), with the values of the constants introduced, becomes

$$\phi = \frac{\theta}{\theta_1} = \frac{e^{-nX} + [(1-m)/(1+m)]e^{-2n}e^{nX}}{1 + [(1-m)/(1+m)]e^{-2n}} \tag{8.2.9}$$

This relation is the generalized temperature distribution, for $X = 0$ to 1.0, as a function of m and n. The definitions of m and n are given in Eqs. (8.2.8) and (8.2.5). Solutions of Eq. (8.2.9) for various values of m and n in the range of practical interest are plotted in Fig. 8.2.4.

The curves in Fig. 8.2.4 indicate that the slope of the temperature curve at the free end of the rod, that is, at $X = 1.0$, is close to zero for most values of m and n. A slope of nearly zero means that the rate of heat flow at the end is near zero and that the convection from the end face may be neglected. If this approximation is made at the outset, in writing the boundary conditions following Eq. (8.2.6), a much simpler solution is obtained. The general solution, Eq. (8.2.7), is the same as before, but the boundary conditions are changed to the following:

At $X = 0$: $\phi = 1$

At $X = 1.0$: $\dfrac{d\phi}{dX} = 0$

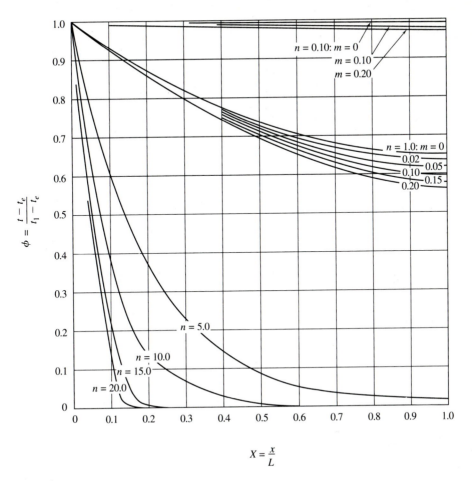

FIGURE 8.2.4
Generalized temperature distributions in uniform-area rods with lateral and end losses. (Calculated by S. A. Shull.)

These conditions give the following simpler approximate solution:

$$\phi = \frac{e^{-nX} + e^{-2n}e^{nX}}{1 + e^{-2n}} \tag{8.2.10}$$

This is, in effect, Eq. (8.2.9), with $m = 0$. Several $m = 0$ curves are shown in Fig. 8.2.4.

It is noted here that the conditions for an adiabatic end-face rod are exactly met by a rod which extends a distance $2L$ between two energy reservoirs of equal temperature. The temperature distribution in the rod is symmetric about its center plane. All the heat coming from either one of the reservoirs is lost on the surface of the rod before the center point is reached. Therefore, this problem is merely an adiabatic end-face rod of length L and is covered by Eq. (8.2.10).

The preceding analysis may be used in the consideration of errors to be expected in the reading of any rodlike thermometric device extending out into a passage through which a fluid flows. If the stem is in good contact with the passage wall and if the fluid and wall are at different temperatures, t_f and t_w, the device will indicate a temperature t_t different from t_f. The preceding analysis applies to this circumstance. In particular, if the temperature-sensitive part of the device is at the end of the rod, the difference between t_t and t_f, that is, the error θ_E, is $\phi_E\theta_1 = \phi_E(t_w - t_f)$. This quantity may be found from Eq. (8.2.9) by taking $X = 1.0$. This error, $(t_t - t_f)$, found as a ratio to $\theta_1 = (t_w - t_f)$, is

$$\frac{t_t - t_f}{t_w - t_f} = \phi_E = \frac{e^{-n} + [(1-m)/(1+m)]e^{-n}}{1 + [(1-m)/(1+m)]e^{-2n}}$$

$$= \frac{1 + [(1-m)/(1+m)]}{e^{n} + [(1-m)/(1+m)]e^{-n}} \tag{8.2.11}$$

If end convection is ignored, $m = 0$ and this equation is further simplified. This result is applied quantitatively by Gebhart (1971), Sec. 11-5, for an air flow in a passage.

EFFECTIVENESS AND EFFICIENCY. The principal approaches to determining these performance quantities are through measured performance, in more complicated geometries, or by calculations in simpler ones. The latter procedure is applied here for the plate fins in Fig. 8.2.3. Recall the assumptions that the fin conduction field is approximated as one dimensional and that h is uniform over the whole external surface.

The temperature variable is again written in terms of $X = x/L$ and $\phi = \theta(x)/(t_0 - t_2)$, where $\theta(x) = t(x) - t_2$. Neglecting contact or bond resistance, at $x = 0$, the conduction into a fin at its base, q_0 per unit of fin length, is

$$q_0 = -k\delta\left(\frac{dt}{dx}\right)_{x=0} = -\frac{k\delta}{L}\left(\frac{d\theta}{dX}\right)_{X=0} = -\frac{k\delta(t_0 - t_2)}{L}\left(\frac{d\phi}{dX}\right)_{X=0} \tag{8.2.12}$$

From Eq. (8.2.9),

$$\left(\frac{d\phi}{dX}\right)_0 = -\frac{n\{1 - [(1-m)/(1+m)]e^{-2n}\}}{1 + [(1-m)/(1+m)]e^{-2n}} \tag{8.2.13}$$

where n and m are defined in Eqs. (8.2.5) and (8.2.8). The P/A ratio for this fin is approximately $2/\delta$ and

$$q_0 = \sqrt{2kh_2\delta}\,(t_0 - t_2)\frac{1 - [(1-m)/(1+m)]e^{-2n}}{1 + [(1-m)/(1+m)]e^{-2n}} \tag{8.2.14}$$

A simple estimate of the increase in overall conductance due to the fins may be made if the plate fin is assumed to be very thin or if its thermal conductivity is very high. Then one may assume that the temperatures of the bare surface and of the base of the fin are both t_0. The overall conductance per unit area of flat surface will be found in terms of the size and spacing of the fins for this condition. The heat dissipation for a typical sector of the outside surface, consisting of one space and one fin, is written in terms of an equivalent outside conductance h_2':

$$q = h_2'(s + \delta)(t_0 - t_2) = h_2 s(t_0 - t_2) + q_0$$

$$= h_2 s(t_0 - t_2) + \sqrt{2kh_2\delta}\,(t_0 - t_2)\frac{1 - [(1 - m)/(1 + m)]e^{-2n}}{1 + [(1 - m)/(1 + m)]e^{-2n}} \quad (8.2.15)$$

Therefore,

$$h_2' = h_2\frac{s}{s + \delta} + \frac{\sqrt{2kh_2\delta}}{s + \delta}\frac{1 - [(1 - m)/(1 + m)]e^{-2n}}{1 + [(1 - m)/(1 + m)]e^{-2n}} \quad (8.2.16)$$

Or, for thin fins and end convection neglected,

$$h_2' = h_2 + \frac{\sqrt{2kh_2\delta}}{s}\frac{1 - e^{-2n}}{1 + e^{-2n}} \quad (8.2.17)$$

The overall conductance in terms of h_2' and the wall thickness Δ is

$$U_0 = \frac{1}{1/h_1 + \Delta/k + 1/h_2'} \quad (8.2.18)$$

This conductance is higher than that for a bare wall because h_2' is larger than h_2.

The effectiveness of extended surface, defined as the heat dissipation rate q_0 divided by the dissipation rate of the bare surface at t_0 covered by the fins, is, for the rectangular fin,

$$e = \frac{q_0}{q_0'} = \frac{q_0}{h_2\delta(t_0 - t_2)} = \sqrt{\frac{2k}{h_2\delta}}\frac{1 - [(1 - m)/(1 + m)]e^{-2n}}{1 + [(1 - m)/(1 + m)]e^{-2n}} \quad (8.2.19)$$

The efficiency of the fin, defined as the ratio of q_0 to the rate of heat dissipation that would result if the whole fin surface were at t_0, is

$$\eta = \frac{q_0}{h_2(2L + \delta)(t_0 - t_2)} = \frac{\sqrt{2kh_2\delta}}{h_2(2L + \delta)}\frac{1 - [(1 - m)/(1 + m)]e^{-2n}}{1 + [(1 - m)/(1 + m)]e^{-2n}}$$

$$(8.2.20)$$

If δ is small compared with L and if end convection is neglected, that is, $m = 0$, we have

$$\eta = \sqrt{\frac{\delta}{2L}} \sqrt{\frac{k}{hL}} \frac{1 - e^{-2n}}{1 + e^{-2n}} = \frac{1}{n} \frac{1 - e^{-2n}}{1 + e^{-2n}} \qquad (8.2.21)$$

The latter quantity, efficiency, is a very valuable criterion and is used to assess the relative merits of various arrangements of extended surface, for example, of the rectangular fin considered previously, compared to a fin of triangular section. Many cross-sectional shapes have been analyzed both for straight fins on flat surfaces and for circular fins on the outside of tubes. Information is available concerning optimum profiles. See, for example, Kern and Kraus (1972). Several results are given in Sec. 8.2.4.

8.2.3 Radiation and Other Effects at High Temperatures

The common use of a combined convection and radiation effect, as $h = h_c + h_r$, is often not a sufficiently accurate procedure. This arises when the temperature difference along a fin, Δt, is comparable to the absolute surface temperature level, T, of the material or of its immediate surroundings. This condition often arises with heat dissipation from a fin at high temperature levels. Then the local radiant emission component along the fin must be modeled in terms of the fourth power of the local absolute temperature level of the fin material T, as $\epsilon\sigma T^4$, where ϵ is the local surface emissivity and σ is the Stefan–Boltzmann constant.

This effect is evaluated here in the simpler circumstance in which convection is not present, as a parallel process. This is often applicable for finlike structures used in space devices to dissipate energy rejection from, for example, energy reactors or solar cell arrays.

Rectangular fins, as shown in Fig. 8.2.3, are analyzed. The changing internal heat conduction along the fin is again related to the local surface loss, as in the differential equation in $t(x)$, Eq. (8.2.3b). For a rectangular fin $A = \delta$ and $P = 2$. However, the local radiant emission from the surface by radiation, which is $2h\,dx(t - t_e)$ for convection, is now $2\epsilon\sigma T^4\,dx$, where $T = T(x)$ is the local absolute temperature of the fin material. Therefore, the second term in Eq. (8.2.3b) becomes $2\epsilon\sigma T^4$. This is again an energy sink. However, there may be a radiation incident on the surface, due to adjacent fins or to the presence of solar radiation or to other nearby radiation emitters. This input is modeled in this simple formulation as $\alpha q_s''$, where α is the surface absorptivity or absorptance and q_s'' is the surface heat flux loading. This additional term in the equation is then $+2\alpha q_s''$. Local heat generation q''' in the fin material has a similar effect and is modeled in a similar way. This additional term is $q'''\delta$. These are both local sources. The resulting equation for $T(x)$ is

$$\frac{d^2T}{dx^2} - \frac{2\epsilon\sigma T^4}{k\delta} + \frac{2\alpha q_s''}{k\delta} + \frac{q'''}{k} = 0 \qquad (8.2.22)$$

The boundary condition at $x = 0$ will be $T(0) = T_1$. Equation (8.2.22) is nonlinear in $T(x)$. For a simpler analysis, the other boundary condition will be taken as no heat loss from the end face of the fin, that is, at $x = L$. This amounted to $m = 0$ in the analysis of the generalized geometry in Fig. 8.2.2. This condition is applied here in the form of $dT/dx = 0$ at $x = L$. The resulting solutions will then be in terms of T_1 and $T_L = T(L)$.

The first result given in the following discussion is for a radiating fin with no radiation loading from any adjacent fins or from the environment. This is, $q_s'' = 0$. Also, the q''' effect will not be included.

Equation (8.2.22) is nonlinear and the formulation is transformed in terms of $p = dT/dx$ as

$$p\frac{dp}{dT} = \frac{2\sigma\epsilon T^2}{k\delta} \tag{8.2.23}$$

Using the condition $dT/dx = 0$ at $x = L$ results in

$$\frac{dT}{dx} = -2\left(\frac{\sigma\epsilon}{5k\delta}\right)^{1/2}(T^5 - T_L^5)^{1/2} \tag{8.2.24}$$

The solution, see Kern and Kraus (1972), is expressed in terms of a beta function. The heat conduction rate q_0, at $x = 0$, and the fin efficiency η are calculated as

$$q_0 = \sqrt{4\sigma\epsilon k\delta(T_1^5 - T_L^5)/5} \tag{8.2.25}$$

and the fin efficiency is

$$\eta = \frac{2\left[(T_L/T_1)^3 - (T_L/T_1)^8\right]^{1/2}}{\sqrt{20\sigma\epsilon L^2 T_L^3/k\delta}} \tag{8.2.26}$$

where q_0 is per unit length along the wall and η is defined as before for convection, immediately preceding Eq. (8.2.20).

For any specified values of $T(0) = T_1$ and L, the preceding results are in terms of the unknown temperature T_L, for ϵ, L, k, and δ given. Any particular calculation is carried out in terms of Fig. 8.2.5. The denominator in Eq. (8.2.26), called B, is the fin parameter. This is proportional to $\sqrt{T_L^3}$, which is unknown. However, a value of T_L may be guessed and B calculated. The matching value of T_1/T_L on the curve will not, in general, be the value of T_L initially assumed. Agreement is obtained through the iteration of this procedure. When T_1/T_L is known, the fin efficiency may be read from Fig. 8.2.6.

The comparable analysis, also including incident radiation from the surroundings as q_s'', includes that additional term in Eq. (8.2.22). Results are given by Kern and Kraus (1972) for $\alpha q_s''$ uniform over the whole fin surface. Again, an approximate analysis is indicated.

There are also applications in which there is internal energy generation, as $q'''\delta$, as included in the formulation in Eq. (8.2.22). For example, this may arise

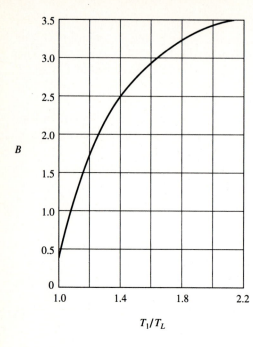

FIGURE 8.2.5

The fin parameter $B = \sqrt{20\sigma\epsilon L^2 T^3/k\delta}$ in Eq. (8.2.26).

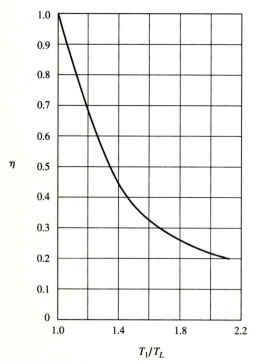

FIGURE 8.2.6

The radiation fin efficiency vs. T_1/T_L.

from electrical energy dissipation, induction heating, and fission particle irradiation. Results are given by Kern and Kraus (1972) for both rectangular and tapered fins.

8.2.4 Tapered Fins

Fins of constant cross section along their length do not yield the highest performance for a given amount of fin material. The internal heat conduction rate decreases due to the lateral losses, as does the temperature gradient. More efficient use of fin material may be achieved if the plate fin cross section $\delta(x)$ decreases with x, as sketched in Fig. 8.2.7. This generalization applies for both convection and radiation surface effects. It might be surmised that the most effective fin shape for convective heat dissipation would be the one which has a uniform temperature gradient along its length.

CONVECTIVE LOSS. Consider only convection from the fin surface. It is assumed that $d\delta/dx$ is sufficiently small so that the total surface area of the element dx thick may be taken as $2\,dx$. The local conduction rate is

$$-\frac{d}{dx}\left(-k\,\delta(x)\frac{dt}{dx}\right)dx = k\frac{d}{dx}\delta(x)\frac{dt}{dx}$$

This is equal to the local convection, $2h\,dx(t - t_e)$, as follows:

$$\frac{d}{dx}\left(\delta(x)\frac{dt}{dx}\right) = \frac{2h(t - t_e)}{k} \tag{8.2.27}$$

Note that x is here measured from the tip of the fin, for convenience in later analysis. The boundary conditions are then: at $x = L$, $t = t_1$; at $x = 0$, $k(dt/dx) = 2h\delta(0)[t(0) - t_e]$, for end convection.

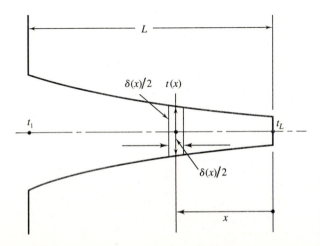

FIGURE 8.2.7
A longitudinal plate fin of tapered cross section of local area $\delta(x)$, per unit length.

A relatively simple example of a tapered fin is a triangular shape, that is, $\delta(x) = \delta_0(x/L)$, where the fin width at the base is then δ_0. The analysis, heat flow, and efficiency are given by Kern and Kraus (1972). Also given are the results for both convex parabolic shapes and for the conditions which will result in fins of the least required amount of material for a given heat dissipation rate. Schneider (1955) considers additional aspects of fin performance. See also Rohsenow et al. (1985) for performance diagrams and the efficiency of many fin shapes.

RADIATION LOSS. Fin shape optimization for heat dissipation by radiation alone was analyzed by Wilkins (1960). The general formulation is shown in Fig. 8.2.7. If environment radiation effects are not included, that is, $T_e = 0$ K, the equation is

$$\frac{d}{dx}\left(\delta(x)\frac{dt}{dx}\right) = \frac{2\epsilon\sigma T^4}{k} \tag{8.2.28}$$

The condition at $x = L$ is taken at t_1. The shape of least volume is determined; see Kern and Kraus (1972). The volume of material in the fin, per unit of lateral length, is

$$V = \int_0^L \delta(x)\, dx \tag{8.2.29}$$

The resulting fin varies in thickness $\delta(x)$ as $x^{7/2}$ as

$$\delta(x) = \frac{3q_L^2}{k\epsilon\sigma T^5}\left(\frac{x}{L}\right)^{7/2} \tag{8.2.30}$$

where q_L is the conduction rate outward from the base of the fin. A consequence of this analysis is that $T(0) = 0$.

8.2.5   Bond or Contact Resistance

This was discussed earlier in this section, as a frequent cause of decreased fin efficiency. Such a resistance, as discussed in Sec. 7.2, interposes a temperature drop Δt_c across the resistive bond, as diagrammed in Fig. 7.2.1. This effect is evaluated here in terms of the pin fin geometry in Fig. 8.2.2. The base surface material and the contacting fin material were postulated, in the foregoing analyses, to have an equal temperature level. With contact resistance R_c, the temperature levels on the two sides are different and are taken as $t_{1,L}$ and $t_{1,R}$, respectively. Therefore,

$$t_{1,L} - t_{1,R} = R_c q_0$$

where R_c is the contact resistance across area A in Fig. 8.2.2., or of area δ, in Fig. 8.2.3, per unit length. This formulation assumes that each $t_{1,L}$ and $t_{1,R}$ is uniform over the contact area. As a consequence, the resistance R_c is directly in series with the fin resistance R_f. Therefore, the total resistance, R_t, between the base material at $t_{1,L} = t_0$ and the environment at t_e is $R_t = R_c + R_f$. The resulting reduced heat transfer rate is $q_{0,c} = (t_0 - t_e)/R_t$. Therefore, the heat transfer rate ratio is

$$\frac{q_{0,c}}{q_0} = \frac{R_f}{R_c + R_f} = \frac{R_f}{R_t} \tag{8.2.31}$$

The resulting reduction in overall fin efficiency is determined from the preceding ratio of resistances. See Kern and Kraus (1972) for an analysis of bond resistance in terms of thermal stresses and expansion.

SUMMARY. The foregoing analyses of the heat transfer performance of appendages to base surfaces indicate the interaction of several design options. The major effects are fin conductivity and geometry, in terms of the heat loss mechanisms from the fins, and the heat supply process on the other side of the bounding base surface. Fin design is an optimization process, in terms of cost and performance considerations. Additional information is given in Rohsenow et al. (1985) page 4-156, Jakob (1949), Schneider (1955), Kern and Kraus (1972), and Kays and London (1964) for compact heat exchangers.

8.3 CONCENTRATED MOVING SOURCES

Metal parts are commonly joined along their edges by welding. These processes rely on intense local heating by an energy source. This source moves along the seam between the two parts, or the source is stationary and the parts move. A local region of melting arises. This region then cools, largely by conduction into the more distant material, and the parts are fused together.

One very common technique uses a tungsten electrode arc, applied to the seam region. The arc is shielded by an inert gas, such as argon, to exclude atmospheric gas components from the region of melting and fusion. Another common process uses a consumable electrode instead. The electrode material is fused and joins the weld region. Another increasingly common process uses very intense and concentrated laser beams. Power levels up to megawatts are in use.

In all of these processes, a liquid pool is formed at the seam. This solidifies as the energy source moves along the seam, as heat is conducted away into the metal parts. An idealization for analysis is a moving concentrated energy source. The space above the surface is often taken as a mirror image of the processes in the material below it. This measure assumes that the local conduction rate heat loss rate in the metal parts is much greater than the surface convective and radiative losses. The conduction field model then becomes a moving concentrated energy source in an infinite region, of twice the strength. In these welding techniques, there is a short starting process. Later, the process becomes quasi-static, if the velocity of the relative motion u of the heating effect, remains constant. Then the process appears to be steady state to an observer moving at the same velocity.

These metal joining processes are analyzed in attempts to improve quality control in the product. The intense local heating may cause very large thermal stresses in the immediate vicinity of the weld. Subsequent cooling and solidification leave other residual stresses in the material. These effects may cause both important distortions and even cracks and failure. Their severity results from a combination of welding speed, energy input level, the mass of the parts, their initial temperature, and the ensuing cooling rate. Analysis quantifies the interaction of these features and may guide the procedures followed in practice.

8.3.1 Quasi-Static Analysis

Moving energy source conduction processes are formulated in Sec. 4.7, recall Fig. 4.7.1. A solution is then given for a propagating plane energy source. The general transformation adds a convection term to the conduction equation, as seen in Eq. (4.7.3). This amounts to translating the region toward the left at a velocity of u. Then a constant-velocity energy source, moving at velocity u, will remain stationary. The formulation is thereby transformed into a coordinate system x, y, and ξ, which moves at the constant velocity u of the heating effect. The x-direction coordinate, relative to the moving source, is then ξ.

This formulation was applied by Rosenthal (1946) to plane, line, and point constant-rate heat sources, propagating at constant velocity u in infinite media. That is, the propagating quasi-static temperature fields $t(\xi)$, $t(\xi, y)$, and $t(\xi, y, z)$, in one, two, and three dimensions, are unchanging in time and are independent of any surface boundary conditions. For example, the first condi-

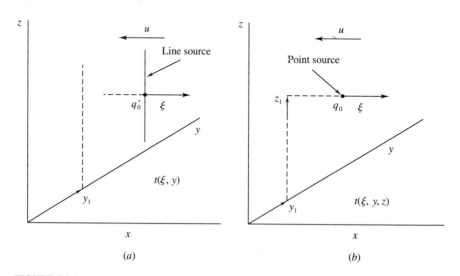

(a) (b)

FIGURE 8.3.1
A vertical line and a point heat source, q_0' and q_0, in (a) and in (b), propagating in an infinite solid at a constant velocity u. Recall Figs. 4.7.1 and 4.7.2 for the formulation in terms of ξ, for a source element dq and for a plane source of strength, q_0'', respectively.

tion, a moving plane source of strength q_0'', in Fig. 4.7.2, was analyzed in Sec. 4.7, using the conditions $t(\xi) \to t_\infty$ as $\xi \to \infty$ and $d\theta/d\xi \to 0$ as $\xi \to \pm\infty$. The solution is given in Eq. (4.7.9). The characteristic temperature difference is $\Delta t_{ch} = q_0''/\rho c u$.

The additional results given here are for the constant heating rate and constant-velocity line and point sources, of strengths q_0' and q_0. In Fig. 4.7.1 a local element dq of a general distributed energy source is seen. For a plane source, dq amounts to q_0'' at $\xi = 0$. For a concentrated line source of strength q_0', $dq = q_0'$. This is the heat dissipation rate at $\xi = 0$, per unit of z-direction source length, at a given location y_1. See Fig. 8.3.1(a). This is a model for the concentrated surface heating of a very thin plate of thickness δ or of a region of very high thermal conductivity, k, in terms of the plate thermal conductance k/δ. The heat loss from the back side is assumed negligible, compared to the input $q_0'\delta$.

For a concentrated point source, q_0 is at a given location y_1 and z_1, at $\xi = 0$. See Fig. 8.3.1(b). In the following two analyses, the two sources will be taken to move along the paths $y = 0$ and $y = z = 0$, respectively, for simplicity in the analysis.

8.3.2 A Uniform Moving Line Source

The temperature field is taken as independent of z. The relevant equation and solution, in terms of $t(\xi, y) - t_\infty = \theta(\xi, y)$, are

$$\frac{\partial^2\theta}{\partial\xi^2} + \frac{\partial^2\theta}{\partial y^2} = -\frac{u}{\alpha}\frac{\partial\theta}{\partial\xi} \tag{8.3.1}$$

$$\theta(\xi, y) = e^{-(u/2\alpha)\xi}f(\xi, y) \tag{8.3.2}$$

where t_∞ is approached at $\xi \to \pm\infty$, the exponential term is the temperature field spatial decay, and $f(\xi, y)$ is the form of the quasi-static two-dimensional temperature field which surrounds the line source. Substituting Eq. (8.3.2) into (8.3.1) results in the following relation for $f(\xi, y)$:

$$\frac{\partial^2 f}{\partial\xi^2} + \frac{\partial^2 f}{\partial y^2} - \left(\frac{u}{2\alpha}\right)^2 f = 0 \tag{8.3.3}$$

The temperature field decays away from the location $\xi = 0$ as given by the exponential term in Eq. (8.3.2). The function $f(\xi, y)$ then is symmetric around $\xi = 0$. Therefore, cylindrical coordinates are used as $r^2 = \xi^2 + y^2$ and Eq. (8.3.3) becomes

$$\frac{d^2 f}{dr^2} + \frac{1}{r}\frac{df}{dr} - \left(\frac{u}{2\alpha}\right)^2 f = 0 \tag{8.3.4}$$

The remote boundary condition requires that $\partial\theta/\partial r$ approaches zero at large ξ. Energy balance considerations require that the rate of outward conduction in the immediate vicinity of the line source is equal to q_0', per unit of source

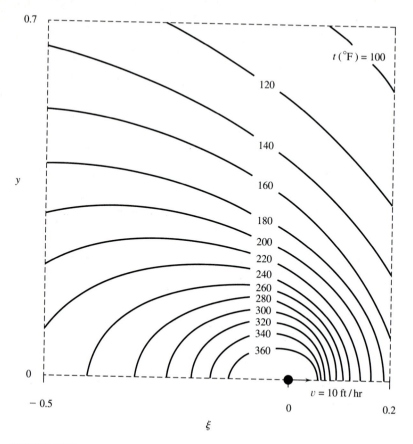

FIGURE 8.3.2

The two-dimensional temperature field around a moving line source, for the conditions $k = 1$ Btu/hr ft °F, $\alpha = 2$ ft^2/hr, $t_\infty = 80$°F, $u = 10$ ft/hr, and $q_0' = 10^3$ Btu/hr ft. The ξ range shown is $-0.5 \leq \xi \leq 0.2$. For these conditions Δt_{ch} is 159°F and $u/2\alpha = 2.5$. [From Schneider (1955).]

length. The resulting solution, see Schneider (1955) for the analysis, is

$$\theta(\xi, r) = \frac{q_0'}{2\pi k} e^{-(u/2\alpha)\xi} K_0\left(\frac{u}{2\alpha} r\right) \tag{8.3.5}$$

where K_0 is the modified Bessel function of the second kind, of zero order and Δt_{ch} is seen to be $q_0'/2\pi k$.

The preceding solution is unbounded at $r = 0$. It also is unbounded for a stationary line source, as may be inferred from Eqs. (2.1.17a) and (2.1.17b). However, isotherms may be plotted elsewhere. An example is shown in Fig. 8.3.2 for particular values of k, α, t_∞, u, and q_0'. The effect of velocity is that the field is not symmetric about $\xi = 0$. A higher velocity would further thin the precursor effect, for $\xi > 0$, and also thicken the trailing region conduction field in the y direction. The y-direction conduction also spreads the temperature field. Recall that the solution for the plane source, q_0'', amounted only to a conduction–precursor region, see Eq. (4.7.9).

For a thick plate, in terms of k/δ small, or for small plate conductance, there would be appreciable conductive resistance in the direction of the line source, that is, in the z direction. The following analysis considers this effect, as for a very thick region, with k/δ small, as a point source.

8.3.3   A Moving Point Source

This is shown in Fig. 8.3.1(b), as a source of strength of say $2q_0$ in an infinite region. This amounts to a source of q_0 in the half-infinite region below or above z_1, if the whole region is initially at t_∞. Then $\theta(\xi, y, z)$ is symmetric about z_1 and $\partial\theta/\partial z = 0$ there. This is a model for an insulated surface subject to a point source moving along the surface, with conduction into the region below z_1. This is formulated as $z = 0$ at z_1 and z is taken positive downward.

The differential equation is Eq. (8.3.1), with the additional term $\partial^2\theta/\partial z^2$. This is as given in Eq. (4.7.3). The transformation is now in terms of $r = \sqrt{\xi^2 + y^2 + z^2}$. The solution, from that of Rosenthal (1946), is

$$\theta(\xi, r) = \frac{q_0}{2\pi kr} e^{-u(\xi+r)/2\alpha} \tag{8.3.6}$$

The temperature at the source, at $r = 0$, is again infinite. This result is the same as that for an embedded sphere at $t_0 \neq t_\infty$, when $u = 0$, as given by Eq. (2.1.39a). It is also the constriction contact resistance result in Sec. 7.3.3. However, a simple characteristic temperature difference does not arise for this moving source, since the source does not have a characteristic dimension.

SUMMARY. The preceding analyses indicate the penetration of idealized concentrated heating effects for several geometries. The conditions are idealized in terms of uniform conductivity and negligible latent heat of the material, in the event that melting occurs. They also neglect fluid circulation in any liquid pool, due to electrical, thermal instability, and surface tension effects. Also, heat loss at the exposed surface by radiative and convective effects is ignored.

These effects and other aspects are receiving increasing attention. Carslaw and Jaeger (1959) formulate the preceding processes, including the effects of radiant loss. Several readily available references are given here. Jhaveri et al. (1962) considered the effects of finite thickness and of surface radiation loss. Atthey (1980), Oreper and Szekely (1984), Davis (1987), and Kou and Wang (1986) analyzed the effects of weld pool fluid motion, whereas Wei and Giedt (1985) considered thermo-capillary flow effects. A symmetric Gaussian energy source was formulated by Nied (1986) for numerical calculations of weld pool geometry. These were in good agreement with video observations. See also Abakians and Modest (1988) for an analysis of evaporative cutting, using a Gaussian-distributed laser beam. Summaries by Lancaster (1980) and Masubuchi (1980) consider the metallurgy and analysis of welding.

Ule et al. (1990) used a semidiscrete numerical technique to model tungsten arc welding. Unequally spaced grids are concentrated near the moving heat source. Property temperature dependence is included. The influence of defects is compared with the idealized results. References are given for other related effects.

Kanouff and Greif (1992) included the Marangoni, Lorentz, and buoyancy effects in numerically modelling a tungsten arc weld pool. Convection decreased the energy losses, increasing the pool size. The Lorentz force increased pool depth, and the Marangoni effect depended on the sign of the surface tension coefficient. These components moderate the efficiency of the process.

8.4 THERMAL STRESSES

The thermal conduction fields, and even those of mass diffusion, often lead to internal stress fields in a region. These result from internal tendencies to expand or contract. Considering thermal conduction, changes in the internal temperature field often result in changes in the internal stress field. This arises from the thermal expansion property of region materials. The inherent thermal expansion of a material when stress free is formulated in terms of a volumetric expansion coefficient β as

$$\rho(t) = \rho(t_r)\left[1 + \beta(t - t_r)\right] \tag{8.4.1a}$$

$$\beta = -\frac{1}{\rho}\left(\frac{\partial \rho}{\partial t}\right)_p \tag{8.4.1b}$$

where β is commonly taken as uniform over any particular region of solid. The comparable formulation for the fractional free linear thermal expansion, or unit elongation ϵ, is

$$\frac{\Delta L}{L} = \epsilon = \alpha\,\Delta t \tag{8.4.2}$$

where α is the thermal linear expansion coefficient.

Recall that the conduction equations developed in Chap. 2 and used throughout this book are based on the condition of zero change of the local material level of strain, during the analyzed processes. However, changes in local strain are usually implicit in transient thermal conduction processes. Such geometric distortion is considered in Sec. 8.4.2 here. Nevertheless, the standard conduction equations are used to determine the temperature field in almost all analyses, since the convective effect of changing strain is commonly very small.

Consider a region of an initially stress-free isotropic and homogeneous solid material, uniformly at t_i. If it is very slowly heated, without imposed surface stresses, it may simply expand uniformly in all directions. No internal stresses will arise. However, if the heating results in internal temperature gradients, internal stresses will arise, due to a tendency to internal differential expansion, according to Eq. (8.4.1). For example, rapid heating of the outside

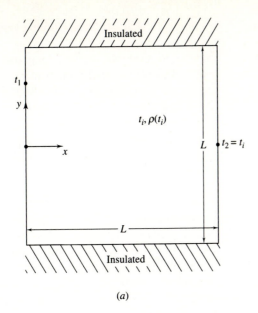

(a)

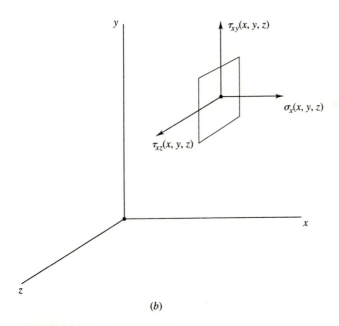

(b)

FIGURE 8.4.1
A long square rod: (a) initially uniformly at t_i with the surfaces at $y = 0$ and L insulated; (b) the general formulation of the internal normal, σ_x, and tangential, τ_{xy} and τ_{xz}, stresses, on a plane normal to the x direction.

surface of a solid sphere would result in internal tensile stresses. Cooling would have the opposite effect.

Figure 8.4.1(a) shows a more complicated example. It is a long rod of material, of dimensions L by L, initially at a uniform internal temperature t_i. It is idealized as initially at zero stress throughout. The surfaces at $y = 0$ and L are insulated. The initial material density is uniformly $\rho(t_i) = \rho_i$. A conduction transient is considered, as an example of thermally caused stresses.

The temperature of the face at $x = 0$ is suddenly changed to $t_1 > t_i$, while that at the right remains at $t_i = t_2$. The resulting transient temperature distribution is written as $\phi(x, y, \tau) = [t(x, y, \tau) - t_2]/(t_1 - t_2)$. In the eventual steady state, the conduction field is one dimensional, as $\phi = \phi(x/L) = 1 - (x/L)$, since the surfaces at $y = 0$ and L are insulated.

Consider the eventual steady-state temperature field. The left part of the region, for $t_1 > t_2$, would have expanded more than the right part, from the initial zero internal stress, or stress-free, condition. Clearly, the resulting normal stress, in the y direction, as $\sigma_y(x, y)$, would be compressive on the left, balanced by an expansive stress, $\sigma_y(x, y)$, on the right. That is, these regions would be in compression and in expansion, respectively. Nevertheless, the left face, at $x = 0$, would have increased in length, with a local strain $\epsilon(0, y) \neq 0$. The surface normal and shear stresses remain zero, in the absence of any external restraints. However, for example, the shear stress $\tau_{xy}(x, y)$ parallel to the surface, and inside $x = 0$ is zero at $y = 0$ and positive and negative for $y > 0$ and $y < 0$, respectively, because of symmetry around $y = 0$. The boundaries at $y = L/2$ and $-L/2$ have also deformed, due to the internal stress fields, $\sigma_x(x, y)$, $\sigma_y(x, y)$, and $\tau_{xy}(x, y) = \tau_{yx}(x, y)$. These effects arise due to the general and spatially dependent thermal expansion tendencies of the region, as $\Delta\rho = \rho\beta(t - t_2)$.

This example relates to an initially stress-free region, with the eventual one-dimensional steady-state internal conduction temperature condition. If the initial transient $t(x, y, \tau)$ is considered instead, the resulting internal thermal expansion tendency varies at each location and with time as $\Delta\rho = \rho\beta[t(x, y, \tau) - t_2]$.

The following contents of this section consider several simple aspects of thermoelastic effects. The first circumstance is a plate, having an imposed temperature distribution internally, as from a distributed energy source, as q''' uniform. In Sec. 8.4.2 cylindrical and spherical geometries are considered. These involve an additional effect. This arises as an elastic expansion of the whole conduction region, due to thermal expansion. That is, local displacement, as $u(r)$, arises throughout the region, including at the boundaries.

8.4.1   Thermal Stress in a Plane Region

Consider the solid layer in Fig. 8.4.2(a), subject to a distributed energy source, q''', caused, for example, by the internal dissipation of electrical energy. For q''' uniform across the region, the temperature distribution is parabolic and sym-

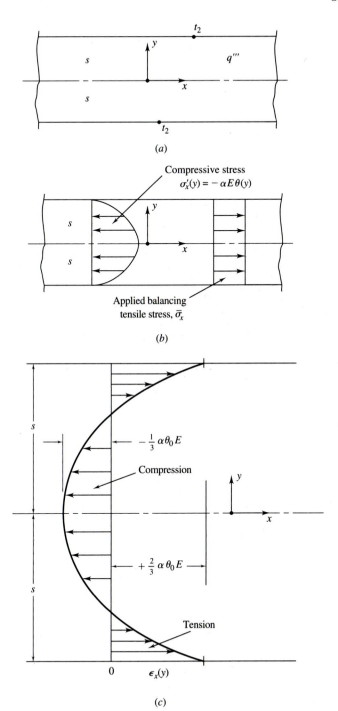

FIGURE 8.4.2
A plate of thickness $2s$: (a) the coordinates; (b) the compressive and balancing stresses $\sigma_x'(y)$ and $\bar{\sigma}_x$; (c) the resulting thermal stress over the region.

metric around the midplane. It is given by Eq. (2.4.8a) and in terms of the characteristic temperature difference, $\Delta t_{ch} = q'''s^2/2k$, is

$$\phi\left(\frac{y}{s}\right) = \frac{t(y) - t_2}{\Delta t_{ch}} = \frac{\theta(y)}{\Delta t_{ch}} = 1 - \left(\frac{y}{s}\right)^2 \qquad (8.4.3)$$

where Δt_{ch} here is the temperature at the midplane, denoted by θ_0, and $\theta(y) = t(y) - t_2$.

If the distribution of $\phi(y)$ were not symmetric, the stresses arising from the internal tendency of thermal expansion would cause moments and the plate would bend. However, for the distribution in Eq. (8.4.3), this would not occur. Instead, the greater thermal expansion tendency around $y = 0$ causes horizontal compressive stresses there as $\sigma_x(y) < 0$, and expansive stresses, $\sigma_x(y) > 0$, at larger $|y| \leq s$. In equilibrium, the sum of these two integrated forces is equal.

No x-direction deformation would occur in a plate of unbounded extent. Therefore, this geometry is constrained. However, the disk and sphere examples following are of finite extent. Then the internal stress fields cause radial changes in the sizes of the material regions. For the infinite plane region the expansion tendency causes compression at small $|y|/s$. That is, $\sigma_x(y) < 0$. At larger y, on each side, a balancing tensile force arises, in terms of $\sigma_x(y) > 0$.

The local longitudinal tendency of expansion is expressed, as in Eq. (8.4.2), by $\alpha\theta(y)$. This tendency is positive over the whole region, $y = -s$ to s, for a positive energy source q''' in Eq. (8.4.3). Then $\theta(y) \geq 0$.

The effects of the preceding tendency of thermal expansion of the region of material would be suppressed by supplying a normal compressive x-direction stress across each x location. The unrestrained linear thermal expansion across the region would have been $\alpha\theta(y)$, where $\theta(y)$ is parabolic as given in Eq. (8.4.3). The postulated balancing local stress $\sigma_x'(y)$ is calculated in terms of Hooke's law, the modulus of elasticity E, and the temperature distribution as

or
$$\epsilon_T(y) = \alpha\theta(y) = -\frac{\sigma_x'(y)}{E} \qquad (8.4.4)$$

$$\sigma_x'(y) = -\alpha E\theta(y)$$

where $\sigma_x'(y)$ is compressive, for $\theta(0) > 0$.

This compressive stress equivalent applies at all x locations in the plate. Recall that the internal temperature has been increased above the stress-free condition, which applied when the material was uniformly at t_2 throughout. The resulting compressive force across the plate, from Eq. (8.4.4), per unit width, is

$$P = \int_{-s}^{+s} \alpha E\theta(y)\, dy \qquad (8.4.5)$$

Since any section of the plate is free from any external or surface forces, an equal total tensile stress is superimposed at each location, as a uniform stress, at an average value $\bar{\sigma}_x$, to balance the preceding force.

$$\bar{\sigma}_x = \frac{1}{2s} \int_{-s}^{s} \alpha E \theta(x) \, dy \tag{8.4.6}$$

Therefore, the thermal stress will be the sum of these two effects, as

$$\sigma_x(y) = \bar{\sigma}_x - \alpha E \theta(x) = \frac{1}{2s} \int_{-s}^{s} \alpha E \theta(x) \, dy - \alpha E \theta(x) \tag{8.4.7}$$

The resulting function $\sigma_x(y)$ may be determined for any given even function of local internal temperature, as $\theta(y)$. For the parabolic form in Eq. (8.4.3), the result from Eq. (8.4.7) is

$$\sigma_x(y) = \frac{2}{3} \alpha \theta_0 E - \alpha \theta_0 E \left(1 - \frac{y^2}{s^2} \right) \tag{8.4.8}$$

This parabolic thermal stress distribution is shown in Fig. 8.4.2(c) (see page 481). The internal higher temperature leads to compressive stresses there. They are opposed, with equal total force, by the layers of tension summed over the two surface layers.

SUMMARY. The preceding method of analysis amounts to internally resisting the tendency to thermal expansion by applying a compressive stress, as in Eq. (8.4.4). This compressive stress field is then balanced, to give zero net force across the region, by superimposing an equal total tensile force. The resulting net internal stress field is then the thermal stress.

The example also shows how thermal cycling may eventually cause surface fatigue failures. Another aspect is that the previous thermal stress result also applies to a long rod of limited length $2W$ in the x direction, but still infinite in the z direction, normal to the figure. For W considerably greater than s, the stress field in the central portion, around $x = 0$, would still be as given in Eq. (8.4.8). This question concerns the effect of the nominally unreasonable approximation of the variable stresses in Eq. (8.4.5), as an average stress $\bar{\sigma}_x$, from Eq. (8.4.6). However, the error of this measure of approximation decreases rapidly away from the edges of such a rod, at $x = \pm W$. Then Eq. (8.4.8) applies accurately to the central region. This kind of approximation, widely used in stress analysis, is called Saint-Venant's principle. It expresses the tendency of local regions of large stress variations to be limited to the material near the boundaries of the region.

8.4.2 Stresses in Radial Geometries

The foregoing section considered a plate, initially stress free, with a uniform distributed energy source q''' in the eventual steady state. The resulting internal stress effect is entirely due to thermal expansion. A simple thermal stress field is determined. In this section, two examples are considered in which additional

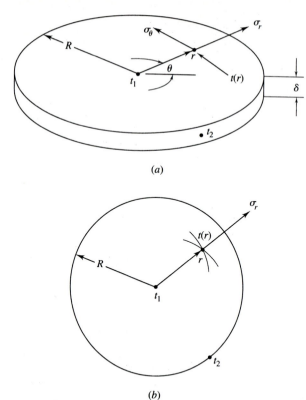

σ_θ

(a)

FIGURE 8.4.3
Radial geometries: (a) a thin
disk of radius R; (b) a solid
sphere of radius R.

(b)

aspects of deformation arise. These are a thin disk and a solid sphere, as shown in Fig. 8.4.3.

A THIN CIRCULAR DISK. See Fig. 8.4.3(a). The steady-state temperature distribution in the material is assumed to depend only on r, as for $\delta \ll R$, for a material of high conductivity k. This analysis also applies for a long cylinder. The temperature distribution in the material is taken as $t(r)$, symmetric about $r = 0$ with central and edge values of $t(0) = t_1$ and $t(R) = t_2$, where $\theta(r) = t(r) - t_2$.

The material is assumed to have been stress free when uniformly at t_2. Stress equilibrium in the region is expressed in terms of the orthogonal normal stresses as follows. Given the symmetry of $\theta(r)$, $\tau_{r\theta} = 0$ and

$$\frac{d\sigma_r}{dr} + \frac{\sigma_r - \sigma_\theta}{r} = 0 \qquad (8.4.9)$$

In the example in Sec. 8.4.1, the strain, and the relative displacement, arose only due to the internal temperature field $\theta(r)$. However, in this geometry the strain arises from a combination of the tendency toward thermal expansion and the local stress arising in the cylindrical geometry. That is, any radial expansion of the material produces circumferential expansion and an additional negative contribution to σ_θ. The result will be an expansion of the volume of the region. The radius R increases. There will be relative radial outward displacement, $u(r)$, with respect to the axis, for $\theta(r) > 0$. In terms of the resulting strains, ϵ_r and ϵ_θ, the formation is

$$\epsilon_r - \alpha\theta(x) = \frac{1}{E}(\sigma_r - \nu\sigma_\theta) \tag{8.4.10}$$

$$\epsilon_\theta - \alpha\theta(x) = \frac{1}{E}(\sigma_\theta - \nu\sigma_r) \tag{8.4.11}$$

where $\nu \approx 0.25$ is Poisson's ratio. This is the lateral deformation arising purely from the application of a normal stress. Equations (8.4.10) and (8.4.11), expressed explicitly in terms of the stresses, are

$$\sigma_r = \frac{E}{(1 - \nu^2)}[\epsilon_r + \nu\epsilon_\theta - (1 + \nu)\alpha\theta(r)] \tag{8.4.12}$$

$$\sigma_\theta = \frac{E}{(1 - \nu^2)}[\epsilon_\theta + \nu\epsilon_r - (1 + \nu)\alpha\theta(r)] \tag{8.4.13}$$

The equilibrium condition, Eq. (8.4.9), in terms of these stress formulations, becomes

$$r\frac{d}{dr}(\epsilon_r + \nu\epsilon_\theta) + (1 - \nu)(\epsilon_r - \epsilon_\theta) = (1 + \nu)\alpha r\frac{d\theta(r)}{dr} \tag{8.4.14}$$

The local radial displacement $u(r)$ is related to the local strain, per unit of radius r, as follows. The angular strain ϵ_θ arises from circumferential expansion.

$$\frac{du}{dr} = \epsilon_r \quad \text{and} \quad \frac{u}{r} = \epsilon_\theta \tag{8.4.15}$$

In terms of u, Eq. (8.4.14) becomes a differential equation for $u(r)$ in terms of $\theta(x)$ as

$$\frac{d}{dr}\left[\frac{1}{r}\frac{d(ru)}{dr}\right] = (1 + \nu)\alpha\frac{d\theta(r)}{dr} \tag{8.4.16}$$

The solution, in terms of the local temperature, $\theta(r)$, is

$$u(r) = \frac{(1 + \nu)}{r}\alpha\int_0^r \theta(r)r\,dr + C_1 r + \frac{C_2}{1} \tag{8.4.17}$$

Using this result, along with Eq. (8.4.15), the stress distributions may be determined entirely in terms of α, E, r, ν, and $\theta(r)$. The term C_2/r must be excluded to achieve $u(0) = 0$. At the outer edge, $\sigma_r = 0$. The stress field is then

$$\sigma_r = \alpha E\left[\frac{1}{R^2}\int_0^R \theta(r)r\,dr - \frac{1}{r^2}\int_0^r \theta(r)r\,dr\right] \qquad (8.4.18)$$

$$\sigma_\theta = \alpha E\left[\frac{1}{R^2}\int_0^R \theta(r)r\,dr + \frac{1}{r^2}\int_0^r \theta(r)r\,dr - \theta(r)\right] \qquad (8.4.19)$$

where $\sigma_r(0)$ and $\sigma_\theta(0)$ are finite for $t(0)$, a given temperature value t_1. This results in $\phi(r) = \theta(r)/(t_1 - t_2)$.

A SOLID SPHERE. See Fig. 8.4.3(b). The region is stress free when at t_2 and the later imposed temperature $t(r)$ is again symmetric. Internal equilibrium is then expressed in a manner similar to that in Eq. (8.4.9), as

$$\frac{\partial \sigma_r}{\partial r} + \frac{2}{r}(\sigma_r - \sigma_t) = 0 \qquad (8.4.20)$$

where σ_t includes the two tangential components implied in Fig. 8.4.3(b) (see page 484). See Timoshenko and Goodier (1987). The strain is related to σ_r and σ_t as

$$\epsilon_r - \alpha\theta(x) = \frac{1}{E}(\sigma_r - 2\nu\sigma_t) \qquad (8.4.21)$$

$$\epsilon_t - \alpha\theta(x) = \frac{1}{E}\left[\sigma_t - \nu(\sigma_r + \sigma_t)\right] \qquad (8.4.22)$$

Displacement $u(r)$ again arises, related to ϵ_r and ϵ_t as

$$\epsilon_r = \frac{du}{dr} \quad \text{and} \quad \epsilon_t = \frac{u}{r} \qquad (8.4.23)$$

The stresses σ_r and σ_t are again calculated and the displacement equation in $u(r)$ is

$$\frac{d}{dr}\left[\frac{1}{r^2}\frac{d(r^2 u)}{dr}\right] = \frac{1+\nu}{1-\nu}\alpha\frac{d\theta(r)}{dr} \qquad (8.4.24)$$

This has the solution

$$u(r) = \frac{(1+\nu)}{(1-\nu)}\frac{\alpha}{r^2}\int_0^r \theta(r)r^2\,dr + C_1 r + \frac{C_2}{r^2} \qquad (8.4.25)$$

The two tangential stresses σ_θ and σ_ϕ are equal. The stress field is then

$$\sigma_r = \frac{2\alpha E}{(1-\nu)}\left[\frac{1}{R^3}\int_0^R \theta(r)r^2\,dr - \frac{1}{r^3}\int_0^r \theta(r)r^2\,dr\right] \qquad (8.4.26)$$

$$\sigma_\theta = \frac{\alpha E}{(1-\nu)}\left[\frac{2}{R^3}\int_0^R \theta(r)r^2\,dr + \frac{1}{r^3}\int_0^r \theta(r)r^2\,dr - \theta(r)\right] \qquad (8.4.27)$$

Timoshenko and Goodier (1987) cite an interesting observation concerning the stress distribution in this geometry. The average temperature over the material of the sphere from $r = 0$ to r is calculated as

$$\theta(r)_m = \int_0^r 4\pi r^2 \theta(r)\, dr \Bigg/ \frac{4\pi r^3}{3} = \frac{3}{r^3} \int_0^r r^2 \theta(r)\, dr \qquad (8.4.28)$$

Therefore, the radial stress σ_r at radius r is proportional to the difference between the mean temperature of the whole sphere and $\theta(r)_m$ given previously. Therefore, given the function $\theta(r)$, the stresses may be simply calculated.

SUMMARY. These results are for the thermal stresses in two additional kinds of geometries. They are relatively simple conditions, but do allow definite estimates of the magnitudes of such effects. Many other commonly important geometries have also been considered. In addition to the summaries of Timoshenko and Goodier (1987) and Parkus (1960), reference is made to Boley and Weiner (1970), Hetnarski (1986, 1987, 1989), and Harvey (1980).

8.5 DIFFUSION SUBJECT TO RANDOM DISTURBANCES AND EFFECTS

The bulk of all formulation, analysis, and experimentation concerning diffusion processes proceeds with the presumption that the component physical inputs and resulting effects are principally deterministic. For example, the specification of geometry, bounding conditions, and characteristics of the diffusion region are assumed sufficient to specify any given heat or mass transport result. In analysis, if the equation and boundary conditions are given, a solution is assumed to be formulated.

However, other kinds of processes sometimes arise which must be or are best understood by postulating some random input mechanism, to either simulate or to account for appreciable effects which are not simply deterministic. An example of this is noncontinuum transport between a spherical surface of diameter D and a gas at low density. This kind of transport arises when the gas mean free molecular path length λ is appreciable compared to D. Then the statistical properties of molecular motion must be taken into account. Other examples are random macroscopic inputs in natural processes or in devices, wherein, for example, a bounding surface is moved in a random way. Some attention has been given to such effects. The two kinds of processes considered here are the transport effect arising from the random motion of solid boundaries in contact with a fluid and the random displacement of a growing film of liquid condensate on a cold surface.

Moving containers of fluids are commonplace. They may translate and rotate, either steadily, periodically, in a transient, or randomly. The first consideration here is of a uniform body of fluid in a container which, in turn, is subject to random perturbations of both its linear and angular velocities, that is, of its translation and rotational rates. The transmission of linear velocity fluctuations is largely by normal, or pressure, stresses at the solid–fluid interfaces. Internal equilibration of internal fluid motion by pressure waves is rapid in all but very large containers. However, changes in container angular motion are transmitted only by shear stresses, and initially only at the interface. This effect then propagates inward, to the contained fluid, relatively very slowly. This is the first kind of mechanism considered here. The second example is the random displacement of a growing surface film of vapor condensate.

Early studies of such transport were related to the operation of fluid-filled devices on a nominally ballistic trajectory in space. The buoyancy-driven motions may be very small. Then the lower limit of diffusion heat transfer through the fluid would be pure conduction. However, such devices are subject to many velocity and rotational disturbances due to attitude control measures, engine burns, vibrations, relative motions of components of the spacecraft, motion of occupants, and particle impacts. The actual motion of the device is often modeled as a zero average gravity trajectory, without rotation, perturbed by imposed random velocity and angular motion fluctuations. It might be modeled as a random distribution of both the time interval between instantaneous or abrupt fluctuations, τ_c, as well as the magnitude of the fluctuation. See Gebhart (1963).

In fact, on-board accelerometers in space devices have repeatedly documented such fluctuations, in terms of an "effective" gravity g_e, as related to velocity fluctuations. The characteristic magnitude has been $g_e = O(10^{-6}g)$, where g is the terrestrial level of gravity. Such fluctuations are called "g-jitter." In a gas, such effects may cause compressive temperature fluctuations adjacent to surfaces. These may augment heat transfer. See, for example, Spradley (1974). A review of such effects and of some on-board Apollo heat transfer data are given by Grodzka and Bannister (1972, 1974).

These effects are also discussed by Shyy (1989), relevant to the movement of spacecraft occupants. At an altitude of 170 km, the resulting microgravity noise is about $g_e = 3 \times 10^{-5}g$ and $10^{-3}g$, when asleep and active, respectively.

8.5.1   The Formulation for a Surface in a Fluid

However, enhanced heat transfer, beyond the pure conduction rate, also arises from angular perturbations. The mechanism may often be very different. A sequence of abrupt orientation changes are not rapidly propagated across an appreciable contained fluid volume. Consider a cup of coffee sitting on a horizontal surface, as the fluid container. Continuing, quick, small, and purely angular rotations of the cup leave the main body of the fluid almost motionless. The cup rotates with respect to the bulk of the coffee.

As a model of the resulting heat transfer effects, consider a heated object whose area A is uniformly at t_0, well away from the wall, immersed in a fluid at t_∞. If the object is attached to the randomly moving enclosing container, it moves in the same random way. In the absence of gravity and any container translational acceleration, this object would lose heat to the fluid only by conduction. Any appreciable subsequent random angular perturbations of the container would repeatedly move this object away from its developing conduction field, into new fluid at t_∞. After each abrupt movement, spaced randomly at time intervals τ_c apart, a new transient conduction field would begin to propagate outward from the object, into the adjacent fluid. This particular transient temperature field development would, in turn, cease at the next movement of the container. Then a new transient would begin, at a new location in the fluid. This kind of phenomenon would occur for any random sidewise motion which removes the heating source geometry from its developing outward transient conduction field.

The resulting average heat transfer rate is the average of the rates during these successive conduction transients. This average is calculated as follows. If $Q(\tau_c)$ is the total heat transferred in the specific time interval τ_c, the average heat transfer rate during this interval is $Q(\tau_c)/\tau_c$. Then the long-term average heat transfer rate from the surface, $\overline{Q}$, written as the average of the value of $Q(\tau_c)/\tau_c$, over all of the subsequent intervals, τ_c, is

$$\overline{Q} = \int_0^\infty \frac{Q(\tau_c)}{\tau_c} f(\tau_c)\, d\tau_c = \overline{q}''A \qquad (8.5.1)$$

where $f(\tau_c)$ is the probability distribution of the variable time interval τ_c, where individual intervals τ_c lie in the elapsed time range from 0 to ∞. The instantaneous heat flux from the surface into the fluid during a given transient episode is $q''(\tau_c)$.

Now τ_c is written as a "disturbance" Fourier number, as Fo $= F = \alpha\tau_c/s^2$, where s is a characteristic length. Then Eq. (8.5.1), for average flux $\overline{Q}'' = \overline{Q}/A$, becomes

$$\overline{q}'' = \frac{\overline{Q}}{A} = \frac{\alpha}{s^2} \int_0^\infty \frac{Q(F)f(F)\, dF}{AF} \qquad (8.5.2)$$

where α is the thermal diffusivity of the liquid. The Nusselt number, based on the average heat transfer rate, is then

$$\text{Nu} = \frac{hs}{k} = \frac{\overline{q}''s}{(t_0 - t_\infty)k} = \frac{\alpha}{sk(t_0 - t_\infty)} \int_0^\infty \frac{Q(F)f(F)\, dF}{AF} \qquad (8.5.3)$$

where k is the thermal conductivity of the fluid and h is the effective average random convection coefficient.

It is convenient to replace F by $P = F/F_m = \tau_c/\tau_m$, where τ_m and $F_m = \alpha\tau_m/s^2$ are the mean, or most probable, values of τ and F, distributed as the function $f(F)$. Then

$$\text{Nu} = \frac{\alpha}{skF_m(t_0 - t_\infty)} \int_0^\infty \frac{Q(P)f(P)\, dP}{AP} \qquad (8.5.4)$$

The random events, now spaced in terms of the random time interval P, as $f(P)$, may arise from one effect or from a combination of many independent effects. If each individual effect follows an exponential distribution $f(P) = e^{-P}$, then the probability distribution of the sum of $n + 1$ such simultaneous effects is the following gamma distribution $f_n(P)$. See Parzen (1960).

$$f_n(F/F_m) = f_n(P) = [(n + 1)/n!] P^n e^{-nP} \tag{8.5.5}$$

This probability approaches the Gaussian as n increases and becomes 1 as $n \to \infty$; see Parzen (1960). The relation for Nu for any choice of n becomes

$$\text{Nu}_n = \frac{\alpha[(n + 1)/n!]}{skF_m(t_0 - t_\infty)} \int_0^\infty \frac{Q(P)P^{n-1}e^{-nP}\, dP}{A} = \frac{hs}{k} = \frac{\bar{q}_s'' s}{(t_0 - t_\infty)k} \tag{8.5.6}$$

The value of Nu_n may be calculated for any kind of heating element. Geometry determines $Q(\tau_c) = Q(P)$, for any number of combined disturbance effects, $n + 1$.

8.5.2 Calculated Heat Conduction During a Time Interval

The preceding relation may be used to calculate either an effective random "convection coefficient," h, the average total heat flow, or the average surface heat flux. This is done by evaluating the transient temperature field generated in the fluid during the time interval τ_c. This is calculated in the following subsections for three geometries.

A PLANE SURFACE. The first process taken is the highly idealized circumstance of a flat surface as shown in Fig. 8.5.1(a). The solid region at t_0 is suddenly in contact with ambient fluid at t_∞. Conduction begins. It is assumed that the surface temperature remains at t_0, as for a relatively large solid region conductivity.

The extent of the heat transfer surface is taken as large, compared to the effective penetration distance $\delta(\tau_c)$ of the temperature field in the fluid, as for relatively small values of τ_c. Recall the integral method formulation in Sec. 4.8.1. The solution of the semiinfinite plane region transient, in terms of $\phi = \theta/\theta_0 = (t - t_\infty)/(t_0 - t_\infty)$, is erfc $x/2\sqrt{\alpha\tau}$, as given in Sec. 4.2.3. From that solution the instantaneous heat flux at $x = 0$ is

$$q''(\tau) = -k\left(\frac{\partial t}{\partial x}\right)_{x=0} = \frac{k\theta_0}{\sqrt{\pi\alpha\tau}} = \frac{k\theta_0}{s\sqrt{\pi F}} \tag{8.5.7}$$

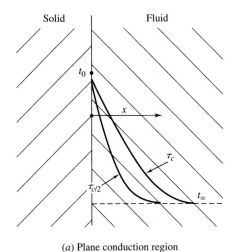

Solid   Fluid

(a) Plane conduction region

(b) Cylindrical or spherical region

FIGURE 8.5.1
A solid region at t_0 in an extensive ambient fluid at t_∞: (a) a flat surface with temperature distributions shown at $\tau = 0$, $\tau = \tau_c/2$, and τ_c; (b) a solid cylindrical or spherical element, subject to the same conditions.

The resulting value of $Q(\tau)/A$ during the time interval $\tau = 0$ to τ_c is

$$\frac{Q(\tau_c)}{A} = \frac{2sk\theta_0}{\alpha}\sqrt{\frac{F_m P}{\pi}} = \frac{2sk\theta_0}{\alpha}\sqrt{\frac{F}{\pi}} \tag{8.5.8}$$

This result will be used in Eq. (8.5.6) to determine h_n.

A SPHERICAL SURFACE. For a sphere maintained at a surface temperature of t_0, suddenly in contact with another region at t_∞, the solution for $\phi =$

$(t - t_\infty)/(t_0 - t_\infty)$ is Eq. (4.3.10) and

$$q''(\tau) = \frac{k\theta_0}{s}\left[1 + \frac{1}{\sqrt{\pi F}}\right] \tag{8.5.9}$$

Then

$$\frac{Q(\tau_c)}{A} = \frac{2sk\theta_0}{\alpha}\left[\frac{F}{2} + \sqrt{\frac{F}{\pi}}\right] \tag{8.5.10}$$

This is seen to amount to the plane conduction region result, with the addition of $F/2$, for the spherical region conduction effect.

A CYLINDRICAL SURFACE. The solution for this circumstance is Eq. (4.3.2), in terms of Bessel functions. Jaeger (1942) gave series for $Q(\tau)/A$ in terms of small and large values of F. Jaeger and Clarke (1942) give a tabulation. Goldenberg (1956) integrated the series of Jaeger. Jakob (1949) tabulates a

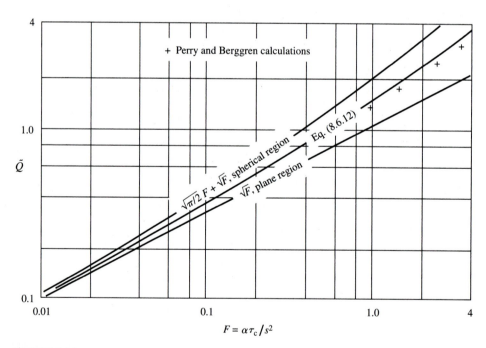

FIGURE 8.5.2
The amount of heat conduction into the fluid $Q(\tau_c)/A$ in the time interval $\tau = 0$ to τ_c, in terms of $\tilde{Q} = 2\sqrt{\pi}\, Q(\tau_c)/2\,Aks\theta_0$ vs. $F = \alpha\tau_c/s^2$.

numerical evaluation by Perry and Berggren, shown as points on Fig. 8.5.2. None of these results is in a convenient form, since $Q(\tau)/A$ must be integrated in Eq. (8.5.6). The data of Jakob, for $F \leq 3.5$, may be approximately fitted by

$$\frac{Q(\tau_c)}{A} = \frac{sk\theta_0}{\alpha}\left(0.4F^{0.95} + 2\sqrt{\frac{F}{\pi}}\right) \tag{8.5.11}$$

However, a simpler suggestion of Jaeger (1942) will be used. The value of $Q(\tau_c)/A$ is taken as the average of the plane and spherical region values, as

$$\frac{Q(\tau_c)}{A} = \frac{2sk\theta_0}{\alpha}\left[\frac{F}{4} + \sqrt{\frac{F}{\pi}}\right] \tag{8.5.12}$$

SUMMARY. The three foregoing expressions for the heat conduction per unit of surface area, from $\tau = 0$ to τ_c, are plotted in Fig. 8.5.2 as a function of $F = \alpha\tau_c/s^2$. The differences between the three geometries are small at small times. Then the effect of surface curvature is small. The specific calculations for the cylinder are seen to be in close agreement with the approximation in Eq. (8.5.12).

8.5.3 Calculation of the Nusselt Number

The general formulation is in Eq. (8.5.1), in terms of $Q(\tau_c)$ and any probability distribution of τ_c, as $f(\tau_c)$. The Nusselt number is expressed in terms of $Q(F)$ and $F = \alpha\tau_c/s^2$ in Eq. (8.5.3). In terms of $P = F/F_m$, where F_m is the most probable value of F, Eq. (8.5.4) applies. Finally, introducing the n-dependent gamma probability distribution in Eq. (8.5.5), Eq. (8.5.6) results. Thereby, Nu_n is determined from the integration in Eq. (8.5.6) when $Q(\tau_c)$, as $Q(P)$, is known, as in Eqs. (8.5.8), (8.5.10), and (8.5.12), for the three geometries.

The resulting values of Nu_n and h_n are in terms of n and the mean, or most probable, time interval, as $F_m = \alpha\tau_m/s^2$, or τ_m. The result for the plane region is

$$Nu_n = C_n/\sqrt{F_m} \quad \text{and} \quad h_n = C_n\sqrt{\rho ck/\tau_m} \tag{8.5.13}$$

where

$$C_n = \frac{\sqrt{n}\,(2n - 1)(2n - 3)\ldots 3 \cdot 1}{n!2^{n-1}} \tag{8.5.14}$$

Thus h_m is dependent on the random process as $1/\sqrt{\tau_m}$.

Also, the preceding value of C_n is very insensitive to the value of n. For example, for $n = 1, 2, 3$, and ∞, $C_n = 1, 1.061, 1.083$, and 1.128. This is only

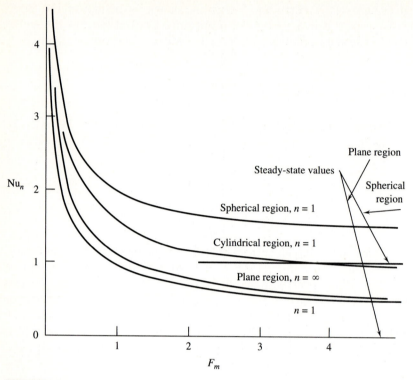

FIGURE 8.5.3
The value of Nu_n as a function of $F_m = \alpha\tau_m/s^2$, with the asymptotes shown for the spherical and plane regions.

about a 13% change over the whole range. The similar results for the spherical and cylindrical regions are

$$Nu_n = 1 + \frac{C_n}{\sqrt{F_m}} \qquad (8.5.15)$$

and

$$Nu_n = \frac{1}{2} + \frac{C_n}{\sqrt{F_m}} \qquad (8.5.16)$$

Examples of these three results are plotted in Fig. 8.5.3. Note that Eqs. (8.5.13) and (8.5.15) go to the proper limits of 0 and 1 as $\tau_m \rightarrow \infty$, as they must. See Sec. 2.1.5. However, Eq. (8.5.16) does not go to zero. This arises due to the approximation made in Eq. (8.5.12), which includes the effect of the spherical geometry.

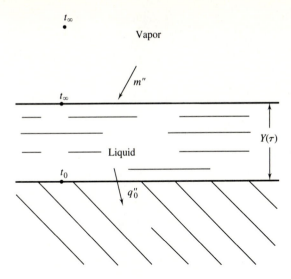

FIGURE 8.5.4
A growing condensate film subject to random removal by disturbances.

8.5.4 The Effect of Random Events on Condensate Films

The other transport process to be considered is the filmwise condensation of a saturated vapor at t_∞ on a surface maintained at $t_0 < t_\infty$. It is assumed that there is no liquid motion induced by vapor velocity, and that the random disturbances, spaced at time interval τ_c, completely clear the surface of an otherwise stagnant liquid film. A new film then forms and grows. Its instantaneous thickness is $Y(\tau)$. Assuming that $Y(\tau)$ remains small compared to any radius of curvature of the solid surface, all geometries may be considered to be a flat surface, as shown in Fig. 8.5.4. The instantaneous rate of mass addition to the liquid film per unit area is taken as m''.

The original conduction assumption of Nusselt is used. It was justified by Sparrow and Gregg (1959a) for small $c_p(t_0 - t_\infty)/h_{fg}$, for gravity-drained films. It is assumed that the conduction process through the liquid film may be treated as steady-state conduction through a slab. Refer to Sec. 8.6.1. Then the thermal flux is written as

$$q'' = (k/Y)(t_0 - t_\infty) = (k/Y)\theta_0 = m''h_{fg} \qquad (8.5.17)$$

where h_{fg} is the latent heat released in condensation. The time rate of change of Y is related to m'' as follows:

$$\rho(dY/d\tau) = m'' \qquad (8.5.18)$$

From Eqs. (8.5.17) and (8.5.18),

$$Y\,dY = (k\theta_0/\rho h_{fg})\,d\tau$$

This is integrated from $\tau = 0$ to $\tau = \tau_c$ and Q/A is found:

$$Y_c^2 = \left(\frac{2k\theta_0}{\rho h_{fg}}\right)\tau_c$$

$$\frac{Q(\tau_c)}{A} = h_{fg}\rho Y_c = (2\rho k h_{fg}\theta_0\tau_c)^{1/2} \tag{8.5.19}$$

$$= \left(\frac{2\rho s^2 k h_{fg}\theta_0 F_m}{\alpha}P\right)^{1/2}$$

where F_m and P are defined as before. The average Nusselt number is found from Eq. (8.5.3) for the probability distribution of Eq. (8.5.5):

$$\text{Nu}_n = \left(\frac{2h_{fg}}{c_p\theta_0 F_m}\right)^{1/2}\frac{n^{n+1}}{n!}\int_0^\infty P^{n-1/2}e^{-nP}\,dP$$

or $\tag{8.5.20}$

$$\text{Nu}_n = \left(\frac{\pi h_{fg}}{c_p\theta_0}\right)^{1/2}\frac{C_n}{\sqrt{F_m}}$$

where C_n is again given as in Eq. (8.5.14). All properties in the preceding relation apply to the liquid phase.

The dependence on F_m in Eq. (8.5.20) is similar to that arising in a single-phase environment. However, the dimensionless parameter $\pi h_{fg}/c_p\theta_0$ is very large for many condensation conditions of practical importance. For example, for mercury vapor at $300°\text{C}$, condensing on a surface at $270°\text{C}$, the value is 225. In such a circumstance Eq. (8.5.20) predicts a high value for the heat transfer rate. Note that the rate in the absence of disturbances approaches zero, as for the flat surface in contact with a single-phase fluid, in Eq. (8.5.13).

8.5.5 Summary

The preceding results indicate that random impulses may have a large effect on transport rates in circumstances where forced fluid motion is not assured. Although only simple heat conduction and vapor condensation processes were treated specifically, the same considerations apply for many other types of transport processes, including analogous diffusion processes. In particular, the conduction solutions presented apply to mass diffusion processes that might arise, for example, in transpiration cooling or in metabolic respiration. If the mole fraction of the diffusing chemical species is small, the same solutions are applicable, if the thermal diffusivity in the Fourier number is replaced by the mass, or chemical diffusivity.

The effects of disturbances are shown to be very large in a condensation process. Boiling might be analyzed similarly. The result of condensation suggests that it might be more desirable to intentionally introduce closely spaced disturbances into a condenser, rather than to arrange for the drainage of extensive condensate films by, for example, controlling the motion of the vapor.

The resulting optimum arrangements of condensing surface undoubtedly would be different.

The principal assumptions that support the formulation of and methods of analysis are that there are continuing disturbances that are appreciable compared to any steady "body" forces present. Also, these disturbances are assumed to cause relative displacements that are the order of size of the various transfer surfaces of interest. Detailed consideration of any proposed design would permit the assessment of these assumptions for that design.

Several studies of these kinds of effects on transport have related to crystal growth. Dressler (1981) analyzes the consequence of a step change of gravity. Thevenard and Hadid (1991) analyze the g-jitter-caused motion effects on crystal growth inhomogeneities, accounting for the effects of the frequency of the fluctuations. The results of Mathioulakis and Grignon (1991) are an example of a statistical interpretation the transport effects of the irregular, random, and unrepetitive instabilities which arise in a flow field. Rahman et al. (1991) analyze thin film flow behavior with and without gravity and indicate the role of the Froude number in flow regimes, including hydraulic jumps. A study of nucleate pooling boiling, by Kenning (1990), indicates the role of spatial and independent temporal effects arising from the statistical nature of random nucleation site behavior. This is related to boiling hysteresis. Stużalec (1991) formulates the finite-element modeling of the effects of random thermal properties, and the effects arising from probabilistic distributions of temperature. See also Tahjadi and Ottino (1991) for results concerning the breakup of freely suspended droplets, and for other related studies. These studies are an indication of the growing interests in stochastic modeling.

8.6 CONDUCTION MODELS OF CONVECTION

Many processes arise in fluids in which the nominally convective transport mechanism is actually evaluated by approximating it as pure conduction heat transfer. Examples include heating a laminar flow in a long tube and cooling oil flows in bearings. The particular example shown in Fig. 8.6.1(a) is the Nusselt (1916) theory of laminar filmwise condensation of a vapor on a cold surface. The vapor is at its saturation temperature, t_{sat}, adjacent to a surface maintained uniformly at $t_0 < t_{sat}$. The falling laminar film of vapor condensate, of local thickness $\delta(x)$, is continuously augmented by vapor condensation at the liquid–vapor interface, that is, at $y = \delta(x)$.

This model amounts to a liquid region between $y = 0$ and $\delta(x)$. It has a temperature of t_0 on one face and $t_v = t_{sat}$ on the other. For steady laminar liquid flow, the local heat flux $q''(x)$ might be estimated as pure conduction across the liquid layer, as simply

$$q''(x) = \frac{k_l}{\delta(x)}(t_v - t_0) \quad \text{or} \quad h_x = \frac{q''(x)}{(t_v - t_0)} = \frac{k_l}{\delta(x)} \quad (8.6.1)$$

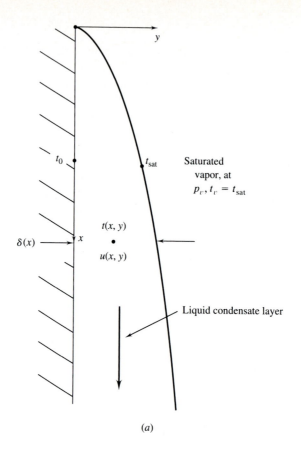

(a)

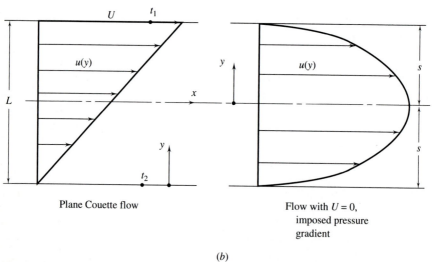

Plane Couette flow

Flow with $U = 0$,
imposed pressure
gradient

(b)

FIGURE 8.6.1
Conductive–convective processes: (a) laminar filmwise condensation of a saturated vapor on a cooled surface; (b) laminar internal plane flows.
498

However, the variation $\delta(x)$ is initially unknown. It results from controlling conditions, such as $(t_v - t_0)$, k_l, and the viscosity of the liquid μ_l. Such analysis is given in Sec. 8.6.1, along with estimates of the resulting corrections required to improve the accuracy of the result.

Another kind of process considered here, in Sec. 8.6.2, is the plane layer internal flow in Fig. 8.6.1(b). Important examples of this kind of geometry include a plane layer with one boundary moving with respect to the other, a Couette flow. The flow might be driven only by the velocity U of the upper surface, by viscous drag. There may also be an imposed pressure gradient in the fluid in the layer. This would produce an added flow effect in the passage. The other example, shown on the right in Fig. 8.6.1(b), is a flow passage with stationary plane boundaries. For both kinds of flows, results are also given for the effects of internal fluid viscous dissipation. This results in an equivalent distributed internal energy source q'''. Uniform boundary conditions are used.

Section 8.6.3 concerns round tubes. Flow is driven by a pressure gradient. The conditions at the inlet may be developed or undeveloped, in terms of both the temperature and the velocity distributions. Boundary conditions of heat flux and of temperature are applied.

8.6.1 Laminar Films

A gravity-drained liquid condensate film is shown in Fig. 8.6.1(a). The vapor is taken as in a saturated state. That is, $t_v = t_{\text{sat}}$ is the vapor phase equilibrium temperature at the ambient pressure p_v. If the surface temperature t_0 is less than $t_v = t_{\text{sat}}$, there is condensation and subsequent conduction through the gravity-drained liquid layer, to the surface. The liquid–vapor interface is taken to be at t_{sat}. That is, it is assumed that there is no thermal resistance there, as might arise with the presence of a noncondensible gas in the vapor or by a contaminating surfactant. Therefore, the simplest formulation of the heat transfer is conduction across the liquid layer, of local thickness $\delta(x)$. The resulting layer conductance, in terms of the liquid conductivity, k_l, is approximated as

$$h(x) = k_l/\delta(x) \qquad (8.6.2)$$

This simple formulation applies quite accurately for thin films. However, in general, there is also convective transport of the liquid in the film. This occurs as the liquid film is cooled below t_v, as it flows downward. The parameter which evaluates this effect is $c_{p,l}(t_v - t_0)/h_{fg}$, the subcooling parameter, where h_{fg} is the latent heat of phase change. The analysis here evaluates $q''(x)$, based on Eq. (8.6.1). Subcooling, convection, and other effects are then discussed.

The preceding model also applies for laminar film boiling, as shown in Fig. 8.6.2. This would arise, for example, for a surface at t_0 greater than an adjacent liquid, at saturation temperature $t_v = t_{\text{sat}}$. Then the heat flux would be outward across the vapor layer, to the vapor–liquid interface. The vaporization occurs there, with the absorption of the latent heat h_{fg}, per unit mass of vapor formed.

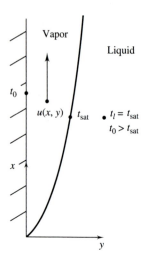

FIGURE 8.6.2
Laminar film boiling of a saturated liquid.

The heat transfer is approximated as conduction across a vapor layer $\delta(x)$ of conductivity, k_v, as

$$h(x) = k_v/\delta(x) \qquad (8.6.3)$$

Then the buoyancy effect in the film, for $\rho_v < \rho_l$, is upward. Therefore, the vapor film flow is upward, from a leading edge at the bottom of the surface.

Both the condensation and vaporization processes may also be analyzed in a similar way even if the ambient fluid is not at the phase equilibrium temperature $t_v = t_{\text{sat}}$, for the pressure level p_v. For the condensation of a superheated vapor, at $t_v > t_{\text{sat}}$, in the present model, there would be an additional conductive layer in the vapor. In this layer, the remote temperature $t_v > t_{\text{sat}}$ is reduced to t_{sat} at the phase interface. Condensation occurs there, with a heat deposition of $c_{p,v}(t_{\text{sat}} - t_v) + h_{fg}$, per unit mass of condensate formed. For the vaporization of a subcooled liquid, that is, for the ambient liquid at $t_l < t_{\text{sat}}$, at pressure level p_v, an outward conduction flux in the liquid heats the adjacent liquid to t_{sat} at the interface. The required heat flux through the vapor, conducted out to the interface, is then $c_{p,l}(t_l - t_{\text{sat}}) + h_{fg}$, per unit mass of vapor formed.

The following analysis is for the simpler circumstance of $t_v = t_{\text{sat}}$ and for condensation, where $t_0 < t_{\text{sat}}$. The formulation is shown for a surface inclined at an angle ζ from horizontal in Fig. 8.6.3. Inclination decreases the downward buoyancy force from $g(\rho_l - \rho_v)$ to $(g \sin \zeta)(\rho_l - \rho_v)$.

LAMINAR FILM CONDENSATION ANALYSIS. The general formulation and a local region, dx by $\delta(x)$, are shown in Fig. 8.6.3. The analysis, originally by Nusselt (1916), amounts to determining the local tangential liquid velocity distribution $u(x, y)$ which is consistent with the buoyancy force $g(\rho_l - \rho_v)\sin \zeta$ and with the local surface shear stress $\tau(x, 0)$ which opposes the liquid flow

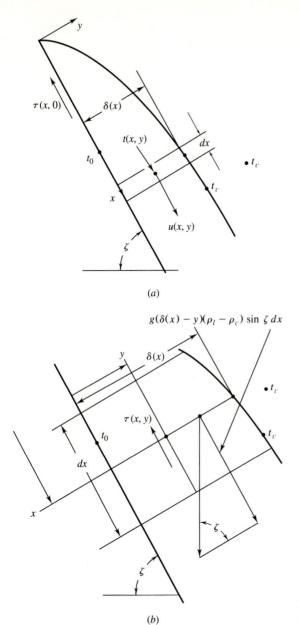

$g(\delta(x) - y)(\rho_l - \rho_v) \sin \zeta \, dx$

FIGURE 8.6.3
A gravity-drained condensate film.

(b)

effect. The film flow rate at x is then matched to the condensation rate on the outer film surface at $\delta(x)$, summed from $x = 0$ to x. The buoyancy force is evaluated as the difference between the hydrostatic pressure gradient in the liquid, $g\rho_l \sin \zeta$, and that in the vapor, $g\rho_v \sin \zeta$.

A force balance is written as follows for the liquid region, of thickness dx, from y to $y = \delta(x)$ in Fig. 8.6.3(b). The viscous drag of the vapor at $y = \delta(x)$ is neglected.

$$(g \sin \zeta)(\rho_l - \rho_v)[\delta(x) - y] \, dx = \tau(x, y) \, dx = \mu_l \frac{du}{dy} \, dx \quad (8.6.4)$$

where du/dy is the local velocity gradient. This becomes

$$du = (g \sin \zeta)(\rho_l - \rho_v)[\delta(x) - y] \, dy/\mu_l \quad (8.6.5)$$

This is integrated from $y = 0$, where $u = 0$, to y.

$$u(x, y) = (g \sin \zeta)(\rho_l - \rho_v)[y \, \delta(x) - y^2/2] \quad (8.6.6)$$

Thus $u(x, y)$ is parabolic in y. The value $\delta(x)$ is next determined in terms of the local condensation rate. The conduction rate through the liquid film layer dx is $dq = k_l(t_v - t_0) \, dx/\delta(x)$; recall Eq. (8.6.1). The local liquid mass flow rate in the film is denoted as $\dot{M}(x)$. The condensation rate at the vapor–liquid interface, of area dx, is then $d\dot{M}$. The latent heat liberation rate there is $dq = h_{fg} \, d\dot{M}$. Therefore,

$$dq = \frac{k_l(t_v - t_0) \, dx}{\delta(x)} = h_{fg} \, d\dot{M}$$

or

$$dM = \frac{k_l(t_v - t_0) \, dx}{h_{fg} \, \delta(x)}$$

$$\quad (8.6.7)$$

Also, from Eq. (8.6.6), $\dot{M}$ is evaluated as

$$\dot{M} = \int_0^{\delta(x)} \rho u \, dy = \frac{\rho_l(\rho_l - \rho_v)g \sin \zeta}{3\mu_l}[\delta(x)]^3$$

and

$$d\dot{M} = \frac{\rho_l(\rho_l - \rho_v)g \sin \zeta}{\mu}[\delta(x)]^2 \, d[\delta(x)]$$

$$\quad (8.6.8)$$

Elimination of $d\dot{M}$ between Eqs. (8.6.7) and (8.6.8) results in the following differential equation for $\delta(x)$:

$$[\delta(x)]^3 \, d[\delta(x)] = \frac{k\mu_l(t_v - t_0)}{\rho_l(\rho_l - \rho_v)h_{fg}g \sin \zeta} \, dx \quad (8.6.9)$$

Integration from $x = 0$, where $\delta(x) = 0$, results in the local film thickness variation

$$[\delta(x)]^4 = \frac{4k_l\mu_l(t_v - t_0)x}{\rho_l(\rho_l - \rho_v)h_{fg}g \sin \zeta}$$

or

$$\delta(x) = \left[\frac{4k_l\mu_l(t_v - t_0)\dot{x}}{\rho_l(\rho_l - \rho_v)h_{fg}g \sin \zeta} \right]^{1/4}$$

$$\quad (8.6.10)$$

Neglecting the convection of thermal energy in the liquid layer, Eq. (8.6.1) evaluates the local heat transfer coefficient h_x from Eq. (8.6.10) as

$$h(x) = \frac{k_l}{\delta(x)} = \left[\frac{\rho_l(\rho_l - \rho_v)k_l^3 h_{fg} g \sin \zeta}{4\mu_l(t_v - t_0)x} \right]^{1/4} \tag{8.6.11}$$

This is rewritten as a local Nusselt number as

$$\frac{h_x x}{k} = \mathrm{Nu}_x = \left[\frac{(\rho_l - \rho_v)c_p x^3 g \sin \zeta}{4\nu_l k_l} \right]^{1/4} \left[\frac{h_{fg}}{c_p(t_v - t_0)} \right]^{1/4} \tag{8.6.12}$$

where c_p applies for the liquid. Averaging h_x in Eq. (8.6.11), from $x = 0$ to the plate length $x = L$, gives

$$\mathrm{Nu} = \frac{hL}{k_l} = 0.943 \left[\frac{(\rho_l - \rho_v)c_p L^3 g \sin \zeta}{\nu_l k_l} \right]^{1/4} \left[\frac{h_{fg}}{c_p(t_v - t_0)} \right]^{1/4} \tag{8.6.13}$$

From Eq. (8.6.10), $\delta(x) \propto \sqrt[4]{x}$. Therefore, $q''(x)$ from Eq. (8.6.1) varies as $1/\sqrt[4]{x}$ and $q''(0)$ is unbounded. This is another example of the lack of realism which arises in conduction solutions when two temperature levels coincide in space. However, for $q''(x) \propto 1/\sqrt[4]{x}$, the integral from $x = 0$ to any value of $x = L$ is bounded, in contrast to many other conduction region solutions.

OTHER ASPECTS OF FILM FLOWS. The principal approximations in the preceding analysis are that the energy effect of liquid subcooling is not accounted for, increasing liquid momentum effects are neglected in the force balance, and the heat transfer rate is calculated as pure conduction across the liquid layer. Bromley (1952) and Rohsenow (1956) determined that the replacement of h_{fg} in the preceding results by $h_{fg} + 0.68c_{pl}(t_v - t_0)$ improved their accuracy. A boundary layer analysis by Sparrow and Gregg (1959a, 1959b) dispensed with all three of the preceding approximations. Specific results were given for $\mathrm{Pr} = (c_p \mu/k)_l = 0.003$ to 100. The Nusselt result, in Eq. (8.6.11), was found to apply for $c_{pl}(t_v - t_0)/h_{fg} < 0.01$ for water. Large corrections arise at larger values. See Gebhart (1971). Cylindrical and other geometries have also been studied.

As discussed previously, a similar analysis applies for the inverse process of film vaporization of a liquid, at t_{sat}, for example, adjacent to a surface at $t_0 > t_{\mathrm{sat}}$. See Fig. 8.6.2. If a film is established and maintained, the lighter vapor film flows upward adjacent to the hot surface, in a film. Then the same mechanisms apply. Common examples of film boiling are the immersion of a hot metal object in a cryogenic liquid, or in oil for quenching.

Falling liquid film processes are also commonly used in mass transfer processes, to promote chemical component adsorption of a gas into the liquid phase. The gas may also be only one component of a gaseous mixture which must be removed, for example, ammonia from air into water. The liquid film may be accelerative as before, or simply a one-dimensional viscosity-controlled flow. See Edwards et al. (1979) for several examples.

In general, sufficiently vigorous film flows may become unstable and undergo transition to turbulence. This regime results in much more effective mixing and heat transfer. The relevant parameter which indicates turbulence is the film Reynolds number. It is defined as $\mathrm{Re}_f(x) = 4\dot{M}(x)/\mu_l$, where $\dot{M}(x)$ is the local condensate flow rate, per unit of layer width. $\mathrm{Re}_f(x) = 1800$ is commonly taken as the upper limit of laminar liquid film flow.

8.6.2 Flow Between Parallel Surfaces

This flow is called Couette. Two parallel surfaces are separated by a fluid layer. In the configuration in Fig. 8.6.1(b) at the left, one surface is in steady motion U. In the other, both are at rest. In the first, the velocity distribution $u(y)$ is linear, as shown in Fig. 8.6.1(b), for $dp/dx = 0$. This flow is a simple idealization of an oil or air bearing. A more complicated similar flow arises when a pressure gradient dp/dx is also applied to the fluid layer, as with forced lubrication. If the bounding surfaces are at different temperatures, t_1 and t_2, the conduction is simply one dimensional, in both circumstances. The flux is $q'' = k(t_1 - t_2)/\delta$, where $\delta = L$ and $2s$, in the two geometries. However, in both of these conditions, the velocity gradients in the liquid layer, du/dx, may be sufficiently great that the viscous dissipation of flow energy creates an important temperature effect. This amounts to a distributed source of strength q'''.

For uniform conditions at each of the two boundaries, as in a long passage, the processes are one dimensional. Two kinds of boundary conditions are considered in the following discussion. The first is the condition at left in Fig. 8.6.1(b), with a velocity applied to a moving top surface. The other circumstance is for $u(s) = u(-s) = 0$, as in the figure to the right. In both configurations, a pressure gradient dp/dx applied in the passage would cause forced flow through the passage.

Also, in each flow, important viscous dissipation commonly arises in vigorous flows. This effect is evaluated in terms of the relevant conduction equation. In one dimension, this is

$$k\frac{d^2t}{dy^2} + q'''(y) = k\frac{d^2t}{dy^2} + \mu\left(\frac{du}{dy}\right)^2 = 0 \tag{8.6.14}$$

where (du/dy) is the local velocity gradient, at y. In the examples here, the cause of this effect is viscosity. For the geometry at the left in Fig. 8.6.1(b), but with a pressure gradient, there are two interacting sources of this energy. One source of the dissipation of mechanical energy into thermal energy is the decreasing flow-work downstream. This is in terms of the imposed pressure gradient $(-dp/dx)$, with an unchanging downstream velocity distribution. The other effect arises from the imposed motion at the boundary of a viscous fluid.

The resulting mechanical energy input is $F \times U$ where F is the shear stress at the interface. This is evaluated as $-U\mu(du/dy)$ at $x = L$ and is $\mu U^2/L$ per unit area. If du/dy is uniform across the region, as for $dp/dx = 0$, this amounts to a distributed thermal source $q''' = \mu U^2/L^2$. This energy is conducted to the bounding surfaces at $y = 0$ and L, depending on the temperature levels maintained there.

FLOW, ONE SURFACE IN MOTION AT VELOCITY U. The flow is assumed caused by a combination of the motion of the top surface and the imposed pressure gradient in the fluid layer. This gradient will aid the flow for $dp/dx < 0$ and oppose it for $dp/dx > 0$, assuming $U > 0$. The pressure gradient and viscous force, as $\mu(du/dy)$, are related as

$$\frac{dp}{dx} = \mu \frac{d}{dy}\left(\frac{du}{dy}\right) \tag{8.6.15}$$

The boundary conditions for $u(y)$ are $u(L) = U$ and $u(0) = 0$. The local flow velocity $u(y)$ is

$$u(y) = \frac{L^2}{2\mu}\left(-\frac{dp}{dx}\right)\frac{y}{L}\left(1 - \frac{y}{L}\right) + \frac{y}{L}U \tag{8.6.16a}$$

$$= N\frac{y}{L}\left(1 - \frac{y}{L}\right) + \frac{y}{L}U \tag{8.6.16b}$$

where $N = (-dp/dx)L^2/2\mu$ is a dimensionless comparison of the pressure gradient and the viscous effects. This distribution is plotted in Fig. 8.6.4 for $N/U = -2, -1, 0, 1$, and 2. For $N/U = 0$, the heat flux across the layer is $q''(x) = $ constant $= k(t_1 - t_2)/L$.

VISCOUS DISSIPATION, ONE SURFACE IN MOTION. The preceding analysis and the simple result, assume that there is no energy source q''' in the region. However, viscous effects dissipate energy directly to the fluid, as formulated in Eq. (8.6.14). This energy is conducted to the boundaries. The value of du/dy in $\mu(du/dy)^2 = q'''(y)$ is calculated from Eq. (8.6.16a) as

$$\left(\frac{du}{dy}\right) = \frac{L}{2\mu}\left(-\frac{dp}{dx}\right)\left[1 - \frac{2y}{L}\right] + \frac{U}{L} \tag{8.6.17}$$

This is seen to be composed of pressure and applied shear force components.

For $(dp/dx) = 0$, that is, no pressure forcing of flow in the passage, the energy equation (8.6.14) becomes simply

$$k\frac{d^2t}{dy^2} = -\mu\left(\frac{du}{dy}\right)^2 = -\mu\left(\frac{U}{L}\right)^2$$

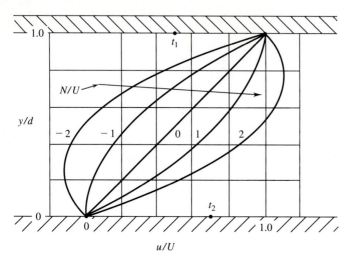

FIGURE 8.6.4
The velocity distributions in Couette flow subject to a pressure gradient, for both aiding, $N > 0$, and opposing, $N < 0$, effects.

This yields

$$t(y) = -\frac{\mu U^2}{kL^2} y^2 + B_1 y + B_2$$

The conditions $t(0) = t_2$ and $t(L) = t_1$ result in

$$\frac{t(y) - t_2}{t_1 - t_2} = \phi\left(\frac{y}{L}, \mathrm{Pr}, \mathrm{Ec}\right) = \frac{\mathrm{Ec}\,\mathrm{Pr}}{2}\left(\frac{y}{L}\right)\left(1 - \frac{y}{L}\right) + \frac{y}{L} \quad (8.6.18)$$

where $\mathrm{Ec} = U^2/c_p(t_1 - t_2)$ is the Eckert number and $c_p\mu/k$ is the Prandtl number. For $\mathrm{Pr} = 0$, that is, for $\mu = 0$, there is no dissipation effect and $\phi = y/L$. Figure 8.6.5 indicates an increasing effect with increasing $\mathrm{Ec}\,\mathrm{Pr}\,\alpha$ $\mu U^2/k(t_1 - t_2)$. This kind of conduction, with $q''' \neq 0$, is very similar to the simple conduction analyses in Sec. 2.4.1.

STATIONARY BOUNDING SURFACES, AT t_i. Figure 8.6.6 is an example of plane Poiseuille flow. In the following example, both surfaces are assumed to be at the same temperature t_1. Viscous dissipation in the fluid results in an energy source $q'''(y)$. There is no moving shear-force energy effect. The energy equation again is

$$k\frac{d^2t}{dy^2} = -\mu\left(\frac{du}{dy}\right)^2 = -q'''(y) \quad (8.6.19)$$

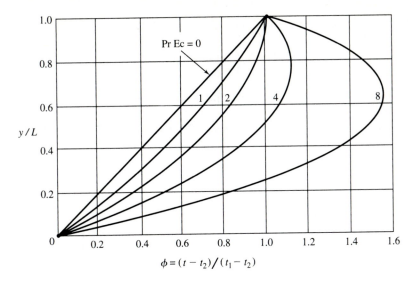

FIGURE 8.6.5
The temperature distribution across a fluid layer, at surface temperatures of t_1 and t_2, accounting for viscous dissipation.

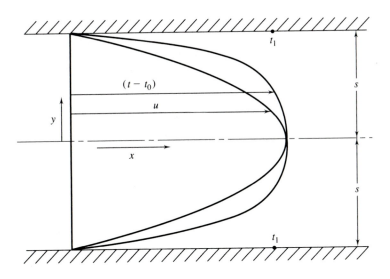

FIGURE 8.6.6
Flow in a passage between stationary surfaces.

The flow velocity distribution is parabolic as

$$u(y) = \frac{s^2}{2\mu}\left(-\frac{dp}{dx}\right)\left[1 - \left(\frac{y}{s}\right)^2\right] = u(0)\left[1 - \left(\frac{y}{s}\right)^2\right] \quad (8.6.20)$$

The temperature condition at $x = s$ and $-s$ is $t = t_1$. A uniform pressure gradient, $dp/dx < 0$, drives the flow, as seen in Eq. (8.6.20). The local dissipation rate q''' is again $\mu(du/dy)^2$ and the result from Eq. (8.6.20) is

$$\mu\left(\frac{du}{dy}\right)^2 = \frac{y^2}{\mu}\left(\frac{dp}{dx}\right)^2 \quad (8.6.21)$$

The resulting temperature distribution is

$$t(y) - t_1 = \frac{s^4}{12\mu k}\left(\frac{dp}{dx}\right)^2\left[1 - \left(\frac{y}{s}\right)^4\right] = \Delta t_{ch}\left[1 - \left(\frac{y}{s}\right)^4\right] \quad (8.6.22)$$

where $\Delta t_{ch} = t_0 - t_1$, where $t_0 = t(0)$, and

$$\frac{t(y) - t_1}{t_0 - t_1} = \phi = 1 - \left(\frac{y}{s}\right)^4 \quad (8.6.23)$$

Therefore, the velocity distribution is parabolic and the temperature distribution is a fourth-degree parabola, as compared in Fig. 8.6.6. The measure of the viscous dissipation effect is Δt_{ch}. This temperature effect arises entirely from viscous dissipation, the conversion of flow energy into a thermal effect. Conduction is equally outward on both sides.

UNEQUAL BOUNDING SURFACE TEMPERATURES. The other result given here is for the top and bottom bounding surfaces both at rest, at t_1 and t_2, respectively. Then conduction flux, for $q''' = \mu(du/dy)^2 = 0$ is $q'' = k(t_1 - t_2)/2s$. With viscous dissipation, the parameter Ec Pr again arises and the temperature distribution becomes

$$\begin{aligned}
\frac{t(y) - t_2}{t_1 - t_2} &= \frac{1}{2}\left(1 + \frac{y}{s}\right) + \frac{s^4}{12k\mu}\left(\frac{dp}{dx}\right)^2\left[1 - \left(\frac{y}{s}\right)^4\right] \\
&= \frac{1}{2}\left(1 + \frac{y}{s}\right) + \frac{\mu[u(0)]^2}{3k(t_1 - t_2)}\left[1 - \left(\frac{y}{s}\right)^4\right] \\
&= \frac{1}{2}\left(1 + \frac{y}{s}\right) + \frac{\text{Pr Ec}}{3}\left[1 - \left(\frac{y}{s}\right)^4\right]
\end{aligned} \quad (8.6.24)$$

The local heat flux is calculated as

$$q''(y) = -k\frac{dt}{dy} = -\frac{k(t_1 - t_2)}{2s}\left[1 - \frac{8\,\text{Pr Ec}}{3}\left(\frac{y}{s}\right)^3\right] \quad (8.6.25)$$

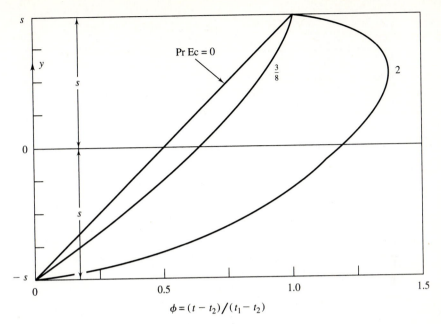

$$\phi = (t - t_2)/(t_1 - t_2)$$

FIGURE 8.6.7
The effect of viscous dissipation, in terms of the parameter $8\,\mathrm{Pr}\,\mathrm{Ec}/3$, for fixed boundaries and $t_1 > t_2$.

The first term in the preceding equation is conduction, as for $q''' = 0$. The second term is the viscous dissipation effect. At $y = s$, $q''(s) = 0$ for $K = 8\,\mathrm{Pr}\,\mathrm{Ec}/3 = 1$. For $t_1 > t_2$, $q''(s)$ is negative, or downward, for $K < 1$. It is positive for $K > 1$. That is, the viscous dissipation effect balances the flux, for $K = 1$. For $K < 1$, conduction is inward at the top. The general behavior is seen in Fig. 8.6.7, for several values of $\mathrm{Pr}\,\mathrm{Ec}$. See Schlichting (1970) for additional considerations.

SUMMARY. The preceding solutions indicate the consequences of viscous dissipation in several kinds of flow and surface boundary conditions. Other conditions also arise. Viscous dissipation is often an important consideration in design. The following section considers processes in circular tubes, which may also often be approximated accurately as one-dimensional conduction.

8.6.3 Flow in Circular Passages

The foregoing conduction analyses of nominally convective circumstances indicate that very simple solutions arise in films and plane layers. These formulations are useful and are sufficiently accurate estimates in many applications. The conditions considered previously in detail are examples of many such opportunities in modeling convective transport.

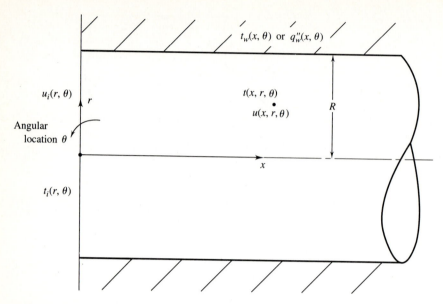

FIGURE 8.6.8
Laminar flow in a circular tube.

There are many models of convection in circular passages which many also be analyzed in relatively simple form. Figure 8.6.8 shows a tube with steady internal flow, subject to general inlet conditions $u_i(r, \theta)$ and $t_i(r, \theta)$. The downstream internal surface conditions are either temperature $t_w(x, \theta)$ or imposed heat flux $q_w''(x, \theta)$, where θ is the angular location around the axis x. Here, uniform fluid properties c_p, ρ, μ, and k are assumed at the outset.

Many specific features of the general formulation postulated in Fig. 8.6.8 have been analyzed in detail. However, for a uniform inlet velocity condition $u_i(r, \theta) = u_i$, the internal flow $u(x, r, \theta)$ eventually develops downstream into a parabolic form. In terms of average velocity V, this distribution is

$$u(r) = 2V\left[1 - \left(\frac{r}{R}\right)^2\right] = \frac{R}{4\mu}\left(-\frac{dp}{dx}\right)\left[1 - \left(\frac{r}{R}\right)^2\right] \qquad (8.6.26)$$

Cross-flows arise in the entry region, to result in this simple downstream flow. It is driven by an imposed pressure gradient $(-dp/dx)$.

Such a downstream flow development is accompanied by a development of the temperature field, locally $t(x, r, \theta)$. Its form would become symmetric, as $t(x, r)$, for symmetric imposed surface conditions t_w or q_w''. However, for either $t_w > t_i$ or $q_w'' > 0$, the local average or cup mixing temperature, $t_m(x)$, would change downstream, as

$$t_m(x) = \frac{1}{\pi R^2 V}\int_0^R 2\pi u(r)t(r)r\,dr \qquad (8.6.27)$$

Many of the important processes of downstream development have been analyzed; see, for example, Kays and Crawford (1980) for an initial consideration of these matters. In applying examples of conduction analysis here, several of the simplest mechanisms are considered. Axial conduction, arising from $\partial t/\partial x = 0$, is neglected, as is permissible for $(R/L)^2 \ll 1$, where L is the heated length. Both the velocity and temperature fields are taken as developed. That amounts to using both Eq. (8.6.26) for $u(r)$ and assuming that the spatial cross-stream form of $t(x, r)$ remains the same downstream, as t_m increases, as with heating. This would also apply far downstream for many kinds of entry conditions, as at $x = 0$ in Fig. 8.6.8.

Three results are given here. The first is for the input wall flux taken uniform downstream, as q''_w. The second is for the imposed flux $q''_w(\theta)$ varying in a periodic way around the wall, as $q''_w(\theta) = q''(1 + b \cos \theta)$. Reynolds (1960, 1963) considered arbitrary angular surface condition variations. The third condition here is for a uniform tube surface temperature downstream, with an initially developed inlet velocity distribution. The temperature field is then governed by

$$\frac{1}{r}\frac{\partial}{\partial r}\left(r\frac{\partial t}{\partial r}\right) = \frac{u}{\alpha}\frac{\partial t}{\partial x} \tag{8.6.28}$$

where $u(r)$ in the convection term on the right is as given by Eq. (8.6.26).

A GENERAL ANALYSIS. Consider a locally uniform wall temperature, which may vary downstream at $t_w(x)$. The following temperature excess ratio is defined, in terms of $t_m(x)$, $t(x, r)$, and $\phi(x, r)$:

$$\phi(x, r) = \frac{t_w(x) - t(x, r)}{t_w(x) - t_m(x)} \tag{8.6.29}$$

Wall conditions which lead to simple results are determined by calculating $\partial t(x,r)/\partial x$ as

$$\frac{\partial t(x, r)}{\partial x} = \frac{dt_w(x)}{dx} - \phi\frac{dt_w(x)}{dx} - t_w(x)\frac{\partial \phi}{\partial x} + \frac{\partial \phi}{\partial x}t_m(x) + \phi\frac{\partial t_m(x)}{\partial x} \tag{8.6.30}$$

However, for a developed temperature profile, $\phi(x, r)$ becomes $\phi(r)$ and

$$\frac{\partial t(x, r)}{\partial x} = \frac{dt_w(x)}{dx} - \phi(r)\frac{dt_w(x)}{dx} + \phi(r)\frac{dt_m(x)}{dx} \tag{8.6.31}$$

This relation evaluates $\partial t(x, r)/\partial x$, in terms of the gradients of $t_w(x)$ and $t_m(x)$. This result is compared with Eq. (8.6.28), to find temperature and flux conditions for which the derivative $\partial t(x, r)/\partial x$ reduces that relation to an ordinary differential equation in terms of $t(r)$ or $\phi(r)$. Two examples are seen

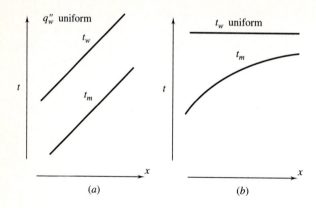

FIGURE 8.6.9
Developed distributions downstream: (*a*) uniform heat flux input q_s''; (*b*) uniform wall temperature t_w.

in Fig. 8.6.9. They are uniform downstream values of q_w'' and t_w, respectively. Then

$$q_w'': \qquad \frac{\partial t}{\partial x} = \frac{dt_w}{dx} = \frac{dt_m}{dx} = \text{constant} \qquad (8.6.32)$$

$$t_w: \qquad \frac{\partial t}{\partial x} = \phi(r)\frac{dt_m}{dx} \qquad (8.6.33)$$

The flux condition is fully developed. See Fig. 8.6.9(*a*). The temperature difference $t_w - t_m$ remains the same downstream. Also, the local temperature distribution, $t(x, r)$, remains of the same shape downstream. The local peak to wall temperature difference $t(x, 0) - t_w(x)$ also remains the same even though both t_w and t_m increase downstream. Equation (8.6.28) is then

$$\frac{u}{\alpha}\frac{\partial t}{\partial x} = \frac{u}{\alpha}\frac{dt_m}{dx} = \frac{2V}{\alpha}\left[1 - \left(\frac{r}{R}\right)^2\right]\frac{dt_m}{dx} = \frac{1}{r}\left(r\frac{dt}{dr}\right) \qquad (8.6.34a)$$

with conditions

$$t(x, R)_w = t(x) \quad \text{and} \quad \frac{dt}{dr} = 0 \quad \text{at } r = 0 \qquad (8.6.34b)$$

The surface temperature condition, t_w in Eq. (8.6.33), is for the local temperature distribution assumed to be developed, in the sense of unchanging shape downstream. Of course, the magnitude of the difference $t(x, 0) - t_w$ will decrease, as the temperature difference and resulting heat flux both attenuate downstream. See Fig. 8.6.9(*b*). Equations (8.6.29) and (8.6.33) result in

$$\frac{2V}{\alpha}\left[1 - \left(\frac{r}{R}\right)^2\right]\phi(r)\frac{dt_m}{dx} = \frac{1}{r}\frac{d}{dr}\left(r\frac{dt}{dr}\right) \qquad (8.6.35a)$$

with conditions

$$t(x, R) = t_0 \quad \text{and} \quad \frac{dt}{dr} = 0 \quad \text{at } r = 0 \qquad (8.6.35b)$$

UNIFORM WALL HEAT FLUX. Since dt_m/dx is constant, Eq. (8.6.34) yields

$$t_w - t(r) = \frac{VR^2}{8\alpha}\left(\frac{dt_m}{dx}\right)\left[3 + \left(\frac{r}{R}\right)^4 - 4\left(\frac{r}{R}\right)^2\right] \qquad (8.6.36)$$

It remains to relate dt_m/dx to the input flux q_w''. First, $t_m(x)$ is determined from $t(r)$ as

$$t_m = \frac{1}{\pi R^2 V}\int_0^R 2\pi u(r) t(r) r \, dr = t_w - \frac{11R^2}{96}\left(\frac{2V}{\alpha}\right)\frac{dt_m}{dx} \qquad (8.6.37)$$

or
$$\frac{dt_m}{dx} = (t_w - t_m)\frac{96}{11R^2}\frac{\alpha}{2V} \qquad (8.6.38)$$

An energy balance of input, $2\pi R q''$, to the energy increase in the fluid, per unit length, is

$$2\pi R q_w'' = \rho c_p V \pi R^2 \, dt_m/dx \qquad (8.6.39)$$

Eliminating dt_m/dx and solving for $q_w''/(t_w - t_m)$ yields

$$h = \frac{48}{11}\frac{k}{D} \quad \text{or} \quad \mathrm{Nu} = \frac{hD}{k} = \frac{48}{11} = 4.364 \qquad (8.6.40)$$

or q', the heat transfer per unit tube length, is

$$q' = h\pi D \,\Delta t = 13.71k(t_w - t_m) \qquad (8.6.41)$$

Since $(t_w - t_m)$ is independent of x, q' is also.

PERIODIC VARIATION OF THE IMPOSED FLUX. The circumferential variation is taken as a single periodic component, as $q_s''(\theta) = q''(1 + b\cos\theta)$. Angular direction conduction requires an additional term, $(1/r^2)(\partial^2 t/\partial\theta^2)$, on the right side of Eq. (8.6.28). The variation $q_s''(\theta)$ results in a θ-varying temperature difference $t_w - t_m$ around the circumference. Therefore, the Nusselt number is θ dependent and is written as Nu_θ. The value is found to be

$$\mathrm{Nu}_\theta = \frac{1 + b\cos\theta}{\dfrac{11}{48} + \dfrac{b}{2}\cos\theta} \qquad (8.6.42)$$

This becomes the previous result, $48/11$, for $b = 0$.

UNIFORM WALL SURFACE TEMPERATURE DOWNSTREAM. This is sometimes called the Graetz problem. The downstream trends are seen in Fig. 8.6.9(b). Using the result in Eq. (8.6.33), Eq. (8.6.28) becomes

$$\frac{2V}{\alpha}\left[1 - \left(\frac{r}{R}\right)^2\right]\phi(r)\frac{dt_m}{dx} = \frac{1}{r}\frac{d}{dr}r\frac{dt}{dr} \qquad (8.6.43)$$

where dt_m/dx is not a constant, but instead is a function of x. Therefore, Eq. (8.6.42), as a function in r, formally applies for a fixed value of dt_m/dx, as at any location along the bottom curve in Fig. 8.6.9(b).

Although the developed circumstance for the flux boundary condition was very simple, the standard analytical treatment required for a temperature condition is very complicated. The general relation, Eq. (8.6.28) is treated as a single function in terms of r and x. A separation of variables is postulated as

$$t(x, r) = X(x)R(r)$$

In terms of the separation constant $1/\lambda^2$, the equations for $X(x)$ and $R(r)$ become

$$\frac{dX(x)}{dx} = -\frac{k}{2\lambda^2\rho c_p V}X(x) \tag{8.6.44}$$

$$\frac{d^2R(r)}{dr^2} + \frac{1}{r}\frac{dR(r)}{dr} + \frac{1}{\lambda^2}\left[1 - \left(\frac{r}{s}\right)^2\right]R(r) = 0 \tag{8.6.45}$$

Equation (8.6.44) determines $X(x)$ as exponential in x. The solution of Eq. (8.6.45) is expressed as a series. Graetz (1885) gave the first solution, in terms of $\phi = (t_0 - t_m)/(t_0 - t_s)$ as a function of $c_p\dot{m}/kL$, where $\dot{m}$ is the mass flow rate, L is the length, and t_m, t_s, and t_0 are the local mean, the tube surface, and the centerline temperatures. Figure 8.6.10, from Jakob (1949), compares the Graetz solutions for the parabolic and the uniform velocity distribution, or plug flow. Also shown is the Lévêque (1928) solution, at larger values of $c_p\dot{m}/kL$.

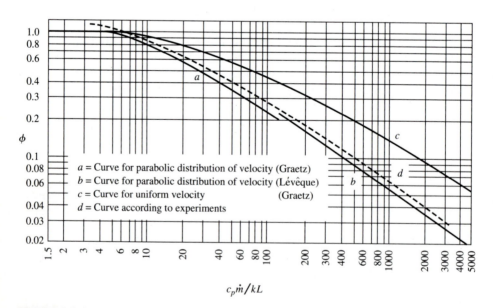

FIGURE 8.6.10
Laminar heat transfer downstream. Comparison of calculations and measurements. [From Jakob (1949).]

The experimental results are in good agreement, for $c_p \dot{m}/kL$ greater than about 8, for both developed flows a and b. In Fig. 8.6.10, $\phi = \theta_m/\theta_0$, where θ_0 is the inlet temperature excess and θ_m is the downstream temperature level.

Extensions and improvements of analysis, including numerical calculations, have increased the value of the results. They also include the developing entry length. See Jakob (1949) for the early analysis and Kays and Crawford (1980) for a recent summary.

SUMMARY. The preceding tube flow heat transfer analyses given are similar to those in Secs. 8.6.1 and 8.6.2, in that heat transfer is determined from a conduction formulation. Streamwise conduction is also ignored. The only added feature is the convection term in the last formulation in this section. Such models are very useful in many other applications and in making initial simple estimates of performance.

PROBLEMS

8.2.1. A round steel rod of 1 in. diameter and 12 in. length is attached to a wall at 100°F. It extends out into air at 60°F. The convection coefficient to the air is 0.5 Btu/hr ft² °F.
(a) Calculate the temperature at the end of the rod and at $x = 6$ in.
(b) Calculate the total heat loss from the rod.
(c) Calculate the end loss.

8.2.2. Find the temperature at the free end and at the middle of a horizontal brass rod 10 cm long and 0.5 cm in diameter. The other end of the rod is maintained at 60°C, and the rod is surrounded by air at 20°C. Assume that the surface coefficient is constant over the whole exposed surface. The surface coefficient is 10 W/m² °C for air and natural convection. Estimate the radiation effect h_r as a linear term, for surrounding surfaces also at 20°C.

8.2.3. Well-mixed combustion gases at 1 atm and 825°C are to flow from a large combustion chamber into a 15-cm-ID pipe whose wall temperature is not to exceed 275°C. Just inside the entrance to the pipe a solid-steel strut of 1 cm diameter is to be placed across the passage, perpendicular to the direction of flow. The strut is expected to be in good thermal contact with the pipe wall at both ends where it contacts the pipe. The average velocity of the gases in the passage is to be 50 ms⁻¹. Determine the maximum temperature to which the metal of the strut will be subjected. For the properties of the combustion gases, use those of air at the same temperature, to calculate the convection coefficient.

8.2.4. For the circumstances of Prob. 8.2.3, the suggestion is made that a very small hole be drilled for the length of the strut and that a thermocouple junction be placed at the center to measure the gas temperature. Estimate the thermometer error for the gas conditions given in Prob. 8.2.2 if the strut is 0.1-cm-diameter steel. How may the design be further modified to reduce this error?

8.2.5. Develop an expression for the overall conductance, per unit length, of a tube having extended surface on the outside. The inside and outside surface areas of the tube per unit length may be taken as A_i and A_0. Neglect the thermal resistance of the metal.

8.2.6. A 1/16-in.-thick aluminum plate is provided on one side with 1/16-by-1/2-in.-high rectangular aluminum fins spaced at 1/2 in. on center. The finned side is in contact with CO_2 at 60°F, and the surface coefficient is to be 3.0 Btu/hr ft^2 °F. The fluid on the plain side is water at 100°F, and the surface coefficient will be 20.0 Btu/hr ft^2 °F. Find the effectiveness and efficiency of the fins.

8.2.7. For the circumstance described in the preceding problem, compute the rate of heat transfer per square foot of area for the bare plate and for the finned plate, assuming that the bare surface and fin base have the same temperature.

8.2.8. Plot the variations of effectiveness and efficiency vs. length for 0.3-cm-thick copper fins for a surface coefficient of 30 W/m^2 °C. Also include the curve for efficiency, if end convection is ignored.

8.2.9. A single long rectangular fin of 0.3 cm thickness and 10 cm height is to be fabricated of surface-coated magnesium, $\epsilon = 0.9$, to dissipate heat to space. The base surface temperature will be 400 K. Therefore, the incident radiation from space may be assumed to have a negligible effect on heat dissipation. Determine the temperature at the outer edge of the fin, the fin efficiency, and the heat dissipation rate.

8.2.10. For the circumstance in Prob. 8.2.9, find the height of fin necessary to dissipate 600 W per meter of length, if the base temperature is 400 K.

8.2.11. Find the fin temperature distribution and the heat transfer, for fin designs:

$$\delta(x) = cx^{1/2}$$

$$\delta(x) = cx^2$$

where c is a constant. The thickness of the fins is $\delta(x)$, x is measured from the tip, and b is the width at the base.

8.2.12. For a plate fin subject to convection on each side, determine the fin thickness variation from the root to the tip which will result in a linear internal temperature gradient in the fin material. Neglect end-face heat convection.

8.2.13. Consider the optimum radiative fin configuration, for the conditions in Prob. 8.2.9. Calculate the base area per unit fin length and the fin efficiency.

8.2.14. For the circumstance in Prob. 8.2.6, consider the effect of bond resistance at the base of the fin. Take the bond temperature to be 200°F.
   (*a*) Calculate the fin efficiency, η, in the absence of bond resistance.
   (*b*) Determine η for rms microfinishes of 10 and 120 in.$\times 10^{-6}$ at bond pressures of 10 and 200 psi.

8.2.15. For the circumstance in Prob. 8.2.6, assume that the fin material is uniformly irradiated, as by fission particles or by microwave radiation.
   (*a*) Determine the effect of this on base area heat dissipation, q_0, if the irradiation energy effect over the fin volume is small, compared to the base area loss.
   (*b*) Calculate the effect on fin efficiency.

8.3.1. Consider a point energy source moving along a relatively thin surface, in comparison with a plane source moving in an infinite region.
   (*a*) For the example in Fig. 8.3.2, plot $\theta/\Delta t_{ch}$ at $y = 0$, where $\theta = t - t_\infty$.
   (*b*) Also plot the result for the plane source for the same value of Δt_{ch} and explain the principal differences.

8.3.2. Make the same comparison as in the preceding problem, but for relatively thin and very thick regions, in terms of k/δ. Use the conditions in Fig. 8.3.2 and comment on the differences of the distributions (at $z = 0$).

8.3.3. A point source q_0 moves along the surface of a thick steel plate, at 25°C, with a velocity u.
  (a) Plot the distance ahead of the source at which the melting temperature is reached, as a function of q_0 (in W). Neglect the latent heat of fusion of steel and assume a velocity of 1 cm/min.
  (b) Repeat part (a) for the distance behind the source at which resolidification occurs. Plot these results on the same curve as in part (a).

8.4.1. An extensive steel wall of 8 cm thickness is subject to fission-particle irradiation, with a resulting approximately uniform internal heating rate of q'''. Assume that the wall was initially stress free at 60°C. The wall surfaces are maintained at 60°C. $\alpha = 6.5 \times 10^{-6}$ °F^{-1} and $E = 30 \times 10^6$ psi.
  (a) Determine the strength of the distributed source which would result in a maximum internal wall temperature of 360°C.
  (b) For this condition, in steady state, calculate the maximum tension and compression thermal stresses in the wall.

8.4.2. A large plane layer of electrical conductor has a uniform internal Joulean energy dissipation rate of q'''.
  (a) Determine the maximum value of the internal thermal stress, as a function of the relevant parameters which apply.
  (b) If the layer is steel of 10 mm thickness, determine the maximum value of q''', in W/cm^3, for the highest local strain to be 10^{-4}, where $\nu = 0.3$. $E = 30 \times 10^6$ psi and $\alpha = 6.5 \times 10^{-6}$ °F^{-1}.

8.4.3. For an aluminum wall, with the conditions in Prob. 8.4.1, determine the maximum values of the actual internal compressive and expansive strains which arise in the wall. $\alpha = 25 \times 10^{-6}$ °C^{-1} and $E = 10 \times 10^6$ psi.

8.4.4. Integrated circuitry is configured on both sides of 10-cm-diameter silicon wafers, of 500 μm thickness. The circuitry amounts to a thickness of only several microns and the design uniform thermal dissipation rate in the circuitry is 10 W/cm^2 on each side. The combined surface coefficient is h, on each side, to ambient air at 25°C. $\alpha = 3 \times 10^{-6}$ °C^{-1}, and $E = 20 \times 10^6$ psi.
  (a) Determine the required value of h so that the circuitry temperature t_e will not exceed 95°C, in an ambient at 25°C.
  (b) For a uniform surface temperature condition, $t_s - t_e$, determine the general dependence of the thermal stresses, σ_r and σ_θ, on radial location.
  (c) Determine the location and magnitude of the maximum values of σ_r and σ_θ.
  (d) Evaluate these values if the stress-free condition applies for a temperature level of 25°C.

8.4.5. A solid sphere of copper, of radius R, is in a stress-free state at t_2. It is internally heated to a center temperature of t_1. $\alpha = 16.6 \times 10^{-6}$ °C^{-1} and $E = 17 \times 10^6$ psi.
  (a) Determine the normal stress distribution for a uniform internal temperature.
  (b) Also determine $\sigma_r(r)$ for a parabolic temperature distribution, as $\phi = 1 - (r/R)^2$.
  (c) For a uniform temperature distribution, and $t_2 = 100$°C, and $t_1 = 300$°C, for $R = 2$ cm, calculate the maximum value of σ_r over the region.

8.5.1. In a materials processing experiment in an earth-orbiting space laboratory, spherical crystals of 0.1 mm diameter are growing while suspended in a large supersaturated water solution of the same material. The solidification thermal effect will result in a temperature difference of 1°C between the water solution, at 20°C, and the growing crystals. The density ratio of the crystals and water solution is 4.2. Due to spacecraft control and other perturbations of the experiment, random displacements of 1 mm occur. Their most probable time interval is 5 s.

(*a*) Calculate the relevant Fourier number.

(*b*) Calculate the Nusselt number and the heat transfer rate q between the solution and a crystal, if the disturbance events are exponentially distributed.

(*c*) Calculate the maximum range of q over all probability distributions.

8.5.2. Repeat part (*a*) of Prob. 8.5.1 over the crystal diameter range from 10^{-2} to 0.5 mm and explain the variations of q and Nu.

8.5.3. In an orbiting laboratory, water vapor at 1 atm condenses on a small cooled test surface maintained at 90°C. It is of 0.2 cm diameter. The spatial perturbations of the experimental module are of the order of 0.5 cm.

(*a*) Taking the most probable time interval as 10s, calculate the relevant Fourier and Nusselt numbers.

(*b*) Determine the heat transfer rate, as flux q'' and as q.

(*c*) Compare this result with the flux determined from the Nusselt theory, for the surface vertical, at standard terrestrial gravity.

8.6.1. Saturated steam at 7.5 psia condenses in a filmwise manner on a vertical surface of 1 ft height, which is maintained at 160°F.

(*a*) Calculate the average value of h_x and the minimum value of h_x which arises.

(*b*) Determine the total heat transfer rate per unit of plate width.

(*c*) Repeat parts (*a*) and (*b*) for an angle of inclination of 45° and compare the results.

8.6.2. A plate 60 cm high and inclined at 45° is maintained at 95°C. The plate is in contact with saturated steam at 1 atm. Estimate the theoretical average surface coefficient and the local coefficient at distances of 3 and 60 cm from the upper edge of the plate. Are these values valid for both the upper and lower plate surfaces? Discuss.

8.6.3. For the circumstance given in Prob. 8.6.1, calculate the theoretical film thickness and film–vapor interface velocity at a position halfway down the plate and at the lower edge.

8.6.4. For the conditions in Prob. 8.6.1, determine the length of vertical surface which would likely result in laminar film transition to turbulence, at the lower edge, taking $Re_f = 4\Gamma/\mu_l = 1800$ there.

8.6.5. For the conditions in Prob. 8.6.1 determine:

(*a*) The local Reynolds number at the bottom of a vertical surface of 1 ft height.

(*b*) The expression for the local Nusselt number.

(*c*) The required height for incipient turbulence at the bottom.

8.6.6. Consider laminar filmwise vaporization of a saturated liquid adjacent to a vertical surface at $t_0 > t_{sat}$. Using the approximations and methods of the Nusselt analysis, develop a theory for the process.

    (*a*) Determine the general velocity distribution.

    (*b*) Calculate the expression for the vapor layer thickness.

    (*c*) Determine the average liquid film maximum velocity $u[x, \delta(x)]$ and the local Nusselt number.

8.6.7. Consider a Couette flow between parallel plates, with the top one in steady parallel motion. Including the effects of an imposed pressure gradient $(-dp/dx)$ and viscous dissipation:

    (*a*) Write the relevant differential equations for the velocity and temperature distributions.

    (*b*) Determine the general solution for temperature difference $t_1 - t_2$ of the two surfaces, in terms of ϕ.

8.6.8. For the Couette flow viscous dissipation solution, for one surface in motion and for $dp/dx = 0$, determine:

    (*a*) The fraction of the heat flow which goes to each boundary, for $Pr\, Ec \geq 2$.

    (*b*) Plot the maximum temperature in the region for a light oil, $Pr = 1000$, with a gap width of 0.5 cm, and $t_1 - t_2 = 10°C$, over a velocity range from 0 to 1 m/s.

8.6.9. Consider flow through a plane passage, with both surfaces stationary and at $20°C$. For a gap width of 0.5 cm and a light oil, as $Pr = 1000$.

    (*a*) Plot the maximum velocity across the passage as a function of the imposed pressure gradient, in atmospheres per meter length, over the range of 0 to 1.

    (*b*) Over this same range, determine the maximum temperature.

    (*c*) Determine the total viscous dissipation rate, per meter of length, over the same range.

8.6.10. Water flows between stationary surfaces, spaced at 0.2 cm, with temperatures of $25°C$ and $20°C$ along the top and bottom surfaces

    (*a*) Including the effect of viscous dissipation, calculate the midplane velocity $u(0)$ at which the heat flux at the top surface is zero.

    (*b*) What is the flux at the bottom surface for this condition?

    (*c*) Calculate the value of $u(0)$ necessary for the maximum water temperature to be $30°C$.

8.6.11. Water flows at an average velocity V in a smooth tube of diameter D.

    (*a*) For $D = 1$ cm and $V = 40$ cm/s, and developed flow, determine the maximum velocity and the pressure gradient.

    (*b*) For a flux heating condition of 1000 W per meter of tube length, determine the convection coefficient and the local heat transfer temperature difference, for developed flow.

8.6.12. For the conditions in the preceding problem, the heat flux input is taken to vary periodically around the tube wall as $q''(1 + 5\cos\theta)$, with an average value of 1000 W/m length. Perform the calculations requested in Prob. 8.6.11, part (*b*).

REFERENCES

Abakians, H., and M. F. Modest (1988) *Trans. ASME, J. Heat Transfer*, **110**, 924.

Atthey, D. R. (1980) *J. Fluid Mech.*, **98**, 787.

Boley, B. A., and J. H. Weiner (1960) *Theory of Thermal Stresses*, John Wiley and Sons, New York.

Bromley, L. A. (1952) *Ind. Eng. Chem.*, **44**, 2966.

Carslaw, H. S., and J. C. Jaeger (1959) *Conduction of Heat in Solids*, Oxford Univ. Press, Oxford p. 58.

Davis, S. H. (1987) *Ann. Rev. Fluid Mech.*, **19**, 403.

Dressler, R. F. (1981) *J. Crystal Growth*, **54**, 523.

Edwards, D. K., V. E. Denny, and A. F. Mills (1979) *Transfer Processes*, Hemisphere, New York.

Gardner, K. A. (1945) *Trans. ASME*, **67**, 621.

Gebhart, B. (1963) *AIAA J.*, **1**, 380.

Gebhart, B. (1971) *Heat Transfer*, 2nd ed., McGraw-Hill, New York.

Goldenberg, H. (1956) *Proc. Phys. Soc. (London)*, **69B**, 256.

Graetz, L. (1885) *Ann. d. Phys. (N.F.)*, **18**, 79.

Grodzka, P. G., and T. C. Bannister (1972) *Science*, **176**, 506.

Grodzka, P. G., and T. C. Bannister (1974) Natural Convection in Low-*g* Environments, AIAA Paper 74-156.

Harvey, J. F. (1980) *Pressure Component Construction*, Van Nostrand Reinhold, Princeton, New Jersey.

Hetnarski, R. B. (1986, 1987, 1989) *Thermal Stresses*, I, II, III, Elsevier, New York.

Jaeger, J. C. (1942) *Proc. Roy. Soc. Edinburgh*, **61A**, 223.

Jaeger, J. C., and M. Clarke (1942) *Proc. Roy. Soc. Edinburgh*, **61A**, 229.

Jakob, M. (1949) *Heat Transfer*, John Wiley and Sons, New York, Vol. 1, p. 269.

Jhaveri, P., W. G. Moffat, and C. M. Adams, Jr. (1962) *Welding J.* **41**, 125.

Kanouff, M., and R. Greif (1992) *Int. J. Heat Mass Transfer*, **35**, 967.

Kays, W. M., and M. E. Crawford (1980) *Convection Heat and Mass Transfer*, McGraw-Hill, New York.

Kays, W. M., and A. L. London (1964) *Compact Heat Exchangers*, McGraw-Hill, New York.

Kenning, D. B. R. (1990) *Ninth International Heat Transfer Conference*, Jerusalem, Israel, G. Hetsroni, Ed., Vol. 3

Kern, D. Q., and A. D. Kraus (1972) *Extended Surface Heat Transfer*, McGraw-Hill, New York.

Kou, S., and W. H. Wang (1986) *Met. Trans.*, **17A**, 2265.

Lancaster, J. F. (1980) *Metallurgy of Welding*, 3rd ed., Allen and Unwin,

Lévêque, J. (1928) *Ann. des Mines*, **13**, 201, 305, and 381.

Masubuchi, K. (1980) *Analysis of Welded Structures*, Pergamon, Oxford.

Mathioulakis, M., and M. Grignon (1991) *Int. Comm. Heat Mass. Transfer*, **18**, 141.

Murray, W. M. (1938) *J. Appl. Mech.*, **5**, A78.

Nied, H. A. (1986) Adv. Welding Sci. and Technology, Proc. Int. Conf. Welding Res., Gatlinburg, Tennessee, 18–22 May 1986, Publ. ASM International.

Nusselt, W. (1916) *Z. d. ver. Deutsch. Ing.*, **60**, 541 and 569.

Oreper, G. M., and J. Szekely (1984) *J. Fluid Mech.*, **147**, 53.

Parkus, H. (1962) *Handbook of Engineering Mechanics*, W. Flügge, Ed., McGraw-Hill, New York, Chap. 43.

Parzen, E. (1960) *Modern Probability Theory and Its Applications*, John Wiley and Sons, New York.

Rahman, M. M., W. L. Hankey, and A. Faghri (1991) *Int. J. Heat Mass Transfer*, **34**, 103.

Reynolds, W. C. (1960) *Trans ASME, Ser C.*, **82**, 108.

Reynolds, W. C. (1963) *Int. J. Heat Mass Transfer*, **6**, 445.

Rohsenow, W. M. (1956) *Trans ASME*, **78**, 1645.

Rohsenow, W. M., J. P. Hartnett, and E. N. Ganić (1985) *Handbook of Heat Transfer Fundamentals*, McGraw-Hill, New York.

Rosenthal, D. (1946) *Trans. ASME*, **68**, 849.

Schlichting, H. (1970) *Boundary Layer Theory*, 7th ed., McGraw-Hill, New York.

Schneider, P. J. (1955) *Conduction Heat Transfer*, Addison-Wesley, Reading, MA.

Shyy, W. (1989) *Int. Comm. Heat Mass Transfer*, **16**, 713.

Sparrow, E. M., and J. L. Gregg (1959a) *J. Heat Transfer*, **81**, 13.

Sparrow, E. M., and J. L. Gregg (1959b) *J. Heat Transfer*, **81**, 291.

Spradley, L. W. (1974) Thermoacoustic Convection of Fluids in Low Gravity, AIAA Paper 74-76.

Stużalec, A. (1991) *Int. J. Heat Mass Transfer*, **34**, 55.

Tahjadi, M. and J. m. Ottino (1991), J. Fluid Mech. 232, 191.

Thevenard, D., and H. B. Hadid (1991) *Int. J. Heat Mass Transfer*, **34**, 2167.

Timoshenko, S. P., and J. N. Goodier (1987) *Theory of Elasticity*, 3rd ed., McGraw-Hill, New York.

Ule, R. L., Y. Joshi, and E. B. Sedy (1990) *Metall. Trans B*, **21B**, 1033.

Wei, P. S., and W. H. Giedt (1985) *Welding J.*, **64**, 2515.

Wilkins, J. E., Jr. (1960) *J. Soc. Ind. Appl. Math.*, **9**, 630.

CHAPTER
9

NUMERICAL
ANALYSIS

9.1 NUMERICAL PROCEDURES

The preceding analysis and results, for both heat conduction and mass diffusion, have mostly been based on considering transport regions as being continua. That is, the representations of the regions ascribe definite physical conditions to every point in the region and over its boundaries.

These techniques are primarily based on generating solutions in mathematical form. The relevant differential equation, subject to idealized boundary and initial conditions, often results in a definite solution. Thereby, many quite complicated processes are reduced to relatively simple analytical results. These permit the calculation of distributions of temperature and concentration, as well as heat and mass flux rates. The results indicate the basic parameters, such as the Fourier and Biot numbers, which govern the transport.

On the other hand, these analytic solutions are based on the approximate Fourier and Fick formulations which relate heat and mass transfer fluxes to temperature and concentration gradients. These procedures often result in unrealistic performance predictions. The resulting effects include both unbounded local heat and mass flux and unbounded transport across boundary surfaces. They also predict unbounded fluxes and infinite propagation velocities, when abruptly changed bounding conditions are used. Several examples of such behavior are discussed in Secs. 3.1.1 and 4.2.1, for steady state and for one-dimensional transients.

Another much more important limitation of analytical techniques is that solutions commonly result only for very simple or idealized geometric regions and initial and boundary conditions. For example, each bounding surface condition usually is restricted to a fixed value of one of the principal coordinates, natural to the geometry of interest. Also, any particular such bounding condition is of assumed form or is taken as uniform in some sense; as an idealized temperature, flux, convection, or combined condition.

9.1.1 The Finite-Difference Representation

The preceding kinds of idealizations are very often not sufficiently representative or accurate models of most actual processes in engineering applications and in design. Numerical methods provide a very suitable and convenient alternative. These methods may avoid the unreasonable consequences which commonly arise in purely mathematical formulations. They may also conveniently accommodate irregular region geometries and more realistic imposed conditions. They also often very simply incorporate distributed sources, variable properties, and other effects, beyond the range of convenient or analytical treatment.

However, additional approximation arises in replacing the classical differential equations and bounding conditions by a spatial network or grid of representative points. The effects of this approximation may be reduced by using a finer grid, say Δx. This was shown in Sec. 3.9 for multidimensional steady state. Also, the permissible time interval of calculation $\Delta \tau$ is sometimes limited by calculational stability considerations, as analyzed in Secs. 4.9 and 5.3 for transients.

Sections 3.9, 4.9, and 5.3 concern the preliminary aspects of the numerical formulations and calculation procedures. This is done in terms of regions in Cartesian coordinates, with idealized bounding conditions. The finite-difference equations, the FDE, are determined for a uniform and square grid. This is in terms of the simplest basis for the estimates of first and second derivatives, as seen in Fig. 3.9.1. Typical resulting finite-difference results are Eqs. (3.9.10), (5.5.3), and (5.5.5), for grid locations internal to the conduction region. Surface point conditions, determined there, include the results in Eqs. (3.9.25), (3.9.27), and (3.9.29) in steady state and in Eqs. (4.9.13), (5.3.9), and (5.3.12) for transients. These apply at surface points away from edges and corners.

Those results are an initial view of the kinds of methodologies used in converting the most important partial differential equations, the PDE, and bounding conditions, BC, into the FDE and BC. Also, the basic properties of forward, central, and backward differences, in transients, were discussed. The simplest explicit and implicit calculation procedures for transients are demonstrated. The requirements for the stability of numerical calculations are determined.

This chapter does not review in any detail the material presented in Secs. 3.9, 4.9, and 5.3. However, it is based directly on that simpler and more direct methodology as an initial treatment of the many basic principles of the numerical analysis of diffusion. It is suggested that this earlier material be reviewed as preparation for the more detailed considerations in this chapter.

9.1.2   The Finite-Element Representation

This method was originally developed primarily for calculations in structural elements and in applications of solid-body mechanics. It is in increasingly wider use in heat transfer analysis. The basis is still the representation of the region of calculation by finite subdivisions. Again, a continuum conduction region is replaced by a spaced grid of representative points. These points, or nodes, are the specific locations at which the solution is locally anchored, for either steady-state processes or for transient response.

However, given the method used in finite-element analysis, there is much flexibility in the choice of the grid used to represent a continuous region. Therefore, irregular regions and complicated boundary conditions may be analyzed almost as simply as very regular ones. This difference of flexibility is apparent in comparing a finite-difference representation, as in Fig. 3.9.5 or 3.9.9, with the finite-element representation of an irregular geometry in Fig. 9.6.1. The practical advantage of the finite-element method is obvious, in this aspect of the method.

The emphasis in the finite-difference calculation is on individual grid points. The finite element itself instead comprises a group of closely associated nodal points. A typical finite element at the surface is identified in Fig. 9.6.1. It is shown as a triangle in two dimensions and may be a tetrahedron in three.

The analysis and calculation procedures are also very different. The subunit is the finite element, which interrelates all of the constituent nodal points. The assembly of elements is then the global formulation over the whole region. The region of the element satisfies the differential equation on the average and the global formulation does the same for the whole region.

This is done in these regions by postulating the approximate transient temperature field $t_a(x, y, z, \tau)$, throughout the entire region, as

$$t(x, y, z, \tau) \approx t_a(x, y, z, \tau) = \sum_{m=1}^{M} p_m(x, y, z) F_m(\tau) \qquad (9.1.1)$$

where M is the total number of nodes, p_m is the shape or interpolation function at point m, and $F_m(\tau)$ is the local amplitude.

Given a specification of the p_m, the method numerically determines the solution over the region, considering steady state. At each step of the calculation, the error between the current solution and the differential equation, called the residual ϵ, is determined, over the region. This guides the solution toward higher accuracy. This process results in a simultaneous solution over the whole region. This methodology is discussed in detail in Sec. 9.6.

9.1.3   The Use of the Methods

The following sections in this chapter describe some of the more elaborate and effective procedures of estimate and calculation technique which are in common use. Section 9.2 concerns estimates of derivatives and also formulations in cylindrical and spherical coordinates. Truncation errors are discussed for several kinds of derivative estimates. Approximation errors, convergence, and stability are considered, along with higher-order derivative estimates.

Section 9.3 concerns principally transients. Again, truncation errors are given and numerical stability limits are analyzed. A number of explicit–implicit methods are given. Procedures in multidimensional transients are discussed.

Section 9.4 considers additional aspects of finite-difference modeling and analysis. The matrix and iterative methods of calculating steady-state temperature fields are discussed, including overrelaxation for iterative methods. Surface point consideration on cylindrical boundaries is analyzed, including stability considerations. Calculation and formulation techniques are discussed for non-conforming boundaries.

Section 9.5 considers numerical procedures for conduction in regions having a conductivity and a specific heat which may be variable with temperature, time, and over the region due to material inhomogeneity. The Boltzmann transformation is also discussed.

Section 9.6 concerns the finite-element method technique, as applied to the analysis of conductive fields. The method is formulated, in terms of elements, the approximation, the interpolation function, and the global equation. The calculation procedure is discussed. The method is then applied to one- and two-dimensional steady-state processes.

9.2   FINITE-DIFFERENCE METHODS

The numerical approximation converts the partial differential equation to finite-difference form. The equation considered is as follows, including the source term, for a region described in Cartesian coordinates:

$$\nabla^2 t + \frac{q'''}{k} = \frac{\partial^2 t}{\partial x^2} + \frac{\partial^2 t}{\partial y^2} + \frac{\partial^2 t}{\partial z^2} + \frac{q'''(x, y, z, t, \tau)}{k} = \frac{1}{\alpha}\frac{\partial t}{\partial \tau} \quad (9.2.1)$$

where q''' may be location, temperature, and time dependent. For a three-dimensional process, $\nabla^2 t$ amounts to one term in each of the three coordinate directions. Note that t may also be written in terms of $\theta = t - t_r$ and also in terms of $\phi = (t - t_r)/\Delta t_{\text{ch}}$ where t_r is a reference temperature and Δt_{ch} is a characteristic temperature difference.

9.2.1   First-Order Estimates of Derivatives

The preceding equation is first approximated here as a finite-difference equation (FDE) in Cartesian coordinates. The notation is given in Fig. 9.2.1 for a typical network point i, j, k, internal to the induction region. The network point instantaneous temperature level at point i, j, k, at some time level n, is $t^n_{i,j,k}$.

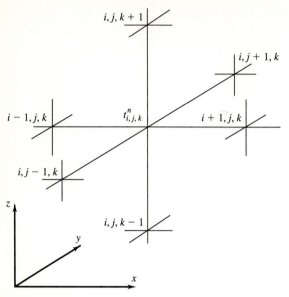

FIGURE 9.2.1
The notation in Cartesian coordinates.

This is the notation used in Secs. 3.9, 4.9, and 5.3 to develop the first-order estimates of the derivatives in Eq. (9.2.1) at grid location i, j, k. Three estimates of $\partial t / \partial \tau$ arose, considering time levels $n - 1$, n, and $n + 1$, as follows:

$$\left[\frac{\partial t}{\partial \tau}\right]^n_{i,j,k} \approx \frac{t^{n+1}_{i,j,k} - t^n_{i,j,k}}{\Delta \tau} \qquad \text{the forward difference} \qquad (9.2.2)$$

$$\left[\frac{\partial t}{\partial \tau}\right]^n_{i,j,k} \approx \frac{t^n_{i,j,k} - t^{n-1}_{i,j,k}}{\Delta \tau} \qquad \text{the backward difference} \qquad (9.2.3)$$

$$\left[\frac{\partial t}{\partial \tau}\right]^n_{i,j,k} \approx \frac{t^{n+1}_{i,j,k} - t^{n-1}_{i,j,k}}{2 \Delta \tau} \qquad \text{the central difference} \qquad (9.2.4)$$

The first estimate calculates the temperature distribution at the next time level $n + 1$ from that at the present time n. The second uses the past and present time levels $n - 1$ and n. The central difference uses time levels $n + 1$ and $n - 1$. Therefore, the forward difference results in an FDE which calculates the future temperature distribution, at time $n + 1$, explicitly from the present one, at time level n. The backward difference uses temperatures at time level $n - 1$ to calculate those at time level n. This is an implicit method, since all of the unknown $t^n_{i,j,k}$ must be calculated simultaneously. This numerical procedure is more demanding. However, it was shown in Sec. 3.9 that the stringent limit on the magnitude of the time step $\Delta \tau$, associated with the forward difference, does not arise in this implicit method. The explicit and implicit methods were compared in Sec. 4.9. Finally, the central difference, originally formulated by Richardson (1910), is not commonly used in the form of Eq. (9.2.4), because of the calculational instabilities which would often arise.

There is only one first-order centered estimate of each of the spatial second-derivative terms, in Eq. (9.2.1), if only temperatures at time level n are used, along with only three adjacent equally spaced grid points. The second-derivative estimate, of $\partial^2 t/\partial x^2$ at point $t_{i,j,k}^n$, is also in terms of $t_{i+1,j,k}^n$ and $t_{i-1,j,k}^n$. A second derivative is the spatial rate of change of the first derivative $\partial t/\partial x$. The difference of the slopes in the region i to $i+1$ and $i-1$ to i are differenced and divided by Δx to yield the three first-order estimates at i, j, k as

$$\left[\frac{\partial^2 t}{\partial x^2}\right]_{i,j,k}^n \approx \left[\frac{t_{i+1,j,k}^n - t_{i,j,k}^n}{\Delta x} - \frac{t_{i,j,k}^n - t_{i-1,j,k}^n}{\Delta x}\right]\frac{1}{\Delta x}$$

$$= \frac{t_{i+1,j,k}^n + t_{i-1,j,k}^n - 2t_{i,j,k}^n}{(\Delta x)^2} \tag{9.2.5}$$

Similar estimates for $\partial^2 t/\partial y^2$ and $\partial^2 t/\partial z^2$, for uniform Δy and Δz, are

$$\left[\frac{\partial^2 t}{\partial y^2}\right]_{i,j,k}^n \approx \frac{t_{i,j+1,k}^n + t_{i,j-1,k}^n - 2t_{i,j,k}^n}{(\Delta y)^2} \tag{9.2.6}$$

$$\left[\frac{\partial^2 t}{\partial z^2}\right]_{i,j,k}^n \approx \frac{t_{i,j,k+1}^n + t_{i,j,k-1}^n - 2t_{i,j,k}^n}{(\Delta z)^2} \tag{9.2.7}$$

The final approximation to be made is of the proper value of $q'''(x, y, z, \tau)$. Recall that q''' may also depend on the local temperature. This would be the value of $t_{i,j,k}^n$, at x, y, z. Therefore

$$q''''^n \approx q_{i,j,k}'''^n \tag{9.2.8}$$

If the distributed source strength is not constant or linearly dependent on $t_{i,j,k}^n$, the numerical calculations may not converge. Approximate linearization methods are given by Patankar (1980).

The foregoing approximations are applied in Eq. (9.2.1) for a one-dimensional transient, $t(x, \tau)$. A forward difference is used for time and Eq. (9.2.5) estimates $\partial^2 t/\partial x^2$:

$$\frac{t_{i+1}^n - t_{i-1}^n - 2t_i^n}{(\Delta x)^2} + q_i'''^n = \frac{1}{\alpha}\frac{(t_i^{n+1} - t_i^n)}{\Delta \tau} \tag{9.2.9}$$

The collection of estimates, in Eqs. (9.2.2)–(9.2.8), are applied to whatever form of Eq. (9.2.1) applies in a particular application. Examples of such applications are in Sec. 3.9 for multidimensional steady-state processes, in Sec. 4.9 for one-dimensional transients, and in Sec. 5.3 for multidimensional transients. These three sections also give the numerical formulations of the boundary conditions which apply on the surfaces of regions subdivided by a network. Steady state and transients are considered in one, two, and three dimensions.

9.2.2 Cylindrical and Spherical Regions

Formulations are given here for these two geometries, for the cylindrical and polar spherical coordinates, also implied in Eq. (9.2.1). Equations (1.2.20) and (1.2.21) for constant and uniform conductivity are given as follows for $t(r, \theta, z, \tau)$ and $t(r, \phi, \theta, \tau)$, where θ and ϕ are measured in the planes in Fig. 1.2.2:

$$\frac{\partial^2 t}{\partial r^2} + \frac{1}{r}\frac{\partial t}{\partial r} + \frac{1}{r^2}\frac{\partial^2 t}{\partial \theta^2} + \frac{d^2 t}{dz^2} + \frac{q'''}{k} = \frac{1}{\alpha}\frac{\partial t}{\partial \tau} \qquad \text{cylindrical} \quad (9.2.10)$$

$$\frac{\partial^2 t}{\partial r^2} + \frac{2}{r}\frac{\partial t}{\partial r} + \frac{1}{r^2 \sin\theta}\frac{\partial}{\partial \theta}\left(\sin\theta \frac{\partial t}{\partial \theta}\right) + \frac{1}{r^2 \sin^2\theta}\frac{\partial^2 t}{\partial \phi^2} + \frac{q'''}{k}$$

$$= \frac{1}{\alpha}\frac{\partial t}{\partial \tau} \qquad \text{spherical} \qquad\qquad\qquad (9.2.11)$$

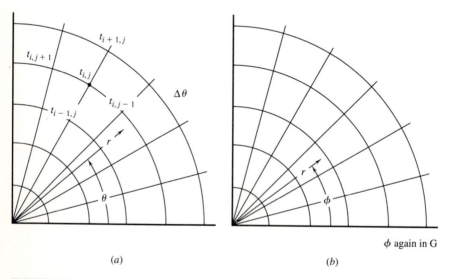

ϕ again in G

(a)　　　　　　　　　　　(b)

FIGURE 9.2.2
The representation of two-dimensional temperature fields in cylindrical and in spherical regions: (a) in r, θ coordinates where i is the axis location at $r = 0$; (b) in r, ϕ where ϕ is the azimuthal angle in polar spherical coordinates.

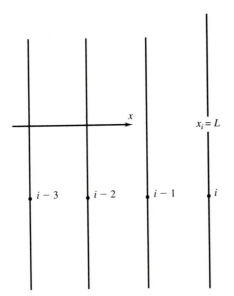

$x_i = L$

$i - 3$ $\quad$ $i - 2$ $\quad$ $i - 1$ $\quad$ i

FIGURE 9.2.3
Estimates of surface heat flux.

The general numerical method subdivisions are shown in Figs. 3.9.2 and 3.9.3, respectively, for three-dimensional analysis.

THE CYLINDRICAL REGION. Central-difference estimates of $\partial^2 t / \partial r^2$, $\partial t / \partial r$, and $\partial^2 t / \partial \theta^2$ are used for $t(r, \theta, \tau)$, in the notation in Figure 9.2.2(a), where

$$\frac{\partial t}{\partial r} \approx \frac{t_{i+1,j} - t_{i-1,j}}{2\,\Delta r}$$

The resulting FDE, in explicit form at time level n, is

$$\frac{1}{(\Delta r^2)}\left[\left[1 - \frac{1}{2i}\right]t_{i-1,j}^n + \left[1 + \frac{1}{2i}\right]t_{-(i-1),j}^n - 2t_{i,j}^n\right]$$

$$+ \frac{1}{i^2(\Delta r\,\Delta\theta)^2}\left(t_{i,j+1}^n + t_{i,j-1}^n - 2t_{i,j}^n\right) + \frac{q_{i,j}'''^n}{k} = \frac{\left(t_{i,j}^{n+1} - t_{i,j}^n\right)}{\alpha\,\Delta\tau} \quad (9.2.12)$$

The first terms of the spatial truncation are of order $(\Delta r)^2$ and $(\Delta\theta)^2$. The differential equation, Eq. (9.2.10), has terms in $1/r$ and $1/r^2$. A question arises concerning how the region around $i = 0$ ($r = 0$) should be modeled. A procedure is given by Özisik (1980) for replacing the cylindrical system in this region by a locally Cartesian one.

THE SPHERICAL REGION. Central differences are used for estimates of $\partial^2 t / \partial r^2$ and $\partial t / \partial r$ for $t(r, \tau)$. The result is

$$\frac{1}{(\Delta r)^2}\left[\left(1 + \frac{1}{i}\right)t_{i+1,j}^n + \left(1 - \frac{1}{i}\right)t_{i-1,j}^n - 2t_{i,j}^n\right]$$

$$+ \frac{q_{i,j}'''^{n}}{k} = \frac{t_{i,j}^{n+1} - t_{i,j}^n}{\alpha\,\Delta\tau} \tag{9.2.13}$$

The first spatial truncation error incurred in Eq. (9.2.13) is $O(\Delta r)^2$. Equation (9.2.11) also has terms in both $1/r$ and $1/r^2$. Özisik (1980) also indicates procedures for replacing the spherical coordinates at and around $r = 0$ by a local Cartesian grid arrangement. It also applies for three-dimensional processes. See also Jaluria and Torrance (1986).

9.2.3 Truncation Errors in Cartesian Coordinates

The foregoing first-order estimates of first and second derivatives are approximate. Consider the error, in terms of the first derivative, $\partial t / \partial \tau$, approximated as $(t_{i,j,k}^{n+1} - t_{i,j,k}^n)/\Delta\tau$. The error is seen in writing $t_{i,j,k}^{n\pm1}$ as a Taylor series expansion about $t_{i,j,k}^n$ as

$$t_{i,j,k}^{n\pm1} = t_{i,j,k}^n \pm \Delta\tau\left(\frac{\partial t}{\partial\tau}\right)_{i,j,k}^n + \frac{(\Delta\tau)^2}{2!}\left(\frac{\partial^2 t}{\partial\tau^2}\right)_{i,j,k}^n \pm \frac{(\Delta\tau)^3}{3!}\left(\frac{\partial^3 t}{\partial\tau^3}\right)_{i,j,k}^n$$

$$+ \frac{(\Delta\tau)^4}{4!}\left(\frac{\partial^4 t}{\partial\tau^4}\right)_{i,j,k}^n \pm \cdots + (-1)^m \frac{(\Delta\tau)^m}{m!}\left(\frac{\partial^m t}{\partial\tau^m}\right)_{i,j,k}^{n+\xi} \tag{9.2.14}$$

The comparable relation for the expansion of, as an example, $t_{i,j,k}^n$ in x, is

$$t_{i\pm1,j,k}^n = t_{i,j,k}^n \pm \Delta x\left(\frac{\partial t}{\partial x}\right)_{i,j,k}^n + \frac{(\Delta x)^2}{2!}\left(\frac{\partial^2 t}{\partial x^2}\right)_{i,j,k}^n \pm \frac{(\Delta x)^3}{3!}\left(\frac{\partial^3 t}{\partial x^3}\right)_{i,j,k}^n$$

$$+ \frac{(\Delta x)^4}{4!}\left(\frac{\partial^4 t}{\partial x^4}\right)_{i,j,k}^n \pm \cdots + (-1)^m \frac{(\Delta x)^m}{m!}\left(\frac{\partial^m t}{\partial x^m}\right)_{i+\xi,j,k}^n \tag{9.2.15}$$

The last term in each series is the remainder of the series, where this derivative is evaluated at some point and in the interval between i and $i + 1$ or i and $i - 1$. The remainder is of the order of Δx^m, which decreases toward zero as the mesh sizes $\Delta\tau$ or Δx are made smaller. This expansion also assumes that derivatives through the range $m + 1$ are continuous.

The series in Eq. (9.2.14) contains the three first-derivative estimates in Eqs. (9.2.2)–(9.2.4). Equation (9.2.15), written as an expansion in x, contains the second derivative. Arranging these expansions in those forms, the truncation errors of the series may be estimated as

$$\left(\frac{\partial t}{\partial \tau}\right)^n_{i,j,k} = \frac{t^{n+1}_{i,j,k} - t^n_{i,j,k}}{\Delta \tau} - \frac{\Delta \tau}{2}\left(\frac{\partial^2 t}{\partial \tau^2}\right)^{n+\xi} \qquad \text{forward difference}$$

(9.2.16)

$$\left(\frac{\partial t}{\partial \tau}\right)^n_{i,j,k} = \frac{t^n_{i,j,k} - t^{n-1}_{i,j,k}}{\Delta \tau} + \frac{\Delta \tau}{2}\left(\frac{\partial^2 t}{\partial \tau^2}\right)^{n+\xi}_{i,j,k} \qquad \text{backward difference}$$

(9.2.17)

$$\left(\frac{\partial t}{\partial \tau}\right)^n_{i,j,k} = \frac{t^{n+1}_{i,j,k} - t^{n-1}_{i,j,k}}{2\,\Delta \tau} - \frac{(\Delta \tau)^2}{6}\left(\frac{\partial^3 t}{\partial \tau^3}\right)^{n+\xi}_{i,j,k} \qquad \text{central difference}$$

(9.2.18)

$$\left(\frac{\partial^2 t}{\partial x^2}\right)^n_{i,j,k} = \frac{t^n_{i+1,j,k} + t^n_{i-1,j,k} - 2t^n_{i,j,k}}{(\Delta x^2)} - \frac{(\Delta x)^2}{12}\left(\frac{\partial^4 t}{\partial x^4}\right)^n_{i+\xi,j,k} \qquad (9.2.19)$$

where, for the second terms in each result, the magnitude of ξ is $0 < \xi < 1$, $-1 < \xi < 0$, $-1 < \xi < 1$, and $-1 < \xi < 1$, respectively.

Comparing Eqs. (9.2.16)–(9.2.19) with Eqs. (9.2.2)–(9.2.5), for example, the several series truncation errors (TE) are apparent. For the forward and backward time derivative estimates, the neglected term is proportional to $\Delta \tau$. For the central difference and the second derivative, the neglected terms are proportional to $(\Delta \tau)^2$ and $(\Delta x)^2$, respectively. Higher derivatives are commonly of increasingly smaller magnitude.

These local first-derivative estimates use only three subsequent times, $n - 1$, n, and $n + 1$, and at one grid point, i, j, k. Each second-derivative estimate uses only three adjacent points in space, for example, $i - 1, j, k$, i, j, k, and $i + 1, j, k$, at one time. Recall the method of Crank and Nicolson (1947) which uses both the n and $n + 1$ time levels for a second-derivative estimate, in connection with time levels n and $n + 1$ for the time derivative.

Using a larger number of sequential points may result in much higher order numerical accuracy. See Jaluria and Torrance (1986) for formulations using four sequential points for first derivatives and five points for the second derivative. A nine-point second-derivative estimate is given in Eq. (9.2.28). Thereby, the estimates of truncation error may be much further reduced. Additional information is given in Sec. 9.2.5. See also the discussion in Mitchell and Griffiths (1980).

9.2.4 Approximations, Errors, Convergence, and Stability

The preceding considerations indicate how the relevant partial differential equation, for example, Eq. (9.2.1), is approximated by a finite-difference equation. Using these different formulations, methods of calculation, and approximations, different results are expected in the analysis of any given heat conduction process. The component sources of the total error in the approximation are separated for consideration and evaluation. Several of these aspects are discussed next.

TRUNCATION ERROR. A numerical formulation should have the property that the truncation error TE should continue to decrease as the finite intervals of the FDE representation are decreased. These are, for example, Δx, Δy, Δz, and $\Delta \tau$ in the preceding formulation. Then the formulation is consistent. The truncation error in any given numerical analysis is the sum of the errors arising in each of the terms in Eq. (9.2.1) which must be retained in the analysis.

DISCRETIZATION ERROR. This arises as the difference between the exact analytical solution of the PDE and the exact numerical solution $\hat{S}$ of the particular FDE formulation used, for any given circumstance. The FDE solution may be exact, in principal, as by carrying an unlimited number of digits in the calculation. The difference between the two solutions is called the discretization error DE. This is the penalty incurred by replacing the PDE by the FDE. That is, it is the error which results from replacing a continuum formulation, the PDE, by a network of specific points in the region, the FDE. This is the error arising solely from the numerical formulation. The numerical formulation is convergent if the discretization error continually decreases as, for example, Δx, Δy, Δz, and $\Delta \tau$ are decreased.

ROUND-OFF ERROR. The FDE are solved numerically by computer. This calculation is often made simultaneously for all the unknown grid point temperatures, for example, all the $t_{i,j,k}$ in steady state, or for the $t_{i,j,k}^n$ sequentially in a transient. This is not a completely precise procedure. There are approximation and/or round-off errors, RO, in the calculation. Commonly, the number of unknowns is very large and an iterative calculation procedure is used. Additional RO then arise. These are errors in calculation.

STABILITY. The conditions required for the stability of a numerical calculation of transient response were discussed in Sec. 4.9 for one-dimensional processes, and in Sec. 5.3 for several dimensions. The question of stability is in terms of the particular first-derivative estimate, in Eqs. (9.2.2)–(9.2.4), which is to be used. Further consideration of stability is found in Secs. 9.3.2 and 9.3.4.

SUMMARY. The preceding considerations concern the derivative estimates necessary to convert Eq. (9.2.1) into the FDE. Any initial condition is applied at the network points. Boundary conditions are modeled as discussed in Secs. 3.9, 4.9, and 5.3. The use of different grid spacings, Δx, Δy, and Δz, is anticipated in the forms of Eqs. (9.2.5)–(9.2.7). Then ratios of these spacings appear in the FDE. An example in two dimensions is given in Sec. 9.2.5. Finally, the previous formulation applies to steady-state processes, in simply omitting $\partial t / \partial \tau$. Then the stability condition does not apply. For further discussion of the preceding matters, see Jaluria and Torrance (1986), along with other references given there.

9.2.5 Other and Higher-Order Estimates

Consecutive three-point derivative estimates are given as Eqs. (9.2.5)–(9.2.7), for second derivatives. Equations (9.2.2)–(9.2.4) are separate kinds of estimates for first derivatives. Each estimate is centered at a typical grid location i, j, k. The estimates apply at time level n, during the calculation of transient response.

Numerical approximations of higher accuracy may be obtained by including more grid points in the formulation of derivative estimates. One kind of approximation required is spatial, in $\nabla^2 t$, to be applied at each time level τ, or value of n. The other is temporal, in $\partial t / \partial \tau$, at a given location i, j, k, through time. Some of the other possibilities of numerical formulation are:

Spatial derivatives, initially as Eq. (9.2.5), for example.

Three consecutive points $i - 1, i, i + 1$ at two time levels among $n - 1, n, n + 1$, the method of Crank and Nicolson (1947).

More than three consecutive points, at time level n, for example, $i - 2$, $i - 1, i, i + 1$, and $i + 2$.

One-sided estimates as $i, i + 1$, and $i + 2$, in from a boundary at i.

The temporal derivative, initially as Eqs. (9.2.2)–(9.2.4), for example.

Again, more consecutive times, outside of $n - 1, n, n + 1$.

Combining estimates at neighboring locations, for example, at $i - 1$ and $i + 1$, at time level $n - 1, n, n + 1$.

There are several interacting basic kinds of considerations in using various kinds of formulations. A principal aspect is accuracy in the ensuing calculation. That is, the formulation must have acceptably small truncation error in each derivative estimate, as well as in the collection of derivative estimates used in the particular application of the FDE. Another aspect is the nature of the resulting calculational procedure. As seen in the time derivative estimates in Eqs. (9.2.2) and (9.2.3), the method is explicit using time levels $n, n + 1$ and implicit using $n - 1, n$. Other and sometimes more convenient calculational procedures arise for transients. These are discussed in Sec. 9.3. There are other considerations as well, in modeling more complicated boundary conditions. The following brief summary considers several of the most general measures of approximation.

ONE-SIDED ESTIMATES OF SPATIAL DERIVATIVES. These may use internal points near a right face boundary at $x_i = L$. See Fig. 9.2.3. The grid locations to the left, $i - 1, i - 2, \ldots$, might be used to calculate the surface heat flux $-k(\partial t/\partial x)_i$ at i, using enough locations to result in a truncation error of order $(\Delta x)^2$. A Taylor series expansion may be written around location i, for t_{i-1} and t_{i-2}. Using the points $i, i - 1, i - 2$ and requiring a truncation error of order $(\Delta x)^2$, Minkowycz et al. (1988) give the following estimate of the local flux at the surface location i:

$$\left(\frac{\partial t}{\partial x}\right)_{i,j,k}^n \approx \frac{t_{i-2,j,k}^n - 4t_{i-1,j,k}^n + 3t_{i,j,k}^n}{2\,\Delta x} \tag{9.2.20}$$

The comparable result for a left face boundary can be obtained by interchanging the subscripts. The one-sided estimate of the second derivative at i on the boundary, to order Δx^2, uses $i, i - 1, i - 2, i - 3$ as

$$\left(\frac{\partial^2 t}{\partial x^2}\right)_{i,j,k}^n \approx \frac{2t_{i,j,k}^n - 5t_{i-1,j,k}^n + 4t_{i-2,j,k}^n - t_{i-3,j,k}^n}{(\Delta x)^2} \tag{9.2.21}$$

These representations may be much more accurate in representing surface boundary conditions, such as flux loading and convective conditions. Polynomial fitting of a solution near boundaries has a similar advantage. See, for example, Anderson et al. (1984).

UNEQUAL GRID SPACING. Such grids are commonly used for convenience and accuracy in irregular geometries. The notation around i, j is shown in Fig. 9.2.4. Again, Taylor series expansions are used to generate estimates of first and

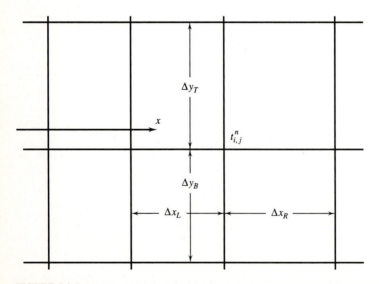

FIGURE 9.2.4
Local unequal grid spacing in two dimensions.

second derivatives, in terms of truncation error level estimates. The simplest second-derivative estimate in the x direction is

$$\left(\frac{\partial^2 t}{\partial x^2}\right)^n_{i,j} = \frac{2}{(\Delta x_L + \Delta x_R)}\left[\frac{t^n_{i+1,j} - t^n_{i,j}}{\Delta x_R} - \frac{t^n_{i,j} - t^n_{i-1,j}}{\Delta x_L}\right] \quad (9.2.22)$$

where the truncation error is of order Δx for Δx_R and Δx_L comparable and becomes of order Δx^2 as $\Delta x_R/\Delta x_L \to 1$. The values of $\partial^2 t/\partial y^2$ and $\partial^2 t/\partial z^2$ are apparent from the preceding equation. Also, for variable grid spacing over the region, a representative value of the spacing must be used for sufficient accuracy in the calculation. See also Minkowycz et al. (1988) and Anderson et al. (1984).

THE CRANK–NICOLSON (1947) METHOD. The time derivative is formulated as a forward difference, n and $n + 1$, as in Eq. (9.2.2). The second derivative is taken, for example, as in Eq. (9.2.5). However, the estimate of $\partial^2 t/\partial x^2$ is determined as the average of its value at n and $n + 1$, see Fig. 9.2.5, as

$$\frac{1}{2}\left[(t^n_{i+1} + t^n_{i-1} - 2t^n_i) + (t^{n+1}_{i+1} + t^{n+1}_{i-1} - 2t^{n+1}_i)\right] = \frac{(\Delta x)^2}{\alpha \Delta \tau}(t^{n+1}_i - t^n_i)$$

$$(9.2.23)$$

This is the one-dimensional form. The two- and three-dimensional forms are apparent as averages of the comparable estimates of $\partial^2 t/\partial y^2$ and $\partial^2 t/\partial z^2$. The truncation error is proportional to $(\Delta x)^2$ and to $(\Delta \tau)^2$. This is an implicit method, in that four terms given previously apply at the future time level $n + 1$. Patankar and Baliga (1978) show physically unreasonable results in calculations using a large time step $\Delta \tau$. On the other hand, a fully implicit method is less accurate at small time steps.

THE DUFORT–FRANKEL (1953) METHOD. The central-difference estimate of $\partial t/\partial \tau$ in Eq. (9.2.4) is an unstable formulation. Stability results if the term t^n_i in the estimate of the second derivative in Eq. (9.2.5) is replaced by the average $(t^{n+1}_i + t^{n-1}_i)/2$. Then the one-dimensional form is

$$t^n_{i+1} + t^n_{i-1} - t^{n+1}_i - t^{n-1}_i = \frac{(\Delta x)^2}{2\alpha \Delta \tau}(t^{n+1}_i - t^{n-1}_i) = \frac{1}{2F}(t^{n+1}_i - t^{n-1}_i)$$

$$(9.2.24)$$

The interrelation of these temperatures is also shown in Fig. 9.2.5. The preceding relation may be rearranged as follows:

$$t^{n+1}_i(1 + 2F) = t^{n-1}_i + 2F(t^n_{i+1} - t^{n-1}_i + t^n_{i-1}) \qquad F = \frac{\alpha \Delta \tau}{(\Delta x)^2} \quad (9.2.25)$$

Thus t^{n+1}_i is known in terms of the present and previous temperature levels, at n and $n - 1$. Therefore, it is an explicit method, as in the simple forward-dif-

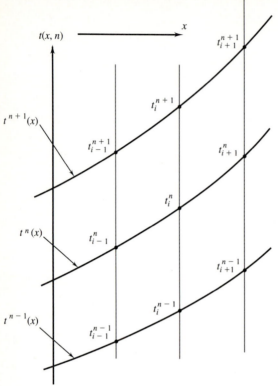

FIGURE 9.2.5
Second derivatives estimated at two time levels.

ference result, where the result was

$$t_i^{n+1} = F(t_{i+1}^n - t_{i-1}^n) + (1 - 2F)t_i^n \qquad (9.2.26)$$

The resulting truncation error is estimated as a sum of contributions, proportional to $(\Delta\tau)^2$, $(\Delta x)^2$, and $\Delta\tau/(\Delta x)^2$. For consistency of the formulation, $\Delta\tau/(\Delta x)^2$ must also approach zero as $\Delta\tau$ and Δx approach zero.

STEADY-STATE PROCESSES. Equation (9.2.1) is given next in its steady-state form, including the distributed source effect. This is the Laplace equation for $q'''(x, y, z, \tau) = 0$, and the Poisson equation otherwise.

$$\nabla^2 t + q'''(x, y, z, \tau) = \frac{\partial^2 t}{\partial x^2} + \frac{\partial^2 t}{\partial y^2} + \frac{\partial^2 t}{\partial z^2} + q'''(x, y, z, \tau) = 0 \quad (9.2.1)$$

The second derivative, for example, $\partial^2 t/\partial x^2$, is the following three-point result:

$$\left(\frac{\partial^2 t}{\partial x^2}\right)_{i,j,k} \approx \frac{t_{i+1,j,k} + t_{i-1,j,k} - 2t_{i,j,k}}{(\Delta x)^2} + \text{TE} \qquad (9.2.27)$$

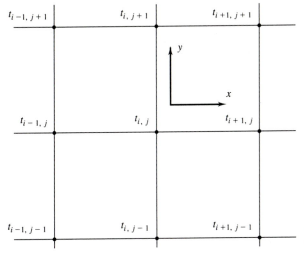

FIGURE 9.2.6
The domain used for the nine-point estimate of the Poisson equation in two dimensions.

where the TE is evaluated in Eq. (9.2.15) and approximated as the one term in Eq. (9.2.19). The error level is $O(\Delta x)^2$, plus $O(\Delta y)^2$ and $O(\Delta z)^2$, in three dimensions. Thus the estimates are based on three, five, and seven points, in one, two, and three dimensions.

The preceding material in this section, for transients, shows that the TE may be reduced by using several time levels to estimate the second derivative at any given time level n. Recall Fig. 9.2.5.

A similar measure in steady state is to use more adjacent intervals to more accurately evaluate second derivatives. A formulation in two dimensions, which uses nine points, is shown in Fig. 9.2.6, for the Laplace equation in two dimensions, for $\Delta x = \Delta y$. The nine points include i, j, the connected four points, and four additional corner points. Estimates of $\partial^2 t/\partial x^2$ and $\partial^2 t/\partial y^2$ at i, j are as given in Eqs. (9.2.5) and (9.2.6). Other estimates in the vicinity of i, j result in the following finite-difference equation:

$$t_{i+1,j+1} + t_{i-1,j+1} + t_{i+1,j-1} + t_{i-1,j-1} + 4t_{i+1,j} + 4t_{i-1,j}$$

$$+ 4t_{i,j+1} + 4t_{i,j-1} - 20t_{i,j} + \frac{q'''_{i,j}(\Delta x)^2}{k} = 0 \qquad (9.2.28)$$

The comparable form of this nine-point equation for $\Delta x \neq \Delta y$ is given by Minkowycz et al. (1988). The TE for the preceding Poisson form is $O(\Delta x)^2 + O(\Delta y)^2$. For the Laplace equation the TE is $O(\Delta x)^6$ for $\Delta x = \Delta y$. See Anderson et al. (1984), Table 3-1, for additional kinds of estimates, along with the associated truncation errors.

SUMMARY. The preceding additional methods demonstrate the range of flexibility which is available, beyond the most apparent estimates given in Eqs. (9.2.2)–(9.2.7). Increased calculation speed and smaller truncation errors may be achieved. Reference is also made to the FDE representation of $\nabla^2 t$ in Cartesian coordinates, described in Jaluria and Torrance (1980), using nine-point estimates of derivatives for a two-dimensional region. Truncation errors of $O(\Delta x)^4$ and $O(\Delta y)^4$ result. Other measures are also discussed. The next section concerns solution methods. Another aspect there is the additional kinds of more complicated formulations which may reduce calculation time.

9.3   TRANSIENT PROCESSES

Section 9.2 considered the estimates of the individual derivatives which arise in the following conduction equation:

$$\nabla^2 t + \frac{q'''}{k} = \frac{1}{\alpha} \frac{\partial t}{\partial \tau} \tag{9.2.1}$$

where $\nabla^2 t$ may be in Cartesian, cylindrical, or spherical coordinates. One-, two-, and three-dimensional regions are discussed. Estimates of the time derivatives are given and explicit and implicit methods arise. The errors of different kinds of estimates are quantified in terms of grid and time intervals of, for example, Δx and $\Delta \tau$. Reliable finite-difference representations are given, in terms of convergence, stability, and consistency considerations.

It is also apparent in these developments that the total error in representing a PDE with an FDE is the sum of the collective errors in estimating each of the derivatives of Eq. (9.2.1) which are retained in any particular application of the FDE. The individual errors include those in terms of Δx, Δy, and Δz, in three dimensions. These are expressed in terms of each other, for example, as $\Delta x = \Delta y = \Delta z$ in a cubical grid.

However, the choice of $\Delta \tau$ is nominally independent of the choices of Δx, Δy, and Δz. For example, in Sec. 4.9, the explicit and implicit methods arose from the particular choice of time levels used to estimate $\partial t / \partial \tau$. Stability limits for the numerical analysis of transients arose when the explicit methods were considered. This effect was found to place a restriction on the maximum value of $\Delta \tau$ for a given choice of grid spacing. The stability limit was also found to be different in one-, two-, and three-dimensional processes. The limiting value was shown in Secs. 4.9 and 5.3 to also depend on the kinds of heat transfer processes arising at the surfaces of conduction regions.

Section 9.3.1 evaluates the overall error incurred in the FDE, compared to the PDE, for a one-dimensional transient in a plane region. Section 9.3.2 considers stability requirements. Section 9.3.3 examines other calculational methods, truncation errors, and stability characteristics. Section 9.3.4 concerns multidimensional and surface effects.

9.3.1   Total Truncation Error

The example here is a plane region transient. The equation is given, followed by the FDE, using the central difference for $\partial^2 t / \partial x^2$ and the forward difference for $\partial t/\partial \tau$. The total truncation error is collected as TE.

$$\frac{1}{\alpha}\frac{\partial t}{\partial \tau} - \frac{\partial^2 t}{\partial x^2} - \frac{q'''}{k} = 0 \qquad (9.3.1)$$

$$\frac{t_i^{n+1} - t_i^n}{\alpha \, \Delta \tau} - \frac{t_{i+1}^n + t_{i-1}^n - 2t_i^n}{(\Delta x)^2} - \frac{q_i'''^n}{k} + \text{TE} = 0 \qquad (9.3.2)$$

where

$$\text{TE} = -\frac{\partial^2 t}{\partial \tau^2}\frac{\Delta \tau}{2\alpha} - \frac{\partial^4 t}{\partial x^4}\frac{(\Delta x)^2}{12} + \cdots \qquad (9.3.3)$$

In this form, the error is $O(\Delta \tau) + O(\Delta x)^2$. The time derivatives in the preceding relation may be eliminated in terms of spatial derivatives. See Anderson et al. (1984) for this method. See also Warming and Hyett (1974). One form of the resulting approximation of Eq. (9.3.1) is the modified result

$$\frac{1}{\alpha}\frac{\partial t}{\partial \tau} - \frac{\partial^2 t}{\partial x^2} - \frac{q'''}{k} = \left[-\frac{\alpha \, \Delta \tau}{2} + \frac{(\Delta x)^2}{12} \right]\frac{\partial^4 t}{\partial x^4}$$

$$+ \left[\frac{\alpha^2(\Delta \tau)^2}{3} - \frac{\alpha \, \Delta \tau (\Delta x)^2}{12} + \frac{(\Delta x)^4}{360} \right]\frac{\partial^6 t}{\partial x^6} + \cdots$$

$$(9.3.4)$$

The order of the TE remains the same, in terms of $\Delta \tau$ and Δx. If the lowest-order term in the spatial derivatives is even, as before, the solution will be principally subject to dissipative errors, which tend to dissipate gradients. In circumstances in which they are of odd order, the error effects will be mostly dispersive. This is a distortion of the relation between any wave-like components which arise. Refer to Anderson et al. (1984).

9.3.2 Stability in Transients

Consider a one-dimensional transient in a plane region with specified temperature boundary conditions. The purely conductive stability constraint is $F = \alpha \, \Delta \tau/(\Delta x^2) \leq 1/2$, for the forward-difference estimate of $\partial t/\partial \tau$. For two- and three-dimensional temperature fields, it is $1/4$ and $1/6$, respectively; see Sec. 5.3. The constraint for an applied temperature-independent heat flux, q_s'', at the surface, or for a uniformly distributed heat source in the region near the surface, q_s''', is the same. See the analysis in Sec. 4.9 along with the heat balance, Eq. (4.9.12), and the coefficient of t_s^{n+1} in Eq. (4.9.13). A convection surface condition resulted in Eq. (4.9.14) and $F \leq 1/2(B + 1)$, where $B = h \, \Delta x/k$ is a grid Biot number. Comparable results in two and three dimensions, including surface and edge geometries, are discussed in Sec. 5.3. Recall that the central-difference estimate of $\partial t/\partial \tau$ results in an unconditionally unstable formulation. This nominally excludes that method, which nevertheless has a smaller truncation error, $O(\Delta\tau)^2 + O(\Delta x)^2$. Thus an attractive formulation is not necessarily reasonable in use. However, the DuFort–Frankel method, discussed in Sec. 9.3.3, uses time levels $n + 1$ and $n - 1$ to evaluate $\partial t/\partial\tau$. The resulting FDE is unconditionally stable.

COMPUTATIONAL STABILITY. A widely used method of determining this stability for linear equations is that of von Neumann; see O'Brien et al. (1950) and Minkowycz et al. (1988). The example considered here is the one-dimensional plane layer forward-difference equation, for $q''' = 0$:

$$\frac{t_i^{n+1} - t_i^n}{\alpha\,\Delta\tau} = \frac{t_{i+1}^n + t_{i-1}^n - 2t_i^n}{(\Delta x)^2} \tag{9.3.5}$$

Stability analysis is applied in terms of the sum δ of all of the errors arising in the numerical solution of the FDE, taking into account the appropriate bounding conditions. The corresponding exact numerical solution $\hat{S}$ is referred to in Sec. 9.2.4. Thus the actual solution at any time level is $\hat{S} + \delta$, where δ is the error. This solution is substituted into Eq. (9.3.5) and the resulting equation for δ is

$$\frac{\delta_i^{n+1} - \delta_i^n}{\alpha\,\Delta\tau} = \frac{\delta_{i+1}^n + \delta_{i-1}^n - 2\delta_i^n}{(\Delta x)^2} \tag{9.3.6}$$

since $\hat{S}$ is defined as satisfying Eq. (9.3.5) exactly. This results because Eqs. (9.2.1) and (9.3.5) are linear. Comparing Eqs. (9.3.5) and (9.3.6) indicates that the growth characteristics of the solution and of the errors are the same.

The test for stability is similar to that used in many other linear formulations. The solution, at each time step in any application, is perturbed by a set of disturbance modes. Here, the form is to be the instantaneous function of the error at time τ, distributed in x as $E(x)$. This function is formulated as

$$E(x) = \sum_n A_n e^{i\beta_n x} \tag{9.3.7}$$

where n designates a harmonic mode, β_n is the wave number, and A_n is the disturbance magnitude. Any given mode, $e^{i\beta x}$, is then considered. A temporal growth rate of the form $e^{\sigma\tau}$ is postulated, where σ may be real or complex. The error at x at time τ is then

$$\delta(x,\tau) = e^{\sigma\tau} e^{i\beta x} \tag{9.3.8}$$

Introducing this form of $\delta(x,\tau)$ into Eq. (9.3.6) results in the following relation at time level τ:

$$e^{\sigma(\tau+\Delta\tau)} e^{i\beta x} - e^{\sigma\tau} e^{i\beta x} = \frac{\alpha\,\Delta\tau\,e^{\sigma\tau}}{(\Delta x)^2} \left[e^{i\beta x(x+\Delta x)} - 2e^{i\Delta x} + e^{i\beta(x-\Delta x)} \right] \tag{9.3.9}$$

Solving for the error at $\Delta(\tau + \Delta\tau, x)$ results in

$$\delta(\tau + \Delta\tau, x) = e^{\sigma(\tau + \Delta\tau)}e^{i\beta x}$$

$$= e^{\sigma\tau}(e^{i\beta x} + Fe^{i\beta(x + \Delta x)} - 2e^{i\beta x} + e^{i\beta(x - \Delta x)}) \quad (9.3.10)$$

The growth, or error amplification, during the time interval $\Delta\tau$, is determined by dividing this result by Eq. (9.3.8). The amplification factor is determined as

$$R = \frac{\delta(\tau + \Delta\tau, x)}{\delta(\tau, x)} = e^{\sigma\Delta\tau} = 1 + 2F[\cos\beta\,\Delta x - 1] \quad (9.3.11)$$

Stability results for the absolute value of $R = \delta(\tau + \Delta\tau, x)/\delta(x, \tau)$ less than or equal to one. That is, a stable component of δ may oscillate around zero but may not increase in magnitude through time. Using this condition results in

$$\left|[1 + 2F[\cos(\beta\,\Delta x) - 1]]\right| \le 1 \quad (9.3.12)$$

This condition is satisfied for $F \le 1/2$.

This result applies for the explicit form in Eqs. (9.3.5) and (9.3.6) used in the preceding analysis. This is also the value surmised in Sec. 4.9, on the basis of intuitive arguments concerning the permissible coefficients of t_i^n in the calculation of t_i^{n+1}.

The present analysis is a specific demonstration of a technique of determining calculational stability limits in transients. It may be applied to additional and more complicated applications. See Anderson et al. (1984). A similar consideration is of multidimensional transients. Another effect would be a linearly temperature-dependent distributed internal energy source, as $q''' \propto t$. This alters the stability limit. Stability conditions may also be assessed from the matrix representation of the transient solution. A general analysis is given by Jaluria and Torrance (1986). The condition is again $F \le 1/2$ for simple assigned constant-temperature boundary conditions, using the explicit method. Also, $F \le 1/4$ and $1/6$ are again found for two- and three-dimensional temperature fields. See also Lapidus and Pinder (1982) and Anderson et al. (1984).

9.3.3 Other Explicit and Implicit Formulations

Section 9.2.5 suggests some additional ways of estimating derivatives to accomplish advantages and increased accuracy in applications of numerical methods. The Crank–Nicolson and DuFort–Frankel finite-difference methods are given as examples of higher-order formulations. Here, a number of related considerations are given, relevant both to accuracy and methodology.

THE SIMPLE IMPLICIT METHOD. This method, discussed in Sec. 4.9.3, does not have a required stability limit. Results are compared with the explicit-method results in Fig. 4.9.4. The formulation, as written in Eq. (4.9.9), for $q_i''' = 0$, is

$$\frac{t_i^{n+1} - t_i^n}{\alpha \, \Delta\tau} = \frac{t_{i+1}^{n+1} + t_{i-1}^{n+1} - 2t_i^{n+1}}{(\Delta x)^2} \qquad (9.3.13)$$

The first error is $O(\Delta\tau) + O(\Delta x)^2$, as for the explicit method. The truncation error is again the same as given in Eq. (9.3.4). The von Neumann stability analysis indicates unconditional stability for $F \geq 0$, for assigned constant temperature surface conditions. Three unknowns, at time level $n + 1$, appear in Eq. (9.3.13). Thus a tridiagonal system of linear algebraic equations must be solved for the t_i^{n+1}, at each time level.

THE CRANK–NICOLSON (1947) METHOD. As seen in Eq. (9.2.34), this method averages the estimates of $\partial^2 t / \partial x^2$ at time levels n and $n + 1$ and uses the forward difference for $\partial t / \partial \tau$; recall Fig. 9.2.5. The truncation error is $O(\Delta\tau)^2 + O(\Delta x)^2$, as

$$\frac{1}{\alpha} \frac{\partial t}{\partial \tau} - \frac{\partial^2 t}{\partial x^2} - \frac{q'''}{k} = \frac{(\Delta x)^2}{12} \frac{\partial^4 t}{\partial x^4} + \left[\frac{\alpha^2 (\Delta\tau)^2}{12} + \frac{(\Delta x)^4}{360} \right] \frac{\partial^6 t}{\partial x^6} + \cdots$$

$$(9.3.14)$$

This is a clear advantage in accuracy. The Keller box method has a comparable truncation characteristic.

   These kinds of methods are special examples of a general combined explicit–implicit formulation. They are discussed by Minkowycz et al. (1988), in terms of a weighting factor between the relative effects of the $n + 1$ and n time levels, in the evaluation of the second-derivative estimate, given the explicit estimate of $\partial t / \partial \tau$. The range of stability, in terms of $F = \alpha \, \Delta\tau / (\Delta x)^2$, depends on the weighting. The examples given have a time component first truncation error of $O(\Delta\tau)^2$.

THE DUFORT–FRANKEL (1953) METHOD. The calculational result of this method was discussed in Sec. 9.2.5, as given in Eq. (9.2.24). The central difference of $\partial t / \partial \tau$ is stabilized by replacing the temperature difference in the transient term $t_i^{n+1} - t_i^n$ by $(t_i^{n+1} + t_i^{n-1})/2$. An explicit method results. The von Neumann analysis indicates that this explicit formulation is unconditionally stable. The truncation error is

$$\text{TE} = \frac{(\Delta x)^2}{12} \frac{\partial^4 t}{\partial x^4} - \left(\frac{\Delta\tau}{\Delta x} \right)^2 \frac{\partial^2 t}{\partial \tau^2} - \frac{(\Delta\tau)^2}{6\alpha} \frac{\partial^3 t}{\partial \tau^3} + \cdots \qquad (9.3.15)$$

THE BARAKAT–CLARK (1966) METHOD. This is an example of an alternating-direction explicit method, ADE. The solution for t_i^{n+1} is divided into two solutions, in p_i^{n+1} and q_i^{n+1}, where t_i^{n+1} is taken as the average of these two, as

$$t_i^{n+1} = \left(p_i^{n+1} + q_i^{n+1}\right)/2 \tag{9.3.16}$$

$$\frac{p_i^{n+1} - p_i^n}{\alpha\,\Delta\tau} = \frac{\left(p_{i+1}^n - p_i^n\right) - \left(p_i^{n+1} - p_{i-1}^{n+1}\right)}{(\Delta x)^2} + \frac{q_i'''}{k} \tag{9.3.17a}$$

$$\frac{q_i^{n+1} - q_i^n}{\alpha\,\Delta\tau} = \frac{\left(q_{i-1}^n - q_i^n\right) - \left(q_i^{n+1} - q_{i+1}^{n+1}\right)}{(\Delta x)^2} + \frac{q_i'''}{k} \tag{9.3.17b}$$

These solutions are each independently generated at each level of time $n + 1$, during the transient. The solutions are generated considering increasing and decreasing i, respectively, across the region. This arises in the form of the two second-derivative estimates seen in Eqs. (9.3.17a) and (9.3.17b), as shown in Fig. 9.3.1. The estimate for $\partial^2 p/\partial x^2$ differs the first derivatives at the two different time levels n and $n + 1$. The estimate for $\partial^2 q/\partial x^2$ does the same, in the other direction.

The method is explicit for both the p_i^{n+1} and q_i^{n+1}. The first calculation is started from the left face, and known boundary conditions there. Therefore, Eq. (9.3.17a) is explicit since that value is known on the right side of the equation. This is seen in arranging Eq. (9.3.17a), for $q_i''' = 0$, as

$$p_i^{n+1} = \frac{F}{1 + F}\left(p_{i+1}^n + p_{i-1}^{n+1}\right) + \frac{(1 - F)}{(1 + F)}p_i^n \tag{9.3.18}$$

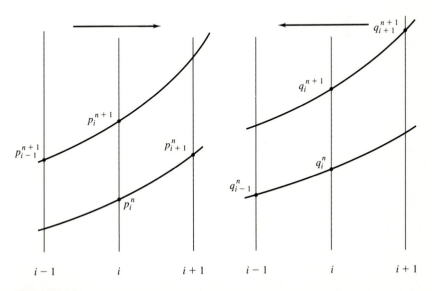

FIGURE 9.3.1
The second derivative estimates in the ADE method: (a) for $\partial^2 p/\partial x^2$; (b) for $\partial^2 q/\partial x^2$.

All terms on the right side of the previous equation are known when calculating toward the right. A similar result arises in Eq. (9.3.17b) when calculating toward the left. The two results, at $n + 1$, are then averaged at each location i, as in Eq. (9.3.16), to yield t_i^{n+1}. The error estimate is $O(\Delta\tau)^2 + O(\Delta x)^2 + O(\Delta\tau/\Delta x)^2$. This method is unconditionally stable.

THE ALTERNATING-DIRECTION IMPLICIT METHOD (ADI). This method, given by Peaceman and Rachford (1955), results in relatively simple calculations for two-dimensional transients. The two spatial-direction second-derivative estimates are each split into two relations as follows, for $\Delta x = \Delta y$:

$$t_{i,j}^{n+1/2} - t_{i,j}^n = \frac{\alpha\,\Delta\tau}{2(\Delta x)^2}\left(\delta_x^2 t_{i,j}^{n+1/2} + \delta_y^2 t_{i,j}^n\right) \tag{9.3.19}$$

$$t_{i,j}^{n+1} - t_{i,j}^{n+1/2} = \frac{\alpha\,\Delta\tau}{2(\Delta x)^2}\left(\delta_x^2 t_{i,j}^{n+1/2} + \delta_y^2 t_{i,j}^{n+1}\right) \tag{9.3.20}$$

where the central-difference operators δ_x and δ_y used previously, for convenience, are defined as

$$\delta_x^2 t_{i,j}^n = t_{i+1,j}^n + t_{i-1,j}^n - 2t_{i,j}^n = \left(t_{i+1,j}^n - t_{i,j}^n\right) - \left(t_{i,j}^n - t_{i-1,j}^n\right) \tag{9.3.21}$$

$$\delta_y^2 t_{i,j}^n = \left(t_{i,j+1}^n - t_{i,j}^n\right) - \left(t_{i,j}^n - t_{i-1,j}^n\right) \tag{9.3.22}$$

For example, the left side of Eq. (9.3.21), divided by $(\Delta x)^2$, is the estimate of $\partial^2 t/\partial x^2$ at grid location i, j at time level n.

Equations (9.3.19) and (9.3.20) intermix x and y three-point second-derivative estimates. In this two-step method, the first step is the solution of a tridiagonal matrix for each j row of grid points. Then a solution is obtained for each i row. The calculations are seen to be implicit. This method has a first truncation error of $O(\Delta\tau)^2$, $O(\Delta x)^2$, and $O(\Delta y)^2$, if $\Delta x \neq \Delta y$. It is unconditionally stable. See Anderson et al. (1984) and Jaluria and Torrance (1986) for other related aspects.

9.3.4 Summary

Multidimensional processes require multiple second-derivative finite-difference estimates. This procedure is discussed in Sec. 3.9 for steady state. The residual equation arose, as Eqs. (3.9.10) and (3.9.11). The relations were determined in Sec. 5.3 for the explicit method, t_i^{n+1}. Second-derivative estimates are in terms of temperature levels at time level n. The stability limits with temperature boundary conditions, for $\Delta x = \Delta y = \Delta z$, were $F \leq 1/4$ and $\leq 1/6$, Eq. (5.3.6). The flat surface point stability limits, with surface convection in one, two, and three dimensions, are given in Sec. 5.3.3 as

$$F_{1D} \leq 1/(2B + 1),\; F_{2D} \leq 1/(2B + 4) \quad \text{and} \quad F_{3D} \leq 1/(2B + 6) \tag{9.3.23}$$

Thus higher dimensionality shortens the maximum permissible computational time step, $\Delta\tau$, for a given $(\Delta x)^2/\alpha$.

Previous considerations have shown that implicit, explicit–implicit, and other measures of derivative estimation yield smaller truncation errors, larger transient steps, and other advantages. The DuFort–Frankel and Barakat–Clark methods simply apply to multidimensions, although the calculational procedures involve more detail. They also remain unconditionally stable. However, the considerable computational advantage of implicit over explicit methods is not retained. Several other alternating-direction implicit methods, are also based on the Crank–Nicolson formulation. Several of these are discussed in Minkowycz et al. (1988). See also Peaceman and Rachford (1955), Douglas (1955), and Douglas and Gunn (1964) for further considerations of ADI methods.

Calculations by Thibault (1985) compared the results of nine different three-dimensional transient formulations, applied to a cubical region. The calculation included the common surface conditions of imposed surface flux, temperature level, and convection. Comparison of the results with exact solutions indicated the limitations and disadvantges of some of the methods, in these applications.

9.4 OTHER ASPECTS OF FINITE DIFFERENCES

The numerical formulations in Secs. 9.2 and 9.3 are for steady state and transient processes. For steady state, the calculation is a simultaneous determination of all unknown grid point temperatures. This often requires the solution of a large number of coupled linear algebraic equations. One method is the direct solution, for all of the unknown temperature levels, $t_{i,j,k}$. However, it is often more convenient or economical to use an iterative method, to obtain successive improved estimates of the $t_{i,j,k}$. Convergence is assured if certain conditions are met. The five- and nine-point formations of the Laplace equation, in two dimensions, satisfy generous conditions. Several schemes are discussed in Sec. 9.4.1.

The calculation for transients is inherently always iterative. It starts from the initial condition, at $\tau = 0$, to determine the transient temperature field response at later times $n + 1, n + 2, \ldots$, throughout the region. These calculations are relatively simple, for the explicit method. However, if an implicit method is used, several adjacent time levels may be involved in more than one term, in each step of the calculation through time. Then the whole instantaneous transient temperature field, at each subsequent time level, must be calculated simultaneously. This is similar to the calculation for a steady-state process. However, the transient calculation must be repeated at each time level during the transient. An exception is the ADE method of Barakat and Clark (1966), formulated in Eqs. (9.3.16) and (9.3.17).

9.4.1   Calculations for Steady State

Many methods and variations have been developed and are in common use. In Sec. 3.9.2 the relaxation procedure was shown by a very simple example of an iteration technique. The initial guesses of the unknown internal grid point temperatures, $t_{i,j}$, are corrected, step by step, as $t_{i,j}^r, t_{i,j}^{r+1}, \ldots$. The procedure is continued to the final level of accuracy necessary to the use of the resulting steady-state result.

The system to be solved includes the unknown temperature levels at all of its internal grid points. For example, consider a long bar of dimension W by H. It is subdivided by an internal grid $\Delta x = \Delta y$, as shown in Fig. 9.4.1. The internal grid locations are at $i = 2, 3, \ldots, I - 1$, for every value of $j = 2, 3, \ldots, J - 1$. The number of these is $(I - 2)(J - 2)$. The boundary conditions are assumed to supply sufficient conditions for a solution. Typical such conditions are given in Sec. 1.4, for temperature and concentration boundary conditions.

THE DIRECT SOLUTION. The solution of the set of linear algebraic equations is simply given for the column vector of unknowns (t), in terms of the square coefficient matrix (A), for the boundary conditions and any distributed energy source, as (B).

$$(A)(t) = (B) \tag{9.4.1}$$

Given (A) and (B), the (t) may be determined by one of the many readily available techniques. However, this procedure is not very commonly used, because of the relatively large amount of required computational effort. This method also does not take advantage of the sparseness of the matrix, which arises for many heat conduction applications.

ITERATIVE TECHNIQUES. The example discussed first here is the Poisson equation in two dimensions. The simplest FDE estimate results in

$$t_{i+1,j} + t_{i-1,j} + t_{i,j+1} + t_{i,j-1} - 4t_{i,j} + \frac{q_{i,j}'''(\Delta x)^2}{k} = 0 \tag{9.4.2}$$

The simple method referred to in Sec. 3.9.2, for $q_{i,j}''' = 0$, supplies initial guesses for the unknowns and then selectively reduced error in the residuals, expressed as $R_{i,j}$ in Eq. (3.9.11). The Gauss–Seidel method of making continuing improvements of the initial guesses is given in the following discussion. It is based on an expression in terms of the temperature term in the FDE which has the largest coefficient. This is $t_{i,j}^r$ in Eq. (9.4.2). The equation is then written in terms of r, which denotes the particular iteration level which is to be advanced to iteration level $r + 1$. For $\Delta x = \Delta y$, at any location i, j,

$$t_{i,j}^{r+1} = \frac{t_{i+1,j}^r + t_{i-1,j}^r + t_{i,j+1}^r - t_{i,j-1}^r}{4} + q_{i,j}''' \frac{(\Delta x)^2}{k} \tag{9.4.3}$$

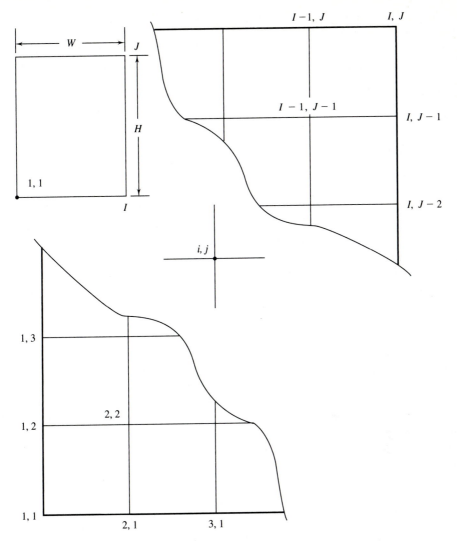

FIGURE 9.4.1
A rectangular conduction region grid array.

The successive calculation of $t_{i,j}^{r+1}$ over the whole region is to sweep across the region in directions of increasing i and j. Then the values of $t_{i,j}^{r}$ are already known at locations $i - 1, j$ and $i, j - 1$.

During the iteration, the convergence rate toward an increasingly accurate result may be improved using the technique called overrelaxation. The simple approach, after the iteration at $r + 1$ is complete, is to merely use these values for the next iteration, which calculates $t_{i,j}^{r+2}$. An improved convergence rate may be obtained by using instead an extrapolation, using the last two values found,

$t_{i,j}^r$ and $t_{i,j}^{r+1}$, as better estimates of the $t_{i,j}^{r+2}$. The relaxation effect is formulated in terms of the parameter ω, in the Laplace equation, as

$$t_{i,j}^{r+1} = (1 - \omega)t_{i,j}^r + \frac{\omega}{4}\left(t_{i+1,j}^r + t_{i-1,j}^r + t_{i,j+1}^r + t_{i,j-1}^r\right) \qquad (9.4.4)$$

Clearly, $\omega = 1$ is the Gauss–Seidel form, in Eq. (9.4.3). The extreme value $\omega = 0$ places all the emphasis on $t_{i,j}^r$. The other extreme, $\omega = 2$, gives comparable weight to all five values. The ranges $0 < \omega < 1$ and $1 < \omega < 2$ are associated with successive under- and overrelaxation, (SUR) and (SOR).

Very large reductions in computer time may be achieved with relaxation. However, there is no general method known for determining the value of ω which would result in the optimized SOR calculation procedure. See, for example, Young (1954) and Frankel (1950). An example is discussed in Minkowycz et al. (1988). The other methods discussed there include the SOR by lines and the alternating-direction implicit method, ADI, using several formulations. An analysis of iterative convergence and rate of convergence is given in Jaluria and Torrance (1986), for a rectangular region with homogeneous boundary conditions. The optimized SOR was shown to be much faster than the Gauss–Seidel method.

9.4.2 Conditions at Other Kinds of Boundaries

The formulations in Secs. 3.9, 4.9, and 5.3 and those given previously here are for regions described in Cartesian coordinates, whose boundaries correspond to fixed values of the coordinates. The equations were also given, in less dimensions, for cylindrical and spherical regions, in Sec. 9.2.2. However, for these two geometries, special considerations arise in determining the surface point equations. These are due to the outward divergence of these regions, as shown in Fig. 9.2.2. The surface point relation for the cylindrical geometry is given, as an example, in the following discussion, in terms of surface–environment interactions.

The circumstance is also considered in which a conduction region boundary does not correspond to a fixed value of any coordinate. The example analyzed is shown in Fig. 9.4.3, in Cartesian coordinates. Analysis methods are given in the following subsections for each of the preceding circumstances.

CYLINDRICAL GEOMETRY SURFACE POINTS. The geometry is shown in Fig. 9.2.2(a). The finite-difference relation for internal conduction points is Eq. (9.2.12). The surface, and the region immediately inside it, are shown in Fig. 9.4.2. A transient surface region energy balance is written, as in Eq. (5.3.7) for two-dimensional Cartesian coordinates and as given in Eq. (5.3.12) for three dimensions. In the following energy balance, the area normal to the radius r is approximated as $R \, \Delta\theta$, not as $(R - \Delta r/2) \, \Delta\theta$, for simplicity in the result. There are five conductive paths which affect the region of t_s. They are from t_1, t_2, t_3, t_4, and t_5. In addition, surface flux q_s'' and convection loading at the surface are

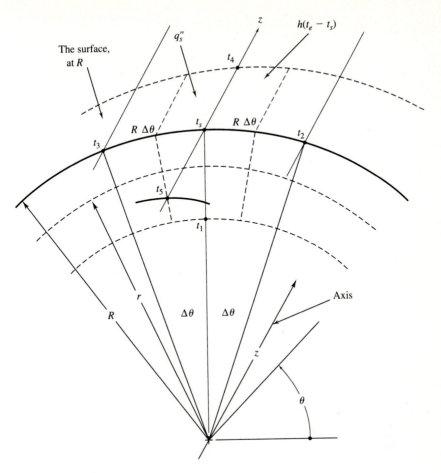

FIGURE 9.4.2
Numerical formulation at curved bounding surfaces of a cylindrical region.

included, as $q_s'' R \Delta\theta \Delta z$ and $h(t_e - t_s) R \Delta\theta \Delta z$. The distributed source q_s''' arises in the volume element $R \Delta\theta \Delta z \Delta r / 2$. The transient term on the right side of the following equation is for a thermal capacity of $\rho c R \Delta\theta \Delta z \Delta r / 2$. The forward-difference energy balance is

$$\frac{kR \Delta\theta \Delta z}{\Delta r}(t_1^n - t_s^n) + \frac{k \Delta r \Delta z}{2R \Delta\theta}(t_2^n - t_s^n) + \frac{k \Delta r \Delta z}{2R \Delta\theta}(t_3^n - t_s^n)$$

$$+ \frac{kR \Delta\theta \Delta r}{2 \Delta z}(t_4^n - t_s^n) + \frac{kR \Delta\theta \Delta r}{2 \Delta z}(t_5^n - t_s^n) + q_s'' R \Delta\theta \Delta z$$

$$+ hR \Delta\theta \Delta z(t_e^n - t_s^n) + \frac{q_s''' R \Delta\theta \Delta z \Delta r}{2} = \frac{\rho c R \Delta\theta \Delta z \Delta r}{2 \Delta\tau}(t_s^{n+1} - t_s^n)$$

$$(9.4.5)$$

Dividing through by $kR \, \Delta\theta \, \Delta z / \Delta r$ yields

$$(t_1^n - t_s^n) + \frac{(\Delta r)^2}{2(R \, \Delta\theta)^2}(t_2^n + t_3^n - 2t_s^n) + \frac{(\Delta r)^2}{2(\Delta z)^2}(t_4^n + t_5^n - 2t_s^n)$$

$$+ \frac{q_s''' \, \Delta r}{k} + \frac{h \, \Delta r}{k}(t_e^n - t_s^n) + \frac{q_s''' \, \Delta r^2}{2k} = \frac{\Delta r^2}{2\alpha \, \Delta\tau}(t_s^{n+1} - t_s^n) \quad (9.4.6)$$

This result indicates grid Biot and Fourier numbers as $B = h \, \Delta r/k$ and $F = \alpha \, \Delta\tau/(\Delta r)^2$. The second term in the preceding equation may be simplified if $\Delta r = R \, \Delta\theta$, as for "square" elements in each z plane. However, the coefficient of the third term remains, since Δz is independent of r. For $\Delta r = R \, \Delta\theta$, and neglecting q_s'' and q_s''', the stability limit based on this surface region equation is

$$F \leq 1/\left[4 + 2B + (\Delta r)^2/(\Delta z)^2\right] \quad (9.4.7)$$

Since all terms in the denominator are positive, the stability limit is reduced by each effect.

NONCONFORMING BOUNDARIES. These arise as conduction region bounding surfaces which are not described in terms of a fixed value of any one of the coordinates which describe the conduction region. A two-dimensional example is shown in Fig. 9.4.3(a), in Cartesian coordinates. The equispaced grid points 1 and 2 lie beyond the surface of the conduction region. Points 0, 3, and 4 lie within. The numerical method, applied internal to the conduction region, uses equally spaced grid points to approximate second derivatives, in both steady-state and transient processes. However, the surface bounding condition, t_s in Fig. 9.4.3(a), applies at points A and B, as t_A and t_B, instead. Therefore, they do not apply for the calculation of t_0, in terms of t_1, t_2, t_3, and t_4.

The temperature levels from which t_0 is to be determined, in terms of the four surrounding temperatures in the internal-region second-derivative estimate, are t_3, t_4, t_A, and t_B. Consider the x direction, in terms of the temperature levels t_3, t_0, and t_A. These are related to the first and second derivatives $\partial t/\partial x$ and $\partial^2 t/\partial x^2$ as follows.

The expansions of t_A and t_3 around t_0 are

$$t_A - t_0 = \left(\frac{\partial t}{\partial x}\right)_0 \lambda_x \, \Delta x + \left(\frac{\partial^2 t}{\partial x^2}\right)_0 \frac{(\lambda_x \, \Delta x)^2}{2} + O\left[(\Delta x)^3\right] \quad (9.4.8a)$$

and

$$t_3 - t_0 = -\left(\frac{\partial t}{\partial x}\right)_0 \Delta x + \left(\frac{\partial^2 t}{\partial x^2}\right)_0 \frac{(\Delta x^2)}{2} + O\left[(\Delta x)^3\right] \quad (9.4.8b)$$

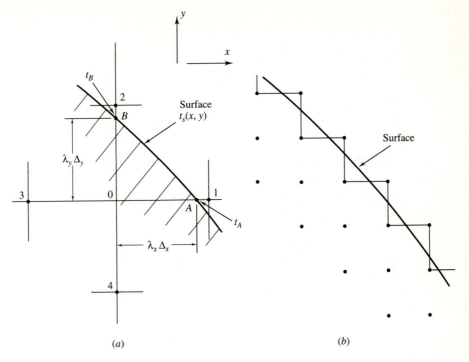

FIGURE 9.4.3
Two methods of accommodating a temperature boundary condition along a nonconforming boundary.

The estimates of the first and second derivatives at location 0 are

$$\left(\frac{\partial t}{\partial x}\right)_0 = \frac{2}{\lambda_x(1 + \lambda_x)\,\Delta x}\left[t_A - t_0(1 - \lambda_x^2) - t_3\lambda_x^2\right] + O\left[(\Delta x)^2\right] \quad (9.4.9a)$$

$$\left(\frac{\partial^2 t}{\partial x^2}\right)_0 = \frac{2}{\lambda_x(1 + \lambda_x)(\Delta x)^2}\left[t_A - t_0(1 + \lambda_x) + t_2\lambda_x\right] + O(\Delta x) \quad (9.4.9b)$$

indicating lower accuracy for the second derivative.

A similar procedure applies in the y direction, to determine $\partial^2 t/\partial y^2$, the other term in $\nabla^2 t = 0$. The result, given in the following equation, expresses the unknown t_0 in terms of t_3 and t_4 and the specified boundary condition temperature levels t_A and t_B. Thereby, the calculation of the internal $t_{i,j}$ values depends on the imposed surface conditions.

$$2\left(\frac{1}{\lambda_x} + \frac{1}{\lambda_y}\right)t_0 = \frac{2t_A}{\lambda_x(1 + \lambda_x)} + \frac{2t_B}{\lambda_y(1 + \lambda_y)} + \frac{2t_3}{1 + \lambda_x} + \frac{2t_4}{1 + \lambda_y} \quad (9.4.10)$$

For $\lambda_x = \lambda_y = 1$, t_0 becomes the average of t_A, t_B, t_3, and t_4 as required. See Anderson et al. (1984) and Özisik (1980) for additional considerations of curved boundary effects.

For an imposed heat flux or convective condition applied at the surface, the temperatures $t_{s,A}$ and $t_{s,B}$ are not known. Instead the normal temperature gradient into the surface becomes the surface condition. The analysis is much more elaborate, as set forth by Forsythe and Wasow (1960) and Fox (1962). See also Allen (1954).

A MODIFIED BOUNDARY. A nominally simpler procedure is to modify the local surface region boundary, as seen in Fig. 9.4.3(b) (see page 551). A specified temperature boundary condition would then be applied along this stepped surface instead. The surface then appears as finned. This implies more effective base surface heat transfer. However, the triangular region results in both increased and decreased resistance across the base area of the triangle. In any case, the modified boundary must continue to conform to the actual surface boundary, in an average way. For greater or lesser angles than that shown, and for curving surfaces, the pitch must change along the surface.

9.5 VARIABLE PHYSICAL PROPERTIES

There are many applications in which the thermal conductivity k and specific heat c vary substantially over the range of imposed temperature conditions. They may also depend on location in an inhomogeneous region. For discussion, Fig. 9.5.1 shows a plane layer, initially at t_i, subject to surface temperatures t_1 and t_2. These may be time dependent. The general conduction equation considered here is

$$\nabla \cdot k \, \nabla t + q''' = \rho c \frac{\partial t}{\partial \tau} \tag{9.5.1}$$

where k and c are known functions of, for example, temperature and location.

The effect of temperature-dependent conductivity $k(t)$ is first considered in Sec. 1.2.3, in terms of the Kirchhoff transformation, to a new dependent variable $V(t)$. Thereby, the nonlinearity, on the left of the preceding equation, is removed. Then the term on the right becomes $(\rho c/k)\partial V/\partial \tau$. The nonlinearity in transients is then completely removed if the temperature dependences of c and k are directly proportional. The term on the right does not appear in steady state. Recall that ρ in Eq. (9.5.1) is formally required to remain a constant at each point in the region, since no convective terms are included in the derivation of the usual conduction equations. This is the constant-strain hypothesis. One-dimensional temperature- and location-dependent conductivity is analyzed in Sec. 2.2. One-dimensional mass diffusion transients are treated in Sec. 6.3.2, for $D = D(C)$. These are analogous to heat conduction, with $k(t)$.

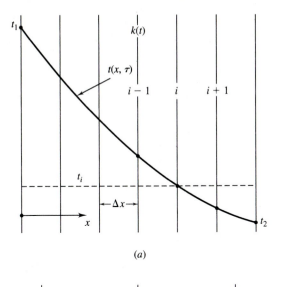

(a)

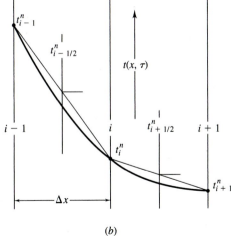

(b)

FIGURE 9.5.1
A conduction transient in a region having k and c possibly both variable with x, y, z, t, and τ: (a) typical location i; (b) a heat flux representation.

Section 9.5.1 concerns the application of numerical analysis to transient conduction, with variable properties. One-dimensional processes, for $k(t)$, are then formulated by the Boltzmann transformation, in Sec. 9.5.2.

9.5.1 Numerical Procedures

The numerical model of Eq. (9.5.1) must accommodate several effects. These relate to the nonlinear conduction term, the distributed source q''', and the specific heat c. The general formulation for the source term is $q''' = q'''(x, y, z, t, \tau)$. The local instantaneous value at time level n is, as before, $q'''^n_{i,j,k}$, at location i, j, k. Within the conditions of the derivation of Eq. (9.5.1), in Sec. 1.2.1, the local instantaneous value of c may be $c(x, y, z, t, \tau)$. The local value is $c^n_{i,j,k}$. For isotropic conduction k is taken as $k(x, y, z, t, \tau)$, denoted hereafter in this section merely as k. A typical instantaneous conduction term in Eq. (9.5.1) is

$$\frac{\partial}{\partial x}\left[k\frac{\partial t}{\partial x}\right] = -\frac{\partial q''_x}{\partial x} \tag{9.5.2}$$

With k uniform and constant, this term was $k\,\partial^2 t/\partial x^2$ and the local estimate of $\partial^2 t/\partial x^2$ was, for example, the three consecutive point central difference, in Eq. (9.2.5), as

$$\left(\frac{\partial^2 t}{\partial x^2}\right)^n_{i,j,k} \approx \frac{t^n_{i+1,j,k} + t^n_{i-1,j,k} - 2t^n_{i,j,k}}{(\Delta x)^2} \tag{9.5.3}$$

For $\nabla \cdot k \nabla t$ in Eq. (9.5.1), the estimate is, instead, of the gradient of the local heat flux component q''_x, and also of q''_y and q''_z, in general. That is, the estimate of the local temperature gradient, $\partial t/\partial x \approx \Delta t/\Delta x$, is first to be multiplied by an appropriate estimate of the local conductivity k. Then this product, as $-q''_k = k(\Delta t/\Delta x)$, is to be estimated, as $-(\Delta q''_x/\Delta x)$, to determine the net conduction effect.

A typical grid location i is shown in Fig. 9.5.1(a) (see page 553), for a one-dimensional transient. The adjacent locations are $i - 1$ and $i + 1$. The transient distribution shown, $t(x, \tau)$, is taken to apply at time level n. At that time, t^n_{i-1}, t^n_i, and t^n_{i+1} are the values at the three consecutive grid locations. Three simple estimates of the local temperature gradient are apparent

$$\left(\frac{\partial t}{\partial x}\right)^n_i \approx \frac{t^n_i - t^n_{i-1}}{\Delta x}, \frac{t^n_{i+1} - t^n_i}{\Delta x}, \text{ and } \frac{t^n_{i+1} - t^n_{i-1}}{2\,\Delta x} \tag{9.5.4}$$

The local flux, as $-q''_x$, related to point i, may be in terms of k evaluated at any of the three grid locations, as $k(t^n_{i-1})$, $k(t^n_i)$, or $k(t^n_{i+1})$. It also might be taken as some average over either of the regions of thickness Δx, denoted for simplicity as $k(t^n_{i-1/2})$ or $k(t^n_{i+1/2})$. All of these procedures relate to an explicit method, in terms of t^n_i, to yield an estimate of t^{n+1}_i.

A direct method of approximation is to evaluate the heat flux in the two layers, between $i + 1$ and i and i and $i - 1$, and take the difference, as seen in Fig. 9.5.1(b) (see page 553). At time level n,

$$-\Delta q''_x = \frac{(k_i + k_{i+1})}{2}\frac{(t^n_{i+1} - t_i)}{\Delta x} - \frac{(k_i + k_{i-1})}{2}\frac{(t_i - t_{i-1})}{\Delta x} \tag{9.5.5}$$

Then $-\Delta q''_x/\Delta x$ is determined, as

$$
\frac{\partial}{\partial x}\left[k\frac{\partial t}{\partial x}\right] \approx -\frac{\Delta q''_x}{\Delta x} = \frac{(k_i + k_{i+1})(t_{i+1} - t_i) - (k_i + k_{i-1})(t_i - t_{i-1})}{2(\Delta x)^2}
$$

$$(9.5.6a)$$

$$
= \frac{(k_i + k_{i+1})t_{i+1} + (k_i + k_{i-1})t_{i-1} - (k_{i+1} + k_{i-1} + 2k_i)t_i}{2(\Delta x)^2}
$$

$$(9.5.6b)$$

$$
= \frac{k_{i+1/2}t_{i+1} + k_{i-1/2}t_{i-1} - (k_{i+1/2} + k_{i-1/2})t_i}{(\Delta x)^2}
$$

$$(9.5.6c)$$

where the two mid-interval averages of k are written as $k_{i+1/2}$ and $k_{i-1/2}$ in the last equation, for simplicity.

The resulting one-dimensional form of Eq. (9.5.1) is then written, in explicit form, as

$$
k^n_{i+1/2}t^n_{i+1} + k^n_{i-1/2}t^n_{i-1} - (k^n_{i+1/2} + k^n_{i-1/2})t^n_i + (\Delta x)^2 q'''^n_i
$$

$$
= \frac{\rho c^n_i(\Delta x)^2}{\Delta \tau}(t^{n+1}_i - t^n_i)
$$

$$(9.5.7)$$

where c^n_i is the instantaneous local value of the specific heat. The two- and three-dimensional forms, with indices i, j, and i, j, k, are apparent. The additional terms are $-\Delta q''_y/\Delta y$ and $-\Delta q''_z/\Delta z$. Multiple grid sizes Δx, Δy, and Δz are easily accommodated.

The simplest implicit form of Eq. (9.5.7) is obtained by changing the superscripts on the temperature and conductivity factors, on the left-hand side, from n to $n + 1$. However, k, q'''_i, and c_i each may be dependent on temperature. Then they are known at time n, and not at $n + 1$, for the time step calculation from n to $n + 1$. They are evaluated as k^n, q'''^n, and c^n_i. See Jaluria and Torrance (1986) for an extrapolation procedure for the conductivities, which avoids this need to use an explicit method.

The preceding procedure in Eq. (9.5.5) amounts to averaging k_{i-1} and k_i, for example, as a representative estimate of the conductive property of the layer of material between locations $i - 1$ and i. That is, k_{i-1} characterizes the half-layer between $i - 1$ and $i - 1/2$ and k_i represents the half-layer between $i - 1/2$ and i. The inverses of k_{i-1} and k_i are resistivities. These resistances may be summed instead, as the total resistance of the two half-layers between $i - 1$ and i as

$$
\frac{1}{2k_{i-1}} + \frac{1}{2k_i} = \frac{1}{2}\left(\frac{1}{k_{i-1}} + \frac{1}{k_i}\right) = \frac{1}{k_{i-1/2}}
$$

$$(9.5.8)$$

where $1/k_{i-1/2}$ denotes the resistance of the whole layer. The conductances of the two whole layers adjacent to i, from Eq. (9.5.8), are

$$k_{i-1/2} = \frac{2k_{i-1}k_i}{k_{i-1} + k_i} \quad \text{and} \quad k_{i+1/2} = \frac{2k_i k_{i+1}}{k_i + k_{i+1}} \tag{9.5.9}$$

These harmonic mean values, two times the product divided by the sum, then replace the arithmetic means used in Eq. (9.5.5). Recall the similar result in Sec. 4.5.2 for a composite barrier. See Patankar (1980), where a similar result for unequal grid spacing is also given. See also the formulation and calculation procedure of Myers (1971) for variable conductivity and also of surface area over a region.

9.5.2 A Transformation for a Semiinfinite Solid

As discussed in Sec. 4.2.3, this geometry also adequately represents early response in finite regions. The relevant equation is

$$\frac{\partial}{\partial x}\left[k(t)\frac{\partial t}{\partial x}\right] = \rho c\frac{\partial t}{\partial \tau} \tag{9.5.10}$$

The Kirchhoff transformation given in Eq. (1.2.14) reduces this to linear form only for $k(t)$ and $c(t)$ proportional to each other, since ρ must remain constant at each location in the conduction region.

The Boltzmann transformation discussed here applies to Eq. (9.5.10). The following analysis assumes that c remains uniform and constant over the conduction region. This transformation, also discussed in Sec. 6.3.2 for variable $D(C)$, combines the space and time variables x and τ as $x/\sqrt{\tau} = \eta$. This classical formulation provides general guidance in analyzing one-dimensional response in many circumstances.

The bounding conditions in x and τ, and the transformed relation from Eq. (9.5.10), in terms of $\phi(x, \tau) = [t(x, \tau) - t_i]/[t(0, \tau) - t_i]$, are

$$\phi(x, 0) = 0 \quad \text{for } x > 0 \tag{9.5.11a}$$

$$\phi(\infty, \tau) = 0 \quad \text{for } \tau > 0 \tag{9.5.11b}$$

$$\phi(0, \tau) = 1 \quad \text{for } \tau > 0 \tag{9.5.11c}$$

$$\frac{d}{d\eta}\left[\alpha\frac{d\phi}{d\eta}\right] + \frac{\eta}{2}\frac{d\phi}{d\eta} = 0 \tag{9.5.11d}$$

where α is a function of t as $k(t)/\rho c$. The two conditions in Eq. (9.5.11a and b) result in the single condition $\phi = 0$ as $x \to \infty$, both at $\tau = 0$ and $\tau > 0$. Therefore, the two conditions of $\phi(\eta)$ required for Eq. (9.5.11d) are

$$\phi(0) = 1 \quad \text{and} \quad \phi \to 0 \quad \text{as } \eta \to \infty \tag{9.5.12}$$

The following additional step transforms Eq. (9.5.11d) into a form suitable for an iterative determination of $\phi(\eta)$:

$$f(\eta) = \frac{1}{\alpha}\frac{d\alpha}{d\phi}\frac{d\phi}{d\eta} + \frac{\eta}{2\alpha} \tag{9.5.13}$$

with the result, for $\phi(\eta)$, of

$$\frac{d^2\phi}{d\eta^2} + f(\eta)\frac{d\phi}{d\eta} = 0 \tag{9.5.14}$$

This form suggests an approximate iterative method of determining $\phi(\eta)$. An initial guess of $\phi(\eta)$, and of $\alpha = \alpha_0$, begins the procedure. An initial guess might be $\phi_i = \mathrm{erfc}(x/2\sqrt{\alpha_0\tau})$. This is the transient solution of Eq. (9.5.11d), for α constant. See Özisik (1980), Sec. 11-3, for the procedure. Also discussed there are several forms of $k(t)$ found by Fujita (1952) to result in analytical solutions of Eq. (9.5.1). These kinds of results are often valuable for estimates of remote region behavior, in various applications.

9.6 THE FINITE-ELEMENT METHOD

This method is very widely used in structural and solid-body mechanics. It is in increasingly wider use in analyzing steady-state and transient heat conduction and mass diffusion. It offers a relatively simple alternative to finite-difference techniques, when the diffusion region is irregular and when more complicated kinds of boundary conditions apply. See, for example, Fig. 9.6.1.

Each finite element is a subvolume of the continuous region in which diffusion occurs. These volumes collectively include the whole volume, to an approximation. Many finite-element shapes may be used (see Fig. 9.6.2). The most common uses are triangles and tetrahedrons, in two and three dimensions, respectively. If these are chosen sufficiently small, they may also accurately represent the regions immediately adjacent to an irregular region boundary as well. The governing conservation equation in integral form is satisfied by applying it to each of the finite elements. The various methods used, in practice, apply in the form of an integral representation of the relevant differential equation, over the region. Common methods are an energy balance, weighted residuals, and variational calculus. See Findlayson (1972), Mitchell and Wait (1977), and Huebner and Thornton (1982) as readily available references.

The procedure in a specific analysis is to subdivide the region into finite elements. Interpolation functions are then taken for the individual elements. These functions represent the element spatial temperature solution, for the number of nodes in each element. The equations for the individual elements are then formulated. These are assembled to represent the whole region. The resulting global equation gives rise to a system of algebraic equations which is solved. For a steady-state process, one set of conditions, over the region, is the solution. In transients, the global formulation provides a set of equations for determining the time dependence.

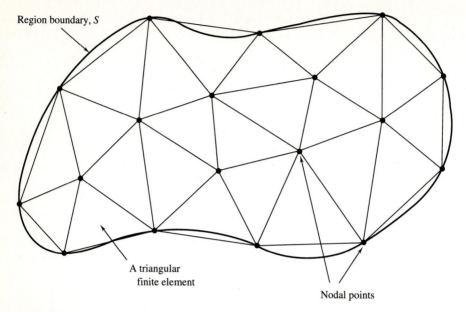

FIGURE 9.6.1
A two-dimensional region, represented by nodal points and triangular finite elements.

The following section concerns the general formulation. This includes representing the regions by finite elements. Such elements are shown for a two-dimensional region in Fig. 9.6.1. The triangles in this geometry commonly are also tetrahedrons in three dimensions. A solution formulation is given in the following section. Section 9.6.2 formulates analysis for one- and two-dimensional regions, for steady-state conduction.

9.6.1   Finite-Element Formulation

A two-dimensional region, subdivided as in Fig. 9.6.1, has a volume V, per unit length normal to the figure. However, the differential equation considered here is of general three-dimensional form, as

$$\nabla \cdot k \nabla t + q''' = \rho c \frac{\partial t}{\partial \tau} \tag{9.6.1}$$

or

$$\rho c \frac{\partial t}{\partial \tau} - \nabla \cdot k \nabla t - q''' = 0 \tag{9.6.2}$$

Equation (9.6.2) is summed over the volume to yield

$$\int_V \left[\rho c \frac{\partial t}{\partial \tau} - \nabla \cdot k \nabla t - q''' \right] dV = 0 \tag{9.6.3}$$

This is zero over V and also over any other volume subdivision ΔV, since the integrand applies at every location in the continuum conduction region. Recall also the assumption implicit in Eq. (9.6.1), that the density at each point in the volume remains constant throughout any process. All other properties may be variable with time.

The finite-element method uses a form of approximation of the temperature field, t_a, for example, as $t_a(x, y, z, \tau) \approx t(x, y, z, \tau)$. This approximation is to satisfy Eq. (9.6.3) over the whole region, as well as over each of the whole set of subregions ΔV which, instead, collectively represent the whole region. The first matter is to divide the region into finite elements. The elements are seen in Fig. 9.6.1, for a two-dimensional region, as triangular elements, each with three nodal points.

The finite-element estimate of, for example, $t(x, y, z, \tau)$, is approximated in terms of the spatial form of the solution at and around the individual nodes. This is done in terms of an interpolation function $p_m(x, y, z)$, as

$$t(x, y, z, \tau) \approx t_a(x, y, z, \tau) = \sum_{m=1}^{M} p_m(x, y, z) F_m(\tau) \qquad (9.6.4)$$

where m denotes the node location, $m = 1, 2, 3, \ldots, M$ and the p_m and F_m apply at each location. Equation (9.6.4) is an expansion of $t_a(x, y, z, \tau)$. It is spatially weighted by p_m, in the summation over all nodal points N. Here $F_m(\tau)$ is the interacting time dependence approximation of $t_a(x, y, z, \tau)$.

The interpolation functions depend on both the types of elements and the number of node points in each element. They are functions which are collectively similar to the anticipated form of the solution. Polynomials are commonly used. Requiring continuity, linear polynomials are used. Then $p_m(x, y, z) = 1$ at node m and becomes zero at immediately neighboring nodes. The amplitudes at each node, $F_m(\tau)$, are linearly interpolated between adjacent nodes. Then the approximation of t in Eq. (9.6.4) varies linearly over each triangle.

THE RESIDUAL EQUATIONS. The approximation, $t_a(x, y, z, \tau)$ in Eq. (9.6.4), is to be substituted in the integral form, Eq. (9.6.3). However, neither the integrand nor the integrals over V or over ΔV, a subvolume, are generally zero, since Eq. (9.6.4) is an approximation. The difference from zero, for either the integrand ΔV or V, is a residual, ϵ. For example,

$$\rho c \frac{\partial t}{\partial \tau} - \nabla \cdot k \nabla t - q''' = \epsilon \qquad (9.6.5)$$

This is similar to the residual which arose at grid points $R_{i,j,k}$, using the finite-difference method. For example, see Eq. (3.9.10). Of course, substituting Eq. (9.6.2) into (9.6.3) yields $\epsilon = 0$, and also

$$\epsilon = \int_V \epsilon \, dV = \int_{\Delta V} \epsilon \, dV = 0 \qquad (9.6.6)$$

since the error ϵ is zero everywhere.

However, substituting the approximating form, t_a, as given in Eq. (9.6.4), does not result in zero residuals, ϵ. That formulation is not an exact solution.

Therefore, the integral in Eq. (9.6.3), over either V or ΔV, results in $\epsilon \neq 0$, in general. A method of approaching a solution is to require that some particular sum of the local values of ϵ, over the whole region, be zero.

Considering nodal points $m = 1, 2, \ldots, M$, a method of weighted residuals is applied as the following integral over the whole conduction region:

$$\int_V W_m \epsilon \, dV = 0 \qquad m = 1, 2, 3, \ldots, M \tag{9.6.7}$$

where the W_m are linearly independent weighting functions. Each integration, for any given m, is over the entire conduction region. That is, there are M simultaneous equations and the residual error is to be zero, as some weighted average.

Many kinds of weighting functions W_m might be used in Eq. (9.6.7). The choice of W_m is substituted into Eq. (9.6.7), along with the formulation of t_a in Eq. (9.6.4), evaluated as in Eq. (9.6.5). The result is a set of equations for the M unknown functions $F_m(\tau)$ in Eq. (9.6.4). These are then solved for the $F_m(\tau)$ and substituted into Eq. (9.6.7), to construct the solution $t_a(x, y, z, \tau)$.

A common choice of the weighting function is the interpolation function, $p_m(x, y, z)$ in Eq. (9.6.4). This choice amounts to the use of what is called the Galerkin method. The global form Eq. (9.6.3), in terms of this approximation, is

$$\int_V p_m(x, y, z) \left[\rho c \frac{\partial t_a}{\partial \tau} - \nabla \cdot k \nabla t_a - q''' \right] dV = 0 \qquad m = 1, 2, 3, \ldots, M \tag{9.6.8}$$

The integral is over the whole region and therefore involves all p_m. The p_m are to be chosen sufficiently small that only the adjacent points are involved, not the more distant ones.

The preceding results are assembled into a formulation for a typical element e consisting of M_e individual nodes. For example, in Fig. 9.6.1, in two dimensions, $M_e = 3$. An element volume is denoted by $(\Delta V)^e$. Interpolation functions p_m are chosen that are nonzero only in the immediate neighborhood of the node m. Then Eq. (9.6.6), for any element, is

$$\int_{(\Delta V)^e} (p_m)^e \left[\rho c \left(\frac{\partial t_a}{\partial \tau} \right)^e - (\nabla \cdot k \nabla t_a)^e - q'''^e \right] dV^e = 0 \qquad m = 1, 2, 3, \ldots, M_e \tag{9.6.9}$$

Note that the p_m^e apply at the nodes of number M_e, while the preceding integral, for any particular element $(\Delta V)^e$, includes only the p_m^e in that element. Also, dV^e is a differential volume element within $(\Delta V)^e$. Therefore, the integration proceeds over the M_e, which are the particular element subset, within $m = 1, 2, 3, \ldots, M$, for the whole region. For the triangular elements in Fig. 9.6.1, $M_e = 3$.

The procedure is to evaluate Eq. (9.6.9) for each of the elements. The overall system of equations is the resulting assembly of the contributions of each element to the solution. Conditions on the p_m are necessary for this assembly to relate to the form of the interpolation function.

THE ELEMENTS. The interpolation functions must satisfy continuity conditions to yield a formulation consistent with the postulate made in expressing t_a as in Eq. (9.6.4). They must also assure convergence to the exact solution as the mesh is refined. These are the compatibility and conformance requirements. These conditions are discussed by Jaluria and Torrance (1986).

Many element geometric forms are shown in Fig. 9.6.2 for one-, two-, and three-dimensional regions, for axisymmetric geometries, and for curved bound-

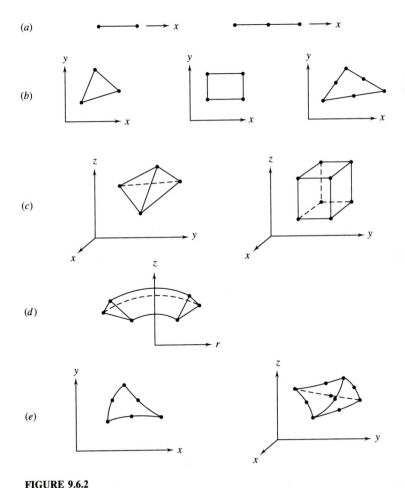

FIGURE 9.6.2
Finite elements: (*a*) one-dimensional line elements; (*b*) two-dimensional elements: three-node triangle, rectangle, six-node triangle; (*c*) three-dimensional elements: tetrahedron, right prism; (*d*) two-dimensional axisymmetric ring element; (*e*) isoparametric elements: triangle, tetrahedron. [From Jaluria and Torrance (1986).]

aries. Elements are specified by their shape, their number, the location of nodes, the variables, and the interpolation functions. Also to be specified are the nature of the imposed conditions at the nodes. The possibilities include temperature level and also a flux loading, as a temperature gradient.

9.6.2   Steady-State Conduction

The preceding formulation is general and applies to both steady-state and transient processes, approximated as in Eq. (9.6.4). The finite-element considerations are similar in steady state, as are the specific procedures and calculations. Equation (9.6.8) in steady state is

$$\int_V p_m[\nabla \cdot k \nabla t + q'''] \, dV = 0 \qquad m = 1, 2, 3, \ldots, M \qquad (9.6.10)$$

where k and q''' may each be location dependent across the conduction region. The function $t(x, y, z)$ is approximated as in Eq. (9.6.4), where $F_m(\tau) = t_m$, the unknown nodal values of t. The boundary conditions are to be satisfied by the interpolation functions p_m. Therefore, the formulation is specific to the particular boundary conditions that the solution must satisfy. For consistency, the first derivatives arising in Eq. (9.6.10) must be continuous.

The order of differentiation in Eq. (9.6.10) is reduced for simplicity by rewriting the conduction term in Eq. (9.6.10) as follows:

$$p_m \nabla \cdot k \nabla t = \nabla \cdot (p_m k \nabla t) - (\nabla p_m) \cdot (k \nabla t) \qquad (9.6.11)$$

Substituting this into Eq. (9.6.10), using the Gauss theorem, gives

$$\int_S p_m(k \nabla t) \cdot \bar{n} \, dS - \int_V (\nabla p_m) \cdot (k \nabla t) \, dV + \int_V p_m q''' \, dV = 0$$

$$m = 1, 2, 3, \ldots, M \qquad (9.6.12)$$

where S denotes a surface integral over the volume of the region and $\bar{n}$ is the unit vector locally normal to the surface. Therefore, the first integral is the sum of the flux over the boundary of the region. This relates the boundary conditions to the conduction volume V. The second term contains only first derivatives. The first term in Eq. (9.6.12) may also be written in terms of local outward surface conduction heat flux $\hat{q}''_s$ as $-\int_S p_m \hat{q}''_s \, dS$.

As in Sec. 9.6.1, the overall formulation in Eq. (9.6.12) may be written instead for a particular element e of volume ΔV^e, having nodes, $m = 1, 2, 3, \ldots, M_e$. Each element may have a nonzero surface heat flux. However, these cancel at element interfaces, in the absence of plane concentrated energy sources, since the two $\bar{n}$ coordinates are opposed. Thus only the region boundary flux appears in the global equations. This flux is accounted for at the external boundaries.

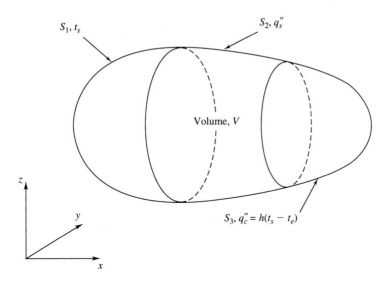

FIGURE 9.6.3
A conduction region with an imposed surface temperature condition over subarea S_1, a flux loading over S_2, and a convection condition to an ambient region at t_e, over S_3.

In order to admit into the formulation a variety of kinds of imposed surface conditions, S is divided into three parts as $S = S_1 + S_2 + S_3$, as idealized in Fig. 9.6.3. The most common conditions, discussed in Sec. 1.4, are: a temperature condition $t_s(x, y, z)$ over S_1; a flux condition $q_s''(x, y, z)$ over S_2; and convection, as $q_c'' = h(t_s - t_e)$, over S_3. Recall that the imposed flux condition includes radiation, as well as an insulating condition, by taking $q_s'' = 0$. In general, the flux is related to the internal temperature gradient, evaluated inward at the surface as $q_s'' = -k(\partial t/\partial n)$. Thus the temperatures over surface regions S_2 and S_3 are initially unknown. The three preceding boundary conditions are written as follows, where q_s'' is positive for an outward flux:

$$t = t_s \qquad \text{on } S_1 \qquad (9.6.13)$$

$$q_s'' = -k\left(\frac{\partial t}{\partial n}\right)_s \qquad \text{on } S_2 \qquad (9.6.14)$$

$$q_c'' = h_c(t_s - t_e) \qquad \text{on } S_3 \qquad (9.6.15)$$

The assembled equations are formed as a combination of Eqs. (9.6.4), and (9.6.12)–(9.6.15), in matrix form, as

$$(A_r)(t) = (F_{q'''}) + (F_{S_1}) + (F_{S_2}) + (F_{S_3}) + (A_{S_3})(t) \qquad (9.6.16)$$

where (A_r) is conductance and (A_{S_3}) arises from the mixed boundary conditions. The components in the global relation, Eq. (9.6.16), are

$$(A_r)_{i,j} = \int_V k \left(\frac{\partial p_i}{\partial x} \frac{\partial p_j}{\partial x} + \frac{\partial p_i}{\partial y} \frac{\partial p_j}{\partial y} + \frac{\partial p_i}{\partial z} \frac{\partial p_j}{\partial z} \right) dV \qquad (9.6.17a)$$

$$(F_{q'''})_i = \int_V p_i q''' \, dV \qquad (9.6.17b)$$

$$(F_{S_1})_i = -\int_{S_1} p_i \hat{q}''_s \, dS \qquad (9.6.17c)$$

$$(F_{S_2})_i = -\int_{S_2} p_i \hat{q}''_s \, dS \qquad (9.6.17d)$$

$$(F_{S_3})_i = \int_{S_3} p_i h t_e \, dS \qquad (9.6.17e)$$

$$(A_{S_3})_{i,j} = \int_{S_3} p_i (h p_j) \, dS \qquad (9.6.17f)$$

where i and j range over $1, 2, 3, \ldots, M$. Equations (9.6.16) and (9.6.17) apply to volume V and its surfaces. They may be used to represent an element when i and j include only the nodal points in the element of number M_e. The preceding formulation, in Eq. (9.6.16) for temperature as (t), contains all of the nodal temperature levels internal to the region and over the boundaries.

The simultaneous relations, in Eq. (9.6.16), are solved by both direct and iteration methods. These include Gauss–Seidel iteration and Gaussian elimination, as discussed in Sec. 9.4. The most efficient methods are commonly those which take advantage of the inherent simplicities of the matrix formulation given previously. The following subsections apply the preceding formulation to one- and two-dimensional processes in steady state, using the Galerkin method.

A PLANE LAYER. A simple example is shown in Fig. 9.6.4(a), where $L = N\Delta x$. For k uniform, where ϕ is the temperature excess ratio, $k \, d^2 t/dx^2 = k \, d^2\phi/dx^2 = 0$, where $\phi = (t - t_2)/(t_1 - t_2)$. The exact solution and heat flux are

$$\phi = 1 - \frac{x}{L} \quad \text{and} \quad q'' = \frac{k}{L}(t_1 - t_2) \qquad (9.6.18)$$

The layer is subdivided into N layers, as for a finite-difference analysis. See Fig. 9.6.4(a). Each of the four elements shown has two nodes. The chosen $p_i(x)$ shown in Fig. 9.6.4(b) have continuity. The approximation of ϕ is

$$\phi = \sum_{i=0}^{N} p_i(x)\phi_i = p_0\phi_0 + p_1\phi_1 + \cdots + p_N\phi_N \qquad (9.6.19)$$

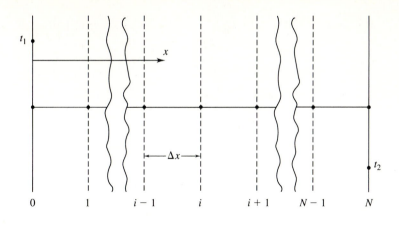

(a)

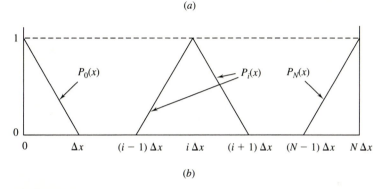

(b)

FIGURE 9.6.4
A steady-state process in a plane layer: (a) the subdivision; (b) the interpolation polynomials.

and Eqs. (9.6.16) and (9.6.17a) simplify to

$$(A_r)(\phi) = (F_{S_1}) \quad \text{and} \quad (A_r)_{i,j} = k\int_0^{N\Delta x} \frac{dp_i}{dx}\frac{dp_j}{dx}\,dx \qquad (9.6.20)$$

In terms of the functions p_i shown in Fig. 9.6.4(b), the results at $i = 0$, i, and $i = N$, in terms of ϕ_i, are

$$\frac{k(\phi_0 - \phi_1)}{\Delta x} = -(\hat{q}_s'')_0 \qquad (9.6.21a)$$

$$\frac{k(\phi_{i+1} + \phi_{i-1} - 2\phi_i)}{(\Delta x^2)} = 0 \qquad (9.6.21b)$$

$$\frac{k(\phi_{N-1} - \phi_N)}{\Delta x} = (\hat{q}_s'')_n = -(\hat{q}_s'')_0 \qquad (9.6.21c)$$

In this simple example, for the conditions in Fig. 9.6.4(a), the temperature

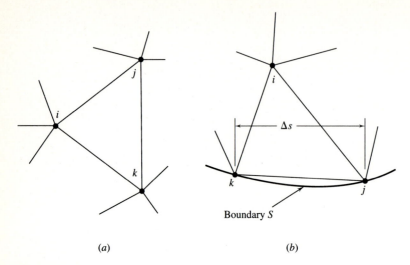

FIGURE 9.6.5
Triangular elements, in two dimensions: (a) internal to the region; (b) adjacent to the surface.

conditions are specified at $x = 0$ and L, that is, at $i = 0$ and N. Therefore, the first and last conditions in Eq. (9.6.21) do not apply. Having a solution for the internal node points throughout, as at locations $i\,\Delta x$, the first and last equations are used to determine $(q''_s)_0$ and $(q''_s)_N$.

A TWO-DIMENSIONAL REGION. The triangular element subdivision is shown in Fig. 9.6.1. A typical completely internal and a surface element are shown in Fig. 9.6.5, where the nodes are i, j, k. The Poisson equation in x and y, with possibly spatially dependent conductivity k, is

$$\frac{\partial}{\partial x}k\frac{\partial t}{\partial x} + \frac{\partial}{\partial y}k\frac{\partial t}{\partial y} + q''' = 0 \qquad (9.6.22)$$

Here $N_e = 3$ and the equation is

$$t^{(e)}(x, y) = \sum_{i=1}^{3} p_i^{(e)}\phi_i^{(e)} \qquad (9.6.23)$$

The polynomials, in the manner of Huebner and Thornton (1983), are

$$p_i^{(e)} = \frac{1}{2\Delta}(a_i + b_i x + c_i y) \qquad (9.6.24a)$$

$$p_j^{(e)} = \frac{1}{2\Delta}(a_j + b_j x + c_j y) \qquad (9.6.24b)$$

$$p_k^{(e)} = \frac{1}{2\Delta}(a_k + b_k x + c_k y) \qquad (9.6.24c)$$

Also

$$a_i = x_j y_k - x_k y_j \qquad b_i = y_j - y_k \qquad c_i = x_k - x_j \qquad (9.6.24d)$$

where the other values of a, b, and c are generated by permuting the subscripts and Δ is the area of the triangle.

The element equations are from Eq. (9.6.17). From Eq. (9.6.24),

$$2\Delta \frac{\partial p_i^{(e)}}{\partial x} = b_i \qquad 2\Delta \frac{\partial p_j^{(e)}}{\partial x} = b_j$$

$$(9.6.25)$$

$$2\Delta \frac{\partial p_i^{(e)}}{\partial y} = c_i \qquad 2\Delta \frac{\partial p_j^{(e)}}{\partial y} = c_j$$

the conductance matrix is

$$(A_k)_{i,j}^{(e)} = \int_{\Delta V^e} k \left(\frac{b_i b_j}{4\Delta^2} + \frac{c_i c_j}{4\Delta^2} \right) dx \, dy \qquad (9.6.26)$$

when k and q''' are uniform over the conduction region, or are approximated as such

$$(A_k)_{i,j}^{(e)} = \frac{k}{4\Delta} (b_i b_j + c_i c_j) \qquad (9.6.27)$$

$$(F_{q'''})_i^{(e)} = \int_{V^{(e)}} p_i q'''^{(e)} \, dx \, dy = -\frac{q'''^{(e)}\Delta}{3} \qquad (9.6.28)$$

The other four matrices in Eq. (9.6.17) relate to the three kinds of surface conditions. They apply to the elements at the boundaries of the conduction region. The example shown in Fig. 9.6.5(b) has a boundary segment approximated as Δs in length. For an imposed temperature t_s, or for a flux condition q_s'', Eqs. (9.6.17c) and (9.6.17d), become

$$(F_{S_1})_i^{(e)} = -\hat{q}_s'' \, \Delta s/2 = (F_{S_2})_i^{(e)} \qquad (9.6.29)$$

A convection condition results in

$$(F_{S_3})_i^{(e)} = ht_e \, \Delta s/2 \quad \text{and} \quad (A_{S_3})^{(e)}(t)^{(e)} = h \, \Delta s \begin{bmatrix} 1/3 & 1/6 & 0 \\ 1/6 & 1/3 & 0 \\ 0 & 0 & 0 \end{bmatrix} \begin{bmatrix} t_i \\ t_j \\ t_k \end{bmatrix}$$

$$(9.6.30)$$

See Jaluria and Torrance (1986) concerning calculation procedures for particular inputs arising in a specific application.

TRANSIENT DIFFUSION. The finite element method is also applied to transient processes, using a Galerkin method for the spatial integrations. Conduction region boundary conditions may again be those of a specified temperature, an imposed surface flux or a convective condition. A combination of these, over the constituent surface of a volume may also be accommodated. Recall Fig. 9.6.3, pg. 563. Transient response is given in terms of conduction and a distributed source, as

$$\frac{\partial t}{\partial \tau} = \alpha \left(\frac{\partial t^2}{\partial x^2} + \frac{\partial t^2}{\partial y^2} \right) + \frac{q'''}{\rho c} \tag{9.6.30}$$

Calculations proceed from the specified initial condition.

Several methods of integration have been used. A matrix system calculation proceeds through time, and over space, commonly using a finite difference method through time. The use of finite elements in time, and modal superposition, are also used. A number of other procedures have also been applied. See Jaluria and Torrance (1986), along with the other references given.

PROBLEMS

9.2.1. Derive an approximation for $\partial t/\partial x$ valid at the end of a three-grid-point interval to $O(\Delta x)^2$. Is there an approximation for $\partial^2 t/\partial x^2$ to the same order of accuracy at this location?

9.2.2. Derive approximations for $\partial t/\partial x$ and $\partial^2 t/\partial x^2$ valid at the middle of a five-grid-point interval.
  (*a*) Obtain expressions that are as accurate as possible for such an interval.
  (*b*) Clearly outline how you would construct the last difference, using a combination of three-point central differences.
  (*c*) Give the truncation errors.

9.2.3. Obtain an expression for $(\partial t/\partial x)_{i,j}$ that is fourth-order accurate.

9.2.4. Consider a nonuniform grid distribution with $(\Delta x)_{i+1,j}/(\Delta x)_{i,j} = a < 1.0$. Obtain finite-difference approximations for the first and the second derivatives, with respect to x, at the grid point (i, j).

9.2.5. A semiinfinite solid transient solution is given in Sec. 4.2.3 as $\phi = \text{erfc } \eta$. Consider the truncation errors in the numerical central-difference formulation for $\partial^2 t/\partial x^2$ and the forward difference in time.
  (*a*) Determine the first term of the TE for approximate choices of $\Delta \tau$ and Δx, in terms of F.
  (*b*) For the time derivative, determine the ratio of the first omitted term compared to the retained one.
  (*c*) Repeat part (*b*) for the spatial derivative.
  (*d*) Find the ratio of these two errors.
  (*e*) Repeat part (*d*) for the backward difference in time.

9.2.6. Make the same determination as in Prob. 9.2.5, parts (*b*), (*c*), and (*d*), for the plane layer transient solution in Eq. (4.2.5), at both $x = 0$ and $x = L/2$.

9.2.7. From the solution referred to in Prob. 9.2.5, calculate the ratio of the first omitted terms in the time and spatial derivatives, for the condition of the maximum permissible time step for stability.

9.2.8. For two-dimensional steady-state conduction in Cartesian coordinates, and $\Delta x = \Delta y$, consider the first component of the truncation errors.
(a) Compare the total error to the retained terms in the three-point second-derivative estimates.
(b) For the conduction field in a long rectangular rod, as in Eq. (3.1.11), calculate the values of the two first truncation terms.
(c) Determine the ratio of the results in parts (a) and (b).

9.2.9. Consider the first truncation error contribution in the geometry for the long square rod condition in Fig. 3.1.2, with a uniform distributed internal heat source q'''. Using the solution in Eq. (3.6.7), find the error at $x = L$ and $y = 0$, compared to the solution, as a ratio.

9.2.10. For a plane region transient with a uniform distributed source q''', determine the first truncation errors, for the conditions in Fig. 4.2.14(a). For $\partial^2 t / \partial x^2$ use the simplest three-point estimate for the forward and backward estimates of $\partial t / \partial \tau$. Compare the sum of the errors with the solution. Make this comparison at $x = L/2$, in terms of F.

9.2.11. Consider the magnitude of the first two truncated terms in the forward-difference estimate of $\partial t / \partial \tau$.
(a) Form the ratio of the second to the first term.
(b) Evaluate this ratio, for the semiinfinite solid solution referred to in Prob. 9.2.5, in the simplest terms.

9.2.12. Repeat Prob. 9.2.11 for the first two truncated terms in the three-point central-difference estimate of $\partial^2 t / \partial x^2$.

9.2.13. Make the determinations as in Probs. 9.2.11 and 9.2.12, for the plane layer transient solution in Eq. (4.2.5), at both $x = 0$ and $x = L/2$.

9.2.14. Consider a long, cylindrical region as in Fig. 3.3.1(b), with the conditions shown.
(a) Determine the central-difference estimates of the second derivatives, and the forward estimate of the first derivative.
(b) Write each of these results as a Taylor series expansion.
(c) Compare the first truncation error in $\partial t / \partial r$, as a ratio of the value to that of each of the other first errors, that is, of the $\partial^2 t / \partial r^2$ and $\partial^2 t / \partial^2 \theta$ terms.

9.2.15. Consider the cylindrical geometry in Fig. 3.3.1(b) and the conditions which led to the solution Eq. (3.3.5).
(a) Determine the first truncation errors for the $\partial t / \partial r$ and $\partial^2 t / \partial r^2$ terms.
(b) Compare these errors to the analytical solution, for values calculated at R and $\theta = 90°$.

9.2.16. Consider the spherical cavity and conditions in Fig. 4.3.3(c).
(a) Determine the first truncation error in $\partial t / \partial r$.
(b) Determine the ratio of this error to the value of $\partial t / \partial r$ from the analytical solution at $r = R$.

9.2.17. The steady-state conduction in a long concrete column is given by Eq. (3.1.11) for $\phi = 0$ on three surfaces and $\phi = 1$ on the fourth. The two temperature levels are 20°C and 0°C.

(a) Calculate the heat flow rate Q across the surface maintained at $\phi = 1$, for $L = H = 1$ m.

(b) Consider calculating $\phi(x, y)$ numerically, using the simplest estimates of $\nabla^2 t = 0$, over the region. Suggest a boundary condition for the two corner point locations on the top face, which have double-valued conditions in the formulation.

(c) Using the postulated condition in part (b), and $\Delta x = \Delta y = L/2$, $L/4$, and $L/8$, calculate the heat flow Q across the top surface, for each of the grid spacings.

(d) Plot Q vs. Δx for the three results and comment on the reasons for the differences, as well as for the different result in part (a).

9.2.18. For the column and conditions in Prob. 9.2.17:

(a) Calculate Q for $\Delta x = \Delta y = L/8$, from the simplest formulation.

(b) Repeat the calculation using the one-sided estimate in Eq. (9.2.21) for both $\partial^2 t/\partial x^2$ and $\partial^2 t/\partial y^2$.

(c) Calculate the heat flux at $x = L/2$ and $y = L$.

(d) Plot the temperature distributions from both results, and from the analytical solution, as $t(L/2, y)$.

(e) Comment on the differences.

9.3.1. For the two-dimensional transient diffusion process, obtain the constraint, if any, on the time step for the Crank–Nicolson explicit method, using the von Neumann analysis.

9.3.2. Consider the Barakat–Clark ADE formulation for a one-dimensional transient.

(a) Indicate each of the first truncation errors which arise in the determination of t_i^{n+1}, as Eq. (9.3.16).

(b) Evaluate this error, as a ratio to the retained term, in the estimate of p_i.

(c) Repeat part (b) for the time derivative.

9.3.3. Consider the truncation error in the surface point convective condition as formulated in Sec. 4.2.2.

(a) Determine the first error in the analytical boundary condition, compared to the approximation in the numerical form used in the heat balance in Eq. (4.9.12).

(b) Repeat this determination for the comparable two-dimensional numerical formulation in Eq. (5.3.7).

9.3.4. For a one-dimensional transient in the plane layer, initially at $24°C$, shown in Fig. 4.2.1(a), several numerical formulations are to be compared with the analytical solution of the transient response after the surface temperature is reduced to $4°C$. The three formulations are the three-point explicit, the Crank–Nicolson, and the DuFort–Frankel methods. The layer is a brick wall of 10 cm thickness. Use $\Delta x = L/8$.

(a) Determine the maximum value of F permissible among these three methods.

(b) Formulate the transient response in the wall, in terms of $\phi = (t - t_0)/(t_i - t_0)$.

(c) Calculate the internal response for the time into the transient in which one of the results, among the three methods, becomes less than $14°C$ at the midplane.

(d) Determine the analytical solution at each grid point at this time.

(*e*) Plot all four distributions of $t(x, \tau)$ at this time, over the range $x = 0$ to L.

(*f*) Determine the rms difference across the region for each of the three numerical results at this time, compared to the analytical solution.

(*g*) Comment on reasons for the differences.

9.3.5. Repeat the considerations for the circumstance in Prob. 9.3.4, if the changed boundary condition is not from t_i to t_0, but a convection condition suddenly imposed with an ambient temperature of 4°C. The Biot number for the convection process is taken as 9. Use the formulation in Fig. 4.2.2 for the analysis. Use the Heisler charts as the analytical solution for comparison.

9.3.6. The semiinfinite solid solution in Eq. (4.2.15) is to be compared with the three numerical method results referred to in Prob. 9.3.4. The plane region is a 1-m-thick concrete wall at 20°C, with soil against one side. The temperature of the other side suddenly drops to 0°C and remains at that level.

(*a*) Determine the time τ_c at which the effect of the temperature change on the exposed face becomes 0.2°C at the concrete–soil interface on the other side. Assume that the soil properties are the same as those of the concrete. Use the analytical solution.

(*b*) Make numerical calculations by each of the three methods, for $0 \leq \tau = \tau_c$ and $\Delta x = 10$ cm, and plot all four results vs. x, at $\tau = \tau_c$.

(*c*) Compare an integral method result with those in part (*b*).

(*d*) Comment on any large differences.

9.4.1. A long steel tube of inside and outside diameters of $D_i = 12$ and $D_0 = 14$ cm has the inner surface maintained at 100°C by two-phase water flowing on the inside. A net thermal radiation loading on the outside surface amounts to a surface flux $q_s''(\theta)_2$. It varies around the surface as $q_s'' = q_{s,a}'' \cos \theta$, where $q_{s,a}'' = 20$ W/cm^2. There are no other outside surface energy effects.

(*a*) Formulate this conduction situation analytically, including all boundary conditions.

(*b*) Write the appropriate surface point equation, for steady state.

(*c*) Taking $\Delta r = 2$ cm and $\Delta \theta = 15°$, determine the outside surface temperature variation.

9.4.2. For two-dimensional time-dependent conduction in a rectangular solid region, the heat input at one wall is given as $q_s = \epsilon \sigma (T^4 - T_a^4) - h(T - T_a)$, where q_s is a constant heat flux, T_a is the absolute ambient temperature, h is the constant convective heat transfer coefficient, ϵ is the surface emissivity, and σ is the Stefan–Boltzmann constant. Obtain the finite-difference form of this boundary condition, employing the energy balance for the region represented by a node at the surface.

9.4.3. A long square duct of 40 by 40 cm is insulated by a covering of magnesia insulation which is formed with a concentric cylindrical outside surface of 100 cm diameter. The outside surface temperature will be 30°C. The insulation occupies all of the region between the outside cylindrical surface and the duct walls, which are at 230°C. The conduction loss rate is to be numerically calculated, per unit of duct length, by the method shown in Fig. 9.4.3(*a*).

(*a*) Develop the finite-difference grid for $\Delta x = \Delta y = 10$ cm.

(*b*) Indicate the number of different values of the set (λ_x, λ_y) necessary for the whole region.

(c) Give their approximate values.

(d) Write the set of grid point equations for each distinctive point over the region of calculation.

(e) Determine the heat loss rate, per unit length, q'.

9.4.4. Repeat the preceding calculation, for q', using a modified outside boundary, as in Fig. 9.4.3(b).

(a) Draw the appropriate resulting grid geometry.

(b) Calculate q' for $\Delta x = \Delta y = 10$ cm.

9.4.5. A long copper rod of triangular cross section, and of equal face widths of 6 cm, carries an electric current. The rod is cooled over its surface to 50°C, by an adjacent circulating liquid.

(a) Generate a suitable numerical grid of the kind in Fig. 9.4.3, for calculating the temperature field over the rod cross section. Use $\Delta x = \Delta y = 1$ cm.

(b) Determine the internal energy generation rate q''' which will result in a maximum internal temperature level of 100°C.

(c) Determine the electrical current in the rod for this condition.

9.4.6. Do the calculation requested in Prob. 9.4.5 if the rod cross section is instead one-half of a round rod of radius 3 cm. Use $\Delta x = \Delta y = 1$ cm.

9.5.1. The local conductivity of an $L = 10$ cm thick plane layer, with surface temperatures of 60°C and 30°C, is $k(x) = 2(1 + x/L)$, in W/cm °C.

(a) Using $\Delta x = 1$ cm, determine the steady-state temperature distribution across the layer, and the heat flux, using harmonic mean conductivities.

(b) Compare this result with that resulting from the averaging procedure in Table 2.3.1.

9.5.2. A barrier consists of three layers, a, b, and c, of equal thickness L. It is at 60°C and 30°C on its left and right surfaces. The conductivity of each material varies linearly with temperature as $k_a(t) = t$, $k_b(t) = 2t$, and $k_c(t) = 2t$, in units of W/cm °C.

(a) Using a uniform grid spacing, $\Delta x = L/4 = 1$ cm, calculate the temperature across the barrier, using the appropriate harmonic mean values at the interfaces.

(b) Determine the heat flux across the barrier.

9.5.3. Consider a transient response in a plane barrier, initially at 30°C, with the left surface suddenly raised to 60°C. The right surface remains at 30°C. The conductivity varies as $k(t) = 2[1 + \alpha(t - 30°C)]$, where $\alpha = 1/60$, in units of W/cm °C and °C. The other properties remain uniform.

(a) Formulate the numerical analysis for determining the transient temperature response, using the forward difference.

(b) Analyze the stability criterion for the $k(t)$ variation given.

(c) Calculate the early temperature response in the layer.

REFERENCES

Allen, D. N. de G. (1954) *Relaxation Methods*, McGraw-Hill, New York.

Anderson, D. A., J. C. Tannehill, and R. H. Pletcher (1984) *Computational Fluid Mechanics and Heat Transfer*, Hemisphere, New York.

Barakat, H. Z., and J. A. Clark (1966) *J. Heat Transfer*, **87–88**, 421.

Crank, J., and P. Nicolson (1947) *Proc. Cambridge Philos. Soc.*, **43**, 50.

Douglas, J. (1955) *J. Soc. Ind. Appl. Math.*, **3**, 42.

Douglas, J., and J. E. Gunn (1964) *Numer. Math.*, **6**, 428.

DuFort, E. C., and S. P. Frankel (1953) *Mathematical Tables and Other Aids to Computation*, 7, 153.

Findlayson, B. A. (1972) *The Method of Weighted Residuals and Variational Principles*, Academic, New York.

Forsythe, G. E., and W. R. Wasow (1960) *Finite Difference Methods for Partial Differential Equations*, John Wiley and Sons, New York.

Fox, L. (1962) *Numerical Solution of Ordinary and Partial Differential Equations*, Pergamon, Oxford.

Frankel, S. P. (1950) *Mathematical Tables and Other Aids to Computation*, Vol. 4, page 65.

Fujita, H. (1952) *Text. Res. J.*, **22**, 757 and 823.

Huebner, K. H., and E. A. Thornton (1982) *The Finite Element Method for Engineers*, 2nd, ed., John Wiley and Sons, New York.

Jaluria, Y., and K. E. Torrance (1986) *Computational Heat Transfer*, Hemisphere, New York.

Lapidus, L., and G. F. Pinder (1982) *Numerical Solution of Partial Differential Equations in Science and Engineering*, Wiley-Interscience, New York.

Minkowycz, W. J., E. M. Sparrow, G. E. Schneider, and R. H. Pletcher (1988) *Handbook of Numerical Heat Transfer*, John Wiley and Sons, New York.

Mitchell, A. R., and D. F. Griffiths (1980) *The Finite Difference Method in Partial Differential Equations*, John Wiley and Sons, New York.

Mitchell, A. R., and R. Wait (1977) *The Finite Element Method in Partial Differential Equations*, John Wiley and Sons, New York.

Myers, G. E. (1971) *Analytical Methods in Conduction Heat Transfer*, McGraw-Hill, New York.

O'Brien, G. G., M. A. Hyman, and S. Kaplan (1950) *J. Math. Phys.*, **29**, 223.

Özisik, M. N. (1980) *Heat Conduction*, Wiley-Interscience, New York.

Patankar, S. V. (1980) *Numerical Heat Transfer and Fluid Flow*, Hemisphere, New York.

Patankar, S. V., and B. R. Baliga (1978) *Numer. Heat Transfer*, **1**, 27.

Peaceman, D. W., and H. H. Rachford (1955) *J. Soc. Ind. Appl. Math.*, **3**, 28.

Richardson, L. F. (1910) *Philos. Trans. Roy. Soc. London, Ser. A*, **210**, 307.

Thibault, J. (1985) *Numer. Heat Transfer*, **8**, 281.

Warming, R. F., and B. J. Hyett (1974) *J. Comp. Phys.*, **14**, 159.

Young, D. (1954) *Trans. Amer. Math. Soc.*, **76**, 92.

APPENDIX
A

CONVERSION
FACTORS

A.1 SI TO OTHER UNITS

Quantity	To convert number of	To	Multiply by
Pressure	N/m^2	Pascal = Pa	1
		$dyne/cm^2$	10
		bar	10^{-5}
		atm (phys)	0.9869×10^{-5}
	bar = $10^5 \ N/m^2$	atm (phys)	0.9869
		kgf/cm^2 (atm abs)	1.0197
		mm Hg	750
Density	kg/m^3	g/cm^3	10^{-3}
Specific volume	m^3/kg	cm^3/g	10^3
Heat capacity	$kJ/kg\,^\circ C$	$kcal/kg\,^\circ C$	0.2388
Thermal conductivity	$W/m\,^\circ C$	$cal/cm\ s\,^\circ C$	2.388×10^{-3}
		$kcal/m\ hr\,^\circ C$	0.86
Viscosity	$N\ s/m^2$ (kg/m s)	$g/cm\ s$ (poise)	10
		$kg\ s/m^2$	0.102
Kinematic viscosity	m^2/s	cm^2/s (stokes)	10^4
Surface tension	N/m	$dyne/cm$ (erg/cm^2)	10^3

Source: From W. M. Rohsenow and J. P. Hartnett, Eds. (1973) *Handbook of Heat Transfer*, McGraw-Hill, New York.

A.2 ENGLISH TO METRIC UNITS

Quantity	To convert number of	To	Multiply by
Length	in.	cm	2.540
	ft	m	0.3048
Area	ft^2	m^2	0.0929
Volume	ft^3	m^3	0.02932
Mass	lbm	kg	0.45359
	slugs	kg	14.594
Force	lbf	newtons	4.4482
Density	lbm/ft^3	kg/m^3	16.02
Work	ft lbf	mkgf	0.1383
	hp hr	mkgf	273.700
Heat	Btu	kcal	0.2520
	Btu	joules	1054.35
	Btu	ft lbf	778.26
	kW hr	Btu	3412.75
Specific heat	$Btu/lbm\,°F$	$cal/g\,°C$	1.000
	$Btu/lbm\,°F$	$W\,s/kg\,°C$	4186.7
Pressure	$lbf/in.^2$ (psi)	kgf/cm^2	0.070309
	psi	atm	0.068046
	psi	bars	0.068948
	psi	$dyne/cm^2$	68947.0
	psi	N/m^2	9894.7
Surface tension	lbf/ft	N/m	14.5937

Source: From W. M. Rohsenow and J. P. Hartnett, Eds. (1973) *Handbook of Heat Transfer*, McGraw-Hill, New York.

A.3 HEAT FLUX, q''

To obtain ↓	Multiply number of the following units by			
	$Btu/ft^2\,hr$	W/cm^2	$kcal/hr\,m^2$	$cal/s\,cm^2$
$Btu/ft^2\,hr$	1	3170.75	0.36865	13,277.26
W/cm^2	3.154×10^{-4}	1	1.63×10^{-4}	4.1868
$kcal/hr\,m^2$	2.7126	8600	1	2.778×10^{-5}
$cal/s\,cm^2$	7.536×10^{-5}	0.2389	36,000	1

Source: From W. M. Rohsenow and J. P. Hartnett, Eds. (1973) *Handbook of Heat Transfer*, McGraw-Hill, New York.

A.4   HEAT TRANSFER COEFFICIENT, *h*

To obtain ↓	Multiply number of the following units by			
	Btu / hr ft^2°F	W / cm^2°C	cal / s cm^2°C	kcal / hr m^2°C
Btu/hr ft^2°F	1	1761	7376	0.20489
W/cm^2°C	5.6785×10^{-4}	1	4.186	1.163×10^{-4}
cal/s cm^2°C	1.356×10^{-4}	0.2391	1	2.778×10^{-5}
kcal/hr m^2°C	4.8826	8600	36000	1

Source: From W. M. Rohsenow and J. P. Hartnett, Eds. (1973). *Handbook of Heat Transfer*, McGraw-Hill, New York.

A.5   THERMAL CONDUCTIVITY, *k*

To obtain ↓	Multiply number of the following units by				
	Btu / hr ft °F	W / cm °C	cal / s cm °C	kcal / hr m °C	Btu in./ hr ft^2°F
Btu/hr ft °F	1	57.793	241.9	0.6722	0.08333
W/cm °C	0.01730	1	4.186	0.01171	1.442×10^{-3}
cal/s cm °C	4.134×10^{-3}	0.2389	1	2.778×10^{-3}	3.445×10^{-4}
kcal/hr m °C	1.488	86.01	360	1	0.1240
Btu in./hr ft^2°F	12	693.5	2903	8.064	1

Source: From W. M. Rohsenow and J. P. Hartnett, Eds. (1973). *Handbook of Heat Transfer*, McGraw-Hill, New York.

A.6   VISCOSITY, μ

To obtain ↓	Multiply number of the following units by				
	lbm / ft hr	lbf s / ft^2	centipoise	kgm / m hr	kgf s / m^2
lbm/ft hr	1	116,000	2.42	0.672	23733
lbf s/ft^2	0.00000862	1	0.00002086	0.00000579	0.2048
centipoise†	0.413	47,880	1	0.278	9807
kgm/m hr	1.49	172,000	3.60	1	35305
kgf s/m^2	0.0000421	4.882	0.0001020	0.0000284	1

†100 centipoise = 1 poise = 1 g/s cm = 1 dyne s/cm^2.
Source: From W. M. Rohsenow and J. P. Hartnett, Eds. (1973). *Handbook of Heat Transfer*, McGraw-Hill, New York.

A.7 KINEMATIC VISCOSITY, ν

To obtain ↓	Multiply number of the following units by			
	$\mathbf{ft^2/hr}$	**stokes**	$\mathbf{m^2/hr}$	$\mathbf{m^2/s}$
ft^2/hr	1	3.875	10.764	38.751
stokes	0.25806	1	2.778	10^4
m^2/hr	0.092903	0.3599	1	3600
m^2/s	0.00002581	10^{-4}	0.0002778	1

Source: From W. M. Rohsenow and J. P. Hartnett, Eds. (1973) *Handbook of Heat Transfer*, McGraw-Hill, New York.

APPENDIX
B

PROPERTIES
OF SOLIDS

B.1 STRUCTURAL AND INSULATING MATERIALS

Structural building materials

Description / composition	Density, ρ, kg / m^3	Thermal conductivity, k, W / m K	Specific heat, c_p, J / kg K
		Typical properties at 300 K	
Building Boards			
Asbestos–cement board	1920	0.58	—
Gypsum or plaster board	800	0.17	—
Plywood	545	0.12	1215
Sheathing, regular density	290	0.055	1300
Acoustic tile	290	0.058	1340
Hardboard, siding	640	0.094	1170
Hardboard, high density	1010	0.15	1380
Particle board, low density	590	0.078	1300
Particle board, high density	1000	0.170	1300
Woods			
Hardwoods (oak, maple)	720	0.16	1255
Softwoods (fir, pine)	510	0.12	1380
Masonry Materials			
Cement mortar	1860	0.72	780
Brick, common	1920	0.72	835
Brick, face	2083	1.3	—
Clay tile, hollow			
1 cell deep, 10 cm thick	—	0.52	—
3 cells deep, 30 cm thick	—	0.69	—
Concrete block, 3 oval cores			
sand/gravel, 20 cm thick	—	1.0	—
cinder aggregate, 20 cm thick	—	0.67	—
Concrete block, rectangular core			
2 cores, 20 cm thick, 16 kg	—	1.1	—
same with filled cores	—	0.60	—
Plastering Materials			
Cement plaster, sand aggregate	1860	0.72	—
Gypsum plaster, sand aggregate	1680	0.22	1085
Gypsum plaster, vermiculite aggregate	720	0.25	—

Structural building materials

Description / composition	Typical properties at 300 K		
	Density, ρ, kg / m^3	Thermal conductivity, k, W / m K	Specific heat, c_p, J / kg K
Blanket and Batt			
Glass fiber, paper faced	16	0.046	—
	28	0.038	—
	40	0.035	—
Glass fiber, coated; duct liner	32	0.038	835
Board and Slab			
Cellular glass	145	0.058	1000
Glass fiber, organic bonded	105	0.036	795
Polystyrene, expanded			
extruded (R-12)	55	0.027	1210
molded beads	16	0.040	1210
Mineral fiberboard; roofing material	265	0.049	—
Wood, shredded/cemented	350	0.087	1590
Cork	120	0.039	1800
Loose Fill			
Cork, granulated	160	0.045	—
Diatomaceous silica, coarse	350	0.069	—
powder	400	0.091	—
Diatomaceous silica, fine powder	200	0.052	—
	275	0.061	—
Glass fiber, poured or blown	16	0.043	835
Vermiculite, flakes	80	0.068	835
	160	0.063	1000
Formed/Foamd-in-Place			
Mineral wool granules with asbestos/inorganic binders, sprayed	190	0.046	—
Polyvinyl acetate cork mastic; sprayed or troweled	—	0.100	—
Urethane, two-part mixture; rigid foam	70	0.026	1045
Reflective			
Aluminum foil separating fluffy glass mats; 10–12 layers; evacuated; for cryogenic applications (150 K)	40	0.00016	—
Aluminum foil and glass paper laminate; 75–150 layers; evacuated; for cryogenic application (150 K)	120	0.000017	—
Typical silica powder, evacuated	160	0.0017	—

Industrial insulation

Description composition	Maximum service temperature, K	Typical density, kg/m³	200	215	230	240	255	270	285	300	310	365	420	530	645	750
										Typical thermal conductivity, k (w / m K), at various temperatures (K)						
Blankets																
Blanket, mineral fiber, metal reinforced	920	96–192									0.038	0.046	0.056	0.078		
metal reinforced	815	40–96									0.035	0.045	0.058	0.088		
Blanket, mineral fiber, glass; fine fiber, organic bonded	450	10				0.036	0.038	0.040	0.043	0.048	0.052	0.076				
organic bonded		12				0.035	0.036	0.039	0.042	0.046	0.049	0.069				
		16				0.033	0.035	0.036	0.039	0.042	0.046	0.062				
		24				0.030	0.032	0.033	0.036	0.039	0.040	0.053				
		32				0.029	0.030	0.032	0.033	0.036	0.038	0.048				
		48				0.027	0.029	0.030	0.032	0.033	0.035	0.045				
Blanket, alumina–silica fiber	1530	48												0.071	0.105	0.150
		64												0.059	0.087	0.125
		96												0.052	0.076	0.100
		128												0.049	0.068	0.091
Felt, semirigid; organic bonded	480	50–125	0.023	0.025	0.026	0.027	0.029	0.035	0.036	0.038	0.039	0.051	0.063			
Felt, laminated; no binder	730	50						0.030	0.032	0.033	0.035	0.051	0.079			
	920	120											0.051	0.065	0.087	
Blocks, Boards, and Pipe Insulations Asbestos paper, laminated and corrugated																
4-ply	420	190								0.078	0.082	0.098				
6-ply	420	255								0.071	0.074	0.085				
8-ply	420	300								0.068	0.071	0.082				

Material	Maximum Service Temperature (K)	Typical Density (kg/m³)	Typical Thermal Conductivity, k (W/m·K), at Various Temperatures
Magnesia, 85%	590	185	0.051, 0.055, 0.061, 0.075, 0.089, 0.104
Calcium silicate	920	190	0.055, 0.059, 0.063, 0.092, 0.098, 0.104
Cellular glass	700	145	0.046, 0.048, 0.051, 0.052, 0.055, 0.058, 0.062, 0.069, 0.079
Diatomaceous silica	1145	345	0.092, 0.098, 0.104
Diatomaceous silica	1310	385	0.101, 0.100, 0.115
Polystyrene, rigid			
Extruded (R-12)	350	56	0.023, 0.022, 0.023, 0.023, 0.025, 0.026, 0.027, 0.029
Extruded (R-12)	350	35	0.023, 0.023, 0.025, 0.025, 0.026, 0.027, 0.029
Molded beads	350	16	0.026, 0.029, 0.030, 0.033, 0.035, 0.036, 0.038, 0.040
Rubber, rigid foamed	340	70	0.029, 0.030, 0.032, 0.033
Insulating Cement			
Mineral fiber (rock, slag, or glass)			
with clay binder	1225	430	0.071, 0.079, 0.088, 0.105, 0.123
with hydraulic setting binder	922	560	0.108, 0.115, 0.123, 0.137
Loose Fill			
Cellulose, wood or paper pulp	—	45	0.036, 0.038, 0.039, 0.042
Perlite, expanded	—	105	0.042, 0.043, 0.046, 0.049, 0.051, 0.053, 0.056
Vermiculite, expanded	—	122	0.056, 0.058, 0.061, 0.063, 0.065, 0.068, 0.071
Vermiculite, expanded	—	80	0.049, 0.051, 0.055, 0.058, 0.061, 0.063, 0.066

Source: From F. P. Incropera and D. P. Dewitt (1991) *Fundamentals of Heat and Mass Transfer*, 3rd ed., John Wiley and Sons, New York.

B.2 OTHER MATERIALS

Description / composition	Temperature, K	Density, ρ, kg / m³	Thermal Conductivity, k, W / m K	Specific heat, c_p, J / kg K
Asphalt	300	2115	0.062	920
Bakelite	300	1300	1.4	1465
Brick, refractory				
Carborundum	872	—	18.5	—
	1672	—	11.0	—
Chrome brick	473	3010	2.3	835
	823		2.5	
	1173		2.0	
Diatomaceous	478	—	0.25	—
silica, fired	1145	—	0.30	
Fire clay, burnt 1600 K	773	2050	1.0	960
	1073	—	1.1	
	1373	—	1.1	
Fire clay, burnt 1725 K	773	2325	1.3	960
	1073		1.4	
	1373		1.4	
Fire clay brick	478	2645	1.0	960
	922		1.5	
	1478		1.8	
Magnesite	478	—	3.8	1130
	922	—	2.8	
	1478		1.9	
Clay	300	1460	1.3	880
Coal, anthracite	300	1350	0.26	1260
Concrete (stone mix)	300	2300	1.4	880
Cotton	300	80	0.06	1300
Foodstuffs				
Banana (75.7% water content)	300	980	0.481	3350
Apple, red (75% water content)	300	840	0.513	3600
Cake, batter	300	720	0.223	—
Cake, fully baked	300	280	0.121	—
Chicken meat, white	198	—	1.60	—
(74.4% water content)	233	—	1.49	
	253		1.35	
	263		1.20	
	273		0.476	
	283		0.480	
	293		0.489	
Glass				
Plate (soda lime)	300	2500	1.4	750
Pyrex	300	2225	1.4	835

Description / composition	Temperature, K	Density, ρ, kg / m^3	Thermal Conductivity, k, W / m K	Specific heat, c_p, J / kg K
Ice	273	920	1.88	2040
	253	—	2.03	1945
Leather (sole)	300	998	0.159	—
Paper	300	930	0.180	1340
Paraffin	300	900	0.240	2890
Rock				
Granite, Barre	300	2630	2.79	775
Limestone, Salem	300	2320	2.15	810
Marble, Halston	300	2680	2.80	830
Quartzite, Sioux	300	2640	5.38	1105
Sandstone, Berea	300	2150	2.90	745
Rubber, vulcanized				
Soft	300	1100	0.13	2010
Hard	300	1190	0.16	—
Sand	300	1515	0.27	800
Soil	300	2050	0.52	1840
Snow	273	110	0.049	—
	500	500	0.190	—
Teflon	300	2200	0.35	—
	400		0.45	—
Tissue, human				
Skin	300	—	0.37	—
Fat layer (adipose)	300	—	0.2	—
Muscle	300	—	0.41	—
Wood, cross gain				
Balsa	300	140	0.055	—
Cypress	300	465	0.097	—
Fir	300	415	0.11	2720
Oak	300	545	0.17	2385
Yellow pine	300	640	0.15	2805
White pine	300	435	0.11	—
Wood, radial				
Oak	300	545	0.19	2385
Fir	300	420	0.14	2720

Source: From F. P. Incropera and D. P. Dewitt (1991) *Fundamentals of Heat and Mass Transfer,"* 3rd ed., *John Wiley and Sons, New York.*

B.3   METALS

	Melting point, K	Properties at 300 K				Properties at various temperatures K k (W / m K) / c_p (J / kg K)									
Composition		ρ kg / m^3	c_p, J / kg K	k, W / m K	$\alpha \times 10^6$, m^2 / s	100	200	400	600	800	1000	1200	1500	2000	2500
Aluminum															
Pure	933	2702	903	237	97.1	302	237	240	231	218					
						482	798	949	1033	1146					
Alloy 2024-T6 (4.5% Cu, 1.5% Mg, 0.6% Mn)	775	2770	875	177	73.0	65	163	186	186						
						473	787	925	1042						
Alloy 195, Cast (4.5% Cu)		2790	883	168	68.2			174	185						
								—	—						
Beryllium	1550	1850	1825	200	59.2	990	301	161	126	106	90.8	78.7			
						203	1114	2191	2604	2823	3018	3227	3519		
Bismuth	545	9780	122	7.86	6.59	16.5	9.69	7.04							
						112	120	127							
Boron	2573	2500	1107	27.0	9.76	190	55.5	16.8	10.6	9.60	9.85				
						128	600	1463	1892	2160	2338				
Cadmium	594	8650	231	96.8	48.4	203	99.3	94.7							
						198	222	242							
Chromium	2118	7160	449	93.7	29.1	159	111	90.9	80.7	71.3	65.4	61.9	57.2	49.4	
						192	384	484	542	581	616	682	779	937	
Cobalt	1769	8862	421	99.2	26.6	167	122	85.4	67.4	58.2	52.1	49.3	42.5		
						236	379	450	503	550	628	733	674		
Copper															
Pure	1358	8933	385	401	117	482	413	393	379	366	352	339			
						252	356	397	417	433	451	480			
Commercial bronze (90% Cu, 10% Al)	1293	8800	420	52	14		42	52	59						
							785	460	545						
Phosphor gear bronze (89% Cu, 11% Sn)	1104	8780	355	54	17		41	65	74						
							—	—	—						
Cartridge brass (70% Cu, 30% Zn)	1188	8530	380	110	33.9	75	95	137	149						
							360	395	425						
Constantan (55% Cu, 45% Ni)	1493	8920	384	23	6.71	17	19								
						237	362								
Germanium	1211	5360	322	59.9	34.7	232	96.8	43.2	27.3	19.8	17.4	17.4			
						190	290	337	348	357	375	395			
Gold	1336	19300	129	317	127	327	323	311	298	284	270	255			
						109	124	131	135	140	145	155			
Iridium	2720	22500	130	147	50.3	172	153	144	138	132	126	120	111		
						90	122	133	138	144	153	161	172		
Iron															
Pure	1810	7870	447	80.2	23.1	134	94.0	69.5	54.7	43.3	32.8	28.3	32.1		
						216	384	490	574	680	975	609	654		
Armco (99.75% pure)		7870	447	72.7	20.7	95.6	80.6	65.7	53.1	42.2	32.3	28.7	31.4		
						215	384	490	574	680	975	609	654		
Carbon steels															
Plain carbon (Mn ≤ 1%, Si ≤ 0.1%)		7854	434	60.5	17.7			56.7	48.0	39.2	30.0				
								487	559	685	1169				
AISI 1010		7832	434	63.9	18.8			58.7	48.8	39.2	31.3				
								487	559	685	1168				
Carbon–silicon (Mn ≤ 1%, 0.1% < Si ≤ 0.6%)		7817	446	51.9	14.9			49.8	44.0	37.4	29.3				
								501	582	699	971				

Composition	Melting point, K	ρ kg/m³	c_p, J/kg K	k, W/m K	$\alpha \times 10^6$, m²/s	100	200	400	600	800	1000	1200	1500	2000	2500
Carbon–manganese–silicon (1% < Mn ≤ 1.65, 0.1% < Si ≤ 0.6%)		8131	434	41.0	11.6			42.2 / 487	39.7 / 559	35.0 / 685	27.6 / 1090				
Chromium steels															
$\frac{1}{2}$Cr–$\frac{1}{4}$Mo–Si (0.18% C, 0.65% Cr, 0.23% Mo, 0.6% Si)		7822	444	37.7	10.9			38.2 / 492	36.7 / 575	33.3 / 688	26.9 / 969				
1 Cr–$\frac{1}{2}$Mo (0.16% C, 1% Cr, 0.54% Mo, 0.39% Si)		7858	442	42.3	12.2			42.0 / 492	39.1 / 575	34.5 / 688	27.4 / 969				
1 Cr–V (0.2% C, 1.02% Cr, 0.15% V)		7836	443	48.9	14.1			46.8 / 492	42.1 / 575	36.3 / 688	28.2 / 969				
Stainless steels															
AISI 302		8055	480	15.1	3.91			17.3 / 512	0.0 / 559	22.8 / 585	25.4 / 606				
AISI 304	1670	7900	477	14.9	3.95	9.2 / 272	12.6 / 402	16.6 / 515	19.8 / 557	22.6 / 582	25.4 / 611	28.0 / 640	31.7 / 682		
AISI 316		8238	468	13.4	3.48			15.2 / 504	18.3 / 550	21.3 / 576	24.2 / 602				
AISI 347		7978	480	14.2	3.71			15.8 / 513	18.9 / 559	21.9 / 585	24.7 / 606				
Lead	601	11340	129	35.3	24.1	39.7 / 118	36.7 / 125	34.0 / 132	31.4 / 142						
Magnesium	923	1740	1024	156	87.6	169 / 649	159 / 934	153 / 1074	149 / 1170	146 / 1267					
Molybdenum	2894	10240	251	138	53.7	179 / 141	143 / 224	134 / 261	126 / 275	118 / 285	112 / 295	105 / 308	98 / 330	90 / 380	86 / 459
Nickel															
Pure	1728	8900	444	90.7	23.0	164 / 232	107 / 383	80.2 / 485	65.6 / 592	67.6 / 530	71.8 / 562	76.2 / 594	82.6 / 616		
Nichrome (80% Ni, 20% Cr)	1672	8400	420	12	3.4			14 / 480	16 / 525	21 / 545					
Inconnel X-750 (73% Ni, 15%Cr, 6.7% Fe)	1665	8510	439	11.7	3.1	8.7 / —	10.3 / 372	13.5 / 473	17.0 / 510	20.5 / 546	24.0 / 626	27.6 / —	33.0 / —		
Niobium	2741	8570	265	53.7	23.6	55.2 / 188	52.6 / 249	55.2 / 274	58.2 / 283	61.3 / 292	64.4 / 301	67.5 / 310	72.1 / 324	79.1 / 347	
Palladium	1827	12020	244	71.8	24.5	76.5 / 168	71.6 / 227	73.6 / 251	79.7 / 261	86.9 / 271	94.2 / 281	102 / 291	110 / 307		
Platinum															
Pure	2045	21450	133	71.6	25.1	77.5 / 100	72.6 / 125	71.8 / 136	73.2 / 141	75.6 / 146	78.7 / 152	82.6 / 157	89.5 / 165	99.4 / 179	
Alloy 60Pt–40Rh (60% Pt, 40% Rh)	1800	16630	162	47	17.4			52 / —	59 / —	65 / —	69 / —	73 / —	76 / —		
Rhenium	3453	21100	136	47.9	16.7	58.9 / 97	51.0 / 127	46.1 / 139	44.2 / 145	44.1 / 151	44.6 / 156	45.7 / 162	47.8 / 171	51.9 / 186	
Rhodium	2236	12450	243	150	49.6	186 / 147	154 / 220	146 / 253	136 / 274	127 / 293	121 / 311	116 / 327	110 / 349	112 / 376	
Silicon	1685	2300	712	148	89.2	884 / 259	264 / 556	98.9 / 790	61.9 / 867	42.2 / 913	31.2 / 946	25.7 / 967	22.7 / 992		
Silver	1235	10500	235	429	174	444 / 187	430 / 225	425 / 239	412 / 250	396 / 262	379 / 277	361 / 292			

Composition	Melting point, K	ρ kg / m³	c_p, J / kg K	k, W / m K	$\alpha \times 10^6$, m² / s	Properties at various temperatures K k (W / m K) / c_p (J / kg K)									
						100	200	400	600	800	1000	1200	1500	2000	2500
Tantalum	3269	16600	140	57.5	24.7	59.2	57.5	57.8	58.6	59.4	60.2	61.0	62.2	64.1	65.6
						110	133	144	146	149	152	155	160	172	189
Thorium	2023	11700	11	54.0	39.1	59.8	54.6	54.5	55.8	56.9	56.9	58.7			
						99	112	124	134	145	156	167			
Tin	505	7310	227	66.6	40.1	85.2	73.3	62.2							
						188	215	243							
Titanium	1953	4500	522	21.9	9.32	30.5	24.5	20.4	19.4	19.7	20.7	22.0	24.5		
						300	465	551	591	633	675	620	686		
Tungsten	3660	19300	132	174	68.3	208	186	159	137	125	118	113	107	100	95
						87	122	137	142	145	148	152	157	167	176
Uranium	1406	19070	116	27.6	12.5	21.7	25.1	29.6	34.0	38.8	43.9	49.0			
						94	108	125	146	176	180	161			
Vanadium	2192	6100	489	30.7	10.3	35.8	31.3	31.3	33.3	35.7	38.2	40.8	44.6	50.9	
						258	430	515	540	563	597	645	714	867	
Zinc	693	7140	389	116	41.8	117	118	111	103						
						297	367	402	436						
Zirconium	2125	6570	278	22.7	12.4	33.2	25.2	21.6	20.7	21.6	23.7	26.0	28.8	33.0	
						205	264	300	322	342	362	344	344	344	

Source: From F. P. Incropera and D. P. Dewitt (1990) *Fundamentals of Heat and Mass Transfer*, 3rd ed., John Wiley and Sons, New York.

B.4 SOME NONMETALLIC SOLIDS

Composition	Melting point, K	Properties at 300 K				Properties at various temperatures, K k (W/m K)/c_p (J/kg K)									
		ρ, kg/m^3	c_p, J/kg K	k, W/m K	$\alpha \times 10^6$ m^2/s	100	200	400	600	800	1000	1200	1500	2000	2500
Aluminum oxide, sapphire	2323	3970	765	46	15.1	450	82	32.4	18.9	13.0	10.5				
						—	—	940	1110	1180	1225				
Aluminum oxide, polycrystalline	2323	3970	765	36.0	11.9	133	55	26.4	15.8	10.4	7.85	6.55	5.66	6.00	
						—	—	940	1110	1180	1225	—	—	—	
Beryllium oxide	2725	3000	1030	272	88.0			196	111	70	47	33	21.5	15	
								1350	1690	1865	1975	2055	2145	2750	
Boron	2573	2500	1105	27.6	9.99	190	52.5	18.7	11.3	8.1	6.3	5.2			
						—	—	1490	1880	2135	2350	2555			
Boron fiber eoxy (30% vol)composite	590	2080													
k, ∥ to fibers				2.29		2.10	2.23	2.28							
k, ⊥ to fibers				0.59		0.37	0.49	0.60							
c_p			1122			364	757	1431							
Carbon Amorphous	1500	1950	—	1.60	—	0.67	1.18	1.89	2.19	2.37	2.53	2.84	3.48		
						—	—	—	—	—	—	—	—		
Diamond, type IIa insulator	—	3500	509	2300		10000	4000	1540							
						21	194	853							
Graphite, pyrolytic	2273	2210													
k, ∥ to layers				1950		4970	3230	1390	892	667	534	448	357	262	
k, ⊥ to layers				5.70		16.8	9.23	4.09	2.68	2.01	1.60	1.34	1.08	0.81	
c_p			709			136	411	992	1406	1650	1793	1890	1974	2043	
Graphite fiber epoxy (25% vol) composite	450	1400													
k, heat flow ∥ to fibers				11.1		5.7	8.7	13.0							
k, heat flow ⊥ to fibers				0.87		0.46	0.68	1.1							
c_p			935			337	642	1216							
Pyroceram, Corning 9606	1623	2600	808	3.98	1.89	5.25	4.78	3.64	3.28	3.08	2.96	2.87	2.79		
						—	—	908	1038	1122	1197	1264	1498		
Silicon carbide	3100	3160	675	490	230			—	—	—	87	58	30		
								880	1050	1135	1195	1243	1310		
Silicon dioxide, crystalline (quartz)	1883	2650													
k, ∥ to c axis				10.4		39	16.4	7.6	5.0	4.2					
k, ⊥ to c axis				6.21		20.8	9.5	4.70	3.4	3.1					
c_p			745			—	—	885	1075	1250					
Silicon dioxide, polycrystalline (fused silica)	1883	2220	745	1.38	0.834	0.69	1.14	1.51	1.75	2.17	2.87	4.00			
						—	—	905	1040	1105	1155	1195			
Silicon nitride	2173	2400	691	16.0	9.65	—	—	13.9	11.3	9.88	8.76	8.00	7.16	6.20	
						—	578	788	937	1063	1155	1226	1306	1377	
Sulfur	392	2070	708	0.206	0.141	0.165	0.185								
						403	606								
Thorium dioxide	3573	9110	235	13	6.1			10.2	6.6	4.7	3.68	3.12	2.73	2.5	
								255	274	285	295	303	315	330	
Titanium dioxide, polycrystalline	2133	4157	710	8.4	2.8			7.01	5.02	3.94	3.46	3.28			
								805	880	910	930	945			

Source: From F. P. Incropera and D. P. Dewitt (1991) *Fundamentals of Heat and Mass Transfer*, 3rd ed., John Wiley and Sons, New York.

B.5 THERMAL PROPERTIES OF SEVERAL GLASS PRODUCTS

For specific heat in J/kg K, multiply values by 4184. For thermal conductivity in W/m K, multiply values in cal/cm s °C by 418.4.

Material	Specific heat			Thermal conductivity, cal / cm s °C × 10^4			
	25°C	500°C	100°C	− 100°C	0°C	100°C	400°C
Fused silica	0.173	0.268	0.292	25.0	31.5	35.4	
7900	0.18	0.24	0.29	24	30	34	
7740	0.17	0.28		21	26	30	
1723	0.18	0.26			29	33	
0311 (chemically strengthend)	0.21	0.28			27	29	35
Soda-lime window glass	0.190	0.300	0.333	19	24	27	
Heavy flint, 80% PbO, 20% SiO$_2$				10	12	14	
Foamglass insulation	0.20			(0.97)	1.3	1.73	(2.81)
Fibrous glass					(0.8)		
9606 glass-ceramic	0.185	0.267	0.311		90	86	75
9608 low-expansion glass-ceramic	0.195	0.286			48	51	55

Notes:
Parentheses indicate extrapolated values.

Specific heat increases with temperature and approaches zero at 0°K. There are no critical temperatures or phase changes. Thermal conductivity increases with temperature and is very high for glass ceramics.

Source: From "Glass," and J. R. Hutchins III and R. V. Harrington, *Kirk-Othmer Encyclopedia of Chemical Technology*, Vol. 10, p. 598. Copyright © 1966 by John Wiley and Sons, Inc. Reprinted by permission.

B.6 THERMAL CONDUCTIVITY OF METALS AT CRYOGENIC TEMPERATURES

As Recommended by National Standard Reference Data System-N.B.S.

Values in this table are in W cm K. To convert to Btu/hr ft °R, multiply the tabular values by 57.818. These data apply only to metals of purity of at least 99.9%. In the table the third significant figure is for smoothness and is not indicative of the degree of accuracy.

Temperature		Aluminum	Cadmium	Chromium	Copper	Gold	Iron	Lead	Magnesium	Molybdenum
K	°R									
1	1.8	7.8	48.7	0.401	28.7	4.4	0.75	27.7	1.30	0.146
2	3.6	15.5	89.3	0.802	57.3	8.9	1.49	42.4	2.59	0.292
3	5.4	23.2	104	1.20	85.5	13.1	2.24	34.0	3.88	0.438
4	7.2	30.8	92.0	1.60	113	17.1	2.97	22.4	5.15	0.584
5	9	38.1	69.0	1.99	138	20.7	3.71	13.8	6.39	0.730
6	10.8	45.1	44.2	2.38	159	23.7	4.42	8.2	7.60	0.876
7	12.6	51.5	28.0	2.77	177	26.0	5.13	4.9	8.75	1.02
8	14.4	57.3	18.0	3.14	189	27.5	5.80	3.2	9.83	1.17
9	16.2	62.2	12.2	3.50	195	28.2	6.45	2.3	10.8	1.31
10	18	66.1	8.87	3.85	196	28.2	7.05	1.78	11.7	1.45
11	19.8	69.0	6.91	4.18	193	27.7	7.62	1.46	12.5	1.60
12	21.6	70.8	5.56	4.49	185	26.7	8.13	1.23	13.1	1.74
13	23.4	71.5	4.67	4.78	176	25.5	8.58	1.07	13.6	1.88
14	25.2	71.3	4.01	5.04	166	24.1	8.97	0.94	14.0	2.01
15	27	70.2	3.55	5.27	156	22.6	9.30	0.84	14.3	2.15
16	28.8	68.4	3.16	5.48	145	20.9	9.56	0.77	14.4	2.28
18	32.4	63.5	2.62	5.81	124	17.7	9.88	0.66	14.3	2.53
20	36	56.5	2.26	6.01	105	15.0	9.97	0.59	13.9	2.77
25	45	40.0	1.79	6.07	68	10.2	9.36	0.507	12.0	3.25
30	54	28.5	1.56	5.58	43	7.6	8.14	0.477	9.5	3.55
35	63	21.0	1.41	5.03	29	61	6.81	0.462	7.4	3.62
40	72	16.0	1.32	4.30	20.5	5.2	5.55	0.451	5.7	3.51
45	81	12.5	1.25	3.67	15.3	4.6	4.50	0.442	4.57	3.26
50	90	10.0	1.20	3.17	12.2	4.2	3.72	0.435	3.75	3.00
60	108	6.7	1.13	2.48	8.5	3.8	2.65	0.424	2.74	2.60
70	126	5.0	1.08	2.08	6.7	3.58	2.04	0.415	2.23	2.30
80	144	4.0	1.06	1.82	5.7	3.52	1.68	0.407	1.95	2.09
90	162	3.4	1.04	1.68	5.14	3.48	1.46	0.401	1.78	1.92
100	180	3.0	1.03	1.58	4.83	3.45	1.32	0.396	1.69	1.79

Temperature

K	°R	Nickel	Niobium	Platinum	Silver	Tantalum	Tin	Titanium	Tungsten	Zinc	Zirconium
1	1.8	0.64	0.251	2.31	39.4	0.115		0.0144	14.4	19.0	0.111
2	3.6	1.27	0.501	4.60	78.3	0.230		0.0288	28.7	37.9	0.223
3	5.4	1.91	0.749	6.79	115	0.345	297	0.0432	42.6	55.5	0.333
4	7.2	2.54	0.993	8.8	147	0.459	181	0.0576	55.6	69.7	0.442
5	9	3.16	1.23	10.5	172	0.571	117	0.0719	67.1	77.8	0.549
6	10.8	3.77	1.46	11.8	187	0.681	76	0.0863	76.2	78.0	0.652
7	12.6	4.36	1.67	12.6	193	0.788	52	0.101	82.4	71.7	0.748
8	14.4	4.94	1.86	12.9	190	0.891	36	0.115	85.3	61.8	0.837
9	16.2	5.49	2.04	12.8	181	0.989	26	0.129	85.1	51.9	0.916
10	18	6.00	2.18	12.3	168	1.08	19.3	0.144	82.4	43.2	0.984
11	19.8	6.48	2.30	11.7	154	1.16	14.8	0.158	77.9	36.4	1.04
12	21.6	6.91	2.39	10.9	139	1.24	11.6	0.172	72.4	30.8	1.08
13	23.4	7.30	2.46	10.1	124	1.30	9.3	0.186	66.4	26.1	1.11
14	25.2	7.64	2.49	9.3	109	1.36	7.6	0.200	60.4	22.4	1.13
15	27	7.92	2.50	8.4	96	1.40	6.3	0.214	54.8	19.4	1.13
16	28.8	8.15	2.49	7.6	85	1.44	5.3	0.227	49.3	16.9	1.12
18	32.4	8.45	2.42	6.1	66	1.47	4.0	0.254	40.0	13.3	1.08
20	36	8.56	2.29	4.9	51	1.47	3.2	0.279	32.6	10.7	1.01
25	45	8.15	1.87	3.15	29.5	1.36	2.22	0.337	20.4	6.9	0.85
30	54	6.95	1.45	2.28	19.3	1.16	1.76	0.382	13.1	4.9	0.74
35	63	5.62	1.16	1.80	13.7	0.99	1.50	0.411	8.9	3.72	0.65
40	72	4.63	0.97	1.51	10.5	0.87	1.35	0.422	6.5	2.97	0.58
45	81	3.91	0.84	1.32	8.4	0.78	1.23	0.416	5.07	2.48	0.535
50	90	3.36	0.76	1.18	7.0	0.72	1.15	0.401	4.17	2.13	0.497
60	108	2.63	0.66	1.01	5.5	0.651	1.04	0.377	3.18	1.71	0.442
70	126	2.21	0.61	0.90	4.97	0.616	0.96	0.356	2.76	1.48	0.403
80	144	1.93	0.58	0.84	4.71	0.603	0.91	0.339	2.56	1.38	0.373
90	162	1.72	0.563	0.81	4.60	0.596	0.88	0.324	2.44	1.34	0.350
100	180	1.58	0.552	0.79	4.50	0.592	0.85	0.312	2.35	1.32	0.332

B.7   PROPERTIES OF CRYOGENIC INSULATION

Description of Advantages

Class

1. *Liquid and vapor shields.* Very low temperature, valuable, or dangerous liquids, such as helium or fluorine, are often shielded by an intermediate cryogenic liquid or vapor container that must in turn be insulated by one of the following methods.
2. *Multilayer reflecting shields.* Foil or aluminized plastic alternated with paper-thin glass- or plastic-fiber sheets; lowest conductivity, low density, and heat storage; good stability; minimum support structure.
3. *Opacified evacuated powders.* Contain metallic flakes to reduce radiation; conform to irregular shapes.
4. *Evacuated dielectric powders.* Very fine powders of low-conductivity adsorbent; moderate vacuum requirement; minimum fire hazard in ozygen.
5. *Vacuum flasks (Dewar).* Tight shield-space with highly reflecting walls and high vacuum; minimum heat capacity; rugged; small thickness.
6. *Gas-filled powders.* Same powders as Class 4 but with air or inert gas; low cost; easy application; no vacuum requirement.
7. *Expanded foams.* Very light foamed plastic; inexpensive; minimum weight but bulky; self-supporting.
8. *Porous fiber blankets.* Blanket material of fine fibers, usually glass; minimum cost and easy installation but not an adequate insulation for most cryogenic applications.

Insulation properties

Class	Descriptive name	Approximate density $\dfrac{\text{lbm}}{\text{ft}^3}$	$\dfrac{\text{kg}}{\text{m}^3}$	Approximate specific heat $\dfrac{\text{Btu}}{\text{lbm}\,°\text{F}}$	$\dfrac{\text{kJ}}{\text{kg K}}$	Range of mean conductivities $\dfrac{\text{Btu}}{\text{hr ft}\,°\text{F}}$	$\dfrac{\text{mW}}{\text{m K}}$	Interspace pressure, mm Hg†
2	Multilayer	5	80	0.22	0.92	0.000023–0.00012	0.04–0.2	10^{-4}
3	Opacified powder	7	110	0.23	0.96	0.00015–0.0004	0.26–0.7	10^{-4}
4	Evacuated powder	6	100	0.25	1.05	0.00057–0.00115	1.0–2.0	10^{-4}
5	Vacuum flask	—	—	—	—	0.0029	5.0	10^{-6}
6	Gas-filled powder	6	100	0.25	1.05	0.001–0.004	1.7–7.0	760
7	Expanded foam	2	30	0.4	1.67	0.0029–0.020	5.0–35	760
8	Fiber blanket	8	130	0.5	2.09	0.02–0.026	33–45	760

†N m² multiply by 133.32.

Structural support

For those insulating materials and constructions requiring structural support, the relative strengths, weights, heat capacities, and conductivities and the supporting materials are important.

Material	Tensile yield strength S, $-$ 1000's psi‡	Density, ρ		Specific heat, c_p		Mean thermal conductivity k,§, 20–300 K		Relative		
		$\dfrac{\text{lbm}}{\text{ft}^3}$	$\dfrac{\text{kg}}{\text{m}^3}$	$\dfrac{\text{Btu}}{\text{lbm}\,°\text{F}}$	$\dfrac{\text{kJ}}{\text{kg K}}$	$\dfrac{\text{Btu}}{\text{hr ft}\,°\text{F}}$	$\dfrac{\text{W}}{\text{m K}}$	$\dfrac{S}{k}$	$\dfrac{\rho}{k}$	$\dfrac{c_p\rho}{S}$
Aluminum alloy	50	170	2720	0.22	0.92	50	86	1	3	0.75
"K" Monel	100	520	8330	0.13	0.54	10	17	10	52	0.68
Stainless steel	100	500	8010	0.12	0.50	5.4	9.3	10	93	0.60
Titanium alloy	100	625	10010	0.06	0.25	3.5	6.1	29	180	0.37
Nylon	15	70	1120	0.4	1.67	0.17	0.29	88	41	1.9
Teflon	2	120	1920	0.25	1.05	0.14	0.24	14	86	15.0

‡For MN m^2 multiply tabulated values in 1000's psi by 6.8948.

§For solid members. Perforation and lamination used to reduce conduction.

Source: Compiled from several sources.

PROPERTIES OF LIQUIDS

C.1 PURE WATER AT ATMOSPHERIC PRESSURE

t, °C	ρ, kg / m³	$\mu \times 10^3$, kg / m s	$\nu \times 10^6$, m² / s	k, W / m K	$\beta \times 10^5$, K^{-1}	c_p, W s / kg K	Pr
0	999.84	1.7531	1.7533	0.5687	− 6.8143	4209.3	12.976
5	999.96	1.5012	1.5013	0.578	1.5985	4201.0	10.911
10	999.70	1.2995	1.2999	0.5869	8.7902	4194.1	9.286
15	999.10	1.1360	1.1370	0.5953	15.073	4188.5	7.991
20	998.20	1.0017	1.0035	0.6034	20.661	4184.1	6.946
25	997.05	0.8904	0.8930	0.6110	20.570	4180.9	6.093
30	995.65	0.7972	0.8007	0.6182	30.314	4178.8	5.388
35	994.03	0.7185	0.7228	0.6251	34.571	4177.7	4.802
40	992.21	0.6517	0.6565	0.6351	38.53	4177.6	4.309
45	990.22	0.5939	0.5997	0.6376	42.26	4178.3	3.892
50	988.04	0.5442	0.5507	0.6432	45.78	4179.7	3.535
60	983.19	0.4631	0.4710	0.6535	52.33	4184.8	2.965
70	977.76	0.4004	0.4095	0.6623	58.40	4192.0	2.534
80	971.79	0.3509	0.3611	0.6698	64.13	4200.1	2.201
90	965.31	0.3113	0.3225	0.6759	69.62	4210.7	1.939
100	958.35	0.2789	0.2911	0.6807	75.00	4221.0	1.729

Source: From D. J. Kukulka (1981). Thermodynamic and Transport Properties of Pure and Saline Water, M. S. Thesis, State University of New York at Buffalo.

C.2 OTHER COMMON LIQUIDS

Common name	Density, kg / m^3	Specific heat, kJ / kg K	Viscosity, N s / m^2	Thermal conductivity, W / m K	Freezing point, K	Latent heat of fusion, kJ / kg	Boiling point, K	Latent heat of evaporation, kJ / kg	Coefficient of cubical expansion, K^{-1}
Acetic acid	1049	2.18	0.001155	0.171	290	181	391	402	0.0011
Acetone	784.6	2.15	0.000316	0.161	179.0	98.3	329	518	0.0015
Alcohol, ethyl	785.1	2.44	0.001095	0.171	158.6	108	351.46	846	0.0011
Alcohol, methyl	786.5	2.54	0.00056	0.202	175.5	98.8	337.8	1100	0.0014
Alcohol, propyl	800.0	2.37	0.00192	0.161	146	86.5	371	779	
Ammonia (aqua)	823.5	4.38		0.353					
Benzene	873.8	1.73	0.000601	0.144	278.68	126	353.3	390	0.0013
Bromine		0.473	0.00095		245.84	66.7	331.6	193	0.0012
Carbon dissulfide	1261	0.992	0.00036	0.161	161.2	57.6	319.40	351	0.0013
Carbon tetrachloride	1584	0.866	0.00091	0.104	250.35	174	349.6	194	0.0013
Castor oil	956.1	1.97	0.650	0.180	263.2		334.4	247	
Chloroform	1465	1.05	0.00053	0.118	209.6	77.0	447.2	263	0.0013
Decane	726.3	2.21	0.000859	0.147	243.5	201	489.4	256	
Dodecane	754.6	2.21	0.001374	0.140	247.18	216	307.7	372	
Ether	713.5	2.21	0.000223	0.130	157	96.2		800	0.0016
Ethylene glycol	1097	2.36	0.0162	0.258	260.2	181	470	180§	
Fluorine refrigerant R-11	1476	0.870‡	0.00042	0.093‡	162		297.0	165§	
Fluorine refrigerant R-12	1311	0.971‡		0.071‡	115	34.4	243.4	232§	
Fluorine refrigerant R-22	1194	1.26‡		0.086‡	113	183	232.4		

596

Glycerine	1259	2.62	0.950	0.287	264.8	200	563.4	974	0.00054
Heptane	679.5	2.24	0.000376	0.128	182.54	140	371.5	318	
Hexane	654.8	2.26	0.000297	0.124	178.0	152	341.84	365	
Iodine		2.15			386.6	62.2	457.5	164	
Kerosene	820.1	2.09	0.00164	0.145				251	
Linseed oil	929.1	1.84	0.0331		253		560		
Mercury		0.139	0.00153	0.131	234.3	11.6	630	295	0.00018
Octane	698.6	2.15	0.00051		216.4	181	398	298	0.00072
Phenol	1072	1.43	0.0080	0.190	316.2	121	455		0.00090
Propane	493.5	2.41‡	0.00011		85.5	79.9	231.08	428§	
Propylene	514.4	2.85	0.00009		87.9	71.4	225.45	342	
Propylene glycol	965.3	2.50	0.042		213		460	914	
Sea water	1025	3.76–4.10			270.6				
Toluene	862.3	1.72	0.000550	0.133	178	71.8	383.6	363	
Turpentine	868.2	1.78	0.001375	0.121	214		433	293	0.00099

†At 1.0 atm pressure (0.101325 MN/m^2), 300 K, except as noted.

‡At 297 K, liquid.

§At 0.101325 MN, saturation temperature.

Source: Reprinted with permission from R. C. Weast, Ed., *Handbook for Tables for Applied Engineering Science.* Copyright © 1970, CRC Press, Inc., Boca Raton, FL.

C.3 LIQUID METALS

Metal (melting point, °F)	Temperature °F	°C	Specific gravity (cal / g °C)	Specific heat	Thermal conductivity Btu / hr ft °F	cal / s cm °C†	Absolute viscosity lbm / ft s	centipoise
Aluminum	1250	677	2.38	0.259				
(1220)	1300	704	2.37	0.259	60.2	0.249	1.88×10^{-3}	2.8
	1350	732	2.36	0.259	63.4	0.262	1.61×10^{-3}	2.4
	1400	760	2.35	0.259	54.3	0.266	1.34×10^{-3}	2.0
	1450	788	2.34	0.259	69.9	0.289	1.08×10^{-3}	1.6
Bismuth	600	316	10.0	0.0345	9.5	0.039	1.09×10^{-3}	1.62
(520)	800	427	9.87	0.0357	9.0	0.037	9.0×10^{-4}	1.34
	1000	538	9.74	0.0369	9.0	0.037	7.4×10^{-4}	1.10
	1200	649	9.61	0.0381	9.0	0.037	6.2×10^{-4}	0.923
Cesium	83	28	1.84	0.060	10.6	0.044		
(83)	150	66					3.84×10^{-4}	0.571
	250	121					2.95×10^{-4}	0.439
	350	177					2.47×10^{-4}	0.368
	400	204					2.30×10^{-4}	0.343
Lead	700	371	10.5	0.038	9.3	0.038	1.61×10^{-3}	2.39
(621)	850	454	10.4	0.037	9.0	0.037	1.38×10^{-3}	2.05
	1000	538	10.4	0.037	8.9	0.036	1.17×10^{-3}	1.74
	1150	621	10.2	0.037	8.7	0.036	1.02×10^{-3}	1.52
	1300	704	10.1		8.6	0.035	9.20×10^{-4}	1.37
Lithium	400	204	0.506	1.0	24.	0.10	4.0×10^{-4}	0.595
(355)	600	316	0.497	1.0	23	0.095	3.4×10^{-4}	0.506
	800	427	0.489	1.0	22.	0.090	3.7×10^{-4}	0.551
	1200	649	0.471				2.9×10^{-4}	0.432
	1800	942	0.442				2.8×10^{-4}	0.417
Magnesium	1250	677	1.55	0.318				
(1203)	1301	705	1.53	0.320				
	1350	732	1.49	0.322				
Mercury	50	10	13.6	0.033	4.8	0.020	1.07×10^{-3}	1.59
(−38)	200	93	13.4	0.033	6.0	0.025	8.4×10^{-4}	1.25
	300	149	13.2	0.033	6.7	0.028	7.4×10^{-4}	1.10
	400	204	13.1	0.032	7.2	0.030	6.7×10^{-4}	0.997
	600	316	12.8	0.032	8.1	0.033	5.8×10^{-4}	0.863
Tin	500	260	6.94	0.058	19	0.079	1.22×10^{-3}	1.82
(449)	700	371	6.86	0.060	19.4	0.080	9.8×10^{-4}	1.46
	850	454	6.81	0.062	19	0.079	8.5×10^{-4}	1.26
	1000	538	6.74	0.064	19	0.079	7.6×10^{-4}	1.13
	1200	649	6.68	0.066	19	0.079	6.7×10^{-4}	0.997
Zinc	600	316	6.97	0.123	35.4	0.146		
(787)	850	454	6.90	0.119	33.7	0.139	2.10×10^{-3}	3.12
	1000	538	6.86	0.116	33.2	0.137	1.72×10^{-3}	2.56
	1200	649	6.76	0.113	32.8	0.136	1.39×10^{-3}	2.07
	1500	816	6.74	0.107	32.6	0.135	9.83×10^{-4}	1.46

†For W/cm °C multiply by 4.184.

Source: From R. C. Weast, ED. (1970) *Handbook of Tables for Applied Engineering Science*, CRC Press, Boca Raton, FL.

APPENDIX
D

PROPERTIES
OF GASES
AND VAPORS

D.1 AIR IN SI UNITS

Temperature			Properties							
K	°C	°F	ρ	c_p	c_p/c_v	μ	k	Pr	h	V_s
100	− 173.15	− 280	3.598	1.028		6.929	9.248	0.770	98.42	198.4
110	− 163.15	− 262	3.256	1.022	1.4202	7.633	10.15	0.768	108.7	208.7
120	− 153.15	− 244	2.975	1.017	1.4166	8.319	11.05	0.766	118.8	218.4
130	− 143.15	− 226	2.740	1.014	1.4139	8.990	11.94	0.763	129.0	227.6
140	− 133.15	− 208	2.540	1.012	1.4119	9.646	12.84	0.761	139.1	236.4
150	− 123.15	− 190	2.367	1.010	1.4102	10.28	13.73	0.758	149.2	245.0
160	− 113.15	− 172	2.217	1.009	1.4089	10.91	14.61	0.754	159.4	253.2
170	− 103.15	− 154	2.085	1.008	1.4079	11.52	15.49	0.750	169.4	261.0
180	− 93.15	− 136	1.968	1.007	1.4071	12.12	16.37	0.746	179.5	268.7
190	− 83.15	− 118	1.863	1.007	1.4064	12.71	17.23	0.743	189.6	276.2
200	− 73.15	− 100	1.769	1.006	1.4057	13.28	18.09	0.739	199.7	283.4
205	− 68.15	− 91	1.726	1.006	1.4055	13.56	18.52	0.738	204.7	286.9
210	− 63.15	− 82	1.684	1.006	1.4053	13.85	18.94	0.736	209.7	290.5
215	− 58.15	− 73	1.646	1.006	1.4050	14.12	19.36	0.734	214.8	293.9
220	− 53.15	− 64	1.607	1.006	1.4048	14.40	19.78	0.732	219.8	297.4

Temperature			Properties							
K	°C	°F	ρ	c_p	c_p/c_v	μ	k	Pr	h	V_s
225	−48.15	−55	1.572	1.006	1.4046	14.67	20.20	0.731	224.8	300.8
230	−43.15	−46	1.537	1.006	1.4044	14.94	20.62	0.729	229.8	304.1
235	−38.15	−37	1.505	1.006	1.4042	15.20	21.04	0.727	234.9	307.4
240	−33.15	−28	1.473	1.005	1.4040	15.47	21.45	0.725	239.9	310.6
245	−28.15	−19	1.443	1.005	1.4038	15.73	21.86	0.724	244.9	313.8
250	−23.15	−10	1.413	1.005	1.4036	15.99	22.27	0.722	250.0	317.1
255	−18.15	−1	1.386	1.005	1.4034	16.25	22.68	0.721	255.0	320.2
260	−13.15	8	1.359	1.005	1.4032	16.50	23.08	0.719	260.0	323.4
265	−8.15	17	1.333	1.005	1.4030	16.75	23.48	0.717	265.0	326.5
270	−3.15	26	1.308	1.006	1.4029	17.00	23.88	0.716	270.1	329.6
275	−1.85	35	1.235	1.006	1.4026	17.26	24.28	0.715	275.1	332.6
280	6.85	44	1.261	1.006	1.4024	17.50	24.67	0.713	280.1	335.6
285	11.85	53	1.240	1.006	1.4022	17.74	25.06	0.711	285.1	338.5
290	16.85	62	1.218	1.006	1.4020	17.98	25.47	0.710	290.2	341.5
295	21.85	71	1.197	1.006	1.4018	18.22	25.85	0.709	295.2	344.4
300	26.85	80	1.177	1.006	1.4017	18.46	26.24	0.708	300.2	347.3
305	31.85	89	1.158	1.006	1.4015	18.70	26.63	0.707	305.3	350.2
310	36.85	98	1.139	1.007	1.4013	18.93	27.01	0.705	310.3	353.1
315	41.85	107	1.121	1.007	1.4010	19.15	27.40	0.704	315.3	355.8
320	46.85	116	1.103	1.007	1.4008	19.39	27.78	0.703	320.4	358.7
325	51.85	125	1.086	1.008	1.4006	19.63	28.15	0.702	325.4	361.4
330	56.85	134	1.070	1.008	1.4004	19.85	28.53	0.701	330.4	364.2
335	61.85	143	1.054	1.008	1.4001	20.08	28.90	0.700	335.5	366.9
340	66.85	152	1.038	1.008	1.3999	20.30	29.28	0.699	340.5	369.6
345	71.85	161	1.023	1.009	1.3996	20.52	29.64	0.698	345.6	372.3
350	76.85	170	1.008	1.009	1.3993	20.75	30.03	0.697	350.6	375.0
355	81.85	179	0.9945	1.010	1.3990	20.97	30.39	0.696	355.7	377.6
360	86.85	188	0.9805	1.010	1.3987	21.18	30.78	0.695	360.7	380.2
365	91.85	197	0.9672	1.010	1.3984	21.38	31.14	0.694	365.8	382.8
370	96.85	206	0.9539	1.011	1.3981	21.60	31.50	0.693	370.8	385.4
375	101.85	215	0.9413	1.011	1.3978	21.81	31.86	0.692	375.9	388.0
380	106.85	224	0.9288	1.012	1.3975	22.02	32.23	0.691	380.9	390.5
385	111.85	233	0.9169	1.012	1.3971	22.24	32.59	0.690	386.0	393.0
390	116.85	242	0.9050	1.013	1.3968	22.44	32.95	0.690	391.0	395.5
395	121.85	251	0.8936	1.014	1.3964	22.65	33.31	0.689	396.1	398.0
400	126.85	260	0.8822	1.014	1.3961	22.86	33.65	0.689	401.2	400.4
410	136.85	278	0.8608	1.015	1.3953	23.27	34.35	0.688	411.3	405.3
420	146.85	296	0.8402	1.017	1.3946	23.66	35.05	0.687	421.5	410.2
430	156.85	314	0.8207	1.018	1.3938	24.06	35.75	0.686	431.7	414.9
440	166.85	332	0.8021	1.020	1.3929	24.45	36.43	0.684	441.9	419.6
450	176.85	350	0.7342	1.021	1.3920	24.85	37.10	0.684	452.1	424.2
460	186.85	368	0.7677	1.023	1.3911	25.22	37.78	0.683	462.3	428.7
470	196.85	386	0.7509	1.024	1.3901	25.58	38.46	0.682	472.5	433.2
480	206.85	404	0.7351	1.026	1.3892	25.96	39.11	0.681	482.8	437.6
490	216.85	422	0.7201	1.208	1.3881	26.32	39.76	0.680	493.0	422.0

Temperature			Properties							
K	°C	°F	ρ	c_p	c_p/c_v	μ	k	Pr	h	V_s
500	226.85	440	0.7057	1.030	1.3871	26.70	40.41	0.680	503.3	446.4
510	236.85	458	0.6919	1.032	1.3861	27.06	41.06	0.680	513.6	450.6
520	246.85	476	0.6786	1.034	1.3851	27.42	41.69	0.680	524.0	454.9
530	256.85	494	0.6658	1.036	1.3840	27.78	42.32	0.680	534.4	459.0
540	266.85	512	0.6535	1.038	1.3829	28.14	42.94	0.680	544.7	463.2
550	276.85	530	0.6416	1.040	1.3818	28.48	43.57	0.680	555.1	467.3
560	286.85	548	0.6301	1.042	1.3806	28.83	44.20	0.680	565.5	471.3
570	296.85	566	0.6190	1.044	1.3795	29.17	44.80	0.680	575.9	475.3
580	306.85	584	0.6084	1.047	1.3783	29.52	45.41	0.680	586.4	479.2
590	316.85	602	0.5980	1.049	1.3772	29.84	46.01	0.680	596.9	483.2
600	326.85	620	0.5881	1.051	1.3760	30.17	46.61	0.680	607.4	486.9
620	346.85	656	0.5691	1.056	1.3737	30.82	47.80	0.681	628.4	494.5
640	366.85	692	0.5514	1.061	1.3714	31.47	48.96	0.682	649.6	502.1
660	386.85	728	0.5347	1.065	1.3691	32.09	50.12	0.682	670.9	509.4
680	406.85	764	0.5189	1.070	1.3668	32.71	51.25	0.683	692.2	516.7
700	426.85	800	0.5040	1.075	1.3646	33.32	52.36	0.684	713.7	523.7
720	446.85	836	0.4901	1.080	1.3623	33.92	53.45	0.685	735.2	531.0
740	466.85	872	0.4769	1.085	1.3601	34.52	54.53	0.686	756.9	537.6
760	486.85	903	0.4643	1.089	1.3580	35.11	55.62	0.687	778.6	544.6
780	506.85	944	0.4524	1.094	1.3559	35.69	56.68	0.688	800.5	551.2
800	526.85	950	0.4410	1.099	1.354	36.24	57.74	0.689	822.4	557.8
850	576.85	1070	0.4152	1.110	1.349	37.63	60.30	0.693	877.5	574.1
900	626.85	1160	0.3920	1.121	1.345	38.97	62.76	0.696	933.4	589.6
950	676.85	1250	0.3714	1.132	1.340	40.26	65.20	0.699	989.7	604.9
1000	726.85	1340	0.3529	1.142	1.336	41.53	67.54	0.702	1046	619.5
1100	826.85	1520	0.3208	1.161	1.329	43.96			1162	648.0
1200	926.85	1700	0.2941	1.179	1.322	46.26			1279	675.2
1300	1026.85	1580	0.2714	1.197	1.316	48.46			1398	701.0
1400	1126.85	2060	0.2521	1.214	1.310	50.57			1518	725.9
1500	1220.85	2240	0.2353	1.231	1.304	52.61			1640	749.4
1600	1326.85	2420	0.2206	1.249	1.299	54.57			1764	772.6
1800	1526.85	2780	0.1960	1.288	1.288	58.29			2018	815.7
2000	1726.85	3140	0.1764	1.338	1.274				2280	855.5
2400	2126.85	3860	0.1467	1.575	1.238				2853	924.4
2800	2526.85	4580	0.1245	2.259	1.196				3599	983.1

†Symbols and units: K, absolute temperature, degrees Kelvin; °C, temperature, degrees Celsius; °F, temperature, degrees Fahrenheit; ρ, density, kg/m³; c_p, specific heat capacity, kJ/kg K; c_p/c_r, specific heat capacity ratio, dimensionless; μ, viscosity [for N s/m² (= kg/m s) multiply tabulated values by 10^{-6}]; k, thermal conductivity, 10^3 W/m K; Pr, Prandtl number, dimensionless; h, enthalpy, kJ/kg; V_s, sound velocity, m/s.

Source: From R. C. Weast, Ed. (1970) *Handbook of Tables for Applied Engineering Science*, CRC Press, Boca Raton, FL.

D.2   AIR IN ENGLISH UNITS

Temperature			Properties							
K	°R	°F	ρ	c_p	c_p/c_v	μ	k	Pr	h	V_s
100	180	−280	0.2247	0.2456		0.0466	0.00534	0.770	42.3	651
110	198	−262	0.2033	0.2440	1.4202	0.0513	0.00586	0.768	46.7	685
120	215	−244	0.1858	0.2430	1.4166	0.0559	0.00638	0.766	51.1	717
130	234	−226	0.1711	0.2423	1.4139	0.0604	0.00690	0.763	55.5	747
140	252	−208	0.1586	0.2418	1.4119	0.0648	0.00742	0.761	59.8	776
150	270	−190	0.1478	0.2414	1.4102	0.0691	0.00793	0.758	64.2	804
160	288	−172	0.1384	0.2411	1.4089	0.0733	0.00844	0.754	68.5	831
170	306	−154	0.1301	0.2408	1.4079	0.0774	0.00895	0.750	72.9	856
180	324	−136	0.1228	0.2406	1.4071	0.0815	0.00946	0.746	77.2	882
190	342	−118	0.1163	0.2405	1.4064	0.0854	0.00996	0.743	81.5	906
200	360	−100	0.1104	0.2404	1.4057	0.0892	0.01045	0.739	85.8	930
205	369	−91	0.1078	0.2403	1.4055	0.0911	0.01070	0.738	88.0	941
210	378	−82	0.1051	0.2403	1.4053	0.0930	0.01095	0.736	90.2	953
215	387	−73	0.1027	0.2403	1.4050	0.0949	0.01119	0.734	92.3	964
220	396	−64	0.1003	0.2402	1.4048	0.0967	0.01143	0.732	94.5	976
225	405	−55	0.0981	0.2402	1.4046	0.0986	0.01168	0.731	96.7	987
230	414	−46	0.0959	0.2402	1.4044	0.1004	0.01191	0.729	98.8	998
235	423	−37	0.0939	0.2402	1.4042	0.1022	0.01215	0.727	101.1	1008
240	432	−28	0.0919	0.2401	1.4040	0.1039	0.01239	0.725	103.1	1019
245	441	−19	0.0901	0.2401	1.4038	0.1057	0.01263	0.724	105.3	1030
250	450	−10	0.0882	0.2401	1.4036	0.1074	0.01287	0.722	107.5	1040
255.4	459.7	0	0.0865	0.2401	1.4034	0.1092	0.01310	0.721	109.6	1051
260	468	8	0.0848	0.2401	1.4032	0.1109	0.01334	0.719	111.8	1061
265	477	17	0.0832	0.2402	1.4030	0.1126	0.01357	0.717	114.0	1071
270	486	26	0.0817	0.2402	1.4029	0.1143	0.01380	0.716	116.1	1081
275	495	35	0.0802	0.2402	1.4026	0.1160	0.01403	0.715	118.3	1091
280	504	44	0.0787	0.2402	1.4024	0.1176	0.01426	0.713	120.4	1101
285	513	53	0.0774	0.2402	1.4022	0.1192	0.01448	0.711	122.6	1111
290	522	62	0.0760	0.2403	1.4020	0.1208	0.01472	0.710	124.8	1120
295	531	71	0.0747	0.2403	1.4018	0.1224	0.01494	0.709	126.9	1130
300	540	80	0.0735	0.2404	1.4017	0.1241	0.01516	0.708	129.1	1140
305	549	89	0.0723	0.2404	1.4015	0.1257	0.01539	0.707	131.3	1149
310	558	98	0.0711	0.2405	1.4013	0.1272	0.01561	0.705	133.4	1158
315	567	107	0.0700	0.2405	1.4010	0.1287	0.01583	0.704	135.6	1167
320	576	116	0.0689	0.2406	1.4008	0.1303	0.01606	0.703	137.7	1177
325	585	125	0.0678	0.2407	1.4006	0.1319	0.01627	0.702	139.9	1186
330	594	134	0.0668	0.2407	1.4004	0.1334	0.01649	0.701	142.1	1195
335	603	143	0.0658	0.2408	1.4001	0.1349	0.01670	0.700	144.2	1204
340	612	152	0.0648	0.2409	1.3999	0.1364	0.01692	0.699	146.4	1213
345	621	161	0.0639	0.2410	1.3996	0.1379	0.01713	0.698	148.6	1221
350	630	170	0.0630	0.2411	1.3993	0.1394	0.01735	0.697	150.7	1230
355	639	179	0.0621	0.2411	1.3990	0.1409	0.01758	0.696	152.9	1239
360	648	188	0.0612	0.2412	1.3987	0.1423	0.01779	0.695	155.1	1247

Temperature			Properties							
K	°R	°F	ρ	c_p	c_p/c_v	μ	k	Pr	h	V_s
365	657	197	0.0604	0.2313	1.3984	0.1437	0.01800	0.694	157.3	1256
370	666	206	0.0595	0.2415	1.3981	0.1452	0.01820	0.693	159.4	1264
375	675	215	0.0588	0.2416	1.3978	0.1465	0.01841	0.692	161.6	1273
380	684	224	0.0580	0.2417	1.3975	0.1479	0.01862	0.691	163.8	1281
385	693	233	0.0572	0.2418	1.3971	0.1494	0.01883	0.690	166.0	1289
390	702	242	0.0565	0.2420	1.3968	0.1508	0.01904	0.690	168.1	1298
395	711	251	0.0559	0.2421	1.3964	0.1522	0.01925	0.689	170.3	1306
400	720	260	0.0551	0.2422	1.3961	0.1536	0.01945	0.689	172.5	1314
410	738	278	0.0537	0.2425	1.3953	0.1563	0.01985	0.688	176.9	1330
420	756	296	0.0525	0.2428	1.3946	0.1590	0.02026	0.687	181.2	1346
430	774	314	0.0512	0.2432	1.3938	0.1617	0.02066	0.686	185.6	1361
440	792	332	0.0501	0.2435	1.3929	0.1643	0.02106	0.684	190.0	1377
450	810	350	0.0490	0.2439	1.3920	0.1670	0.02144	0.684	194.4	1392
460	828	368	0.0479	0.2443	1.3911	0.1695	0.02183	0.683	198.8	1.407
470	846	386	0.0469	0.2447	1.3901	0.1719	0.02222	0.682	203.2	1421
480	864	404	0.0459	0.2451	1.3892	0.1744	0.02260	0.681	207.6	1436
490	882	422	0.0450	0.2456	1.3881	0.1769	0.02298	0.680	212.0	1450
500	900	440	0.0441	0.2460	1.3871	0.1794	0.02335	0.680	216.4	1464
510	918	458	0.0432	0.2465	1.3861	0.1818	0.02373	0.680	220.8	1478
520	936	476	0.0424	0.2469	1.3851	0.1842	0.02409	0.680	225.3	1492
530	954	494	0.0416	0.2474	1.3840	0.1867	0.02445	0.680	229.7	1506
540	972	512	0.0408	0.2479	1.3829	0.1891	0.02482	0.680	234.2	1520
550	990	530	0.0400	0.2484	1.3818	0.1914	0.02518	0.680	238.7	1533
560	1008	548	0.0393	0.2490	1.3806	0.1937	0.02554	0.680	243.1	1546
570	1026	566	0.0386	0.2495	1.3795	0.1960	0.02589	0.680	247.6	1559
580	1044	584	0.0380	0.2500	1.3783	0.1983	0.02624	0.680	252.1	1572
590	1062	602	0.0373	0.2506	1.3772	0.2005	0.02659	0.680	256.6	1585
600	1080	620	0.0367	0.2511	1.3760	0.2027	0.02694	0.680	261.1	1597
620	1116	656	0.0355	0.2522	1.3737	0.2071	0.02762	0.681	270.2	1622
640	1152	692	0.0344	0.2533	1.3714	0.2115	0.02829	0.682	279.3	1647
660	1188	728	0.0334	0.2545	1.3691	0.2156	0.02896	0.682	288.4	1671
680	1224	764	0.0324	0.2556	1.3668	0.2198	0.02962	0.683	297.6	1695
700	1260	800	0.0315	0.2568	1.3646	0.2239	0.03026	0.684	306.8	1718
720	1296	836	0.0306	0.2579	0.3623	0.2279	0.03089	0.685	316.1	1742
740	1332	872	0.0298	0.2591	1.3601	0.2320	0.03151	0.686	325.4	1764
760	1368	908	0.0290	0.2602	1.3580	0.2359	0.03214	0.687	334.8	1787
780	1404	944	0.0282	0.0213	1.3559	0.2398	0.03275	0.688	344.1	1808
800	1440	980	0.0275	0.2624	1.354	0.2435	0.03337	0.689	353.6	1830
850	1530	1070	0.0259	0.2653	1.349	0.2529	0.03485	0.693	377.3	1883
900	1620	1160	0.0245	0.2678	1.345	0.2618	0.03627	0.696	401.3	1934
950	1710	1250	0.0231	0.2704	1.340	0.2705	0.03768	0.699	425.5	1985
1000	1800	1340	0.0220	0.2728	1.336	0.2790	0.03903	0.702	450.0	2032
1100	1980	1520	0.0200	0.2774	1.329	0.2954			499.5	2126
1200	2160	1700	0.0184	0.2817	1.322	0.3108			549.8	2215

Temperature			Properties							
K	**°R**	**°F**	**ρ**	**c_p**	**c_p / c_v**	**μ**	**k**	**Pr**	**h**	**V_s**
1300	2340	1880	0.0169	0.2860	1.316	0.3256			600.9	2300
1400	2520	2060	0.0157	0.2900	1.310	0.3398			652.7	2381
1500	2700	2240	0.0147	0.2940	1.304	0.3535			705.3	2459
1600	2880	2420	0.0138	0.2984	1.299	0.3667			758.6	2535
1800	3240	2780	0.0122	0.3076	1.288	0.3917			867.7	2676
2000	3600	3140	0.0110	0.3196	1.274				980.5	2807
2400	4320	3860	0.0092	0.3760	1.238				1226.8	3033
2800	5040	4550	0.0078	0.5396	1.196				1547.3	3225

†Symbols and units: K, degrees Kelvin; °R, degrees Rankine; °F, degrees Fahrenheit; ρ, density, lbm/ft^3; c_p, specific heat capacity, Btu/lbm R° = cal/g K; c_p/c_r, specific heat capacity ratio dimensionless; μ, viscosity (for lbm/ft multiply by 10^{-4}); k, thermal conductivity, Btu/hr ft °R: Pr, Prandtl number, dimensionless; h, enthalpy, Btu/lbm (for cal/g multiply by 0.5555); V_s, sound velocity, ft/s.

Source: From R. C. Weast, Ed. (1970) *Handbook of Tables for Applied Engineering Science*, CRC Press, Boca Raton, FL.

D.3 OTHER GASES AND VAPORS

Substance	Temperature °C	Temperature °F	ρ, g / cm^3	c_v, cal / g K	k, cal / s cm K	μ, centipoise
Ammonia	0	32	9.56×10^{-4}	0.52	5.23×10^{-9}	9.18×10^{-3}
	20	68	8.94×10^{-4}	0.52	5.69×10^{-5}	9.82×10^{-3}
	50	122	8.11×10^{-4}	0.52	6.48×10^{-5}	1.09×10^{-2}
	100	212	7.02×10^{-4}	0.53		1.28×10^{-2}
	200	392	6.20×10^{-4}			1.64×10^{-2}
	300	572	5.12×10^{-4}			1.99×10^{-2}
Argon	-13	9	1.87×10^{-3}	0.125	3.74×10^{-3}	2.04×10^{-2}
	-3	37	1.81×10^{-3}	0.125	3.87×10^{-3}	2.11×10^{-2}
	7	45	1.74×10^{-3}	0.125	3.99×10^{-3}	2.17×10^{-2}
	27	81	1.62×10^{-3}	0.125	4.22×10^{-3}	2.30×10^{-2}
	77	171	1.39×10^{-3}	0.124	4.79×10^{-3}	2.59×10^{-2}
	227	441	9.74×10^{-4}	0.124	6.31×10^{-5}	3.37×10^{-2}
	727	1341	4.87×10^{-4}	0.124	1.02×10^{-4}	5.42×10^{-2}
	1227	2241	3.25×10^{-4}	0.124	1.31×10^{-4}	7.08×10^{-2}
	1727	3141	2.43×10^{-4}	0.124		
Butane	0	32	2.59×10^{-3}	0.3802	3.16×10^{-5}	6.84×10^{-3}
	100	212	1.90×10^{-3}	0.4842	5.60×10^{-5}	9.26×10^{-5}
	200	392	1.50×10^{-3}	0.5865	8.70×10^{-5}	1.17×10^{-2}
	300	572	1.24×10^{-3}	0.6721	1.24×10^{-4}	1.40×10^{-2}
	400	752	1.05×10^{-3}	0.7474	1.66×10^{-4}	1.64×10^{-2}
	500	932	9.16×10^{-4}	0.8131	2.15×10^{-4}	1.87×10^{-2}
	600	1112	8.12×10^{-4}	0.8704	2.69×10^{-4}	2.11×10^{-2}
Carbon dioxide	-13	9	2.08×10^{-3}	0.1944	3.25×10^{-3}	1.31×10^{-2}
	-3	27	2.00×10^{-3}	0.1967	3.42×10^{-3}	1.36×10^{-2}
	7	45	1.93×10^{-3}	0.1989	3.60×10^{-3}	1.40×10^{-2}
	17	63	1.86×10^{-3}	0.2012	3.78×10^{-5}	1.45×10^{-2}
	27	81	1.80×10^{-3}	0.2035	3.96×10^{-5}	1.49×10^{-2}
	77	171	1.54×10^{-3}	0.2146	4.89×10^{-5}	1.72×10^{-2}
	227	441	1.07×10^{-3}	0.2424	8.01×10^{-5}	2.32×10^{-2}
	727	1341	5.36×10^{-4}	0.2946		3.89×10^{-2}
	1227	2241	3.57×10^{-4}	0.3166		
Carbon monoxide	-13	9	1.31×10^{-3}	0.2489	5.30×10^{-5}	1.59×10^{-2}
	-3	27	1.27×10^{-3}	0.2489	5.49×10^{-5}	1.64×10^{-2}
	7	45	1.22×10^{-3}	0.2489	5.67×10^{-5}	1.69×10^{-2}
	17	63	1.18×10^{-3}	0.2489	5.85×10^{-5}	1.74×10^{-2}
	27	81	1.14×10^{-3}	0.2489	6.03×10^{-5}	1.79×10^{-2}
	77	171	9.75×10^{-4}	0.2493	6.89×10^{-5}	2.01×10^{-2}
	227	441	6.82×10^{-4}	0.2542	9.23×10^{-5}	2.61×10^{-2}
	727	1341	3.41×10^{-4}			4.17×10^{-2}
	1227	2241	2.27×10^{-4}			5.44×10^{-2}
Ethane	0	32	1.342×10^{-3}	0.3934	4.52×10^{-5}	8.60×10^{-3}
	100	212	9.83×10^{-4}	0.4938	7.59×10^{-5}	1.14×10^{-2}
	200	392	7.76×10^{-4}	0.5947	1.13×10^{-4}	1.41×10^{-2}
	300	572	6.40×10^{-4}	0.6854	1.55×10^{-4}	1.68×10^{-2}
	400	752	5.45×10^{-4}	0.7676	2.04×10^{-4}	1.93×10^{-2}
	500	932	4.74×10^{-4}	0.8405	2.57×10^{-4}	2.20×10^{-2}
	600	1112	4.20×10^{-4}	0.9045	3.15×10^{-4}	2.45×10^{-2}
Ethanol	100	212	1.49×10^{-3}	0.403	5.50×10^{-3}	1.08×10^{-2}
	200	392	1.18×10^{-3}	0.480	8.39×10^{-3}	1.37×10^{-2}

Substance	Temperature °C	°F	ρ, g/cm³	c_v, cal/g K	k, cal/s cm K	μ, centipoise
	300	572	9.74×10^{-4}	0.554	1.19×10^{-4}	1.67×10^{-2}
	400	752	8.28×10^{-4}	0.624	1.59×10^{-4}	1.97×10^{-2}
	500	932	7.20×10^{-4}	0.691	2.05×10^{-4}	2.26×10^{-2}
Helium	-240	-400	1.463×10^{-3}		8.43×10^{-5}	3.74×10^{-3}
	-129	-200	3.38×10^{-4}		2.22×10^{-4}	1.19×10^{-2}
	0	32	3.68×10^{-4}	1.23	3.40×10^{-4}	1.86×10^{-2}
	20	68	1.67×10^{-4}	1.24	3.55×10^{-4}	1.94×10^{-2}
	40	104	1.56×10^{-4}	1.24	3.70×10^{-4}	2.03×10^{-2}
	49	120		1.24	3.76×10^{-4}	2.06×10^{-2}
Hydrogen	-13	9	9.44×10^{-3}	3.373	3.86×10^{-4}	8.14×10^{-3}
	-3	37	9.10×10^{-5}	3.388	3.98×10^{-4}	8.35×10^{-3}
	7	45	8.77×10^{-5}	3.400	4.11×10^{-4}	8.55×10^{-3}
	27	81	8.47×10^{-5}	3.410	4.22×10^{-4}	8.76×10^{-3}
	77	171	8.19×10^{-3}	3.418	4.34×10^{-4}	8.96×10^{-3}
	227	441	7.02×10^{-3}		4.91×10^{-4}	9.94×10^{-3}
	727	1341	4.912×10^{-3}	3.467	6.50×10^{-4}	1.26×10^{-2}
Methane	0	32	7.16×10^{-4}	0.5172	7.33×10^{-3}	1.04×10^{-2}
	100	212	5.25×10^{-4}	0.5848	1.11×10^{-4}	1.32×10^{-2}
	200	392	4.14×10^{-4}	0.6704	1.52×10^{-4}	1.59×10^{-2}
	300	572	3.42×10^{-4}	0.7584	1.96×10^{-4}	1.83×10^{-2}
	400	752	2.91×10^{-4}	0.8430	2.43×10^{-4}	2.07×10^{-2}
	500	932	2.53×10^{-4}	0.9210	2.91×10^{-4}	2.29×10^{-2}
	600	1112	2.24×10^{-4}	0.9919	3.43×10^{-4}	2.52×10^{-2}
Nitrogen	-13	9	1.31×10^{-3}	0.2488	5.52×10^{-5}	
	-3	37	1.27×10^{-3}	0.2487	5.71×10^{-5}	
	7	45	1.22×10^{-3}	0.2487	5.89×10^{-5}	
	27	81	1.18×10^{-3}	0.2487	6.06×10^{-5}	
	77	171	1.14×10^{-3}	0.2487	6.24×10^{-5}	1.79×10^{-2}
	227	441	9.75×10^{-3}	0.2490	7.11×10^{-3}	2.00×10^{-2}
	727	1341	6.82×10^{-3}	0.2524	9.49×10^{-2}	2.57×10^{-2}
Oxygen	-13	9	1.50×10^{-3}	0.2188	5.60×10^{-3}	1.85×10^{-2}
	-3	37	1.45×10^{-3}	0.2190	5.80×10^{-3}	1.90×10^{-2}
	7	45	1.39×10^{-3}	0.2193	5.98×10^{-3}	1.96×10^{-2}
	27	81	1.35×10^{-3}	0.2195	6.22×10^{-3}	2.01×10^{-2}
	77	171	1.30×10^{-3}	0.2198	6.40×10^{-3}	2.06×10^{-2}
	227	441	1.11×10^{-3}	0.2221	7.33×10^{-3}	2.32×10^{-2}
	727	1341	7.80×10^{-3}	0.2324	9.97×10^{-3}	2.99×10^{-2}
Propane	0	32	1.97×10^{-3}	0.3701	3.62×10^{-3}	7.50×10^{-3}
	100	212	1.44×10^{-3}	0.4817	6.26×10^{-3}	1.00×10^{-2}
	200	392	1.14×10^{-3}	0.5871	9.56×10^{-3}	1.25×10^{-2}
	300	572	9.39×10^{-4}	0.6770	1.34×10^{-4}	1.40×10^{-2}
	400	752	7.99×10^{-4}	0.7550	1.78×10^{-4}	1.72×10^{-2}
	500	932	6.94×10^{-4}	0.8237	2.28×10^{-4}	1.94×10^{-2}
	600	1112	6.16×10^{-4}	0.8831	2.83×10^{-4}	2.18×10^{-2}

†Symbols and conversion factors: ρ, density in g/cm³ (for lb/ft³ multiply by 62.428, for kg/m³ multiply by 1000); c_p, specific heat, cal/g K (for Btu/lb °R multiply by 1, for J/kg K multiply by 4184.0); k, thermal conductivity, cal/s cm K (for W/m K multiply by 418 4, for Btu/hr ft °R multiply by 241.9); μ, absolute viscosity in centipoises (for lb/ft hr multiply by 2.419, for N s/m² multiply by 0.001).

Source: Reprinted with permission from R. C. Weast, Ed., *Handbook of Tables for Applied Engineering Science*, Copyright © 1970, CRC Press, Inc., Boca Raton, FL.

D.4   PROPERTIES OF STEAM

Pres-sure, psia	Satu-rated vapor	Temperature, °F						
		32	**200**	**400**	**600**	**800**	**1000**	**1200**
Viscosity, lb / hr ft								
0	—	0.023	0.031	0.041	0.050	0.059	0.067	0.074
500	0.054	—	—	—	0.059	0.073	0.073	0.080
1000	0.070	—	—	—	0.069	0.074	0.080	0.086
1500	0.082	—	—	—	0.082	0.082	0.087	0.092
2000	0.094	—	—	—	—	0.086	0.094	0.097
2500	0.108	—	—	—	—	0.101	0.101	0.104
3000	0.116	—	—	—	—	0.110	0.108	0.110
3500	—	—	—	—	—	0.119	0.114	0.116
Thermal conductivity, Btu / hr ft °F								
0	—	0.0092	0.0133	0.0184	0.0238	0.0292	0.0347	
250	0.0211	—	—	—	0.0247	0.0206	0.0349	
500	0.0251	—	—	—	0.0260	0.0302	0.0352	
1000	0.0316	—	—	—	0.0301	0.0314	0.0357	
1500	0.0379	—	—	—	0.0376	0.0332	0.0364	
1750	0.0408	—	—	—	—	0.0343	0.0368	
2000	0.0445	—	—	—	—	0.0355	0.0372	

Source: Reprinted with permission from B. Gebhart, *Heat Transfer*, Copyright © 1971, McGraw-Hill, New York. Also from J. Keenan and F. Keyes, *Thermodynamic Properties of Steam*. Copyright © 1955, John Wiley and Sons, New York.

APPENDIX
E

DIFFUSIVITIES
OF CHEMICAL
SPECIES

E.1 THE TEMPERATURE DEPENDENCE OF WATER VAPOR DIFFUSION INTO AIR

Temperature, °C	Diffusion constant, D		$\left(\dfrac{\mu}{\rho D}\right)$†
	ft^2 / hr	cm^2 / s	
0	0.844	0.218	0.608
10	0.898	0.232	0.610
20	0.952	0.246	0.612
30	1.01	0.260	0.614
40	1.06	0.275	0.615
50	1.12	0.290	0.616
60	1.18	0.305	0.618
70	1.24	0.321	0.619
80	1.30	0.337	0.619

†The values of $\mu/\rho D$ were calculated using the viscosity and density of dry air. Thus the values apply only when the diffusing water vapor is very dilute.

Source: Reprinted with permission from R. C. Weast, Ed., *Handbook of Tables for Applied Engineering Science.* Copyright © 1970, CRC Press, Inc., Boca Raton, FL.

E.2 GASES AND VAPORS INTO AIR

Substance	Diffusion constant, D, ft^2 / hr		Diffusion constant, D, cm^2 / s		$\left(\dfrac{\mu}{\rho D}\right)$†	
	0°C	25°C	0°C	25°C	0°C	25°C
H_2	2.37	2.76	0.611	0.712	0.217	0.216
NH_3	0.766	0.886	0.198	0.229	0.669	0.673
N_2	0.691		0.178		0.744	
O_2	0.689	0.80	0.178	0.206	0.744	0.748
CO_2	0.550	0.635	0.142	0.164	0.933	0.940
CS_2	0.36	0.414	0.094	0.107	1.41	1.44
Methyl alcohol	0.513	0.615	0.132	0.159	1.00	0.969
Formic acid	0.509	0.615	0.131	0.159	1.01	0.969
Acetic acid	0.411	0.515	0.106	0.133	1.25	1.16
Ethyl alcohol	0.394	0.461	0.102	0.119	1.30	1.29
Chloroform	0.352		0.091		1.46	
Diethylamine	0.342	0.406	0.0884	0.105	1.50	1.47
n-Propyl alcohol	0.329	0.387	0.085	0.100	1.56	1.54
Propionic acid	0.328	0.383	0.0846	0.099	1.57	1.56
Methyl acetate	0.325	0.387	0.0840	0.100	1.58	1.54
Butylamine	0.318	0.391	0.0821	0.101	1.61	1.53
Ethyl ether	0.304	0.360	0.0786	0.093	1.69	1.66
Benzene	0.291	0.341	0.0751	0.088	1.76	1.75
Ethyl acetate	0.277	0.330	0.0715	0.085	1.85	1.81
Toluene	0.274	0.325	0.0709	0.084	1.87	1.83
n-Butyl alcohol	0.272	0.348	0.0703	0.090	1.88	1.17
i-Butyric acid	0.263	0.313	0.0679	0.081	1.95	1.90
Chlorobenzene		0.283		0.073		2.11
Aniline	0.236	0.279	0.0610	0.072	2.17	2.14
Xylene	0.228	0.275	0.059	0.071	2.25	2.17
Amyl alcohol	0.228	0.271	0.0589	0.070	2.25	2.20
n-Octane	0.195	0.232	0.0505	0.060	2.62	2.57
Naphthalene	0.199	0.20	0.0513	0.052	2.58	2.96

†Based on $\mu/\rho = 0.1325$ cm^2/s for air at 0°C and 0.1541 cm^2/s for air at 25°C; applies only when the diffusing gas or vapor is very dilute.

Source: Reprinted with permission from R. C. Weast, Ed., *Handbook of Tables for Applied Engineering Science*. Copyright © 1970, CRC Press, Inc., Boca Raton, FL.

E.3 MEASURED DIFFUSION COEFFICIENTS IN GASES AT 1 ATM

Gas pair	T, K	cm^2 / s
Air–CH_4	273.0	0.196
Air–C_2H_5OH	273.0	0.102
Air–CO_2	265.2	0.142
	317.2	0.177
Air–H_2	273.0	0.611
Air–D_2	296.8	0.565
Air–H_2O	289.1	0.282
	298.2	0.260
	312.6	0.277
	333.2	0.3050
Air–He	276.2	0.6242
Air–O_2	273.0	0.1775
Air–n-hexane	294	0.080
Air–n-heptane	294	0.071
Air–benzene	298.2	0.096
Air–toluene	299.1	0.0860
Air–chlorobenzene	299.1	0.074
Air–aniline	299.1	0.074
Air–nitrobenzene	298.2	0.0855
Air–2-propanol	299.1	0.099
Air–butanol	299.1	0.087
Air–2-butanol	299.1	0.089
Air–2-pentanol	299.1	0.071
Air–ethylacetate	299.1	0.087
CH_4–Ar	298	0.202
CH_4–He	298	0.675
CH_4–H_2	298.0	0.726
CH_4–H_2O	307.7	0.292
CO–N_2	295.8	0.212
^{12}CO–^{14}CO	373	0.323
CO–H_2	295.6	0.7430
CO–D_2	295.7	0.5490
CO–He	295.6	0.7020
CO–Ar	295.7	0.1880
CO_2–H_2	298.0	0.6460
CO_2–N_2	298.2	0.165
CO_2–O_2	293.2	0.160
CO_2–He	298	0.612
CO_2–Ar	276.2	0.1326
CO_2–CO	296.1	0.1520
CO_2–H_2O	307.5	0.202
CO_2–N_2O	298.0	0.117
CO_2–SO_2	263	0.064
$^{12}CO_2$–$^{14}CO_2$	312.8	0.125
CO_2–propane	298.0	0.0863
CO_2–ethyleneoxide	298.0	0.0914
H_2–N_2	297.2	0.779
H_2–O_2	273.2	0.697

Gas pair	T, K	cm^2 / s
H_2–D_2	288.2	1.24
H_2–He	298.2	1.132
H_2–Ar	287.9	0.828
H_2–Xe	341.2	0.751
H_2–SO_2	285.5	0.525
H_2–H_2O	307.1	0.915
H_2–NH_3	298	0.783
H_2–acetone	296	0.424
H_2–ethane	298.0	0.537
H_2–n-butane	287.9	0.361
H_2–n-hexane	288.7	0.290
H_2–cyclohexane	288.6	0.319
H_2–benzene	311.3	0.404
H_2–SF_6	286.2	0.396
H_2–n-heptane	303.2	0.283
H_2–n-decane	364.1	0.306
N_2–O_2	273.2	0.181
	293.2	0.22
N_2–He	298	0.687
N_2–Ar	293	0.194
N_2–NH_3	298	0.230
N_2–H_2O	307.5	0.256
N_2–SO_2	263	0.104
N_2–ethylene	298.0	0.163
N_2–ethane	298	0.148
N_2–n-butane	298	0.096
N_2-isobutane	298	0.0905
N_2–n-hexane	288.6	0.076
N_2–n-octane	303.1	0.073
N_2–2,2,4-trimethylpentane	303.3	0.071
N_2–n-decane	363.6	0.084
N_2–benzene	311.3	0.102
O_2–He (He trace)	298.2	0.737
(O_2 trace)	298.2	0.718
O_2–He	298	0.729
O_2–H_2O	308.1	0.282
O_2–CCl_4	296	0.075
O_2–benzene	311.3	0.101
O_2–cyclohexane	288.6	0.075
O_2–n-hexane	288.6	0.075
O_2–n-octane	303.1	0.071
O_2–2,2,4-trimethylpentane	303.0	0.071
He–D_2	295.1	1.250
He–Ar	298	0.742
He–H_2O	298.2	0.908
He–NH_3	297.1	0.842
He–n-hexane	417.0	0.1574
He–benzene	298.2	0.384
He–Ne	341.2	1.405
He–methanol	42.2	1.032
He–ethanol	298.2	0.494

Gas pair	T, K	cm^2/s
He–propanol	423.2	0.676
He–hexanol	423.2	0.469
Ar–Ne	303	0.327
Ar–Kr	303	0.140
Ar–Xe	329.9	0.137
Ar–NH_3	295.1	0.232
Ar–SO_2	263	0.077
Ar–n-hexane	288.6	0.066
Ne–Kr	273.0	0.223
Ethylene–H_2O	307.8	0.204
Ethane–n-hexane	294	0.0375
N_2O–propane	298	0.0860
N_2O–ethyleneoxide	298	0.0914
NH_3–SF_6	296.6	0.1090
Freon-12–H_2O	298.2	0.1050
Freon-12–benzene	298.2	0.0385
Freon-12–ethanol	298.2	0.0475

Source: From Cussler (1985).

E.4 DIFFUSION OF SOLUTES INTO DILUTE WATER SOLUTIONS AT 20°C

Substance	Diffusion constant, D†		Schmidt number, $(\mu/\rho D)$¶	Substance	Diffusion constant, D†		Schmidt number, $(\mu/\rho D)$¶
	English‡	Metric§			English‡	Metric§	
H_2	19.8	5.13	196	H_2SO_4	6.70	1.73	581
O_2	6.97	1.80	558	NaOH	5.84	1.51	666
CO_2	6.85	1.77	568	NaCl	5.22	1.35	744
NH_3	6.81	1.76	571	Ethyl alcohol	3.87	1.00	1005
N_2	6.35	1.64	613	Acetic acid	3.41	0.88	1140
Acetylene	6.04	1.56	644	Phenol	3.25	0.84	1200
Cl_2	4.72	1.22	824	Glycerol	2.79	0.72	1400
HCl	10.2	2.64	381	Sucrose	1.74	0.45	2230
HNO_3	10.1	2.6	390				

†The following relationship may be used to estimate the effect of temperature on the diffusion constant: $D_1/D_2 = (T_2/T_2)(\mu_2/\mu_1)$, where T is temperature, degrees Kelvin, and μ is the solution viscosity, centipoises. The diffusion constant varies with concentration because of the changes in viscosity and the degree of ideality of the solution.

†For English units in ft^2/hr multiply by 10^{-5}.

§For metric units in m^2/s multiply by 10^{-9}.

¶Based on $\mu/\rho = 0.01005 \ cm^2/s$ for water at 20°C; applies only for dilute solutions.

Source: Reprinted with permission from R. C. Weast, Ed., *Handbook of Tables for Applied Engineering Science.* Copyright © 1970, CRC Press, Inc., Boca Raton, FL.

E.5   DIFFUSION COEFFICIENTS AT INFINITE DILUTION IN WATER AT 25°C

Solute	cm^2/s ($\times 10^5$)
Argon	2.00
Air	2.00
Bromine	1.18
Carbon dioxide	1.92
Carbon monoxide	2.03
Chlorine	1.25
Ethane	1.20
Ethylene	1.87
Helium	6.28
Hydrogen	4.50
Methane	1.49
Nitric oxide	2.60
Nitrogen	1.88
Oxygen	2.10
Propane	0.97
Ammonia	1.64
Benzene	1.02
Hydrogen sulfide	1.41
Sulfuric acid	1.73
Nitric acid	2.60
Acetylene	0.88
Methanol	0.84
Ethanol	0.84
1-Propanol	0.87
2-Propanol	0.87
n-Butanol	0.77
Benzyl alcohol	0.821
Formic acid	1.50
Acetic acid	1.21
Propionic acid	1.06
Benzoic acid	1.00
Glycine	1.06
Valine	0.83
Acetone	1.16
Urea	$(1.380 - 0.0782C + 0.00464C^2)$†
Sucrose	$(0.5228 - 0.265C)$†
Ovalbumin	0.078
Hemoglobin	0.069
Urease	0.035
Fibrinogen	0.020

†C in moles per liter.

Source: From Cussler (1985).

E.6 DIFFUSION COEFFICIENTS AT INFINITE DILUTION IN NONAQUEOUS SOLUTIONS, AT 25°C UNLESS INDICATED OTHERWISE

Solute	Solvent	$cm^2 / s \ (\times 10^5)$
Acetone	Chloroform	2.35
Benzene		2.89
n-Butyl acetate		1.71
Ethyl alcohol (15°)		2.20
Ethyl ether		2.14
Ethyl acetate		2.02
Methyl ethyl ketone		2.13
Acetic acid	Benzene	2.09
Aniline		1.96
Benzoic acid		1.38
Cyclohexane		2.09
Ethyl alcohol (15°)		2.25
n-Heptane		2.10
Methyl ethyl ketone (30°)		2.09
Oxygen (29.6°)		2.89
Toluene		1.85
Acetic acid	Acetone	3.31
Benzoic acid		2.62
Nitrobenzene (20°)		2.94
Water		4.56
Carbon tetrachloride	n-Hexane	3.70
Dodecane		2.73
n-Hexane		4.21
Methyl ethyl ketone (30°)		3.74
Propane		4.87
Toluene		4.21
Benzene	Ethyl alcohol	1.81
Camphor (20°)		0.70
Iodine		1.32
Iodobenzene (20°)		1.00
Oxygen (29.6°)		2.64
Water		1.24
Carbon tetrachloride		1.50
Benzene	n-Butyl alcohol	0.988
Biphenyl		0.627
p-Dichlorobenzene		0.817
Propane		1.57
Water		0.56
Acetone (20°)	Ethyl acetate	3.18
Methyl ethyl ketone (30°)		2.93
Nitrobenzene (20°)		2.25
Water		3.20
Benzene	n-Heptane	3.40

Source: From Cussler (1985).

E.7 SOLID-STATE DIFFUSION COEFFICIENTS

Material	t, °C	cm^2 / s
Hydrogen in iron	10	1.66×10^{-9}
	50	11.4×10^{-9}
	100	124×10^{-9}
Hydrogen in nickel	85	1.16×10^{-8}
	165	10.5×10^{-8}
Carbon monoxide in nickel	950	4×10^{-8}
	1050	14×10^{-8}
Aluminum in copper	850	2.2×10^{-9}
Uranium in tungsten	1727	1.3×10^{-11}
Cerium in tungsten	1727	95×10^{-11}
Yttrium in tungsten	1727	1820×10^{-11}
Tin in lead	285	1.6×10^{-10}
Gold in lead	285	4.6×10^{-6}
Gold in silver	760	3.6×10^{-10}
Antimony in silver	20	3.5×10^{-21}
Zinc in aluminum	500	2×10^{-9}
Silver in aluminum	50	1.2×10^{-9}
Bismuth in lead	20	1.1×10^{-16}
Aluminum in copper	20	1.3×10^{-30}
Cadmium in copper	20	2.7×10^{-15}
Carbon in iron	800	1.5×10^{-8}
	1100	45×10^{-8}
Helium in SiO_2	20	4.0×10^{-10}
	500	7.8×10^{-8}
Hydrogen in SiO_2	200	6.5×10^{-10}
	500	1.3×10^{-8}
Helium in Pyrex	20	4.5×10^{-11}
	500	2×10^{-8}

Source: From Cussler (1985).

APPENDIX
F

THE
ERROR
FUNCTION

$\dfrac{x}{2\sqrt{\alpha\tau}}$	$f\left(\dfrac{x}{2\sqrt{\alpha\tau}}\right)$	$\dfrac{x}{2\sqrt{\alpha\tau}}$	$f\left(\dfrac{x}{2\sqrt{\alpha\tau}}\right)$	$\dfrac{x}{2\sqrt{\alpha\tau}}$	$f\left(\dfrac{x}{2\sqrt{\alpha\tau}}\right)$
0.00	0.00000	0.76	0.71754	1.52	0.96841
0.02	0.02256	0.78	0.73001	1.54	0.97059
0.04	0.04511	0.80	0.74210	1.56	0.97263
0.06	0.06762	0.82	0.75381	1.58	0.97455
0.08	0.09008	0.84	0.76514	1.60	0.97635
0.10	0.11246	0.86	0.77610	1.62	0.97804
0.12	0.13476	0.88	0.78669	1.64	0.97962
0.14	0.15695	0.90	0.79691	1.66	0.98110
0.16	0.17901	0.92	0.80677	1.68	0.98249
0.18	0.20094	0.94	0.81627	1.70	0.98379
0.20	0.22270	0.96	0.82542	1.72	0.98500
0.22	0.24430	0.98	0.83423	1.74	0.98613
0.24	0.26570	1.00	0.84270	1.76	0.98719
0.26	0.28690	1.02	0.85084	1.78	0.98817
0.28	0.30788	1.04	0.85865	1.80	0.98909
0.30	0.32863	1.06	0.86614	1.82	0.98994
0.32	0.34913	1.08	0.87333	1.84	0.99074
0.34	0.36936	1.10	0.88020	1.86	0.99147
0.36	0.38933	1.12	0.88679	1.88	0.99216
0.38	0.40901	1.14	0.89308	1.90	0.99279
0.40	0.42839	1.16	0.89910	1.92	0.99338
0.42	0.44749	1.18	0.90484	1.94	0.99392
0.44	0.46622	1.20	0.91031	1.96	0.99443
0.46	0.48466	1.22	0.91553	1.98	0.99489
0.48	0.50275	1.24	0.92050	2.00	0.995322
0.50	0.52050	1.26	0.92524	2.10	0.997020
0.52	0.53790	1.28	0.92973	2.20	0.998137
0.54	0.55494	1.30	0.93401	2.30	0.998857
0.56	0.57162	1.32	0.93806	2.40	0.999311
0.58	0.58792	1.34	0.94191	2.50	0.999593
0.60	0.60386	1.36	0.94556	2.60	0.999764
0.62	0.61941	1.38	0.94902	2.70	0.999866
0.64	0.63459	1.40	0.95228	2.80	0.999925
0.66	0.64938	1.42	0.95538	2.90	0.999959
0.68	0.66378	1.44	0.95830	3.00	0.999978
0.70	0.67780	1.46	0.96105	3.20	0.999994
0.72	0.69143	1.48	0.96365	3.40	0.999998
0.74	0.70468	1.50	0.96610	3.60	1.000000

APPENDIX
G

LAPLACE
TRANSFORM
PAIRS

$L[t(x, y, z, \tau)] = \bar{t} = \int_0^\infty e^{-p\tau} t(x, y, z, \tau)\, d\tau$ where n, a, ω, α, h, and q are constants

$\bar{t}(p)$	$t(\tau)$
1. $\dfrac{1}{p}$	1
2. $p^{1/(n+1)}$ $\quad n > -1$	$\dfrac{\tau^n}{\Gamma(n+1)}$
3. $\dfrac{1}{p+\alpha}$	$\exp(-\alpha\tau)$
4. $\dfrac{\omega}{p^2 + \omega^2}$	$\sin \omega\tau$
5. $\dfrac{p}{p^2 + \omega^2}$	$\cos \omega\tau$
6. $\exp\left(-\sqrt{\dfrac{p}{a}}\, x\right)$ $\quad \begin{matrix} x > 0 \\ a > 0 \end{matrix}$	$\dfrac{x}{2\sqrt{\pi a \tau^3}} \exp\left(\dfrac{-x^2}{4a\tau}\right)$
7. $\dfrac{1}{\sqrt{p/a}} \exp\left(-\sqrt{\dfrac{p}{a}}\, x\right)$ $\quad \begin{matrix} x > 0 \\ a > 0 \end{matrix}$	$\sqrt{\dfrac{a}{\pi\tau}} \exp\left(\dfrac{-x^2}{4a\tau}\right)$
8. $\dfrac{1}{p} \exp\left(-\sqrt{\dfrac{p}{a}}\, x\right)$ $\quad \begin{matrix} x > 0 \\ a > 0 \end{matrix}$	$\operatorname{erfc} \dfrac{x}{\sqrt{4a\tau}}$
9. $\dfrac{1}{p\sqrt{p/a}} \exp\left(-\sqrt{\dfrac{p}{a}}\, x\right)$ $\quad \begin{matrix} x > 0 \\ a > 0 \end{matrix}$	$2\sqrt{\dfrac{a\tau}{\pi}} \exp\left(\dfrac{-x^2}{4a\tau}\right) - x \operatorname{erfc} \dfrac{x}{\sqrt{4a\tau}}$
10. $\dfrac{1}{p^2} \exp\left(-\sqrt{\dfrac{p}{a}}\, x\right)$ $\quad \begin{matrix} x > 0 \\ a > 0 \end{matrix}$	$\left(\tau + \dfrac{x^2}{2a}\right)\operatorname{erfc}\dfrac{x}{\sqrt{4a\tau}} - x\sqrt{\dfrac{\tau}{\pi a}} \exp\left(\dfrac{-x^2}{4a\tau}\right)$
11. $\dfrac{1}{p^{1+n/2}} \exp\left(\sqrt{\dfrac{p}{a}}\, x\right)$ $\quad \begin{matrix} x > 0 \\ a > 0 \\ n = 0,1,2,\ldots \end{matrix}$	$(4\tau)^{n/2} i^n \operatorname{erfc} \dfrac{x}{\sqrt{4a\tau}}$
12. $\dfrac{\exp\left(-\sqrt{p/a}\, x\right)}{\sqrt{p/a} + h}$ $\quad \begin{matrix} x > 0 \\ a > 0 \end{matrix}$	$\sqrt{\dfrac{a}{\pi\tau}} \exp\left(\dfrac{-x^2}{4a\tau}\right)$ $- ha \exp(hx + ah^2\tau)\operatorname{erfc}\left(\dfrac{x}{\sqrt{4a\tau}} + h\sqrt{a\tau}\right)$
13. $\dfrac{1}{\sqrt{p/a}\left(\sqrt{p/a} + h\right)}$ $\times \exp\left(-\sqrt{\dfrac{p}{a}}\, x\right)$ $\quad \begin{matrix} x > 0 \\ a > 0 \end{matrix}$	$a \exp(hx + ah^2\tau)\operatorname{erfc}\left(\dfrac{x}{\sqrt{4a\tau}} + h\sqrt{a\tau}\right)$

$\bar{t}(p)$	$t(\tau)$
14. $\dfrac{1}{p\left(\sqrt{p/a}+h\right)}$	$\dfrac{1}{h}\,\text{erfc}\,\dfrac{x}{\sqrt{4a\tau}}$
$\times\exp\left(-\sqrt{\dfrac{p}{a}}\,x\right)\qquad\begin{array}{l}x>0\\a>0\end{array}$	$-\dfrac{1}{h}\exp(hx+ah^2\tau)\,\text{erfc}\left(\dfrac{x}{\sqrt{4a\tau}}+h\sqrt{a\tau}\right)$
15. $\dfrac{1}{p\sqrt{p/a}\left(\sqrt{p/a}+h\right)}$	$\dfrac{2}{h}\sqrt{\dfrac{a\tau}{\pi}}\,\exp\left(\dfrac{-x^2}{4a\tau}\right)-\dfrac{1+hx}{h^2}\,\text{erfc}\,\dfrac{x}{\sqrt{4a\tau}}$
$\times\exp\left(-\sqrt{\dfrac{p}{a}}\,x\right)\qquad\begin{array}{l}x>0\\a>0\end{array}$	$+\dfrac{1}{h^2}\exp(hx+ah^2\tau)\,\text{erfc}\left(\dfrac{x}{\sqrt{4a\tau}}+h\sqrt{a\tau}\right)$
16. $\dfrac{1}{(p/a)^{(n+1)/2}\left(\sqrt{p/a}+h\right)}$	$\dfrac{a}{(-h)^n}\exp(hx+ah^2\tau)\,\text{erfc}\left(\dfrac{x}{\sqrt{4a\tau}}+h\sqrt{a\tau}\right)$
$\times\exp\left(-\sqrt{\dfrac{p}{a}}\,x\right)\qquad\begin{array}{l}x>0\\a>0\end{array}\text{A}$	$-\dfrac{a}{(-h)^n}\sum_{r=0}^{n-1}\left(-2h\sqrt{a\tau}\right)^r i^r\,\text{erfc}\,\dfrac{x}{\sqrt{4a\tau}}$
17. $\dfrac{1}{p-\alpha}\exp\left(-\sqrt{\dfrac{p}{a}}\,x\right)\qquad\begin{array}{l}x>0\\a>0\end{array}$	$\dfrac{1}{2}\exp\alpha\tau\left[\exp\left(-x\sqrt{\dfrac{\alpha}{a}}\right)\text{erfc}\left(\dfrac{x}{\sqrt{4a\tau}}-\sqrt{\alpha\tau}\right)\right.$
	$\left.+\exp\left(x\sqrt{\dfrac{\alpha}{a}}\right)\text{erfc}\left(\dfrac{x}{\sqrt{4a\tau}}+\sqrt{\alpha\tau}\right)\right]$
18. $K_0\left(\sqrt{\dfrac{p}{a}}\,x\right)\qquad\begin{array}{l}x>0\\a>0\end{array}$	$\dfrac{1}{2\tau}\exp\left(\dfrac{-x^2}{4a\tau}\right)$
19. $\dfrac{1}{p}\exp\dfrac{x}{p}$	$I_0\left(\sqrt{4x\tau}\right)$
20. $p^{\nu/2}K_\nu\left(x\sqrt{p}\right)$	$\dfrac{x^\nu}{(2\tau)^{\nu+1}}\exp\left(\dfrac{-x^2}{4\tau}\right)$

APPENDIX
H

STANDARD DIMENSIONS OF PIPES AND TUBING

| Nominal OD, in. | Actual OD, in. | | Wall thickness, in. | | | | | |
| | Pipe | Tubing | Standard pipe and tubing | Welded and seamless steel pipe schedule | | | Copper tubing | |
				40	80	160	Type K	Type L
$\frac{1}{8}$	0.405	—	0.068	0.068	0.095			
$\frac{1}{4}$	0.540	—	0.088	0.088	0.119			
$\frac{3}{8}$	0.675	0.500	0.091	0.091	0.126	—	0.049	0.035
$\frac{1}{2}$	0.840	0.625	0.109	0.109	0.147	0.187	0.049	0.040
$\frac{5}{8}$†	—	0.750	—	—	—	—	0.049	0.042
$\frac{3}{4}$	1.050	0.875	0.113	0.113	0.154	0.218	0.065	0.045
1	1.315	1.125	0.133	0.133	0.179	0.250	0.065	0.050
$1\frac{1}{4}$	1.660	1.375	0.140	0.140	0.191	0.250	0.065	0.055
$1\frac{1}{2}$	1.900	1.625	0.145	0.145	0.200	0.281	0.072	0.060
2	2.375	2.125	0.154	0.154	0.218	0.343	0.083	0.070
$2\frac{1}{2}$	2.875	2.625	0.203	0.203	0.276	0.375	0.095	0.080
3	3.500	3.125	0.216	0.216	0.300	0.437	0.109	0.090
$3\frac{1}{2}$	4.000	3.625	0.226	0.226	0.318	—	0.120	0.100
4	4.500	4.125	0.237	0.237	0.337	0.531	0.134	0.110
5	5.563	5.125	0.258	0.238	0.375	0.625	0.160	0.125
6	6.625	6.125	0.280	0.280	0.432	0.718	0.192	0.140
8	8.625	8.125	0.277	0.322	0.500	0.906	0.271	0.200
10	10.75	10.125	0.307	0.365	0.593	1.125	0.338	0.250
12	12.75	12.125	0.330	0.406	0.687	1.312	0.405	0.280
14	14.0	—	—	0.437	0.750	1.406		
16	16.0	—	—	0.500	0.843	1.562		
18	18.0	—	—	0.560	0.937	1.750		
20	20.0	—	—	0.593	1.031	1.937		
24	24.0	—	—	0.687	1.218	2.312		
30	20.0							

†Tubing.

AUTHOR INDEX

623

SUBJECT INDEX